Second Edition

SMART POLYMERS

Applications in Biotechnology and Biomedicine

Second Edition

SMART POLYMERS

Applications in Biotechnology and Biomedicine

Edited by

Igor Galaev
Bo Mattiasson

CRC Press is an imprint of the
Taylor & Francis Group, an **informa** business

CRC Press
Taylor & Francis Group
6000 Broken Sound Parkway NW, Suite 300
Boca Raton, FL 33487-2742

First issued in paperback 2019

ISBN-13: 978-0-8493-9161-3 (hbk)
ISBN-13: 978-0-367-38882-9 (pbk)

Library of Congress Cataloging-in-Publication Data

Smart polymers : applications in biotechnology and biomedicine / edited by Igor Galaev and Bo Mattiasson. -- 2nd ed.
p. cm.
Rev. ed. of: Smart polymers for bioseparation and bioprocessing / edited by Igor Galaev and Bo Mattiasson. c2002.
Includes bibliographical references (p.).
ISBN 978-0-8493-9161-3 (alk. paper)
1. Smart materials. 2. Polymers. 3. Biomolecules--Separation. 4. Biochemical engineering. I. Galaev, Igor. II. Mattiasson, Bo, 1945- III. Smart polymers for bioseparation and bioprocessing.

TA418.9.S62S525 2007
620.1'92--dc22 2007061451

Visit the Taylor & Francis Web site at
http://www.taylorandfrancis.com

and the CRC Press Web site at
http://www.crcpress.com

Table of Contents

Preface

Smart polymers are macromolecules capable of undergoing rapid, reversible phase transitions from a hydrophilic to a hydrophobic microstructure. These transitions are triggered by small shifts in the local environment, such as slight variations in temperature, pH, ionic strength, or the concentration of specific substances like sugars.

Smart polymers have an extensive range of applications, but this book focuses solely on their roles within the fields of bioseparation and biomedicine. Until recently, polymers were considered to be passive participants within these fields. The first edition of this volume outlined an entirely novel approach that advocated a much more active role for smart polymers within the process of bioseparation. This new edition devotes more attention to theories describing the behavior of smart polymers in three states: in solution, as gels, and when grafted to surfaces.

The field of smart polymers has now matured to the stage where there is a clear need for solid quantitative descriptions and reliable guidelines for the development of new smart polymer systems. This edition focuses on smart gels, especially the fast-responding and macroporous gels, as these gels pave the way to the most promising applications of smart polymers, namely drug release and microfluidics.

This volume was written by the leading scientists involved in research on smart polymers and their applications. It offers a comprehensive overview of both the current state of affairs within this research field and the potential for future developments. The book will be of interest to those working on techniques of bioseparation and bioprocessing, polymer chemists developing new smart polymers, and graduate students in biotechnology.

Introduction

The most important components of the living cell — proteins, carbohydrates, and nucleic acids — are polymers. Even lipids, which are of lower molecular weight, could be regarded as methylene oligomers with a polymerization degree of around 20. Furthermore, the spontaneous aggregation of lipids contributes to even larger supramolecular structures. Nature uses polymers both as construction elements and as parts of the complicated cell machinery. The salient feature of functional biopolymers is their all-or-nothing (or at least their highly nonlinear) response to external stimuli. Small changes happen in response to changing conditions until a critical point is reached; then the transition suddenly occurs within a narrow range of the varied parameter, and after the transition is completed, there is no significant further response of the system. This nonlinear response of biopolymers is the end result of many highly cooperative interactions. Despite the weakness of each particular interaction taking place within each separate monomer unit, these interactions, when summed through hundreds and thousands of monomer units, can provide significant driving forces for the processes occurring within such systems.

Guided by a deeper understanding of the cooperative interactions of biopolymers in natural systems, scientists have attempted to mimic this cooperative biopolymer behavior in synthetic systems. Over the past few decades, research has identified a variety of synthetic functional polymers that respond in some desired way to changes in temperature, pH, electric or magnetic fields, or some other parameters. These polymers were originally nicknamed as "stimuli-responsive." The current designation of "smart polymers" was coined to emphasize the similarity of the stimuli-responsive polymers to the natural biopolymers. We have a strong belief that nature has always strived for smart solutions in creating life. Thus it is a worthy goal for scientists to better understand and eventually mimic biological processes in an effort to create novel chemical species and invent new processes.

The field of smart polymers and their applications is developing at a very fast rate. The first product using a thermoresponsive polymer was commercialized in 1996 by Gel Sciences/GelMed (Bedford, Massachusetts). The product, Smart-Gel, is a viscoelastic gel that is soft and pliable at room temperature but becomes much firmer when exposed to body heat; it is being used as a shoe insert to make the shoe conform to the wearer's foot. More recently, a novel suite of temperature-responsive cultureware, RepCell and UpCell, has been marketed by the Japanese company CellSeed (http://www.cellseed.com/index-e.html). The CellSeed products allow the cultivation of anchor-dependent mammalian cells at 37°C, and intact cells or even intact cell layers (cell sheets) can be collected in approximately 10 to 30 min after reducing temperature to 20 to 25°C.

The first edition of this book, which was published in 2002, was the first to cover the application of smart polymers in bioseparation and bioprocessing. A lot has happened since then. Reports on new smart polymer systems and original applications appear, literally, every week.

The "smart" polymer systems considered in this revised and extended edition are those that show a highly nonlinear response to small changes in the microenvironment. Most applications of smart polymers in biotechnology and medicine include biorecognition or biocatalysis, which take place principally in aqueous solutions. Thus only water-compatible smart polymers are considered, and the smart polymers in organic solvents or water/organic solvent mixtures are left beyond the scope of the book. The systems discussed in the book are based on

Soluble/insoluble transitions of smart polymers in aqueous solution

Conformational transitions of macromolecules physically attached or chemically grafted to a surface

Shrinking/swelling of covalently cross-linked networks of macromolecules, i.e., smart hydrogels

The smart polymer systems considered in this book are defined as *macromolecules that undergo fast and reversible changes from hydrophilic to hydrophobic microstructure. These microscopic changes, which are triggered by small changes in the microenvironment, are apparent at the macroscopic level as precipitate formation in solutions of smart polymers, as changes in wettability of the surface to which a smart polymer is grafted, or as dramatic shrinking/swelling of a hydrogel. The changes are fast and reversible, i.e., the system returns to its initial state when the trigger is removed.*

The force behind these transitions (or so-called critical phenomena) is driven by neutralization of charged groups, either by pH shift or addition of oppositely charged polymer; by changes in the efficiency of hydrogen bonding with an increase in temperature or ionic strength; or by cooperative processes in hydrogels and interpenetrating polymer networks. An appropriate balance of hydrophobicity and hydrophilicity in the molecular structure of the polymer is believed to be of key importance in inducing the phase transition.

Compared with the first edition, the current volume devotes more attention to the theory describing the behavior of smart polymers in three states: in solution, as gels, and when grafted to surfaces. The field of smart polymers has matured to the stage where there is a clear need for solid quantitative descriptions and reliable guidelines for the development of new smart polymer systems. This book focuses on smart gels, especially the fast-responding and macroporous gels, as these gels pave the way to the most promising applications of smart polymers, namely drug release and microfluidics.

The editors have done their best to collect, under one cover, chapters written by leading scientists involved in research on smart polymers and their applications. This book is intended to summarize the state of the art and perhaps spark further development in this exciting field of research, mostly by attracting fresh recruits — polymer chemists capable of developing new advanced polymers and biotechnologists ready to use these polymers for new applications.

One of the most frequently used smart polymers is a thermosensitive polymer, poly(*N*-isopropylacrylamide), which is cited throughout this book. At present, in the original literature, there is no consistency in the abbreviation of the name of this polymer. One can come across different notations like PNIPAM, PNIPAAm, polyNIPAAm, polyNIPAM, PNIAAm, and pNIPAm. As editors, we have left these notations as they were used by the authors, despite the temptation to unify the abbreviation throughout the book. The use of these varying notations in different chapters will help the reader to recognize the variations in nomenclature for this polymer when he or she comes across one of the abbreviations used in the original literature. Several chapters provide a list of abbreviations, and the reader is kindly directed to consult these lists.

Igor Galaev

Bo Mattiasson

About the Editors

Igor Yu. Galaev is currently an associate professor in the Department of Biotechnology at Lund University, Sweden. Previously he worked at the Chemical Department of Moscow University and the All-Union (now All-Russian) Research Institute of Blood Substitutes and Hormones (Moscow). He has written more than 160 articles and holds patents in polymer chemistry, bioorganic chemistry, and downstream processing.

Bo Mattiasson is a professor of biotechnology at the Department of Biotechnology and the Center for Chemistry and Chemical Engineering at Lund University. He has published more than 700 papers and edited 7 volumes. He holds several patents within his research interests, including enzyme technology, downstream processing, biosensors, and environmental biotechnology.

Contributors

Matilde Alonso Departamento de Física de la Materia Condensada, Universidad de Valladolid, Valladolid, Spain

Carmen Alvarez-Lorenzo Departamento de Farmacia y Tecnologia Farmaceutica, Universidad de Santiago de Compostela, Santiago de Compostela, Spain

Oleg V. Borisov Institute of Macromolecular Compounds, Russian Academy of Sciences, St. Petersburg, Russia

Deepak Chitkara Department of Pharmaceutics, National Institute of Pharmaceutical Education and Research, Punjab, India

Babur Z. Chowdhry Medway Sciences, University of Greenwich, Kent, United Kingdom

Jeffrey Chuang Boston College, Boston, Massachusetts

Angel Concheiro Departamento de Farmacia y Tecnologia Farmaceutica, Universidad de Santiago de Compostela, Santiago de Compostela, Spain

Ben Corn Department of Medicinal Chemistry, The Hebrew University of Jerusalem, Jerusalem, Israel

Avi Domb Department of Medicinal Chemistry, The Hebrew University of Jerusalem, Jerusalem, Israel

Igor Yu. Galaev Department of Biotechnology, Center for Chemistry and Chemical Engineering, Lund University, Lund, Sweden

Alexander Yu. Grosberg Department of Physics, University of Minnesota, Minneapolis, Minnesota

Alexander E. Ivanov Department of Biotechnology, Center for Chemistry and Chemical Engineering, Lund University, Lund, Sweden

Kong Jilie Chemistry Department, Fudan University, Shanghai, China

Nighat Kausar Medway Sciences, University of Greenwich, Kent, United Kingdom

Sergey V. Kazakov Department of Chemistry and Physical Sciences, Pace University, Pleasantville, New York

Alexei Khokhlov Institute of Organoelement Compounds, Russian Academy of Science, Moscow, Russia

Ashok Kumar Department of Biological Sciences and Bioengineering, Indian Institute of Technology, Kanpur, India

Neeraj Kumar Department of Pharmaceutics, National Institute of Pharmaceutical Education and Research, Punjab, India

Antti Laukkanen Drug Discovery and Development Technology Center, University of Helsinki, Helsinki, Finland

Mu Li Chemistry Department, Fudan University, Shanghai, China

Bo Mattiasson Department of Biotechnology, Center for Chemistry and Chemical Engineering, Lund University, Lund, Sweden

Anupama Mittal Department of Pharmaceutics, National Institute of Pharmaceutical Education and Research, Punjab, India

Jaisree Moorthy University of Wisconsin, Madison, Wisconsin

Oguz Okay Department of Chemistry, Istanbul Technical University, Maslak, Turkey

Rajendra Pawar Department of Medicinal Chemistry, The Hebrew University of Jerusalem, Jerusalem, Israel

Nicholas A. Peppas Departments of Chemical and Biomedical Engineering and Division of Pharmaceutics, The University of Texas at Austin, Austin, Texas

Susana Prieto Departamento Física de la Materia Condensada, Universidad de Valladolid, Valladolid, Spain

Javier Reguera Departamento Física de la Materia Condensada, Universidad de Valladolid, Valladolid, Spain

J. Carlos Rodríguez-Cabello Departamento Física de la Materia Condensada, Universidad de Valladolid, Valladolid, Spain

Martin J. Snowden Medway Sciences, University of Greenwich, Kent, United Kingdom

Sergey Starodubtsev Physics Department, Moscow State University, Moscow, Russia

Heikki Tenhu Laboratory of Polymer Chemistry, University of Helsinki, Helsinki, Finland

Valentina Vasilevskaya Institute of Organoelement Compounds, Russian Academy of Science, Moscow, Russia

Ekaterina B. Zhulina Institute of Macromolecular Compounds, Russian Academy of Sciences, St. Petersburg, Russia

Chapter 1

Phase Transitions in Smart Polymer Solutions and Light Scattering in Biotechnology and Bioprocessing

Sergey V. Kazakov

CONTENTS

Abbreviations

Ab	antibodies
Ab-PMAA	antibody–poly(methacrylic acid) conjugate
AFM	atomic force microscopy
Ag	antigen
Ag-PMAA	antigen–poly(methacrylic acid) conjugate
CC	coexistence curve
dGAPDH	gluceraldehyde-3-phosphate dehydrogenase
DLS	dynamic light scattering
IMAC	immobilized metal affinity chromatography
LCST	lower critical solution temperature
LS	light scattering
PCS	photon correlation spectroscopy
PEC	polyelectrolyte complexes
PMAA	poly(methacrylic acid)
PNIPA	poly(*N*-isopropylacrylamide)
PNIPA-VI	poly(*N*-isopropylacrylamide)-*co*-(*N*-vinylimidazole)
PPC	protein–polyelectrolyte conjugates
PVCL	poly(*N*-vinylcaprolactam)
RGD	Rayleigh-Gans-Debye approximation
SDS	sodium dodecyl sulfate
SLS	static light scattering
SMBV™	supramolecular biovector
TEMED	tetraethylene methylenediamine
UCST	upper critical solution temperature
VCL	*N*-vinyl caprolactam
VI	*N*-vinyl imidazole

1.1 Introduction

Bioseparation, immunoanalysis, immobilized enzyme systems,[1–3] drug delivery and drug targeting systems,[4–6] biodetection, biosensors, and artificial muscles[7–11] are only some of the recently identified biotechnological and biomedical application fields for the so-called smart polymers. Smart polymers, or stimuli-responsive macromolecules (artificial or natural), are thermodynamic systems capable of going reversibly through a phase transition within a certain range of thermodynamic parameters (pressure, temperature, concentration), which are often named as environmental conditions. Although macromolecules can drastically differ in their chemical composition and structure, the behavior of their solutions is universal in the vicinity of phase transition, and especially near the critical point.[12,13] The thermodynamic system exhibits a high susceptibility to the environmental conditions, i.e., infinitesimal changes in thermodynamic parameters (pressure, temperature, concentration, solvent, etc.) cause tremendous (infinite at the critical point) variations in the physicochemical properties of the macromolecules (solubility, structure, shape, size) and their solutions (stability). These significant property changes in the course of phase transition are what determine the application of smart polymers in the biotechnological protocols listed above.

According to the increasingly recognized concept of a gel-like cytoplasm,[14] a cytoplasmic protein-ion-water matrix can operate by the same working principles as a concentrated (even cross-linked) biomacromolecular system. In nature, there are reasons to believe that basic cellular processes (secretion, communication, transport, motility, division, and contraction) are provided through the tiny regulated phase transitions that occur inside the multifunctional, flexible, and dynamic machineries called "cells." In this context, all biotechnological applications of smart polymers, in some sense, can be considered as mimicry of living-matter action. The main argument supporting this point is that the most applicable stimuli-responsive polymers have their "smartness," or working phase transition, at temperatures close to the human body temperature (~37°C) where biomacromolecules (polyaminoacids, polynucleotides, polysaccharides, etc.) can function.

The thermodynamic coordinates of the phase transition — e.g., transition temperature, pressure, concentration, pH and ionic strength of solvent, light intensity, electric field applied, etc. — are specific for different macromolecular solutions, being derived from the competing repulsive and attractive interactions that are characteristic of the chemical composition and structure of the macromolecule as well as the nature of the solvent. As a consequence, in designing a technologically optimal bioprocess, macromolecules with relevant microscopic and macroscopic properties (molecular mass, size, shape, conformation, charge, solubility, specific ligand modification, etc.) should be synthesized to provide phase-transition points at the desired thermodynamic coordinates. Bioseparation processes, such as partitioning of a target substance and impurities (two-phase separation) and recovery of the target from the enriched phase, involve the immiscibility phase transitions.[1–3] For example, the precipitation step employs the addition of a compound that shifts the phase out of the mechanical stability region and results in precipitation of the target substance. Affinity precipitation[1] (pp. 55–77) supposes an involvement of a water-soluble polymer with a number of highly selective ligands specific to the target protein. Thus, knowing how the ligand attachment affects the phase diagrams of pH- or temperature-sensitive polymers or hydrogels is of crucial importance. Bare polymers, ligand-modified polymers, and polymer–protein conjugates need to be characterized in terms of their molecular weight, size, shape, and mutual interactions, i.e., in terms of the parameters that affect phase transition and consequently the purification procedure.

Similar information on phase transition in the smart polymers' solutions is in demand for many biotechnological processes (e.g., controlled-release systems in drug delivery, biodetection, and bioanalysis). The choice of a pertinent experimental method for this type of characterization is decisive. Indeed, the choice is already predetermined by the thermodynamic and statistical nature of the macromolecular solutions: it is static and dynamic light scattering.

Light scattering (LS) is a highly informative method *per se*. The combination of dynamic LS (DLS) with simultaneous measurements of the scattered intensity (static LS, or SLS) in the same experimental setup provides a detailed characterization of microscopic and macroscopic properties of the system.[15,16] Photon correlation spectroscopy (PCS) has recently become a popular technique for studying structural transformations in systems containing proteins, enzymes, polymers, and other macromolecular components.[17–20]

In this chapter, after a brief introduction into the concepts of solubility phase transitions and the theory of light scattering, the works (including ours) on different applications of static (SLS) and dynamic light scattering (DLS) techniques in bioseparation and other biotechnological processes are reviewed. The chapter ends with a view of some of the future trends in the use of light scattering applications in biotechnology and biomedicine.

1.2 Solubility Phase Transitions and Liquid–Liquid Coexistence Curves

On a practical level, all biotechnological processes deal with macromolecular solutions, i.e., systems containing at least two components that can have regions of limited solubility depending on interactions between similar (A-A, B-B) and different molecules (A-B). Knowing the boundaries of the liquid–liquid phase equilibria, or solubility phase diagrams, for smart polymer solutions is a key point for the development of bioprocesses, especially bioseparation. Ideally, if there had been a general theory, one could predict the phase diagram for a polymer of a known chemical structure in different solvents. Unfortunately, nowadays, all existing theoretical approaches, both microscopic[21–28] and phenomenological,[29–34] are unable to analytically represent the so-called liquid–liquid coexistence curve in a wide range of thermodynamic parameters. The most successful ones are only capable of fitting the multiparametric equations into the existing experimental data,[32] but not vice versa, i.e., not by calculating the limiting solubility based on the microscopic (chemical) structure of macromolecules. One of the main challenges in both analytical description and experimental study is the high diversity in shapes of liquid–liquid coexistence curves, especially for polymer solutions.

In Figure 1.1, different types of "temperature–composition" phase diagrams are schematically presented. Typical coexistence curves (CC) with the upper critical solution temperature (UCST, Figure 1.1a) and the lower critical solution temperature (LCST, Figure 1.1b) can be interpreted as parts of the closed-loop CC with two critical temperatures (Figure 1.1c) when the second critical point lies either beneath the melting point (Figure 1.1a) or above the boiling point (Figure 1.1b) of the solution, respectively. It is obvious that Figure 1.1d also shows the closed-loop CC without critical points because the two-phase region expands beyond the liquid phase. References on experimental phase diagrams of this kind, including polymer solutions, can be found in the literature.[31,32] Interestingly, the two-phase region substantially contracts with an increase in the molecular mass of polyethylene glycol in water.[35–37] Another type of CC with two critical points separated by the total solubility region also exists (Figure 1.1e). In this case, the difference between UCST and LCST is governed by properties of the mixture components. For example, an increase in molecular mass of polystyrene in *tert*-butyl acetate

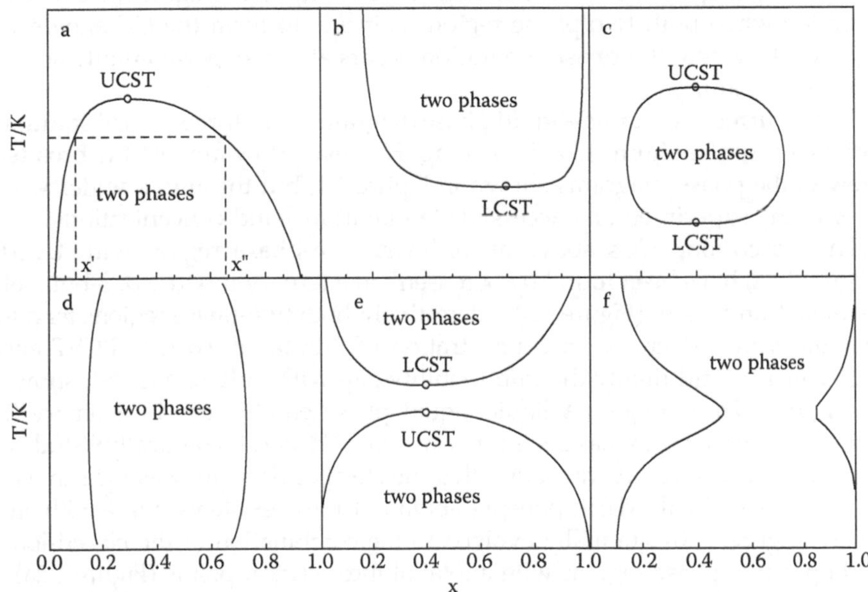

FIGURE 1.1
Schematic representation of liquid–liquid phase diagrams for mixtures with (a) one UCST; (b) one LCST; (c) a closed-loop coexistence curve, UCST > LCST; (d) a closed-loop coexistence curve, where both critical points are beyond the range of measurement; (e) two critical points, UCST < LCST; and (f) two overlapping immiscibility gaps. The frame for each phase diagram is considered as a window of the experimentally accessible temperatures and concentrations.

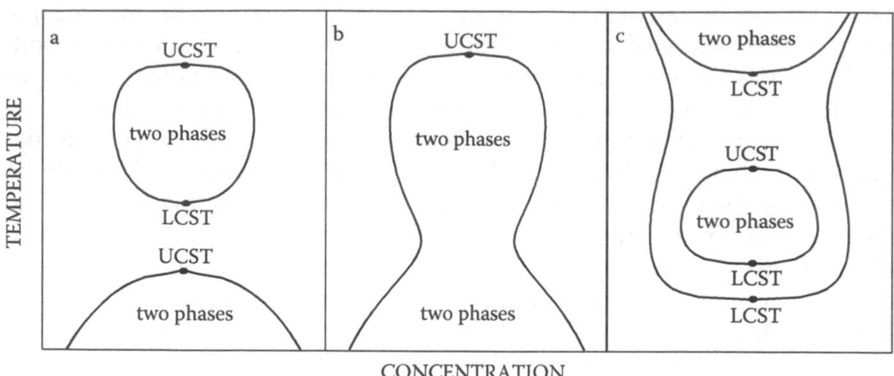

FIGURE 1.2
Schematic representation of liquid–liquid phase diagrams for mixtures with three critical points: (a) closed-loop immiscibility gap lies above CC with one UCST, (b) overlapping immiscibility gaps, (c) hypothetical CCs with three critical points when closed-loop CC lies beneath CC with one LCST. The frame for each phase diagram is considered as a window of the experimentally accessible temperatures and concentrations.

narrows the region of total miscibility,[36] so that one can imagine the extreme situation when both two-phase regions coincide to form the CC shown in Figure 1.1f, where the phase separation occurs at any experimentally accessible temperature.

The occurrence of liquid–liquid phase diagrams with three critical solution points, shown in Figure 1.2a, is striking. However, they are just the broader view of the phase diagrams shown in Figure 1.1, but through a wider window of the experimentally accessible temperatures and concentrations.

The closed-loop CCs above an ordinary two-phase region with UCST (Figure 1.2a) have been found by Sorensen[38] in a ternary mixture of 3-butanol, 2-butanol, and water (Figure 1.3). Interestingly, both two-phase regions tend to expand with a decrease in concentration of 3-butanol, so that LCST and UCST merge, and finally the immiscibility gap with only one UCST, shown in Figure 1.2b, emerges. A liquid–liquid phase equilibrium in a binary 2-butanol/water mixture has been experimentally[39,40] and theoretically[41] studied at different pressures. It was found that the effect of pressure was the same as the addition of the third component (3-butanol): the coexistence curve with one UCST (Figure 1.2b) gradually evolved to the combination of the closed-loop and open two-phase regions with a total of three critical points (Figure 1.2a).

The general thermodynamic consideration of liquid–liquid coexistence curves with several critical points can be found in the literature.[31,32,41] It is noteworthy that immiscibility gaps like those depicted in Figure 1.2c are not forbidden thermodynamically, although there is no experimental evidence. This type of CC must be extremely rare because it supposes to exhibit two LCSTs out of three, which in turn requires highly specific molecular orientations (a negative excess of entropy of mixing) at higher temperatures.[21–28]

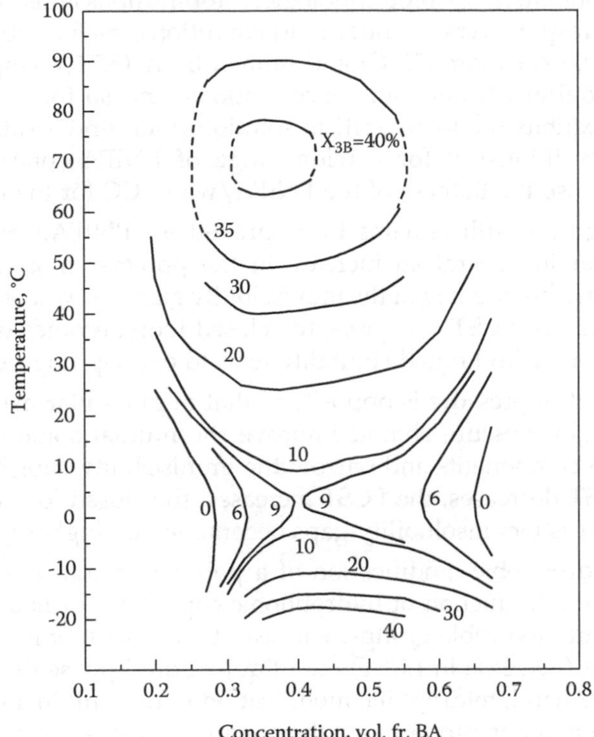

FIGURE 1.3
Cloud-point curves for the 3-butanol/2-butanol/water system. The experimental data were taken from Sorensen.[38] Concentration of butanol (BA) is the sum of volume fractions of both alcohols in aqueous solution. X_{3B} is a volume concentration of 3-butanol.

Therefore, polymer systems and their "smart" properties relevant to their use in bioseparation and bioprocessing procedures can be elucidated by studying the effect of thermodynamic parameters (temperature, pressure, and composition) and additives (solvent, pH, ionic strength, surfactant) on the evolution of immiscibility gaps on the "temperature–composition" plane. Up to now, a vast pool of experimental data has been accumulated on the phase behavior of the smart (both temperature- and ionic-sensitive) polymers that are the most pertinent in biotechnological protocols as phase-forming and precipitation agents, as ligand carriers, and as charged additives. The comprehensive information on distinct smart polymers and their properties is available in a number of original papers and remarkable reviews (see, for instance, the previous and current editions of the present book). The following general observations can be summarized:

1. Practically, not many polymers are "smart." The most popular smart polymer is poly(*N*-isopropylacrylamide) (PNIPA) and its combinations with other copolymers (acrylic acid, *N*-vinylimidazole, etc.).

2. Smart polymers for biotechnological applications must reveal the stimuli-responsiveness under mild conditions, such as atmospheric pressure, near room (25°C) and human body (37°C) temperatures, physiological pH and salt concentrations, and so forth. PNIPA in water exhibits the lower critical solution point under satisfactorily mild conditions[42–46] for a wide range of PNIPA concentrations. (Notice also the flatness of the PNIPA/water CC for the LCST.)
3. Although this still remains to be proved for PNIPA,[45–48] one may state that in general an increase in the polymer molecular mass results in a broadening of the immiscibility gaps,[35–37] where the UCST increases, the LCST decreases, the closed loop becomes wider, and the regions with limited solubility tend to overlap (Figure 1.1e,f).
4. The effect of pressure is opposite to that of molecular mass, i.e., an increase in pressure should improve the mutual solubility of the solution components and narrow the immiscibility gaps,[39–41] where the UCST decreases, the LCST increases, the closed loops contract, and the distinct insolubility gaps become increasingly separated.
5. Any hydrophobic modification of a polymer in aqueous solution, either by introduction of hydrophobic copolymer or neutralization of polymer ionizable groups, will lead to a growth of immiscibility domains (decrease in LCST) on a liquid–liquid phase diagram. On the other hand, hydrophilic modification may result in a loss of the phase-separation properties of the polymer/water system. However, in each case, the balance of polymer–polymer, polymer–solvent, and solvent–solvent interactions should be analyzed to elucidate the shape of the final phase diagram.
6. Presumably, the addition of a solvent may change the liquid–liquid immiscibility pattern and make it similar to those shown in Figure 1.3. However, solvents that are good for the polymer alone do not always increase the polymer solubility in water.[49] Additional entropy effects, such as the capability of forming local structures between water and co-nonsolvent, should be taken into account.[50] For example, in polymer/water/alcohol systems, water–alcohol complexation may reduce the interaction between polymer and water and consequently decrease the LCST, which can be interpreted as a broadening of the immiscibility gap.
7. Ion-sensitive smart polymers can be synthesized on the basis of thermosensitive polymers by copolymerization with monomers containing ionizable groups (carboxyls, imidazolyls, amines, imines, and so on). In aqueous solutions of those polymers, an ionization/deionization (protonation/deprotonation) of the ionizable groups will generate the changes in their "temperature–concentration" solubility phase diagrams.
8. The development of new smart polymer systems for bioseparation and bioprocessing may aim at finding the experimental conditions

bringing the immiscibility gap for the known polymer/water solution to the required point on the temperature–composition diagram. Another option is to modify known polymers or synthesize novel ones (copolymers) with the required immiscibility gap location.

9. Shifting the total solubility phase diagrams for smart polymers by a physical enforcement (light, electric, and magnetic fields) is an alternative way to design new biotechnological protocols.
10. The recently developed hydrogel particles of nanometer scale (nanogels) are of great significance for drug delivery and affinity-ligand carriers. In terms of properties, nanogels are in between polymers and bulk gels. Nanogels have substantial advantages for biotechnology,[51] including their fast response to stimuli and ion-exchange ability, their homogeneous structure, and their high swelling/shrinking capability. For example, the response time for hydrogel nanoparticles of 1 to 1000 nm in size is expected to be less than milliseconds. Bearing in mind the great potential of nanogels in bioseparation, the interactions between those particles with or without ligands and proteins are under study[1] (pp. 191–206). The elucidation of phase-transition boundaries in nanogel suspensions is in demand as well. The potential of employing a relatively new type of nanoparticle (lipobeads, a liposome-hydrogel assembly) as a drug-delivery system has already been reviewed.[52] One can easily imagine how the use of such bicompartmental structures will expand to bioseparation and other biotechnological fields.
11. Liquid–liquid phase transitions exhibit a divergence of the system susceptibility, resulting in significant growth of concentration fluctuations and, as a consequence, an increase in light scattering intensity. Hence, the integral intensity light scattering (SLS) technique is the most informative method to detect the transition itself. Spectral characteristics of scattered light (DLS) are used to monitor the conformational changes of polymers and their aggregates in the course of transition. Being sensitive to the size of scattering particles, the combination of SLS and DLS is a proven and effective approach in testing polymer–polymer, polymer–solvent, and polymer–protein interactions as well as the interactions between their complexes, conjugates, and aggregates.

1.3 Light Scattering in Macromolecular Solutions

The following is a short assessment of our knowledge on the information that can be extracted from SLS and DLS measurements in macromolecular solutions (polymers and biopolymers). It is especially important to be aware of the theoretical restrictions and experimental conditions at which this information can be collected. In the recent decade, the LS techniques have been applied to the more-complex macromolecular systems, such as highly scattering polymeric and colloidal suspensions, viscoelastic nonergodic gel-like systems, and charged and strongly interacting macromolecular solutions. Reviews and monographs concerning either specific theoretical issues or experimental features of light scattering methods are cited in the text. Where possible, these reviews and more recent papers containing references to earlier works are cited in an effort to reduce the number of citations. I regret the impracticability of citing directly all relevant works in the field.

1.3.1 Static Light Scattering

In 1910, Einstein extended the Rayleigh theory of light scattering in gases to liquids and solutions.[19] Scattering of light in any homogeneous medium results from the optical heterogeneities that originate from the thermal molecular motion, i.e., random local variations in density, temperature, and concentration (in solutions) from their average values lead to local fluctuations of the optical dielectric permittivity, $\varepsilon = n^2$. The total intensity of scattered light is proportional to the mean-square deviations of the optical dielectric permittivity, $\langle\Delta\varepsilon^2\rangle$, which is a function of density, ρ, temperature, T, and concentration, x:

$$I = I_0 \frac{\pi^2}{\lambda^4 r^2} V \langle\Delta\varepsilon^2\rangle P(q) \quad \text{(polarized light)} \tag{1.1}$$

where

$$\langle\Delta\varepsilon^2\rangle = \left(\frac{\partial\varepsilon}{\partial\rho}\right)^2_{T,x} \langle\Delta\rho^2\rangle + \left(\frac{\partial\varepsilon}{\partial T}\right)^2_{\rho,x} \langle\Delta T^2\rangle + \left(\frac{\partial\varepsilon}{\partial x}\right)^2_{\rho,T} \langle\Delta x^2\rangle \tag{1.2}$$

λ is the wavelength of radiation; r is the distance from the scattering volume to the detector; I_0 is the intensity of incident light; V is the scattering volume; $P(q)$ is the so-called scattering form factor, the function of size and shape of scattering particles; and q is the scattering vector, given by

$$q = \frac{4\pi n}{\lambda} \sin(\theta/2) \tag{1.3}$$

where n is the refractive index of the medium, and θ is the angle between the incident and scattered beams. The second term in Equation 1.2 is less than 2% of the first one, and concentration fluctuations are much greater than density fluctuations in solutions. Hence, the expression for the intensity of scattered light Equation 1.1 can be reduced to

$$I = I_0 \frac{\pi^2}{\lambda^4 r^2} V \left(\frac{\partial \varepsilon}{\partial x} \right)^2_{\rho,T} \langle \Delta x^2 \rangle P(q) \tag{1.4}$$

Thus, the intensity of light scattered in solution is a function of (a) the difference in refractive indices of the scattering objects (particles, macromolecules, fluctuations, clusters, aggregates, etc.) and surrounding medium (solvent), which is included into the refractive index increment $(\frac{\partial \varepsilon}{\partial x})^2 = 4n^2 (\frac{dn}{dx})^2$; (b) the concentration fluctuations $\langle \Delta x^2 \rangle$; and (c) the size and shape of the scatterers $P(q)$.

1.3.1.1 Fluctuations and Scattering in the Vicinity of Phase Transitions

From the statistical thermodynamics,[53] the measure of concentration fluctuations, $\langle \Delta x^2 \rangle$, is inversely proportional to the first derivative of the chemical potential of the solvent, μ_1, over the concentration:

$$\langle \Delta x^2 \rangle = \frac{RTx_2}{N_A \left(\frac{\partial \mu_1}{\partial x_1} \right)_{p,T}} \tag{1.5}$$

where x_1 and x_2 are the molar fractions of the solvent and solute, respectively, N_A is the Avogadro number, and R is the gas constant. Equation 1.5 clearly shows the potential of light scattering technique for detecting the critical solution point and the line of second-order phase transitions (spinodal) in solutions where concentration fluctuations diverge to infinity due to the thermodynamic definition of the spinodal as $(\frac{\partial \mu_1}{\partial x_1})_{p,T} = 0$. It becomes obvious that different absorbance, transmittance, or turbidity methods traditionally used for detection of phase-transition temperature or pH by monitoring the optical density changes at a fixed wavelength are just modifications of the SLS technique.

1.3.1.2 Scattering in Polymer Systems

For diluted polymer solutions, chemical potential, μ_1, can be expanded into a series of weight fractions, c_2 (grams per ml), as follows

$$\mu_1 = \mu_1^\phi - RTV_1^0 c_2 \left(\frac{1}{M_2} + A_2 c_2 + A_3 c_2^2 + \quad \right) \tag{1.6}$$

where μ_1^ϕ and V_1^0 are the standard chemical potential and molar volume of pure solvent, respectively; M_2 is the molecular mass of polymer; and A_2 and A_3 are termed the second and third virial coefficients, the values of which depend on the binary and ternary interactions, respectively.

Substituting Equations 1.6 and 1.5 into Equation 1.4 and defining the absolute scattering intensity, R_θ, as

$$R_\theta = \frac{IV}{I_0 r^2} \tag{1.7}$$

one can easily obtain

$$\frac{Kc_2}{R_\theta} = \frac{\frac{1}{M_2} + 2A_2c_2 + 3A_3c_2^2 + \cdots}{P(q)} \tag{1.8}$$

where

$$K = 4\pi^2 n^2 \left(\frac{dn}{dc_2}\right)^2 N_A^{-1} \lambda^{-4}$$

Equation 1.8 clearly demonstrates the idea of analyzing the "scattering intensity–concentration" data to determine the molecular mass of the scatterers by using the so-called Debye plot (Figure 1.4):

$$\frac{Kc_2}{R_\theta} = \frac{1}{M_2} + 2A_2c_2 + \tag{1.9}$$

However, it is necessary always to bear in mind a set of assumptions made on the way of deriving Equation 1.9: (a) solution should be sufficiently diluted ($c_2 << 1$) to provide the virial expansion (Equation 1.6) and to neglect

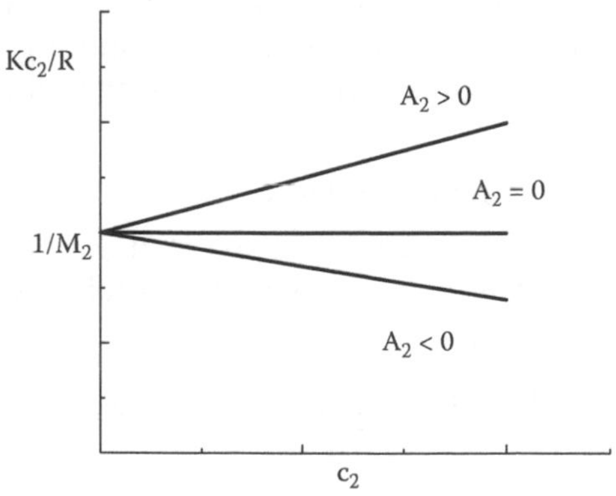

FIGURE 1.4
Kc_2/R_θ as a function of macromolecular concentration c_2: determination of molecular mass M_2 and second virial coefficient A_2; the slopes yield A_2 for different ionic strengths.

the third term in Equation 1.8; (b) the size of scatterers is much less than the wavelength of the scattering light ($d << \lambda$) so that $P(q) = 1$.

In the solutions of polyelectrolytes (polymers, proteins), the second virial coefficient A_2 can be determined as a slope of Kc_2/R-factor vs. the macromolecule's concentration (Figure 1.4). The interaction coefficient, A_2, accounts for (a) the total sum charge on the macroion chain, (b) the effect of the exclusion volume, (c) macroion–macroion interactions, (d) macroion–salt-ion interactions and their complexation, and (e) salt-ion–ion interactions. Depending on the sum of those interactions, the value of A_2 can be negative, positive, or zero, as sketched in Figure 1.4. There is a special case when A_2 is small or, in general, $2A_2c_2 << 1/M_2$:

$$R_\theta \approx Kc_2M_2 \tag{1.10}$$

i.e., the absolute scattering intensity is a measure of the polymer molecular mass and its concentration.

1.3.1.3 Light Scattering from Large Particles

The main difference in scattering from small and large particles is that the electric field of incident light is constant in the bulk of small particles ($d << \lambda/20$), whereas the interference effects within the volume of large particles result in a reduction of the scattering intensity.[54] Those effects may be taken into account by the form factor $P(q)$, which by definition is the ratio of the intensity of light scattered by large particles ($I_{\text{large particles}}$) to the intensity of scattering light without an interference ($I_{\text{no interference}}$):

$$P(q) = \frac{I_{\text{large particles}}}{I_{\text{no inteference}}} \tag{1.11}$$

The Mie method[54–57] is the most precise calculation method for light scattering intensity of large particles. However, calculation by the Mie method can be performed only numerically. The so-called Rayleigh-Gans-Debye (RGD) approximation[16,54–57] is a practically important approach to interpret SLS data for the scatterers with sizes comparable with the wavelength of incident light ($d \sim \lambda$). The Rayleigh-Gans criterion is often expressed using the following inequality:

$$|n - n_0|d << \frac{\lambda}{4\pi} \tag{1.12}$$

where n_0 is the refractive index of the medium. The physical meaning of RGD approximation comprises two assumptions: (a) as light passes through the particle, the phase shift of the light is negligible, and (b) no internal reflections take place. Particularly, the refractive index of the particle should

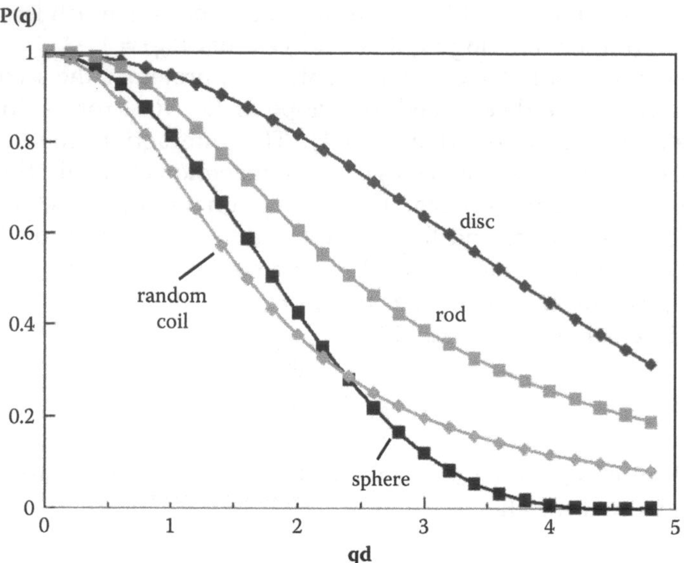

FIGURE 1.5
Scattering form factors for particles of different shape in the RGD approximation.

be close to that of the medium, and the size of the particle should be smaller than the light wavelength (Figure 1.5).[54]

The form factor $P(q)$ is shown for the scatterers of different shapes. The formulae for the different shapes shown in the figure are quite complex.[19,54–57] Nevertheless, for a sphere and a polymer random coil they are not too bad:

$$P(q) = 9\left[\frac{\sin(qd) + (qd)\cos(qd)}{(qd)^3}\right]^2$$

and

$$P(q) = \frac{2}{(qR_g)^4}\left[\exp\left(-q^2R_g^2\right) + (qR_g)^2 - 1\right]$$

respectively. For the small-angle scattering, $qd < 1$, the form factor can be expanded into the series over qd. Keeping two terms of the expansion, both of the above expressions give:

$$P(q) \approx 1 - q^2R_g^2/3 \tag{1.13}$$

where R_g is the radius of gyration of a macromolecule.

Equation 1.13 gives a method for extraction of information on macromolecular sizes in diluted solutions ($c_2 << 1$) by fitting the angular dependence of the light scattering intensity into the equation:

$$\frac{I(q)}{I_0} = 1 - q^2 R_g^2 / 3 \qquad (1.14)$$

The so-called Guinier plot, $I(q)/I_0$ versus q^2, yields the best estimation of the radius of gyration if the condition $qR_g < 1$ is satisfied. Since the macromolecule size is not known *a priori*, a series of Guinier plots should be constructed for different maximum values of q and fitted in until the value of R_g is progressively reduced to the convergent value.

If the shape of macromolecules under study is known or suggested *a priori*, then the value of a single directly measured parameter, the mean square radius of gyration, R_g^2, can be related to the other conformational parameters of the macromolecule in the framework of the corresponding model. For example:

For high-molecular-weight polymers with the identical monomer units, a relation exists between R_g^2 and the mean square end-to-end distance, $\langle h^2 \rangle = 6R_g^2$.

For helical polymers resembling quite rigid rods without kinks, the radius of gyration is given in terms of the length, L, of the rod and its diameter, d, by the relation: $R_g^2 = d^2/8 + L^2/12$.

If the length is large compared with the diameter, R_g^2 reduces to $L^2/12$.

For long and flexible rods, the wormlike coil model was proposed by Porod and Kratky[154] to result in the expression given by $R_g^2 = \frac{1}{3} L^2 p(1 - p + pe^{-1/p})$, where $p = l_p/L$, and l_p is the persistence length, the measure of the flexibility of a polymer chain, increasing as the chain under consideration becomes rigid and less flexible.

For very long and highly flexible rods, $l_p << L$ or $p << 1$, the wormlike coil is referred to as a random coil and $R_g^2 = \frac{1}{3} l_p L$.

In a random coil, the strong interactions between distant residues along the chain can cause a polymer to adopt a unique conformation: the polymer chain coils up into a compact globule with a definite tertiary structure in which each monomer unit has a fixed position and rotation about the bonds of the backbone is severely inhibited. In this case, the radius of gyration is, with high accuracy, equal to the real size of the globule.

The orders of magnitude for the radii of gyration of a polymer in a compact globular state, in a random coil state, and in a rigid rodlike state are 1 to 2 nm, 8 to 10 nm, and 100 to 150 nm, respectively. Those figures demonstrate the potential of light scattering for studying the conformational changes of macromolecules (polymers and biopolymers) in the course of phase transitions.

1.3.1.4 Zimm Plots

Putting Equations 1.8 and 1.13 together, in 1948, Zimm derived the relationship between the concentration, scattering angle, and intensity of the scattered light[58]:

$$\frac{Kc_2}{R_\theta} = \left(\frac{1}{M_2} + 2A_2c_2 + \quad\right)\left(1 + q^2R_g^2/3\right) \tag{1.15}$$

Note that the approximation $(1-x)^{-1} \approx (1+x)$ is used in deriving Equation 1.15. Zimm then proposed the use of a special graphing technique that has been named after him, a Zimm plot: the experimental values of Kc_2/R are plotted against $q^2 + kc_2$, as illustrated in Figure 1.6, to form a grid of points. Herein, k is some constant that can be chosen arbitrarily to make the plot reasonable in scale.

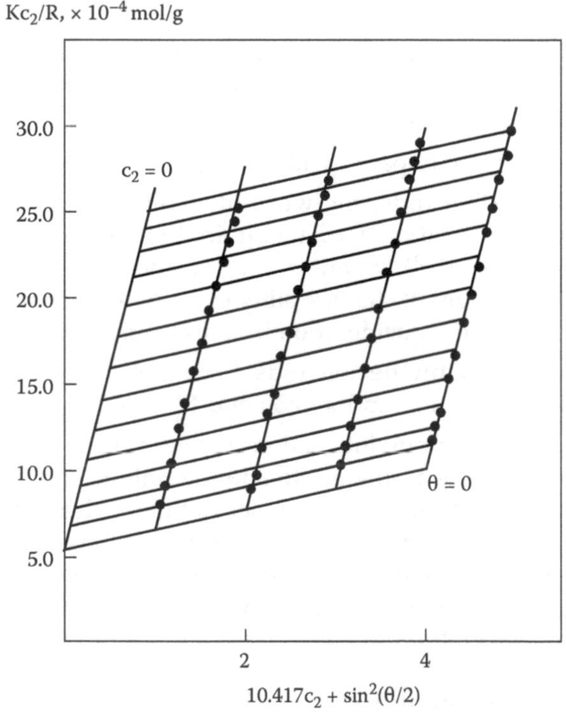

FIGURE 1.6
Zimm plot for the PNIPA aqueous solution at 23°C.

Extrapolation of all points corresponding to different concentrations, c_2, at the identical scattering angle (θ = const) to zero concentration ($c_2 = 0$) yields a straight line with a slope of $R_g^2/3M_2$:

$$\frac{Kc_2}{R_\theta} \approx \frac{1}{M_2}\left(1+q^2R_g^2/3\right) \tag{1.16}$$

Analogously, extrapolating all points at constant c_2, but different scattering angles, to zero scattering angle, $\theta = 0$, one can obtain a straight line with a slope of $2A_2$:

$$\frac{Kc_2}{R_\theta} = \frac{1}{M_2} + 2A_2c_2 \tag{1.17}$$

Two limiting straight lines intercept the y-axis to give an estimate of $1/M_2$ (an example is shown in Figure 1.6).

Three parameters can be extracted from a Zimm plot: M_2, the average molar mass of macromolecules; A_2, the second virial coefficient; and R_g, the radius of gyration. If the preparation is polydisperse, the obtained radius of gyration should be interpreted as z-average. The success of the whole procedure requires the scattering intensity (R_θ) to vary with q significantly. This means that the size of scatterers must be greater than 0.05λ, i.e., the method is inapplicable to macromolecules that are too small. On the other hand, the RGD approximation used in Equation 1.15 puts the practical upper limit of 0.5λ on the radius of gyration. Note that a nonlinear Zimm plot will require keeping higher terms in the concentration and wave vector expansions in Equation 1.15.

1.3.2 Dynamic Light Scattering

As mentioned in Section 1.3.1, the scattering of light in solutions results from the local concentration fluctuations, the strength of which is characterized by $\langle \Delta x^2 \rangle$. However, these concentration heterogeneities are not static or "frozen" clusters. Resulting from the thermal molecular motion, the local concentration fluctuations constantly appear and disappear, causing a sequential modulation of the amplitude of the electric field of the scattered wave. As a result of the modulation, the spectral line of the scattered light is somewhat broadened in comparison with the incident light.

In accord with the Onsager hypothesis,[59,60] the spontaneous local disturbances in solution, the measure of which is the concentration fluctuations, vary in time similar to the macroscopic gradients resulting from the external stimulus. Hence, their behavior can be described with the macroscopic equations of hydrodynamics,[61] resulting in the following form for the autocorrelation function for the concentration fluctuations[62]:

$$G(\tau) = \langle \Delta x(0)\Delta x(\tau) \rangle = \langle \Delta x^2 \rangle \exp(-Dq^2\tau) \tag{1.18}$$

where $\langle \cdots \rangle$ denotes the averaging over a large number of separate time periods of duration τ, D is the diffusion coefficient, and q is the scattering wave vector, given by Equation 1.3.

1.3.2.1 Photon Correlation Spectroscopy

The theory of DLS has been extensively reviewed by Berne and Pecora.[19] In brief, the modulation of the amplitude of the electric field of the scattered wave by the local concentration fluctuations will lead to the same form of the field autocorrelation function as the autocorrelation function for the concentration fluctuations (Equation 1.18):

$$G^{(1)}(\tau) = \langle E^*(0)E(\tau)\rangle = G^{(1)}(0)\exp(-Dq^2\tau) \tag{1.19}$$

where $E(\tau)$ is the representation of the electric field at time τ, $E^*(0)$ is the complex conjugate of the electric field at time zero, and $G^{(1)}(0) = \langle E^2(0)\rangle = I_0$ is the average intensity of the scattered light.

The optical spectrum of the scattered light, $I(\omega)$, is related to the field autocorrelation function $G^{(1)}(\tau)$ by the expression referred to as the Weiner-Kinchine theorem:

$$I(\omega) = \frac{1}{2\pi}\int_0^{\infty} G^{(1)}(\tau)e^{i\omega\tau}d\tau \tag{1.20}$$

where ω is the angular frequency. Performing the integration of Equation 1.19, it can be shown that the exponential field autocorrelation function $G^{(1)}(\tau)$ is related to the spectrum often termed as Lorentzian, whose shape is specified by the equation

$$I(\omega') = I_0 \frac{Dq^2/\pi}{(\omega'-\omega)^2 + (Dq^2)^2} \tag{1.21}$$

It can be readily observed (Figure 1.7a) that the autocorrelation function is equal to $G^{(1)}(0) = I_0$ at $\tau = 0$, falls to zero as τ increases to infinity, and the so-called correlation time is given by $\tau_c = 2\pi\Gamma^{-1}$. The corresponding scattering-intensity spectrum is sketched in Figure 1.7b. $I(\omega')$ has its maximum value at a frequency corresponding to that of incident light $\omega' = \omega$. This maximum value is $I_0/(\pi Dq^2)$, and the half-width of the Lorentzian spectrum at half height is $\Gamma = Dq_2$. Hence, the DLS data can be analyzed either by the spectroscopy of optical mixing (intensity spectrum) or by the photon correlation spectroscopy (correlation function) (PCS).

Nowadays, PCS is the most common way to analyze dynamic light scattering data. There can be two alternatives: (a) in the heterodyne spectroscopy,

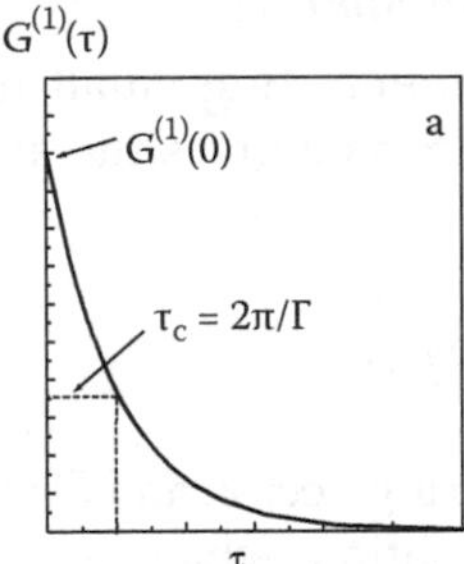

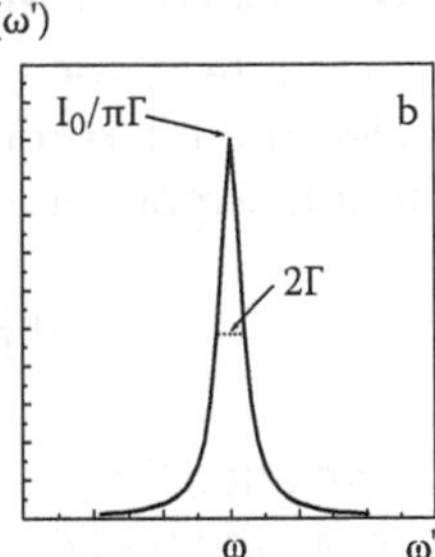

FIGURE 1.7
The field autocorrelation function (a) and corresponding intensity spectrum (b) of the scattered light.

the scattered light makes beats with an external beam, and (b) the homodyne spectroscopy is based on the so-called self-beating effect when the different frequencies present in the scattered light can combine with each other in the absence of an external reference beam. The homodyne version, often called "self-beating spectroscopy," is more feasible experimentally. It is worth realizing that in the homodyne spectroscopy, the intensity autocorrelation function defined as

$$G^{(2)}(\tau) = \langle I(t)I(t+\tau)\rangle = \int_0^{\infty} I(t)I(t+\tau)dt \tag{1.22}$$

is commonly measured, where $I(t)$ is the intensity measured at time t. Like the field autocorrelation function, it reaches the maximum at $\tau = 0$ and falls to zero as τ increases. If the scattering signal is a stationary Gaussian process, the experimentally determined intensity autocorrelation function $G^{(2)}(\tau)$ can be converted to the normalized field autocorrelation function, $g^{(1)}(\tau) = \frac{G^{(1)}(\tau)}{G^{(1)}(0)}$, through the Siegert relation

$$G^{(2)}(\tau) = B(1+\beta^2 |g^{(1)}(\tau)|^2) \tag{1.23}$$

where B is the experimentally determined baseline, and β ($0 < \beta < 1$) is an experimental constant that characterizes the efficiency of optical mixing (coherence).[63]

Note that this consideration is valid only for the so-called ergodic systems, those with the statistically independent optical inhomogeneities resulting from the molecular motion. For these systems, as a consequence, the measured time-averaged parameters (such as the scattered intensity or its autocorrelation function) are equivalent to the ensemble-averaged ones. The problem of nonergodicity in light scattering measurements is discussed in Section 1.3.3.2.

1.3.2.2 DLS Data Analysis: Monodisperse Spheres

In the case of a dilute suspension of noninteracting small monodisperse spherical particles, the interpretation of DLS data gives the autocorrelation function in a simple exponential form:

$$g^{(1)}(\tau) = \exp(-Dq^2\tau) \tag{1.24}$$

where τ is the correlation time, q is the scattering vector, and D is the diffusion coefficient given by the Stokes-Einstein equation, which is a definition for an effective sphere of the so-called hydrodynamic radius, R_h.

$$D = \frac{k_B T}{6\pi\eta R_h} \tag{1.25}$$

where η is the viscosity of the medium, k_B is the Boltzmann constant, and T is the absolute temperature.

1.3.2.3 Effect of Polydispersity

For a mixture of particles of different size, the autocorrelation function $g^{(1)}(\tau)$ is related to the distribution of relaxation rates, $G(\Gamma)$, or to the distribution of relaxation times, $A(\tau)$, through a Laplace transformation:

$$g^{(1)}(t) = \int_0^\infty G(\Gamma)\exp(-\Gamma t)d\Gamma = \int_0^\infty A(\tau)\exp(-t/\tau)d\tau \tag{1.26}$$

where $\Gamma = \tau^{-1}$ is the relaxation rate. To get the particle size distribution, $G(\Gamma)$ or $A(\tau)$ should be recovered from the experimentally determined $g^{(1)}(\tau)$ by solving the integral Equation 1.26. Even though this procedure is the so-called ill-posed problem, i.e., the infinitesimally small experimental errors can result in the infinite divergence of the numerical solution, there are several numerical methods, such as CONTIN[64,65] and REPES,[66] that have been developed for the analysis of the measured autocorrelation function.

For polydisperse systems of noninteracting spherically approximated macromolecules, DLS results are often interpreted using the cumulants analysis, in which the autocorrelation function is expanded in moments about the mean relaxation rate, $\bar{\Gamma}$, given by[19]

$$g^{(1)}(\tau) = \exp(-\bar{\Gamma}\tau)\left[1 + \frac{\mu_2}{2!}\tau^2 - \frac{\mu_3}{3!}\tau^3 + \cdots\right] \tag{1.27}$$

where $\bar{\Gamma} = Dq^2$ is the relaxation rate, and $\mu_n = \int_0^\infty (\Gamma - \bar{\Gamma})^n G(\Gamma)d\Gamma$ is the nth cumulant. The cumulant analysis is effective for a relatively narrow particle size distribution when the condition $\Delta\Gamma/\bar{\Gamma} \leq 1$ is satisfied and no more than three terms of the expansion series (Equation 1.27) are meaningful.

The correlation time is determined as $\tau_c = 2\pi/\bar{\Gamma}$. The diffusion coefficient is calculated from $\bar{\Gamma}$ using $\bar{\Gamma} = Dq^2$. "Z-average hydrodynamic radius" can be expressed using the definition in Equation 1.25:

$$\langle R_h \rangle = \frac{k_B T}{6\pi\eta D} \tag{1.28}$$

The reduced second cumulant $\sqrt{\mu_2}/\bar{\Gamma}$ is referred to as the polydispersity index (PI), which is the dimensionless measure of the size-distribution broadness and is related to the polydispersity of the sample as follows:

$$M_w/M_n \approx 1 + \mathrm{PI} \tag{1.29}$$

where M_w and M_n are the weight average and number average molecular mass of macromolecules, respectively. It is worth noting that the system can be considered as monodisperse if PI $\leq$ 0.1. The value $\mu_3/(\mu_2\bar{\Gamma})$ is the asymmetry index (AI), which is the dimensionless measure of the skewness of the particle size distribution. In the most cases, AI << 1 and is negligible.

1.3.2.4 Effect of Size and Shape of Larger Particles

For larger particles (macromolecules) or larger scattering vector, q, the average diffusion coefficient is weighted by the scattering form factor $P(q)$, since it becomes significant when $R_g q < 1$. Assuming a diluted solution with noninteracting particles, and expanding the form-factor into the series over $R_g q$, one can calculate the apparent diffusion coefficient from the mean relaxation rate using the following relation:

$$D_{app}(q) = \bar{\Gamma}(q)/q^2 \approx D_0\left(1 + CR_g^2 q^2\right) \tag{1.30}$$

where C is the so-called architecture parameter. It is obvious that the combination of the architecture parameter and radius of gyration, CR_g^2, can be found if the apparent diffusion coefficient significantly depends on the scattering angle. To estimate the architecture parameter alone, the radius of gyration is measured using SLS. Furthermore, D_0 related to the equivalent hydrodynamic radius of the particles through the Stokes-Einstein equation (Equation 1.25) is determined from the angular dependence of $D_{app}(q)$. It is important to keep in mind that in the Stokes-Einstein approach, a scattering particle (macromolecule) of any shape is defined as an equivalent sphere with the radius R_h. Therefore, the characterization of conformational changes in the particle shape using DLS data is possible only if the information about the particle shape is available *a priori*. For example, assuming that particles are rigid ellipsoids of revolution, e.g., $R_g = R_h$, the hydrodynamic radius for a prolate or oblate particle is a function of its eccentricity, e[19,67]:

$$R_{\mathrm{h}}^{\mathrm{pro}} = \left[\frac{1}{ae_{\mathrm{pro}}} \ln \frac{a(1+e_{\mathrm{pro}})}{b} \right]^{-1} \tag{1.31a}$$

$$R_{\mathrm{h}}^{\mathrm{ob}} = \left[\frac{1}{be_{\mathrm{ob}}} \tan^{-1} \frac{be_{\mathrm{ob}}}{a} \right]^{-1} \tag{1.31b}$$

where a is the length of the particle symmetry axis (the long axis for a prolate particle or the short axis for an oblate particle), b is the length of the perpendicular axis, and $e_{\mathrm{pro}} = \sqrt{1-b^2/a^2}$ and $e_{\mathrm{ob}} = \sqrt{1-a^2/b^2}$ are two eccentricity parameters for prolate and oblate particles, respectively. For both cases, the zero eccentricity corresponds to spherical particles. The case of a cylindrical rod can be derived from the prolate ellipsoid of revolution within the limit of the large axial ratio.

1.3.2.5 Effect of Concentration

Salt-induced protein precipitation is a key purification step in commercial and laboratory bioseparations.[68–72] For concentrated systems or macromolecular solutions with strong intermolecular interactions, the diffusion coefficient determined by DLS exhibits the concentration dependence.[73, 74]

From the nonequilibrium thermodynamics,[75,76] the translational diffusion coefficient contains both kinetic and thermodynamic factors:

$$D = b_2 x_2 \left(\frac{\partial \mu_2}{\partial x_2} \right)_{p,T} \tag{1.32}$$

where $b_2 = 1/N_A f$ is the mobility of the macromolecules,[77] N_A is the Avogadro number, and f is the particle–solvent friction coefficient. The exact calculation of both mobility and $(\partial \mu_i/\partial x_i)_{p,T}$ in nonideal solutions is an unsolved problem. However, the concentration effect on the diffusion coefficient can be derived by representing the chemical potential (Equation 1.6) and friction coefficient, $f = f_0(1 + k_f c_2 + \cdots)$, as virial expansions in Equation 1.32:

$$D = \frac{RT(1+2A_2M_2c_2 + \quad)}{N_A f_0(1+k_f c_2 + \quad)} \approx \frac{k_B T}{f_0}(1+(2A_2M_2 - k_f - 2v_2)c_2 + \quad) = D_0\left(1+k_d c_2 + \quad\right) \tag{1.33}$$

where f_0 is the friction coefficient at infinite dilution; $D_0 = k_B T/f_0$ is the translational diffusion coefficient at infinite dilution; $k_d = 2A_2M_2 - k_f - 2v_2$ is the so-called hydrodynamic virial coefficient, often termed as the interaction parameter; and v_2 is the partial specific volume of the solute. Note that the approximation $(1 - x)^{-1} \approx (1 + x)$ has been used again in deriving Equation 1.33. Within the framework of the Stokes-Einstein approach (cf. Equation 1.25), the friction coefficient per particle at infinite dilution is $f_0 = 6\pi\eta R_h$.

For proteins, DLS experiments show[72] that the interaction parameter k_d changes from positive to negative with an increase in the salt concentration. This observation indicates the changes in macromolecular interactions from repulsion to attraction due to enhanced screening of Coulombic repulsion by counterions with the increase in ionic strength.

1.3.2.6 "Dynamic" Zimm Plot

Putting Equations 1.30 and 1.33 together, one can derive the relationship between the concentration, scattering angle, and apparent diffusion coefficient, as determined in the DLS technique:

$$D_{\text{app}}(c_2, q) = \bar{\Gamma}(c_2, q)/q^2 \approx D_0\left(1 + k_d c_2 + \cdots\right)\left(1 + CR_g^2 q^2 + \cdots\right) \qquad (1.34)$$

By analogy with the Zimm plot procedure (Section 1.3.1.4), the experimental values of $\bar{\Gamma}(c_2, q)/q^2$ can be plotted against $q^2 + kc_2$ to form a grid of points. This special graphing technique can be named a "dynamic" Zimm plot.

Extrapolation of all points corresponding to different concentrations, c_2, at the identical scattering angle (θ = const) to zero concentration ($c_2 = 0$) yields a straight line with a slope of CR_g^2. Analogously, extrapolating all points at constant c_2 but at different values of the scattering angle to zero scattering angle ($\theta = 0$), one can obtain a straight line with a slope of k_d. Two limiting straight lines intercept at the y-axis to give an estimate of D_0, which inversely relates to the hydrodynamic radius R_h. Two points are noteworthy:

1. The success of the whole procedure requires that $D_{\text{app}}(c_2,q)$ should significantly vary with the scattering angle and concentration, meaning that the size of the scatterers and scattering angle must satisfy the conditions $0.1 < qR_g < 2$, whereas the concentration and interactions must satisfy the conditions $0.1 < k_d c_2 \leq 1$.
2. If the practical upper limits of one or both restrictions are exceeded due to either strong interactions or large sizes, the "dynamic" Zimm plot becomes quite nonlinear, and the higher terms should be kept within the virial expansions over c_2 and q (see Equation 1.34).

1.3.2.7 Some Applications

There are a number of macromolecular systems that cannot be approximated as an ensemble of Brownian spherical particles described by the simple Stokes-Einstein equation. The incomplete list of such systems includes scatterers of an essentially nonspherical shape and macromolecules, diffusion of which is complicated by chemical reactions or conformational transitions. For all listed cases, the electric field autocorrelation function is represented by two terms under the following set of quite strong restrictions: the scatterers are small; their solutions are diluted; and the rotational, chemical, and conformational modes of molecular motion are independent of their position (translational mode)

$$g^{(1)}(\tau) = B_0 \exp(-\Gamma\tau) + B_1 \exp[-(\Gamma + \tau_1^{-1})\tau] \tag{1.35}$$

where $\Gamma = D_T q^2$ is the relaxation rate for the translational mode, and τ_1 is the characteristic relaxation time for the corresponding nontranslational mode. For example, for nonspherical particles,[78] the relaxation time relates to the rotational diffusion coefficient, $\tau_1 = (6D_R)^{-1}$, and the values of B_0 and B_1 are different for distinct models of scatterers (rods, disks, ellipsoids of revolution, etc.).

Changes in concentrations of a reacting mixture and spontaneous fluctuations in the position of the reaction chemical equilibrium will modulate the amplitude of the scattering field, resulting in an additional spectral component with a line width proportional to the reaction rate constant τ_1^{-1} but independent of the scattering angle.[19]

Macromolecules in solution may have different conformations with almost the same energy. Two types of transition can be distinguished between the molecular conformations:

1. Transition from one conformation to another when the properties of the surrounding medium (pH, temperature, ionic strength) change
2. Conformational fluctuations in flexible polymers and biomacromolecules (for example, oscillations of long rigid rods with bending point in the middle). These dynamic transitions are characterized by the relaxation times of the internal modes of motion, τ_i.

The common feature of the aforementioned nontranslational modes is that their characteristic relaxation time, τ_1, is independent of the scattering angle, θ. Therefore, DLS measurements at small angles, when $D_T q^2 \to 0$, allow one to determine the relaxation times, τ_1, whereas the measurements at quite large angles, when $D_T q^2 >> \tau_1$, give the way to recover the translational diffusion coefficient.

The diffusion coefficients for charged and neutral macromolecules of the same size and mass in solution may significantly differ due to (a) the ability of charged molecules to form a cloud of surrounding counterions and (b) the influence of electrostatic intermolecular interactions on the chemical potential. The most viable approach to the problem of charged-macromolecule diffusion[72] seems to be the one based on the concentration dependence of the apparent diffusion coefficient for different ionic strengths, as discussed in Section 1.3.2.5.

Nevertheless, there is an electrophoresis-like method of dynamic light scattering to determine charge-carrying by a macromolecule. If the electric field E is applied to the solution of macromolecules carrying charge Q, the charged molecules begin to move with the velocity of $v = EQ/f$, where f is the friction coefficient, as defined in Equations 1.32 and 1.33. The spectrum of light scattered on moving particles is shifted by the frequency proportional to the velocity, $\omega_0 = qv$. The corresponding field–field autocorrelation function is modulated by the same frequency ω_0:

$$g^{(1)}(\tau) = \exp(-Dq^2\tau)\cos\omega_0\tau \tag{1.36}$$

Practically, Equation 1.36 can be fitted into the $g^{(1)}(\tau)$ extracted from DLS experimental data, and the hydrodynamic radius of the particle, R_h, is calculated from D, whereas the charge on the particle, Q, is calculated from ω_0.

1.3.3 Combining Static and Dynamic Light Scattering

For the macromolecular solutions, the six parameters can be directly extracted from the combination of SLS and DLS (static and dynamic, respectively) Zimm plots, namely:

1. Weight-average molecular mass M_2 as an intercept of the angular and concentration dependences of the scattering-intensity-related quantity $Kc_2/R_\theta(q,c_2)$ at $q \to 0$ and $c_2 \to 0$
2. z-Average radius of gyration, R_g, as the slope of the angular dependence of $Kc_2/R_\theta(q,c_2)$ at $q \to 0$ and $c_2 = \text{const}$
3. The second virial coefficient, A_2, as the slope of the concentration dependence of $Kc_2/R_\theta(q,c_2)$ at $c_2 \to 0$ and $q = \text{const}$
4. Diffusion coefficient D_0 as an intercept of the angular and concentration dependences of the relaxation-rate-related quantity $\Gamma(q,c_2)/q^2$ at $q \to 0$ and $c_2 \to 0$
5. Molecular architecture parameter, CR_g^2, as the slope of the angular dependence of $\Gamma(q,c_2)/q^2$ at $q \to 0$ and $c_2 = \text{const}$
6. The hydrodynamic virial coefficient k_d as the slope of the concentration dependence of $\Gamma(q,c_2)/q^2$ at $c_2 \to 0$ and $q = \text{const}$

Furthermore, two types of interactions — hydrodynamic and thermodynamic — that occur between macromolecules in a solution of a finite concentration can be defined as specific dimensions calculated from the directly measured parameters. The average hydrodynamic radius $R_h = k_BT/6\pi\eta D_0$ characterizes hydrodynamic interactions between a macromolecule and a solvent. It indicates how deeply a particle is drained by the solvent:

a deep draining causes a reduction in R_h; on the other hand, if shallow draining is the only possible alternative, R_h can become much larger than the geometrically defined size of a macromolecule (i.e., the radius of gyration, R_g). The thermodynamic interactions (repulsion or attraction) between macromolecules are characterized by the thermodynamically effective equivalent sphere radius $R_T = (3A_2M_2/16\pi N_A)^{1/3}$, which is the spatial measure of the domain of interpenetration of two macromolecules; in other words, the thermodynamic interactions are characterized by the excluded volume.

Generalized ratios of these differently defined radii can also be derived. The ratio $\rho = R_g/R_h$ compares the range of hydrodynamic interaction with the geometrical dimensions of a molecule and can be theoretically calculated for a particular structure. For example, for a homogeneous sphere, ρ is less than unity; for microgels it can be less than 0.5; for a random coil it changes from 1.5 to 2, depending on polydispersity, solvent, and conditions; and for a rigid rod it is above 2. The ratio $V_T = R_T/R_h$ compares the range of thermodynamic interactions between macromolecules with that of hydrodynamic interactions of macromolecules and solvent. Experiments mostly demonstrate that the ratio is close to unity, indicating that the thermodynamic and hydrodynamic interactions act over similar distances.

Thus, on the one hand, a complete picture of structural features of macromolecules in solution can be obtained by the parallel analysis of all parameters or their combinations as a result of simultaneous SLS and DLS measurements in the same sample under the identical experimental conditions. On the other hand, caution must be used, since this approach is valid under rather strong limitations imposed on the concentration ($2A_2M_2c_2 \leq 0.5$) and scattering angles ($qR_g < 2$) when using the corresponding Equations 1.15 and 1.34. Experimentally, the macromolecular solution should be transparent, without aggregates, not highly concentrated, and without statistical limitations on the molecular motion (ergodic system). In the following, let us consider how the combined SLS and DLS can be applied under the aforementioned complications.

1.3.3.1 Aggregates

In SLS and DLS measurements, many efforts are made to eliminate polymer aggregates from the solutions under study. However, in bioseparation, the aggregation in concentrated macromolecular solutions, especially near the point of solubility phase transition, is an inevitable process preceding precipitation. This makes the characterization of those solutions by light scattering difficult because of at least two problems: (a) the aggregates interact with the nonaggregating component and (b) the aggregates of different conformations (wormlike chains, clusters, and so on) may exist and contribute to the static and dynamic structure factors.

Recent theoretical and experimental light scattering studies of aggregating macromolecular systems[79] show that the contribution of initial polymers and their aggregates can be separated if (a) the size of the aggregates is much greater than the size of the initial polymer and (b) the concentration of aggregates is low in comparison with the concentration of the nonaggregating polymer.

For mixtures of polymer molecules and their aggregates that are quite different in size, the regularization algorithm of inverse Laplace transformation (see Section 1.3.2.3), applied for the autocorrelation function $g^{(1)}(\tau)$ determined from DLS measurements using CONTIN or REPES analysis, results in the bimodal distribution of relaxation times $A(\tau)$, which are easily separated into the fast and slow relaxation modes:

$$\sum_{i\in \text{fast}} A(\tau_i) + \sum_{i\in \text{slow}} A(\tau_i) = 1 \tag{1.37}$$

where the summations are taken over τ_i belonging to each mode. The first cumulant of the corresponding relaxation mode can be calculated to give the fast (Γ_{fast}) and slow (Γ_{slow}) relaxation rates:

$$\Gamma_{\text{fast}} = -\lim_{t\to 0} \frac{d}{dt} \ln\left[\sum_{i\in \text{fast}} A(\tau_i)\exp(-t/\tau_i)\right] = \sum_{i\in \text{fast}} \tau_i^{-1} A(\tau_i) \Big/ \sum_{i\in \text{fast}} A(\tau_i) \tag{1.38a}$$

$$\Gamma_{\text{slow}} = -\lim_{t\to 0} \frac{d}{dt} \ln\left[\sum_{i\in \text{slow}} A(\tau_i)\exp(-t/\tau_i)\right] = \sum_{i\in \text{slow}} \tau_i^{-1} A(\tau_i) \Big/ \sum_{i\in \text{slow}} A(\tau_i) \tag{1.38b}$$

Furthermore, static structure factors for the fast (molecularly dispersed polymer) and slow (aggregates) relaxation modes can be obtained by[79]

$$S_{\text{fast}}(q) = S(q) \sum_{i\in \text{fast}} A(\tau_i) \tag{1.39a}$$

$$S_{\text{slow}}(q) = S(q) \sum_{i\in \text{slow}} A(\tau_i) \tag{1.39b}$$

where $S(q)$ is the total static structure factor obtained from the static light scattering experiment as $Kc_2/R \approx [S(q)]^{-1}$ (for notations, see Equations 1.7 and 1.8). This is an explicit example of how a combination of dynamic and static light scattering methods gains the information on a mixture of a polymer and its aggregates. Once the size of a molecularly dispersed polymer is assumed to be small but the polymer solution is concentrated, $S_{\text{fast}}(q)$ is independent of the scattering angle but is concentration dependent, so that Equation 1.17 can be applied. In contrast, the aggregating component contribution to the static structure factor $S_{\text{slow}}(q)$ is a function of both the concentration and scattering vector given by

$$[S_{\text{slow}}(q,c_2)]^{-1} \approx [S_{\text{slow}}(0,c_2)]^{-1}\left(1 + R_{g,A}^2 q^2/3 + \cdots\right) \tag{1.40}$$

where $R^2_{g,A}$ is the z-average mean-square radius of gyration of the aggregates. If the degree of aggregation is independent of concentration, it can be extracted from the second fitting parameter $[S_{slow}(0,c_2)]^{-1}$.

The z-average hydrodynamic radius of aggregates, $R_{h,A}$, is obtained by extrapolating Γ_{slow} to the zero scattering angle at constant concentration:

$$\lim_{q\to 0}\frac{\Gamma_{slow}}{q^2}=\frac{k_BT}{6\pi\eta R_{h,A}} \tag{1.41}$$

To analyze the shape of the aggregates, the ratio of the radius of gyration to the hydrodynamic radius, $\rho = R_{g,A}/R_{h,A}$, can be calculated for different polymers, concentrations, and experimental conditions. It is known[80] that $\rho = 0.775$ for a sphere with a uniform density, and it increases with increasing axial ratio but decreases with increasing density of the sphere in the central part. Flexible linear chains take values of ρ from 1.2 to 1.5. For stiff polymer chains, ρ can exceed 2.

A more complex system of aggregating macromolecules is a mixture of a molecularly dispersed polymer and aggregates of different size and shape. In this case, the analysis of the field autocorrelation function, $g^{(1)}(\tau)$, by an inverse Laplace transformation (CONTIN or REPES) does not yield a consistent picture of the dynamic behavior of the system. In particular, the number of resolved relaxation modes, their mean rates Γ_I, and their widths vary arbitrarily with the scattering angle. An alternative way of analyzing the measured correlation functions is the use of a so-called model-based approach.[81] In this approach, *a priori* assumptions on the shape of the existing aggregates should be made based on other experimental microscopic or spectroscopic methods. For example, if a hypothetical macromolecular system consists of flexible wormlike chains (fibers), clusters, and spherical formations (or molecularly dispersed polymers), the field autocorrelation function $g^{(1)}(\tau)$ of the three species, as determined from DLS measurements, can be written as follows:

$$g^{(1)}(\tau)=\frac{\gamma_F g_F^{(1)}(\tau)+\gamma_C g_C^{(1)}(\tau)+\gamma_S g_S^{(1)}(\tau)}{\gamma_F+\gamma_C+\gamma_S} \tag{1.42}$$

where $\gamma_i(q)$ are the angular dependent amplitudes of the field autocorrelation functions for fibers (F), clusters (C), and spheres (S), respectively. Even if the cluster and sphere contributions are represented by single-exponential correlation functions $g_{C,S}^{(1)}(\tau)=\exp(-\Gamma_{C,S}\tau)$ with relaxation rates $\Gamma_{C,S} = D_{C,S}q^2$, the fiber contribution will require a more sophisticated model that takes into account the polydispersity distribution of the contour length L and the flexibility (l_p/L) of the wormlike chains, as well as the translational and rotational motion of the fiber aggregates. For example, according to the Schulz-Zimm model, $g_F^{(1)}(\tau)$ is given by[82,83]

$$g_F^{(1)}(\tau) = \frac{\sum_L p(\mathrm{PI}, L_n, L) g_F^{(1)}(\tau, L)}{\sum_L p(\mathrm{PI}, L_n, L)} \tag{1.43}$$

where $p(\mathrm{PI}, L_n, L)$ is the *z*-average size distribution with polydispersity index $\mathrm{PI} = M_w/M_n - 1$, and L_n is the number-average length of the fiber aggregates. To approximate the fiberlike aggregates as stiff cylinders with a low flexibility ($L \approx l_p$), the model of stiff rods was proposed by Pecora[78]:

$$g_F^{(1)}(\tau) = \sum_{n=0}^{N} B_n(2n, qL) \exp\left[-\left(D_F^{(\mathrm{tr})} q^2 + x_n D_F^{(\mathrm{rot})}\right)\tau\right] \tag{1.44}$$

where $x_n = 4n^2 + 2n$. The number of terms as well as the amplitudes B_n depend on the value of qL. The translational and rotational diffusion coefficients depend on the geometry of the rods (length L and diameter d):

$$D_F^{(\mathrm{tr})} = \frac{k_B T}{3\pi\eta L} \ln\left(\frac{L}{d}\right) \tag{1.45a}$$

$$D_F^{(\mathrm{rot})} = \frac{9 D_F^{(\mathrm{tr})}}{L^2} \tag{1.45b}$$

Practically, by recoding the autocorrelation functions for different scattering angles, one can fit the described model into the experimental data to recover the average length L_n, shape, and width of the length distribution represented by the polydispersity index PI as well as the characteristics of the cluster aggregates and spherically approximated polymer.

1.3.3.2 Nonergodic Systems

Concentrated macromolecular solutions with a high ability to aggregate and with strong interactions between macromolecules and their aggregates can form gel-like or even solidlike media (Figure 1.8). Such states precede polymer and protein precipitation and are not rare in bioseparation and bioprocessing. To optimize separation and purification protocols, it is important to characterize the structural and dynamic properties of the precipitating biomacromolecules and smart polymers, especially at the intermediate stage. Light scattering measurements in these dense systems at the transitional

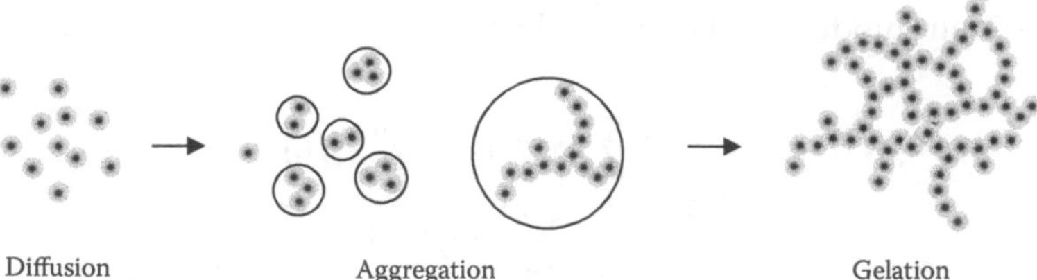

FIGURE 1.8
The illustrative transition from diffusion to aggregation to gelation: attractive interparticle interactions enforce the formation and growth of aggregates into clusters and flexible fibers until they form a solidlike medium filled with a cross-linked network.

steps frequently fail due to strong multiple light scattering and poor transparency (high turbidity). The so-called ergodicity of the scattering medium is a prerequisite for DLS applicability; however, most of the concentrated, highly aggregated, and cross-linked media are nonergodic.

Ergodicity of the medium means that the measured time-averaged quantities $\langle \ldots \rangle_T$ (such as the scattered intensity or its autocorrelation function) are identical to the ensemble-averaged ones $\langle \ldots \rangle_E$. In nonergodic random media, the scatterers are localized near fixed average positions due to the spatial restrictions of the individual particle motion. As a result, time averaging will differ from the ensemble averaging, i.e., $\langle \ldots \rangle_T \neq \langle \ldots \rangle_E$. Experimentally, repeated measurements of the time-averaged parameters on the same nonergodic sample yield a set of different values, making the characteristic value of the medium uncertain. Furthermore, the condition of ergodicity, $\langle \ldots \rangle_T = \langle \ldots \rangle_E$, is required for theoretical interpretation of experimental data, since the ensemble-averaged parameters are commonly calculated theoretically, while time averaging is the common experimental procedure.

In 1989, Pusey and van Megen[84] first proposed the theory and method for DLS in nonergodic media. Good reviews of the methods for proper averaging of the data obtained in DLS experiments can be found in the literature.[85–87] The idea is to collect the autocorrelation functions of light scattered by different parts of the nonergodic sample in the course of its slow motion or rotation in order to perform ensemble averaging of scattered light. The rotation/translation method can be extended to concentrated, turbid suspensions without any principal difficulties.

For a nonergodic medium, the time-averaged scattering intensity $\langle I \rangle_T$ for each chosen sample position at a given scattering vector comprises a fluctuating but position-independent part $\langle I_F \rangle_T$ (sometimes called the dynamic contribution) and a time-independent but position-dependent part $\langle I_S \rangle_E$ (also called the static contribution):

$$\langle I \rangle_T = \langle I_F \rangle_T + \langle I_S \rangle_E \tag{1.46}$$

Obviously, being the ensemble (position)-averaged term of scattered intensity, $\langle I_S \rangle_E$ is a measure of the degree of the spatial inhomogeneities. On the other hand, the time-averaged autocorrelation function extracted from DLS data for a given sample position can, in many cases, be represented by a single exponential function:

$$g^{(1)}(\tau) = \exp(-D_A q^2 \tau) \tag{1.47}$$

where D_A is the apparent diffusion coefficient, which is not identical to the so-called collective diffusion coefficient D, which is the factual dynamic characteristic of a nonergodic medium. It has been shown[88] that D_A and D are related as follows:

$$\frac{\langle I \rangle_T}{D_A} = \frac{2\langle I \rangle_T}{D} - \frac{\langle I_F \rangle_T}{D} \tag{1.48}$$

Experimentally, for each sample position, $\langle I \rangle_T$ is measured from SLS data, and D_A is calculated from DLS data. After taking measurements at a number of sample positions, a linear plot of $\langle I \rangle_T / D_A$ against $\langle I \rangle_T$ can be constructed, and the values of D and $\langle I_F \rangle_T$ are obtained from the slope and intercept, respectively. The measure of the degree of the spatial inhomogeneities in the nonergodic medium is calculated from the following expression:

$$\langle I_S \rangle_E = \langle \langle I \rangle_T(p) - \langle I_F \rangle_T \rangle_E = \langle I \rangle_E - \langle I_F \rangle_T \tag{1.49}$$

where $\langle I \rangle_E = \Sigma_{p=1}^{N} \langle I \rangle_T(p)/N$, $p = 1,2,\ldots,N$, and N is the number of positions measured.

Hence, this is an explicit example of how a combination of static and dynamic light scattering can provide information on the dynamic behavior of a nonergodic system. The variations of the time-independent part, $\langle I_S \rangle_E$, and the time-fluctuating part, $\langle I_F \rangle_T$, of the scattering intensity, as well as the tendency of the collective diffusion coefficient D to change in the proximity of phase transitions, can be studied for concentrated solutions of smart polymers, biopolymers, their aggregates, and their gels, depending on their origin, modifications, immobilized ligands, conditions of preparation and utilization, and surrounding environments. This technique is mainly used by specialists, but it is expected that it eventually will be disseminated into applied research and industrial environments.

1.4 Light Scattering Study on Phase Transitions, Conformational Transformations, and Interactions in Smart Polymer Solutions

1.4.1 Transition-Point Determination (Cloud Point)

The simplest method to detect a point of phase separation in aqueous polymer solutions is referred as the cloud-point method.[89] It is based on a visual observation of an emergence of a milky opalescence at a specific temperature. As was stated above (Section 1.1), the strong increase in the scattering intensity (critical opalescence) results from the grown concentration fluctuations near the critical point, when the size of those fluctuations exceeds the wavelength of the scattered light. The researchers who use a standard UV-VIS spectrophotometer[1,46–48,90,91] (pp. 1–26) to determine the cloud point still measure the relative changes of scattering intensity in the course of nearing the transition point, even if they employ different wavelengths or integrated intensity over a spectral interval. In many cases, the solution may remain transparent, and it may be difficult to determine the phase transition point based on visual observation or even with the use of a UV-VIS apparatus. The light scattering technique is the most precise method of observing the process of phase separation not only in the polymer solutions, but also in binary mixtures of organic solvents.[16] Herein, the static LS allows one to follow the growing fluctuations of concentration by recording the scattering intensity at the scattering angles suitable for the hydrodynamic regime ($qR_g < 1$). The dynamic LS allows one to detect (a) changes in the hydrodynamic size and shape of macromolecules and (b) their aggregation following their precipitation in the course of transition.

Figure 1.9 shows a sharp increase in the scattering intensity at 34.9 ± 0.1°C for the aqueous solution of poly(*N*-isopropylacrylamide) (PNIPA).[15] However, the transition can be rather smooth, resulting in uncertainty in the definition of the transition temperature: is it the very beginning of the intensity increase (point 1 in Figure 1.9a) or the reaching of a plateau (point 2 in Figure 1.9a)? To

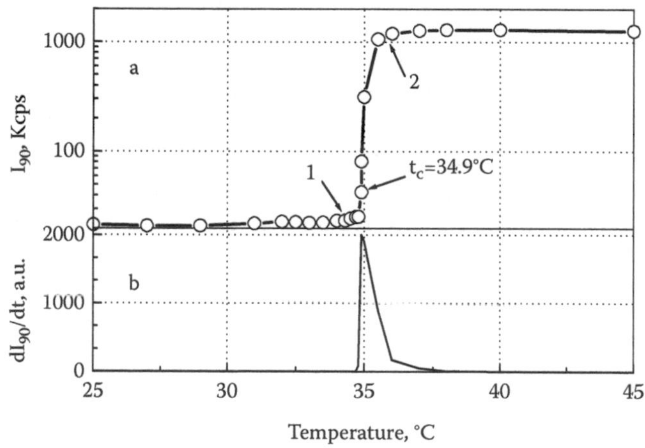

FIGURE 1.9
Determination of the cloud point of PNIPA in aqueous solution by SLS: (a) scattering intensity and (b) its derivative over temperature. PNIPA was synthesized by radical polymerization in aqueous solution at room temperature using a redox pair — ammonium persulfate/tetraethylene methylenediamine (TEMED) — as an initiator, and PNIPA was then purified by three sequential reprecipitations at 40°C.

avoid uncertainty, the temperature corresponding to the maximum rate of the intensity growth can be assigned as the transition temperature. In this case, to differentiate the precipitation curve, a sufficient number of experimental points is required within the transition interval, which is not always experimentally possible.

Although there are other methods for determining the phase-separation point in polymer solutions, e.g., differential scanning calorimetry[44,48] or even nuclear magnetic resonance,[45] the methods that use the light scattering phenomenon (visual observation, UV-VIS, SLS, and DLS) still remain the simplest and most convenient. It is also possible to follow the temperature dependence of the average hydrodynamic size of the polymer chain: the temperature at which R_h starts to increase, i.e., when the individual polymer chains begin to aggregate, can be defined as the LCST.[92]

There is a vast pool of papers published on phase separation in polymer/water solutions. It is interesting to note that a number of synthetic polymers have been reported to exhibit a lower critical solution temperature (LCST) in water. Those polymers have been proposed to model protein denaturation.[46,93–95] There is a strong interest in those polymers with respect to their use for separation systems in biotechnology.[1,2] However, the question is still open as to whether it is crucial to use the LCST polymers in bioseparation or whether the UCST polymers can also be used if they exhibit the transition point within a suitable temperature range.

Water-soluble polymers with LCST, for instance poly(*N*-isopropylacrylamide) (PNIPA) or poly(*N*-vinylcaprolactam) (PVCL), have a rather flat top of the coexistence curve.[42–44,46,92] On a practical level, this means that

1. The phase-separation temperature does not change within a wide range of polymer concentration.
2. All phase-separation temperatures are equal to the LCST within this range.
3. The spinodal and binodal cannot be experimentally distinguished. Along this flattened top of the coexistence curve, the system exhibits properties similar to the critical point because $(\partial^2 G/\partial x^2)_{p,T} = 0$ on the spinodal.

Figure 1.10 summarizes the numerous effectors that shift the position of the flattened part of the coexistence curve to higher or lower temperatures, that narrow or broaden the immiscibility region, and that change the critical temperature and concentration in smart-polymer/water systems.

Besides the study of phase-separation coexistence curves for polymer/water systems, the precipitation curve, as a temperature-dependent light scattering intensity measured over a wide range of temperatures before and after thermoprecipitation, *per se* may be a highly informative method for distinguishing the properties of copolymers of different composition. For example, 1-vinylimidazole copolymer with *N*-vinylcaprolactam was studied as a carrier for immobilized metal affinity precipitation,[96,97] a displacement chromatography

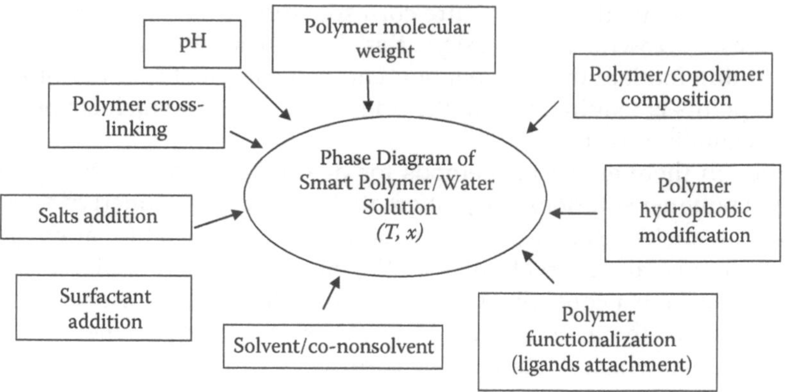

FIGURE 1.10
Different effectors of the shape and parameters of polymer/water phase diagrams as studied by LS techniques.

technique for preparative separation of proteins. In this study, the parent copolymer with molecular mass of 11,700 g/mol (determined by combined SLS and DLS) containing 9:1 molar portions of VCL and VI (11%), respectively, was separated on a Cu^{2+}-IDA-Sepharose column into the fraction (8.5% of VI) purified by IMAC and the nonadsorbing fraction (5% of VI). Interestingly, the precipitation curves for different fractions and parent copolymer were different in terms of the cloud-point temperature and the general shape. Namely, the nonadsorbing fraction of the copolymer exhibited a substantially wider and smoother phase transition beyond 48°C, which is much higher than the temperature for the parent copolymer and its purified fraction.

The scattering-intensity-precipitation-curve method has been further developed for the study of the Cu^{2+}-IDA-Sepharose-bound and -unbound fractions of another so-called proteinlike copolymer, poly(*N*-isopropylacrylamide)-*co*-(*N*-vinylimidazole) (PNIPA-VI), at different pH levels.[98] The molecular weights calculated from Zimm plots of the bound and unbound PNIPA-VI fractions were 1.4×10^5 and 1.4×10^6 g/mol, respectively. The content of *N*-vinylimidazole in both fractions was also different: the bound fraction was somewhat enriched with vinylimidazole (9.1%) compared with the unbound fraction (7.8%). Strikingly, although the unbound fraction contained some amount of vinylimidazole, its transition temperature (~32°C), recorded at the very beginning of the intensity increase (see Figure 1.8), was similar to that of PNIPA independent of pH. Only the sharpness of the transitions decreased at lower pH. In contrast, the transition temperatures for the bound fraction were strongly dependent on pH (Figure 1.11), and this dependence was correlated with pH variations of the hydrodynamic diameter of the bound PNIPA-VI copolymer as determined by DLS. Whether the existence of two fractions is a consequence of the preparation method (polymerization) or fractionation process is still an open question that must be answered to improve the yield of IMAC.

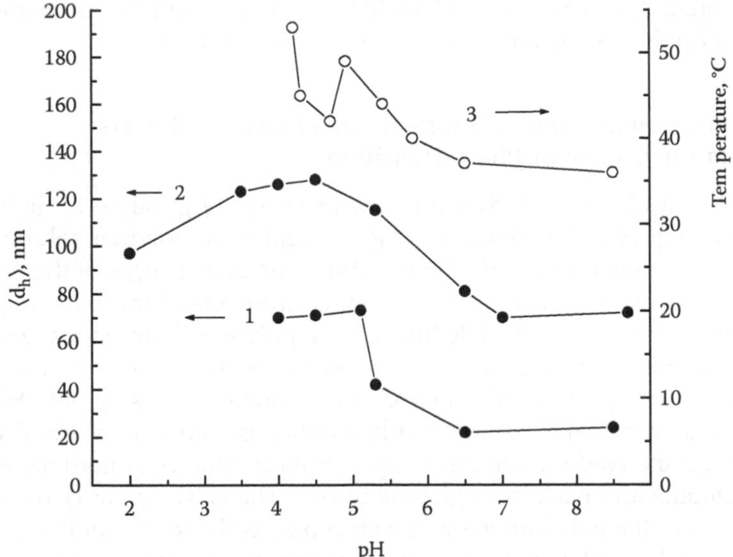

FIGURE 1.11
pH dependences of: (curve 1) the hydrodynamic diameter of the Cu^{2+}-IDA-Sepharose-bound PNIPA-VI copolymer; (curve 2) the hydrodynamic radius of the PNIPA-VI nanogels; and (curve 3) the beginning precipitation temperatures of the bound PNIPA-VI copolymer. For experimental conditions, see Wahlund et al.[98] and Kazakov et al.[99]

In Figure 1.11, the pH-dependence of transition temperatures for the bound PNIPA-VI fraction is compared with the corresponding changes in the hydrodynamic diameter of PNIPA-VI nanogels photopolymerized in a liposomal microreactor.[99] In the PNIPA-VI copolymer solution (curve 1), a decrease in pH from 6.5 to 4.5 (an increase in charge density on the chain) leads to a transition of polymer chains from a compact conformation (random coil) to a highly expanded state (a perturbed random coil).

The behavior of PNIPA-VI nanogels is somewhat similar (curve 2). A decrease in pH from 7 to 4.5 results in their swelling due to imbalance between repulsive electrostatic forces (ionic osmotic pressure) and attractive forces of the polymer network (network swelling pressure), until all imidazolyl groups are protonated at approximately pH 4.5. For the pH range from 4.5 to 2, a further decrease in pH results in an excess of H^+ and Cl^- ions around the ionizable groups, and the nanogels undergo contraction due to the screening effect.[100,101] In contrast, there is no significant change in the size of the hydrogel nanoparticles in the range of pH from 7 to 9, as most imidazolyl groups are already deprotonated.

Figure 1.11 explicitly reveals a correlation between the sizes of both aforementioned systems and their precipitation temperatures (curve 3). A steep drop of the transition temperature, followed by its further growth and eventual smoothing of the total precipitation curves when all imidazolyl groups are protonated, can be ascribed to the severe conformational changes of PNIPA-VI polymeric chains due to the protein unfolding.

1.4.2 Sizes, Shape, and Conformations of Smart Polymers in the Course of Phase Transition

Measurement of the hydrodynamic size or radius of gyration by light scattering techniques in the course of phase separation in polymer solutions has become a common method of following the structural changes of the polymer before and after the transition. The transition from a random coil to a globule has been predicted for a flexible linear homopolymer chain once the solvent quality changes.[102] An elegant work by Wu's research group[103] presents an example of the experimental study of this phenomenon. Using both SLS and DLS measurements for extremely dilute aqueous solutions of PNIPA, they reported on the conformational changes of individual polymer chains from coil to globule and back from globule to coil. The existence of two intermediate states of the polymer chain, the crumpled coil and the molten globule, was revealed by analyzing the temperature dependence of the shape-sensitive ratio $\rho = R_g/R_h$. The coil-to-globule transition of PNIPA in concentrated solutions can also be detected,[15] but caution in analyzing light scattering data is required, since the globules formed above the transition temperature in a concentrated polymer/water solution tend to aggregate and then precipitate.

Static LS measurements (angular and concentration dependences) have been carried out at 23 and 37°C, i.e., before and after the transition characterized by the precipitation curve in Figure 1.9. At 37°C, the aggregates exist in solution as a stable suspension. The SLS data at 23°C are presented as a typical Zimm plot (Figure 1.6). Interestingly, the aggregates formed after transition were 60-fold heavier ($M_2 = 126 \times 10^6$ g/mol) but two times smaller (R_g = 50 nm) than polymer chains in the state of random coils before the transition ($M_2 = 2 \times 10^6$ g/mol, R_g = 100 nm). The second virial coefficient was found to drop from 1.1×10^{-3} to 1.3×10^{-6} mol·ml/g^2, indicating Θ-conditions of the solution of aggregates. The values of the thermodynamic radius $R_T = (3A_2M_2/16\pi N_A)^{1/3}$, estimated before (28 nm) and after (12 nm) the phase transition, indicate strong interactions for both random coils and aggregates of globules.

It has been shown[104] by combined SLS and DLS that conformational states of smart polymer (PNIPA) can be governed by the addition of a surfactant, sodium dodecyl sulfate (SDS). Three areas on the "temperature–surfactant concentration" phase diagram have been discovered wherein the polymer existed as globules, coils, or expanded coils. This suggests that variations in temperature or surfactant concentration can initiate the following conformational transitions: globule-to-coil, globule-to-expanded coil, coil-to-expanded coil, and the corresponding reverse transitions.

It is not straightforward how the conformational transformations of smart polymers in the course of phase transition may contribute to biotechnological implementations. Nevertheless, intuitively, one could expect such conformational changes to be tiny tools for optimization of biotechnological processes. Hopefully, they can enhance protein separation and purification.

1.4.3 Interactions: Polymer–Solvent, Polymer–Polymer, Polymer–Protein

Depending on the charge density, charge distribution, chain length, chain conformation, ionic strength, pH, and distinct ions added, self-association of smart polymers in the course of phase transition can result in soluble polymer/ion complexes and stable aggregates or amorphous precipitates. Dynamic and static light scattering techniques have been very effective in detecting and characterizing these phenomena.[98,105,106] Although the aggregation makes it difficult to analyze polymer behavior in concentrated polymer solutions by light scattering, recent developments of the method,[79,81,107] discussed in Section 1.3.3.1, allow one not only to separate analytically the contribution of aggregated and nonaggregated portions, but also to extract information on the shape of different aggregates.

Another suitable object of study by light scattering is the polyelectrolyte–polyelectrolyte complexes. It is well documented[1,108] (pp. 99–121) that synthetic polymers and natural proteins are able to form either polyelectrolyte complexes (PEC) or protein–polyelectrolyte complexes (PPC) in the course of interaction of the oppositely charged polyions. From a practical point of view, given their property of being reversibly soluble with respect to pH sensitivity, they show great potential for use in bioseparation, biocatalysis/enzymology, bioanalysis,[109,110] immunoassays[1] (pp. 207–229), and for mimicking the regulation of the living cell's metabolism.[111,112] Recently, despite a number of theoretical and experimental problems hindering the acquisition and interpretation of the light scattering data, DLS has become a popular method for the study of structural changes in polyelectrolyte systems comprising polymers, proteins, and their complexes and conjugates.[17,18] It was shown[112] that, even in the case of pronounced aggregation in polyelectrolyte systems, the combination of static and dynamic light scattering was able to facilitate a straightforward interpretation of the light scattering results, providing additional insight into the role played by each component of the conjugate as well as the mechanisms of protein–polymer interactions, conjugate aggregation, and precipitation.

1.4.3.1 Smart-Polymer–Protein Conjugates

Solutions of poly(methacrylic acid) (PMAA), monoclonal antibodies (Ab) from 6C5 clone against denatured rabbit gluceraldehyde-3-phosphate dehydrogenase (dGAPDH), antibody–polymer conjugate (Ab-PMAA), and a mixture of antibody–polymer (Ab-PMAA) and antigen–polymer (Ag-PMAA) conjugates have been analyzed at different pH.[113] This was accomplished by simultaneously recording four light scattering parameters — integral scattering intensity (I_{90}), radius of gyration (R_g), hydrodynamic diameter $\langle d_h \rangle$, and polydispersity index (PI) — for the same sample under the same experimental conditions.

For the PMAA solution, all sets of parameters indicated that, in acidic medium at $pH < 5$, macromolecular chains of this weak polyelectrolyte ($pK \approx 5$) were in the unperturbed random-coil conformation stabilized by the hydrophobic interactions of α-methyl groups and hydrogen bonds between carboxylic and carboxylate groups: –COO…H…OOC–. A pronounced expansion of PMAA chains begins at $pH > 5$, when the deprotonation of carboxylic groups, $COOH \rightarrow COO^-$, results in the mutual repulsion of the COO^- groups.

For the free Ab solution, the detected divergence of all parameters around pH 6, being approached from either the acidic or basic sides, clearly indicated that association of Ab molecules was accompanied by pronounced precipitation. This observation was in agreement with the well-known fact[114] that protein solubility decreases around the isoelectric point, pI. For the antibodies, it was measured to be of pI 5.95.

The solution of Ab–PMAA conjugates exhibited the following striking features:

1. The conjugate precipitated at $pH < 4.8$, which is not the isoelectric point for free Ab but is close to the point of conformational transition of PMAA from an expanded random coil to a compact random coil.
2. Both average sizes (R_g and $\langle d_h \rangle$) of free Ab appeared to be larger than those of the conjugate, suggesting that free Ab molecules are to some extent associated, whereas the presence of a charged polymer covalently attached to the surface of the antibody prevents the assemblage of the protein–polymer conjugates.
3. The average geometrically defined size, R_g, of the Ab–PMAA conjugates detected before precipitation was less than the corresponding size of the expanded coil of the free polymer. This suggests that a strong interaction between the polymer chains and the surface of the protein compresses the extended random coil of polymer up to the surface of the protein globule to form a compact arrangement.
4. On the other hand, the hydrodynamically defined size, $\langle d_h \rangle$, of the Ab–PMAA conjugates exceeded that of the free polymers, indicating the existence of a noticeable hydrodynamic shell of water molecules around the protein–polymer conjugate.
5. In the pH range from 6 to 5, a decrease in R_g and $\langle d_h \rangle$ plus a simultaneous increase in scattering intensity, I_{90}, and polydispersity, PI, of the conjugates unambiguously indicated that the further cooperative shrinkage of the polymer chains around the protein globule preceded the aggregation and precipitation of the conjugates at $pH \approx 4.8$.

Thus, altogether, the observations confirm that the coil conformation of PMAA covalently attached to the antibodies is mainly responsible for the precipitation of Ab–PMAA conjugates in acidic medium.

The antibody–polymer and antigen–polymer conjugates were specifically designed to study their interaction.[112] Consequently, the mixture of Ab-PMAA and dGAPDH-PMAA was prepared and tested with light scattering. In particular, DLS data obtained at pH 7.3 proved the complexation of antibody–polymer and antigen–polymer conjugates by recording a two-fold increase in the average hydrodynamic diameter $\langle d_h \rangle$ for the complexes compared with those for the individual conjugates. The peak of size for the "polymer-Ab–Ag-polymer" complexes in Figure 1.12 shows that the maximum possible association of one antibody with two antigens occurs at about pH 5.5, when PMAA molecules are still in the state of extended random coils. A drop in the average size of the complexes with decreasing pH demonstrates, even more clearly than for the individual conjugates, that the shrinkage of the polymer random coil precedes the precipitation of complexes with the following structure: PMAA-Ag–Ab-PMAA–Ag-PMAA.

Immunoassay technology based on PNIPA-monoclonal human immunoglobulin (Ig) has also been tested by light scattering technique to detect small aggregates, even in cases where the conjugated polymer did not precipitate at elevated temperatures.[115,116]

1.4.3.2 Smart Polymers Modified by Bioaffinity Ligands

Extensive control of protein separation and purification by means of affinity precipitation can be achieved through the use of dynamic light scattering.[117] The affinity precipitation method[1] (pp. 55–77) is based on the selective binding

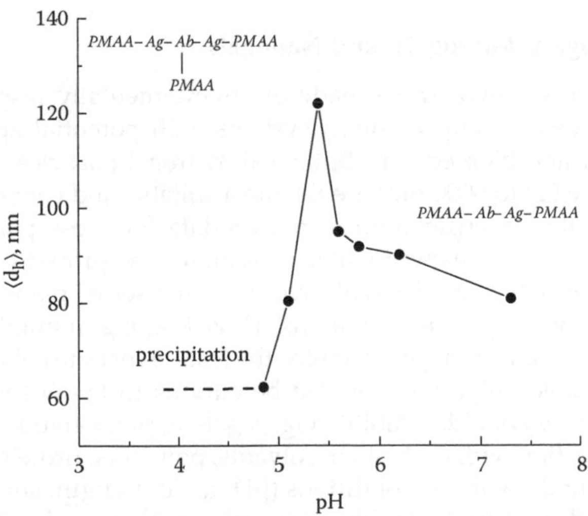

FIGURE 1.12
pH dependence of the z-average hydrodynamic diameter for a mixture of Ab-PMAA and dGAPDH-PMAA conjugates. The hypothetical structures of the conjugate–conjugate complexes are shown in the corresponding pH regions.

of a targeted protein to a smart protein with a controllable solubility. In this work, p-aminophenyl-α-D-glucopyranoside (pAP-α-D-Glu), a ligand specific to Concanavalin A (Con A), was attached (80 ligands/polymer chain) to Eudragit S-100, a 1:2 copolymer of methacrylic acid and methyl methacrylate, which precipitates below pH 5. The occurrence of modification was proved by an increase in the hydrodynamic diameter of the modified polymer compared with the unmodified one. The beginning of aggregation of polymer chains was detected using DLS at pH 5.5 for the unmodified polymer and at pH 6.0 for the modified one. Macroscopic precipitation was observed using SLS at lower pH values of 4.8 and 5.2 for the unmodified and modified polymers, respectively. Moreover, the time-dependent formation of a polymer network was observed for the modified polymer at pH 5.2. Mixing the ligand-modified Eudragit and lectin (Con A) at pH 7.5 also resulted in network formation but at a certain molar ratio (1:2.6) between lectin and polymer. The kinetics of the lectin/polymer complex formation was studied by following the time dependence of the average hydrodynamic diameter.[118] It was shown that the initial rate of the complex formation strongly depends on the Con A/polymer ratio. Furthermore, the kinetics of the complex dissociation initiated by the addition of different sugars has been studied using time-resolved DLS.

The examples presented in this section prove that both static and dynamic light scattering techniques are not only an effective way to control process parameters, but also provide an explicit pattern of polymer–protein interactions at different steps of the process.

1.4.4 Hydrogels, Microgels, and Nanogels

Polymer networks (hydrogels) made of environmentally responsive polymers form another group of smart systems with potential applications in biotechnology and biomedicine. Spherical hydrogel particles of micro- and nanometer size (20 to 5000 nm) are the most suitable and convenient objects for SLS and DLS. Interpretation of the LS data for those particles can be straightforward if the experimental conditions are properly chosen. The combination of SLS and DLS will give the same set of parameters as discussed above for the polymer solutions. Thus, keeping in mind that stimuli-responsive micro- and nanogels combine the features of both colloidal particles and water-soluble polymers, one can be curious to know their swelling/deswelling ability, colloidal stability (aggregation, precipitation), and surface activity (interaction with each other, solvents, polymers, proteins, conjugates and so forth) under various conditions (pH, ionic strength, surfactant, additives, etc.) and modifications (chemical composition, hydrophobic, ligand attachment, etc.). The scheme in Figure 1.9 can be referred to the LS study of the micro- and nanogels as well. Numerous references regarding the DLS study of micro- and nanogels can be found in excellent reviews, monographs, and current literature.[1,42,119,120]

1.4.4.1 Drug-Delivery Systems

Many drug-delivery systems such as polymers, nanoparticles, microspheres, micelles, and liposomes have been developed to prolong the circulation time of certain molecules, to deliver them to the appropriate sites, and to protect them from degradation in the plasma.[121–124] Applications of hydrogel as an element of controlled-release systems have also been well described.[125–129] Herein, a decrease in the hydrogel size to the nanometer scale can solve the problem of the hydrogel's slow response to environmental changes.[130] As a result, over the past three decades, aqueous nanogel suspensions have been one of the most studied colloidal systems by light scattering, with the objectives of improving drug targeting to organs and increasing drug bioavailability across biological membranes. A new type of nanoparticle, lipobeads, has recently been developed.[52] An appropriate assembly of a lipid bilayer on a spherical hydrogel surface combines biocompatible surface properties of liposomes with the mechanical stability and greater loading capacity of a polymer network. This combination broadens the potential of hydrogel–liposome assemblies for pharmaceutical applications, biomimetic sensory systems, controlled-release devices, multivalent receptors, and so on. The bicompartmental structure of hydrogel–liposome capsules may serve as a container for loading various agents such as drugs, fluorescent dyes, or other molecules, and as a functionalization site for the attachment of various ligands, depending on the desired applications.

Two methods of lipobead preparation can be distinguished:

1. Spontaneous formation of lipid layers around nanogels when the preformed hydrogel beads are incubated with liposomes
2. Synthesis of hydrogel within the liposomal interior

The combination of SLS and DLS is the most effective method for controlling the production steps in both cases.

The researchers from Biovector Therapeutics (France) have synthesized[131–134] the so-called Supramolecular Biovector (SMBV™) of ~60-nm diameter, which consists of an ionically charged, cross-linked polysaccharide core surrounded by a lipid membrane. These particles were found to be very efficient as protein carriers for the development of modern nasal vaccines.[134] Kiser et al.[135,136] have made anionic microgels by precipitation polymerization, loaded them with doxorubicin (anticancer agent), and cohomogenized them with hydrating phospholipids (not liposomes) to encapsulate the microgel within a lipid membrane. Rosenzweig and coworkers have mixed and incubated suspensions of 1.6-μm polystyrene beads, phospholipids, and fluorescent dyes to prepare micrometric lipobeads with the ability to sense hydrogen,[137] oxygen,[138] or chloride[139] ions. These beads show potential for use in intracellular measurements. A research group at the University of Toronto has described[140–142] the synthesis of micrometric hydrogel particles with acrylamide-functionalized lipids tethered to their surface as anchors, which promote the assembly of a lipid bilayer around each hydrogel bead when incubated with liposomes. Lipobeads prepared in this way model a cytoskeleton-supported cell membrane. In all aforementioned cases, the spontaneous formation of lipid layers around hydrogels has been confirmed experimentally by a light scattering technique.

The proposed method of lipobead preparation proposed in the literature[143,144] differs from all previously described methods in that the nanoscale (10 to 1000 nm) hydrogel core is prepared via UV-induced polymerization within a liposomal reactor. This method has the advantage of an increased mechanical strength and the enhanced stability of a lipid bilayer that is formed. Recently, in accordance with this protocol, hemoglobin was encapsulated into lipobeads for use as an artificial blood substitute.[145, 146] Once extracted from liposomes, the hydrogel spherical particles (nanogels) have a strong compatibility with the phospholipid bilayer,[99,147] and the phospholipids are self-assembling around the nanogels upon mixing with liposomes, as detected by DLS and AFM. Strikingly, it was found by DLS that if a lipobead is composed of a hydrophobically modified nanogel trapped within a closed lipid bilayer, the hydrophobic chains penetrate into the lipid bilayer and stabilize the liposomal membrane against fusion; thus the lipobeads aggregate reversibly. In contrast, nonanchored PNIPA-VI nanogels within the liposomes were unable to prevent the fusion of lipid bilayers, so that their irreversible aggregation resulted in formation of the so-called giant lipobead, where a number of bare nanogels aggregate within one giant lipid vesicle after the lipid membranes have fused. Two novel drug-delivery systems have been proposed based on the observed reversible and irreversible aggregation of the lipobeads.

1.4.4.2 Frozen Inhomogeneities and Dynamic Fluctuations

It is known that, in hydrogels, so-called frozen inhomogeneities exist on the cross-linked network resulting from quenched-concentration fluctuations in the course of polymerization. Herein, the motion of the chemically or physically cross-linked polymer chains is highly restricted due to topological constrains. As a result, hydrogel becomes a nonergodic random medium, for which time averaging is not equivalent to the ensemble averaging. For the nonergodic systems, a special procedure (see Section 1.3.3.2) is required to distinguish the contributions into the scattering intensity from frozen inhomogeneities, $\langle I_S \rangle_E$, and their dynamic fluctuations, $\langle I_F \rangle_T$, as well as to extract the collective diffusion coefficient from the DLS measurements. The time-independent portion of the light scattered on the frozen inhomogeneities can characterize their concentration, whereas the cooperative diffusion coefficient can characterize the dimension of those inhomogeneities. Currently, the light scattering characterization of hydrogels points to them as nonergodic systems[148] (see also references in Norisuye[148]). The time-independent $\langle I_S \rangle_E$ strongly depends on the sample position,[84,88] the cross-linking density[88,149] and the way of cross-linking, interaction parameters,[150] observation temperature,[150,151] and preparation temperature.[150–152] The $\langle I_S \rangle_E$ can be applied to characterize the kinetics and mechanism of gelation[152] and aggregation. The time-dependent $\langle I_F \rangle_T$ strongly depends on the observation temperature[150] and is almost independent of the preparation temperature[150] and cross-linking density. It was believed that the $\langle I_F \rangle_T$ did not depend on the sample position, but recent results[148] revealed its position dependence.

Besides the characterization of the nonergodicity of polymer networks, it is worthwhile to find the conditions, if any, for the transition from nonergodicity to ergodicity, and vice versa, that can be associated with the point of network breaking or formation. Wu's group in Hong Kong[151] designed a clever "hybrid" hydrogel consisting of swollen PNIPA nanogels (~100 nm) in a concentrated dispersion that was jam-packed into a three-dimensional (3-D) macroscopic hydrogel such that the jammed nanogels physically contacted, whereas inside each particle, the polymer chains were chemically cross-linked. The transition from nonergodic state (3-D hydrogel) to ergodic state (uniform nanogel dispersion) has been observed using DLS and SLS when nanogels collapsed upon temperature elevation from 20 to 30°C, as shown in Figure 1.13.

To some extent, the polymer network in hydrogel resembles the macromolecular solution with strong inter- and intramolecular interactions, so that many of the concentrated, highly aggregating and strongly interacting systems (polymer–polymer, polymer–protein, protein–protein, protein–conjugate, conjugate, etc.) forming networks can be considered as nonergodic. Static and dynamic LS techniques remain to be developed for those systems which are typical in bioseparation and bioprocessing.

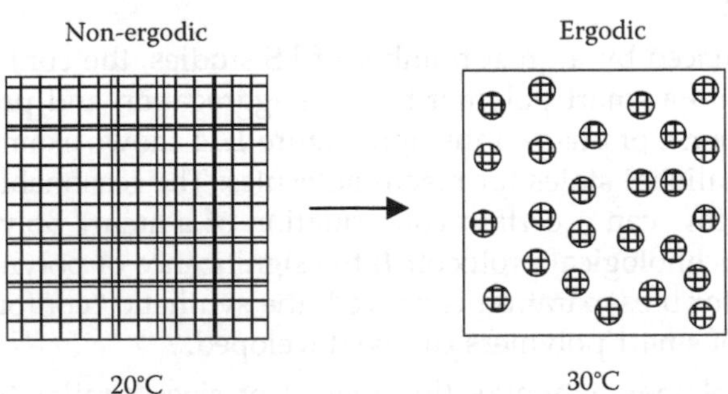

FIGURE 1.13
Nonergodic-to-ergodic transition in "hybrid" hydrogel.

1.5 Closing Remarks: Perspectives on Future Light Scattering Studies in Biotechnology and Biomedicine

The role of light scattering techniques in the past and future developments of smart-polymer systems for biotechnology continues to be in: the determination of phase-transition points; the characterization of polymer properties (molecular mass, size, shape); the identification of interactions of polymers with each other, with ligands, and with biomacromolecules (proteins, DNA, RNA); and the characterization of the properties of their conjugates. Further, the light scattering technique remains an attractive method for controlling all technological steps in drug delivery (permeation, controlled release, targeting, and sensing) and bioseparation because the method is extremely informative, noninvasive, and versatile. The light scattering technique is also a mainstay in the theoretical and experimental development of the more sophisticated interacting systems (polymer–polymer, polymer–protein, protein–protein, conjugate–protein, conjugate–conjugate, networks, nonergodic).

The main goal of this chapter was to highlight some of the hot topics that should be examined in the near future using the combined static and dynamic LS technique to improve bioseparation processes.

1. Once the specific applications of smart polymers in biotechnological protocols have been determined by the significant changes in their microscopic and macroscopic properties in the course of phase transition, the precipitation-curve method (light scattering intensity vs. temperature) will become increasingly important. Not only the transition temperatures, but also the shape of the entire precipitation changes with the external conditions (pH) and polymer structure (different fractions). This correlation between the precipitation-curve shape and the microscopic state of macromolecules should be assessed.
2. As evidenced by a great number of LS studies, the conformational changes of a smart polymer precede aggregation and precipitation in the course of phase transition. Figure 1.14 shows examples of the conformational states for macromolecules. The reasonable question arises: How can a certain conformation of a smart polymer affect the biotechnological protocol? If the significance of polymer conformation for bioseparation is proved, the synthetic "conformational" design of smart polymers can be developed.
3. Smart-polymer networks (hydrogels) of sizes similar to those of polymers, proteins, and conjugates (20 to 500 nm) have been synthesized by different methods. It is intuitively clear that specifically modified smart microgels and nanogels might be quite advantageous as multifunctional agents involved in known biotechnological

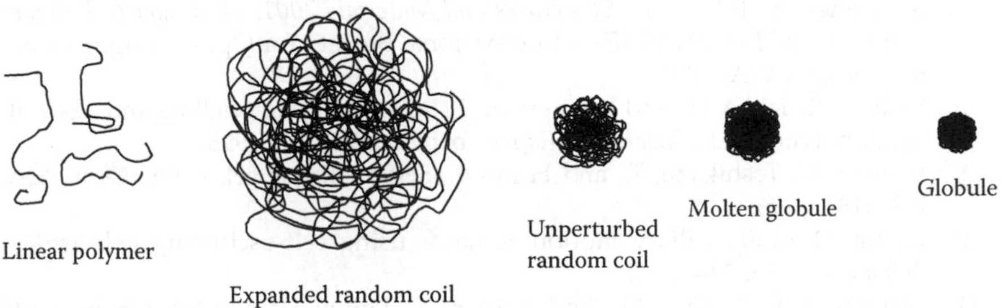

FIGURE 1.14
Some conformational states of a smart polymer.

schemes. The assemblage of a lipid bilayer on a spherical nanogel surface will broaden the potential of those nanocapsules for pharmaceutical applications due to the combination of biocompatible surface properties of liposomes with the mechanical stability and greater loading capacity of a polymer network. The use of combined LS is the most suitable method for studying the interactions between the modified lipobeads, nanogels and polymers, proteins, conjugates, and complexes.

4. Realizing that polymer cross-linked networks are the model for a nonergodic system, it is particularly interesting (for instance, in bioseparation) to examine whether or not the multivalent and highly interacting agents tend to form nonergodic networks. The study of transition from ergodicity to nonergodicity, and vice versa, in biotechnological systems will lead to the development of DLS techniques for this purpose and to the use of modern LS techniques, such as diffusion wave spectroscopy.[87, 153]

References

1. Galaev, I.Y. and Mattiasson, B., Eds., *Smart Polymers for Bioseparation and Bioprocessing*, Taylor & Francis, London and New York, 2002.
2. Galaev, I.Y. and Mattiasson, B., Thermoreactive water-soluble polymers, nonionic surfactants, and hydrogels as reagents in biotechnology, *Enzyme Microb. Technol.*, 15, 354, 1993.
3. Galaev, I.Y., Gupta, M.N., and Mattiasson, B., Use smart polymers for bioseparation, *ChemTech*, 12, 19, 1996.
4. Peppas, N.A., Hydrogels and drug delivery, *Curr. Opin. Colloid Interface Sci.*, 2, 531, 1997.
5. Uhrich, K.E. et al., Polymeric systems for controlled drug release, *Chem. Rev.*, 99, 3181, 1999.
6. Rolland, A., *Pharmaceutical Particulate Carriers: Therapeutic Applications*, Dekker, New York, 1993.

7. Bar-Cohen, Y., Ed., *Smart Structures and Materials 2001: Electroactive Polymer Actuators and Devices*, SPIE — International Society for Optical Engineering, Bellingham, WA, 2001.
8. Yoshida, R., Ichijo, H., and Yamaguchi, T., Self-oscillating swelling and deswelling of polymer gels, *Macromol. Rapid Commun.*, 16, 305, 1995.
9. Yoshida, R., Toshikazu, T., and Hisao, I., Self-oscillating gel, *J. Am. Chem. Soc.*, 118, 5134, 1996.
10. Tabata, O. et al., Ciliary motion actuator using self-oscillating gel, *Sensors Actuators A*, 95, 234, 2002.
11. Giannos, S.A., Dinh, S.M., and Berner, B., Polymeric substitution in a pH oscillator, *Macromol. Rapid Commun.*, 16, 527, 1995.
12. Stanley, H.E., *Introduction to Phase Transitions and Critical Phenomena*, Oxford University, New York, 1973.
13. Anisimov, M.A., *Critical Phenomena in Liquids and Liquid Crystals*, Gordon and Breach, London, 1981.
14. Pollack, G.H., *Cells, Gels, and the Engines of Life*, Ebner & Sons, Seattle, WA, 2001.
15. Kazakov, S.V., Galaev, I.Y., and Mattiasson, B., Characterization of macromolecular solutions by a combined static and dynamic light scattering technique, *Jt. J. Thermophys.*, 23, 161, 2002.
16. Kazakov, S.V. and Chernova, N.I., Static and dynamic light scattering in phase-separating systems, in *Light Scattering and Photon Correlation Spectroscopy*, Pike, E.R. and Abbis, J.B., Eds., Kluwer Academic, Dordrecht, 1997, pp. 401–422.
17. Brown, W., Ed., *Dynamic Light Scattering*, Clarendon, Oxford, 1993.
18. Harding, S.E., Sattelle, D.B., and Bloomfield, V.A., Eds., *Laser Light Scattering in Biochemistry*, Royal Society of Chemistry, London, 1990.
19. Berne, B.J. and Pecora, R., *Dynamic Light Scattering: with Applications to Chemistry, Biology, and Physics*, Dover, New York, 2000.
20. Burchard, W., Solution properties of branched macromolecules, *Adv. Polym. Sci.*, 143, 113, 1999.
21. Walker, J.S. and Ghebremichael, F., Asymmetric coexistence curves in an exactly solvable model of binary liquid mixtures, *Pure Appl. Chem.*, 63, 1381, 1991.
22. Hirschfelder, J., Stevenson, D., and Eyring, H., A theory of liquid structure, *J. Chem. Phys.*, 5, 896, 1937.
23. Barker, J.A. and Fock, W., Theory of upper and lower critical solution temperatures, *Discuss. Faraday Soc.*, 15, 188, 1953.
24. Wheeler, J.C., "Exactly soluble" two-component lattice solution with upper and lower critical solution temperatures, *J. Chem. Phys.*, 62, 433, 1975.
25. Andersen, G.R. and Wheeler, J.C., Directionality dependence of lattice models for solutions with closed loop coexistence curves, *J. Chem. Phys.*, 69, 2082, 1978.
26. Andersen, G.R. and Wheeler, J.C., Theory of lower critical solution point in aqueous mixtures, *J. Chem. Phys.*, 69, 3403, 1978.
27. Huckaby, D.A. and Ballermans, A., An exactly solvable two-component solution having a "closed-loop" phase diagram, *J. Chem. Phys.*, 81, 3691, 1984.
28. Walker, J.S. and Vause, Ch.A., Theory of closed-loop phase diagrams in binary fluid mixtures, *Phys. Lett. A*, 79, 421, 1980.
29. Goldstein, R.E., Phenomenological theory of multiply reentrant solubility, *J. Chem. Phys.*, 83, 1246, 1985.
30. De Pablo, J.J. and Prausnitz, J.M., Thermodynamics of liquid–liquid equilibria including the critical region, *AIChE J.*, 34, 1595, 1988.

31. Kazakov, S.V., Revokatov, O.P., and Chernova, N.I., Reentrant phase transitions: evolution and properties of coexistence curve and new types of multicritical transitions, *Doklady Physics*, 43, 137, 1998.
32. Kazakov, S.V. and Chernova, N.I., Properties of the liquid–liquid coexistence curves with several critical points, *Chem. Eng. Comm.*, 190, 213, 2003.
33. Koningsveld, R., Partial miscibility of multicomponent polymer solutions, *Adv. Colloid Interface Sci.*, 2, 151, 1968.
34. Koningsveld, R., Liquid–liquid equilibria in multicomponent polymer solutions, *Trans. Faraday Soc.*, 49, 144, 1970.
35. Saeki, S. et al., Upper and lower critical solution temperatures in poly(ethylene glycol) solutions, *Polymer*, 17, 685, 1976.
36. Bae, Y.C. et al., Cloud-point curves of polymer solutions from thermooptical measurements, *Macromolecules*, 24, 4403, 1991.
37. Malcolm, G.N. and Rowlinson, J.S., The thermodynamic properties of aqueous solutions of polyethylene glycol, polypropylene glycol and dioxane, *Trans. Faraday Soc.*, 53, 921, 1957.
38. Sorensen, C.M., Cloud-point measurements for the mixture tertiary butyl alcohol, secondary butyl alcohol, and water, *Int. J. Thermophys.*, 9, 703, 1988.
39. Moriyoshi, T. et al., Mutual solubility of 2-butanol+water under high pressure, *J. Chem. Thermodyn.*, 7, 537, 1975.
40. Dolgolenko, W., Uber die untere kritische loslichkeits temperatur zweier flussigkeiten, *Zeitschr. F. Physik. Chemie*, 62, 499, 1908.
41. Kazakov, S.V. and Chernova, N.I., The secondary butanol–water coexistence curve with three critical stratification points, *Russ. J. Phys. Chem.*, 76, 1236, 2002.
42. Schild, H.G., Poly(*N*-isopropylacrylamide): experiment, theory and application, *Prog. Polym. Sci.*, 17, 163, 1992.
43. Heskins, M. and Guillet, J.E., Solution properties of poly(*N*-isopropyl-acrylamide), *J. Macromol. Sci. Chem. A*, 2, 1441, 1968.
44. Boutris, C., Chatzi, E.G., and Kiparissides, C., Characterization of the LCST behavior of aqueous poly(*N*-isopropylacrylamide) solutions by thermal and cloud point techniques, *Polymer*, 38, 2567, 1997.
45. Zheng, X. et al., Phase separation in poly(*N*-isopropyl acrylamide)/water solutions, I: cloud point curves and microgelation, *Polym. J.*, 30, 284, 1998.
46. Taylor, L.D. and Cerankowski, L.D.J., Preparation of films exhibiting a balanced temperature dependence to permeation by aqueous solutions: a study of lower consolute behavior, *Polym. Sci. A: Polym. Chem.*, 13, 2551, 1975.
47. Fujishige, S., Kubota, K., and Ando, I., Phase transition of aqueous solutions of poly(*N*-isopropylacrylamide) and poly(*N*-isopropylmethacrylamide), *J. Phys. Chem.*, 93, 3311, 1989.
48. Schild, H.G. and Tirrell, D.A., Microcalorimetric detection of lower critical solution temperatures in aqueous polymer solutions, *J. Phys. Chem.*, 94, 4352, 1990.
49. Schild, H.G., Muthukumar, M., and Tirrell, D.A., Cononsolvency in mixed aqueous solutions of poly(*N*-isopropylacrylamide), *Macromolecules*, 24, 948, 1991.
50. Winnik, F.M., Ringsdorf, H., and Venzmer, J., Methanol-water as a co-nonsolvent system for poly(*N*-isopropylacrylamide), *Macromolecules*, 23, 2415, 1990.
51. Kazakov, S. et al., Ion concentration of external solution as a characteristic of micro- and nanogel ionic reservoir, *J. Phys. Chem. B*, 110, 15107, 2006.

52. Kazakov, S. and Levon, K., Liposome-hydrogel structures for future pharmaceutical applications, *Curr. Pharm. Design.*, 12, 4713, 2006.
53. Landau, L.D. and Lifshitz, E.M., *Statistical Physics*, Part 1, 3rd ed., Pergamon Press, Oxford, 1986.
54. van de Hulst, H.C., *Light Scattering by Small Particles*, Dover, New York, 1981.
55. Mie, G., Beiträge zur Optik trüber Medien, speziell kolloidaler Metallösungen, *Ann. Physik*, 330, 377, 1908.
56. Kerker, M., *The Scattering of Light and Other Electromagnetic Radiation*, Academic Press, New York, 1969.
57. Bohren, C.F. and Huffman, D.R., *Absorption and Scattering of Light by Small Particles*, Wiley, New York, 1983.
58. Zimm, B., Apparatus and methods for measurement and interpretation of angular variation of light scattering: preliminary results on polystyrene solutions, *J. Chem. Phys.*, 16, 1099, 1948.
59. Mountein, R.D., Spectral distribution of scattered light in a simple fluid, *Rev. Mod. Phys.*, 38, 205, 1966.
60. Landau, L.D. and Lifshitz, E.M., *Fluid Mechanics*, 2nd ed., Butterworth-Heinemann, 1987.
61. Landau, L.D. and Lifshitz, E.M., *Electrodynamics of Continuous Media*, Pergamon Press, 1960.
62. Cummings, H.Z. and Swinney, H.L., Light scattering spectroscopy, *Prog. Optics*, 8, 133, 1970.
63. Kazakov, S.V. and Chernova, N.I., Experimental study and optimization of parameters of the optical heterodyne spectrometer, *Sov. Optics Spectrosc.*, 49, 404, 1980.
64. Provencher, S.W., A constrained regularization method for inverting data represented by linear algebraic or integral equations, *Comp. Phys. Commun.*, 27, 213, 1982.
65. Provencher, S.W., CONTIN: a general purpose constrained regularization program for inverting noisy linear algebraic and integral equations, *Comp. Phys. Commun.*, 27, 229, 1982.
66. Jakes, J., Regularized positive exponential sum (REPES) program: a way of inverting Laplace transform data obtained by dynamic light scattering, *Coll. Czech. Chem. Commun.*, 60, 1781, 1995.
67. Pencer, J. and Hallett, F.R., Effects of vesicle size and shape on static and dynamic light scattering measurements, *Langmuir*, 19, 7488, 2003.
68. Janaky, N. and Lui, X.Y., Protein interactions in undersaturated and supersaturated solutions: a study using light and X-ray scattering, *Biophys. J.*, 84, 523, 2003.
69. Daniel, E.K. et al., Interactions of lysozyme in concentrated electrolyte solutions from dynamic light scattering measurements, *Biophys. J.*, 73, 3211, 1997.
70. Martin, M. and Franz, R., Interactions in undersaturated and supersaturated lysozyme solutions: static and dynamic light scattering results, *J. Chem. Phys.*, 103, 10424, 1995.
71. Tardieu, A. et al., Proteins in solution: from X-ray scattering intensities to interaction potentials, *J. Crystal Growth*, 196, 193, 1999.
72. Li, S., Xing, D., and Li, J., Dynamic light scattering application to study protein interactions in electrolyte solutions, *J. Biol. Phys.*, 30, 313, 2004.
73. Batchelor, G.K., Brownian diffusion of particles with hydrodynamic interactions, *J. Fluid Mech.*, 74, 1, 1976.

74. Felderhof, B.U., Diffusion of interacting Brownian particles, *J. Phys. A: Math. Gen.*, 11, 929, 1978.
75. Croxton, C.A., *Liquid State Physics*, Cambridge University Press, Cambridge, England, 1974.
76. De Groot, S.P. and Mazur, P., *Non-Equilibrium Thermodynamics*, Dover, New York, 1984.
77. Tyrrell, H.J.V. and Harris, K.R., *Diffusion in Liquids*, Butterworth, London, 1984.
78. Pecora, R.J., Spectrum of light scattered from optically anisotropic macromolecules, *J. Chem. Phys.*, 49, 1036, 1968.
79. Kanao, M., Matsuda, Y., and Sato, T., Characterization of polymer solutions containing a small amount of aggregates by static and dynamic light scattering, *Macromolecules*, 36, 2093, 2003.
80. Yamakawa, H., *Helical Wormlike Chains in Polymer Solutions*, Springer Verlag, Berlin, 1997.
81. Furtterer, T. et al., Aggregation of an amphiphilic poly(p-phenylene) in micellar surfactant solutions: static and dynamic scattering, *Macromolecules*, 38, 7443, 2005.
82. Drogemeier, J., Hinssen, H., and Eimer, W., Flexibility of F-actin in aqueous solution: a study on filaments of different average lengths, *Macromolecules*, 27, 87, 1994.
83. Lang, P., Kajiwara, J.K., and Burchard, W., Investigations on the solution architecture of carboxylated tamarind seed polysaccharide by static and dynamic light scattering, *Macromolecules*, 26, 3992, 1993.
84. Pusey, P.N. and van Megen, W., Dynamic light scattering by non-ergodic media, *Physica A*, 157, 705, 1989.
85. Xue, J.Z. et al., Nonergodicity and light scattering from polymer gels, *Phys. Rev. A*, 46, 6550, 1992.
86. Schatzel, K., Accuracy of photon correlation measurements on nonergodic samples, *Appl. Opt.*, 32, 3880, 1993.
87. Scheffold, F. and Schurtenberger, P., Light scattering probes of viscoelastic fluids and solids, *Soft Mater.*, 1, 139, 2003.
88. Joosten, J.G.H., McCathy, J.L., and Pusey, P.N., Dynamic and static light scattering by aqueous polyacrylamide gels, *Macromolecules*, 24, 6690, 1991.
89. Wolf, B.A., Solubility of polymers, *Pure. Appl. Chem.*, 57, 323, 1985.
90. Winnik, F.M., Fluorescence studies of aqueous solutions of poly(*N*-isopropylacrylamide) below and above their LCST, *Macromolecules*, 23, 233, 1990.
91. Winnik, F.M., Effect of temperature on aqueous solutions of pyrene-labeled (hydroxypropyl)cellulose, *Macromolecules*, 20, 2745, 1987.
92. Lau, A.C.W. and Wu, C., Thermally sensitive and biocompatible poly(*N*-vinylcaprolactam) synthesis and characterization of high molar mass linear chains, *Macromolecules*, 32, 581, 1999.
93. Franks, F. and Eagland, D., The role of solvent interactions in protein conformation, *CRC Crit. Rev. Biochem.*, 4, 165, 1985.
94. Robb, I.D., *Chemistry and Technology of Water-Soluble Polymers*, Finch, C.A., Ed., Plenum, New York, 1981.
95. Molyneux, P., *Water-Soluble Polymers: Properties and Behavior*, Vol. I, CRC Press, Boca Raton, FL, 1983.
96. Ivanov, A.E. et al., Thermosensitive copolymer of *N*-vinylimidazole as displacers of proteins in immobilized metal affinity chromatography, *J. Chromatog. A*, 907, 1155, 2001.

97. Ivanov, A.E. et al., Thermosensitive copolymer of N-vinylcaprolactam and 1-vinylimidasole: molecular characterization and separation by immobilized metal affinity chromatography, *Polymer*, 42, 3373, 2001.
98. Wahlund, P.-O. et al., "Protein-like" copolymers: effect of polymer architecture on the performance in bioseparation process, *Macromol. Biosci.*, 2, 33, 2002.
99. Kazakov, S. et al., Poly(*N*-isopropylacrylamide-*co*-1-vinylimidazole) hydrogel nanoparticles prepared and hydrophobically modified in liposome reactors: atomic force microscopy and dynamic light scattering study, *Langmuir*, 19, 8086, 2003.
100. Saunders, B.R., Crowther, H.M., and Vincent, B., Poly[(methyl methacrylate)-*co*-(methacrylic acid)] microgel particles: swelling control using pH, cononsolvency, and osmotic deswelling, *Macromolecules*, 30, 482, 1997.
101. Zhou, S. and Chu, B., Synthesis and volume phase transition of poly(methacrylic acid-*co*-*N*-isopropylacrylamide) microgel particles in water, *J. Phys. Chem. B*, 102, 1364, 1998.
102. Grosberg, A.Y. and Khokhlov, A.R., *Statistical Physics of Macromolecules*, AIP Press, Woodbury, NY 1994.
103. Wu, C. and Wang, X., Globule-to-coil transition of a single homopolymer chain in solution, *Phys. Rev. Lett.*, 80, 4092, 1998.
104. Meewes, M. et al., Coil-globule transitions of poly(N-isopropylacrylamide): a study of surfactant effects by light scattering, *Macromolecules*, 24, 5811, 1991.
105. Li, M. and Wu, C., Self-association of poly(N-isopropylacrylamide) and its complexation with gelatine in aqueous solution, *Macromolecules*, 32, 4311, 1999.
106. Peng, S. and Wu, C., Light scattering study of the formation and structure of partially hydrolyzed poly(acrylamide)/calcium(II) complexes, *Macromolecules*, 32, 585, 1999.
107. Zhang, Y. et al., A light scattering study of the aggregation behavior of fluorocarbon-modified polyacrylamides in water, *Macromolecules*, 29, 2494, 1996.
108. Kabanov, V.A., Physicochemical basis and the prospects of using soluble interpolyelectrolyte complexes (review), *Polym. Sci.*, 36, 143, 1994.
109. Izumrudov, V.A., Galaev, I.Y., and Mattiasson, B., Polycomplexes: potential for bioseparation, *Bioseparation*, 7, 207, 1998.
110. Margolin, A.L. et al., Enzymes in polyelectrolyte complexes: the effect of phase transition on thermal stability, *Eur. J. Biochem.*, 146, 625, 1985.
111. Dainiak, M.B. et al., Conjugates of monoclonal antibodies with polyelectrolyte complexes: an attempt to make an artificial chaperone, *Biochim. Biophys. Acta*, 1381, 279, 1998.
112. Muronetz, V.I. et al., Interaction of antibodies and antigens conjugated with synthetic polyions: on the way of creating an artificial chaperone, *Biochim. Biophys. Acta*, 1475, 141, 2000.
113. Kazakov, S.V. et al., Light scattering study of the antibody-poly(methacrylic acid) and antibody-poly(acrylic acid) conjugates in aqueous solutions, *Macromol. Biosci.*, 1, 157, 2001.
114. Scopes, R.K., *Protein Purification: Principles and Practice*, Springer-Verlag, New York, 1994.
115. Cole, C.A. et al., *Polym. Preprint.*, 27, 237, 1986.
116. Cole, C.A. et al., *ACS Symp. Ser.*, 350, 245, 1987.
117. Linné Larsson, E. et al., Affinity precipitation of Concanavalin A with p-aminophynyl-α-D-glucopyranoside modified Eudragit S-100, I: initial complex formation and build-up of the precipitate, *Bioseparation*, 6, 273, 1996.

118. Linné Larsson, E. et al., Affinity precipitation of Concanavalin A with p-aminophynyl-α-D-glucopyranoside modified Eudragit S-100, II: kinetic studies of the formation and the dissociation of the protein-macroligand complex, *Bioseparation*, 6, 283, 1996.
119. Pelton, R., Temperature-sensitive aqueous microgels, *Adv. Colloid Interface Sci.*, 85, 1, 2000.
120. Gehrke, S.H., Synthesis, equilibrium swelling, kinetics, permeability and application of environmentally responsive gels, *Adv. Polym. Sci.*, 110, 81, 1993.
121. Rolland, A., *Pharmaceutical Particulate Carriers: Therapeutic Applications*, Marcel Dekker, New York, 1993.
122. Kreuter, J., *Colloidal Drug Delivery Systems*, Marcel Dekker, New York, 1994.
123. Woodle, M.C. and Storm, G., *Long Circulating Liposomes: Old Drugs, New Therapeutics*, Springer, Berlin, 1998.
124. Allen, T.M., Liposomal drug formulations: rationale for development and what we can expect for the future, *Drugs*, 56, 747, 1998.
125. Peppas, N.A., *Hydrogels in Medicine and Pharmacy*, CRC Press, Boca Raton, FL, 1987.
126. Park, K., *Controlled Drug Delivery: Challenges and Strategies*, ACS, Washington, DC, 1997.
127. Ende, M. and Mikos, A.G., Diffusion controlled delivery of proteins from hydrogels and other hydrophilic systems, in *Protein Delivery: Physical Systems*, Sanders, L.M. and Hendren, R.W., Eds., Plenum, New York, 1997, pp. 139–165.
128. Lowman, A.M. and Peppas, N.A., in *Encyclopedia of Controlled Drug Delivery*, Mathiowitz, E., Ed., Wiley, New York, 1999, pp. 397–418.
129. Byrne, M.E., Park. K., and Peppas, N.A., Molecular imprinting within hydrogels, *Adv. Drug Delivery Rev.*, 54, 149, 2002.
130. Tanaka, T. and Fillmore, D.J., Kinetics of swelling of gels, *J. Chem. Phys.*, 70, 1214, 1979.
131. De Miguel, I. et al., Synthesis and characterization of supramolecular biovector (SMBV) specifically designed for the entrapment of ionic molecules, *Biochim. Biophys. Acta*, 1237, 49, 1995.
132. Major, M. et al., Characterization and phase behavior of phospholipids bilayers adsorbed on spherical polysaccharidic nanoparticles, *Biochim. Biophys. Acta*, 1327, 32, 1997.
133. De Miguel, I. et al., Proofs of the structure of lipid-coated nanoparticles (SMBV™) used as drug carriers, *Pharm. Res.*, 17, 817, 2000.
134. von Hoegen, P., Synthesis biomimetic supra molecular Biovector™ (SMBV™) particles for nasal vaccine delivery, *Adv. Drug Delivery Rev.*, 51, 113, 2001.
135. Kiser, P.F., Wilson, G., and Needham, D., A synthetic mimic of the secretory granule for drug delivery, *Nature*, 394, 459, 1998.
136. Kiser, P.F., Wilson, G., and Needham, D., Lipid-coated microgels for the triggered release of doxorubicin, *J. Control. Release*, 68, 9, 2000.
137. McNamara, K.P. et al., Synthesis, characterization, and application of fluorescence sensing lipobeads for intracellular pH measurements, *Anal. Chem.*, 73, 3240, 2001.
138. Ji, J. et al., Molecular oxygen-sensitive fluorescent lipobeads for intracellular oxygen measurements in murine macrophages, *Anal. Chem.*, 73, 3521, 2001.
139. Ma, A. and Rosenzweig, Z., Submicrometric lipobead-based fluorescence sensor for chloride ion measurements in aqueous solution, *Anal. Chem.*, 76, 569, 2004.

140. Ng, C.C., Cheng, Y.-L., and Pennefather, P.S., One-step synthesis of a fluorescent phospholipid-hydrogel conjugate for driving self-assembly of supported lipid membranes, *Macromolecules*, 34, 5759, 2001.
141. Ng, C.C., Cheng, Y.-L., and Pennefather, P.S., Properties of a self-assembled phospholipid membrane supported on lipobeads, *Biophys. J.*, 87, 323, 2004.
142. Buck, S. et al., Engineering lipobeads: properties of the hydrogel core and the lipid bilayer shell, *Biomacromolecules*, 5, 2230, 2004.
143. Kazakov, S., Kaholek, M., and Levon, K., Lipobeads and Their Production, pending U.S. Patent 2002, publication no. US2003035842, A1.
144. Kazakov, S. et al., UV-induced gelation on nanometer scale using liposome reactor, *Macromolecules*, 35, 1911, 2002.
145. Patton, J.N. and Palmer, A.F., Photopolymerization of bovine hemoglobin entrapped nanoscale hydrogel particles within liposomal reactors for use as an artificial blood substitute, *Biomacromolecules*, 6, 414, 2005.
146. Patton, J.N. and Palmer, A.F., Engineering temperature-sensitive hydrogel nanoparticles entrapping hemoglobin as a novel type of oxygen carrier, *Biomacromolecules*, 6, 2204, 2005.
147. Kazakov, S., Kaholek, M., and Levon, K., Hydrogel nanoparticles compatible with phospholipid bilayer, *Polym. Preprint*, 43, 381, 2002.
148. Norisuye, T., Tran-Cong-Miyata, Q., and Shibayama, M., Dynamic inhomogeneities in polymer gels investigated by dynamic light scattering, *Macromolecules*, 37, 2944, 2004.
149. Shibayama, M., Norisuye, T., and Nomura, S., Cross-link density dependence of spatial inhomogeneities and dynamic fluctuations of poly(*N*-isopropylacrylamide) gels, *Macromolecules*, 29, 8746, 1996.
150. Shibayama, M., Tanaka, S., and Norisuye, T., Static inhomogeneities and dynamic fluctuations of temperature sensitive polymer gels, *Physica A*, 249, 245, 1998.
151. Zhao, Y., Zhang, G., and Wu, C., Nonergodic dynamics of a novel thermally sensitive hybrid gel, *Macromolecules*, 34, 7804, 2001.
152. Kayaman, N. et al., Structure and protein reparation efficiency of poly(*N*-isopropylacrylamide) gels: effect of synthesis conditions, *J. Appl. Polym. Sci.*, 67, 805, 1998.
153. Scheffold, F. et al., Diffusion-wave spectroscopy of nonergodic media, *Phys. Rev. E*, 63, 61404, 2001.
154. Kratky, O. and Porod, G., Röntgenuntersuchong gelöster Fadenmoleküle, *Rec. Trav. Chim. Pays-Bas.*, 68, 1106, 1949.

Chapter 2

Responsive Polymer Brushes: A Theoretical Outlook

Oleg V. Borisov and Ekaterina B. Zhulina

CONTENTS

Glossary

α	degree of ionization of monomer units in the brush
α_b	degree of ionization of monomer units in the bulk of the solution
$\Delta\Pi$	excess osmotic pressure exerted by mobile ions inside the polyelectrolyte brush
$\Delta\Psi$	excess electrostatic potential inside the polyelectrolyte brush
$\theta(z)$	fraction of monomers complexed with surfactant micelles
μ	$\delta f\{c(z)\}/\delta c$ chemical potential of a monomer unit in the brush

μ_i **(i = A,B)**	chemical potential of solvent molecule of type i in the brush
μ_{bi} **(i = A,B)**	chemical potential of solvent molecule of type i in the bulk of the solution
ξ	correlation length in semidilute polymer solution
π	restoring force (per unit area) in the compressed brush
φ_i	volume fraction of solvent (I = A,B) in the brush
Φ_i	volume fraction of solvent (I = A,B) in the bulk of the solution
χ, χ_A, χ_B	Flory-Huggins parameter of polymer–solvent interactions
χ_{AB}	Flory-Huggins parameter of interactions between solvents A and B
a	monomer unit length
c, $c(z)$	monomer unit number density in the brush
$c_{end}(z)$	number density of the end segments multiplied by s
$c(A)$	monomer unit number density in the amphiphilic brush swollen in solvent A (water)
$c(B)$	monomer unit number density in the amphiphilic brush swollen in solvent B (oil)
c_{bj}	number density of mobile ions of type j in the bulk of the solution
c_{H+}	number density of hydrogen ions in the brush
c_j	number density of mobile ions of type j in the brush
$E(z,z')$	local chain stretching at distance z from the surface, provided that the chain end is localized at distance z'
F, $F(H)$	free energy per chain in the brush
$f\{c(z)\}$	free-energy density in the brush
g	number of monomer units per blob
H	height of the brush
$\mathbf{k_B}$	Boltzmann constant
K	dissociation constant of a monomer unit in polyelectrolyte brush
N	number of monomer units per chain
$\mathbf{n_A}$	number of molecules of solvent A (water) per chain in the amphiphilic brush
$\mathbf{n_B}$	number of molecules of solvent B (oil) per chain in the amphiphilic brush
s	grafting area per chain
T	temperature
u	free energy of complexation per monomer in polymer–surfactant complex (in k_BT units)
v	the second virial coefficient (excluded volume) of a monomer unit
w	the third virial coefficient of a monomer unit

2.1 Introduction

Functional polymer brushlike structures are found in living organisms and artificial biomedical devices. Examples of natural biopolymer brushes include extracellular polysaccharides on bacterial surfaces,[1] neurofilaments and microtubules with associated proteins in neurites,[2,3] aggrecan macromolecules in articular cartilage,[4] and glycomacropeptide "hairs" on milk casein micelles.[5,6]

Modification of artificial and biological surfaces by water-based multifunctional polymers is one of the promising directions in the field of biomaterial engineering and nanomedicine.[7] This strategy allows for direct biological response of artificial materials and biomedical devices operating in contact with biofluids such as serum or plasma. Ultrathin polymer coatings are of particular interest for the surface modification of small (nano) particles. Examples of applications include antithrombogenic, antimicrobial, and anti-inflammatory coatings[7] to enhance compatibility of biomaterials; drug-delivery systems with improved circulation time and targeting for specific cells/tissues;[8–11] and biochips for DNA/RNA, protein, and carbohydrate analysis in medical diagnostics and drug development.[12]

The surface-modification strategies include "grafting from," "grafting to," and molecular assembly/self-organization of polymers at interfaces of two-dimensional, flat substrates/devices and of nanoparticles. The most straightforward way to combine pronounced responsive features with the high stability of a functional polymer coating is to attach chains covalently to the surface by one of the terminal segments, i.e., to create a polymer brush. An inherent feature of a polymer brush with high surface coverage is strong intermolecular interaction that governs chain conformation and ensures stimuli-responsive behavior. Here, the surface-initiated controlled radical polymerization ("grafting from" approach) is most efficient. In contrast to the "grafting to" approach, the "grafting from" strategy (a) avoids the problem of steric hindrance created by the chains already attached to the interface and (b) makes possible incorporation of monomers with different properties at well-defined positions in growing chains.

The possibility of triggering conformational transitions in surface-attached polymeric layers by external physical (temperature, electrical voltage, light, etc.) or chemical (pH, salinity, solvent composition, etc.) fields opens a perspective for design of stimuli-responsive interfaces. Stimuli-induced manipulation of the structure and properties of the interfaces on a nanometer-length scale is of great interest for nanotechnology and bioengineering (design of molecular-resolution templates, data-storage materials, tuning availability of the molecular-recognition sites, etc.). A responsive polymer brush should therefore exhibit either thermosensitive or ionic pH-sensitive features. In the latter case, the chain conformation can be affected by both chemical (pH, salinity) and physical (electrical) fields. In a wider context, specific interactions like binding of a particular solute species to grafted chains can be used to stimulate conformational transitions and to mediate the structure of a polymer brush.

This chapter provides an overview of the current understanding (based on analytical theoretical modeling) of the conformational transitions that can be induced in polymer and polyelectrolyte brushes by various external stimuli.

2.2 Theory of Polymer Brushes

The polymer brush is one of the most extensively explored model systems in polymer science.[13–15] Theoretical models of the polymer brush consider an array of long polymer chains attached by one end onto an impermeable solid surface and immersed in a solvent. The number of monomers per chain is $N >> 1$. The density of grafting is characterized by area per chain s, which is assumed to be sufficiently small to ensure crowding of neighboring polymer chains and predominance of intermolecular interactions over intramolecular ones. For simplicity, we consider the case when the grafted chains are intrinsically flexible (the Kuhn segment is on the order of monomer length a), and the generalization for semiflexible polymers is straightforward. Below, we discuss only the case of a planar (or quasi-planar) surface, i.e., when the curvature radius of the surface R is much larger than the characteristic size of the chains H in the direction normal to the grafting surface ($R >> H$). Generalization of the model for brushes grafted onto spherical or cylindrical surfaces of large curvature, $R << H$, has been elaborated on in a number of works.[16–20]

Below, we consider brushes formed by both neutral (nonionic) and charged polymers (polyelectrolytes). In the case of brushes formed by nonionic polymers, only short-range (van der Waals, excluded volume) interactions between monomers are considered. As long as the volume fraction of monomers in the brush is significantly below unity, only binary monomer–monomer interactions with second virial coefficient $v \sim a^3$ or ternary interactions with third virial coefficient $w \sim a^6$ are relevant. The second virial coefficient v is related to the Flory-Huggins parameter $\chi(T)$ as $va^{-3} = 0.5 - \chi(T)$ and vanishes under theta-solvent conditions, $v(T = \Theta) = 0$. For most nonionic water-soluble polymers (polyethylene glycol [PEG], poly(*N*-isopropylacrylamide) [PNIPAAm], etc.), an increase in temperature leads to inferior solvent strength of water ($\chi(T)$ increases). For example, water is a good solvent for PEG chains at room temperature, whereas theta-solvent conditions are approached at about 100°C.[21] The third virial coefficient wa^{-6} is fairly independent of temperature T and is on the order of unity.

In the case of ionic polymers, we distinguish between strongly and weakly dissociating (pH-sensitive) polyelectrolytes. In the former case, the fraction of charged monomers in a chain α is quenched and is determined by the chain chemical sequence (like, for example, for partially sulfonated polystyrene at low degree of sulfonation). In the latter case, α is affected by the salinity, pH, and polymer concentration in solution. Typical examples are weak polyacid (e.g., poly[acrylic acid]) or polybase (e.g., poly[ethyleneimine]). In both cases, we assume weak charging of the chains ($\alpha \leq 1$). Under these conditions, a number of issues relevant for strongly charged polyelectrolytes (electrostatic stiffening, finite chain extensibility, and possible Manning condensation of counterions) could be safely neglected.

2.2.1 Alexander-de Gennes Scaling Theory

The pioneering theoretical model of the nonionic polymer brush has been proposed by Alexander[22] and de Gennes,[23] who analyzed the conformational properties of flexible neutral polymer chains end-grafted to planar substrates in terms of scaling theory. The scaling analysis is based on the blob model developed originally to describe the structure of semidilute polymer solutions.[24] The solution is envisioned as an array of closely packed blobs. Each blob comprises a chain segment with the dimensions proportional to correlation length ξ, which scales with local polymer concentration c as $\xi \sim c^{-3/4}v^{-1/4}$ or $\xi \sim c^{-1}w^{-1/3}$ under good or theta-solvent conditions, respectively. Within each blob, the chain conformation is not perturbed by interactions with other chains, and the blob size is related to the number of monomers g as $\xi \sim g^{3/5}v^{1/5}$ or $\xi \sim g^{1/2}$ (self-avoiding walk or Gaussian chain statistics) under conditions of good or theta solvent, respectively.

In the framework of the blob model, a polymer brush is envisioned as an array of closely packed blobs of size equal to the lateral separation between the grafts $\xi \approx s^{1/2}$ (Figure 2.1). Each chain constitutes a string of blobs extended in the direction perpendicular to the grafting surface but randomly bent in the lateral direction. The characteristic extension of grafted chains in the direction perpendicular to the surface is $H \sim N/g$, and brush thickness H varies in a good or theta solvent as

$$H \cong Na \cdot (s/a^2)^{-1/3}(v/a^3)^{1/3} \tag{2.1}$$

$$H \cong Na \cdot (s/a^2)^{-1/2}(w/a^6)^{1/4}$$

respectively.

Although the scaling approach does not allow for any numerical factors (as reflected by the sign $\cong$ in Equation 2.1) and does not provide information

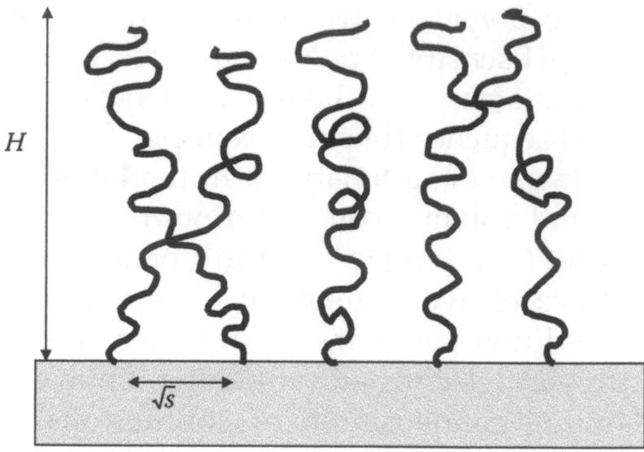

FIGURE 2.1
Polymer brush: s = surface area per grafted chain; H = height of the brush; N = number of monomers in grafted chain; a = size of a monomer.

about internal brush structure, it comes up with correct exponents for power-law dependences of brush thickness H that are confirmed by more refined theories. The most essential prediction is that chains in a planar brush are extended proportionally to their contour length, $H \sim N$, and therefore the average monomer concentration inside the brush $c \cong N/Hs$ is independent of the degree of polymerization N. The blob model of Alexander-de Gennes has been further generalized by Daoud and Cotton[16] and by Zhulina and Birshtein[17–19] to describe the conformations of polymer chains end-grafted to convex spherical and cylindrical surfaces.

2.2.2 Mean-Field Boxlike Model

This model explicitly assumes that the brush comprises uniformly and equally extended chains, that polymer concentration $c = N/sH$ is constant throughout the brush, and that all the chain ends are localized at the edge of the brush. The mean-field brush model is reminiscent of the Flory theory for a polymer coil with excluded-volume interactions.[24] The equilibrium brush thickness follows from the balance of intermolecular repulsive interactions and the conformational entropy penalty for the chains extension:

$$F(H) = F_{\text{int}}(H) + F_{\text{conf}}(H) \tag{2.2}$$

The conformational contribution is taken into account assuming uniform extension and the Gaussian entropic elasticity of the chains:

$$F_{\text{conf}}(H) = \frac{3k_B TH^2}{2Na^2} \tag{2.3}$$

The interaction term comprises relevant contributions (the short-range van der Waals interactions, electrostatic interactions, etc.) that can be expressed (in a mean-field approximation) as a function of polymer concentration c in the brush. For a nonionic brush, the interaction contribution to the free energy can be expressed in terms of virial expansion as

$$F_{int}(H)/k_B T = vNc + wNc^2 + \qquad (2.4)$$

where higher order concentration terms are omitted. Corresponding expressions for the free energy of interactions in ionic polymer brushes are introduced in Section 2.6.

The boxlike mean-field model enables us to calculate the brush thickness (polymer concentration) and its elastic (compression) modulus as functions of environmental conditions. The results for structural properties in terms of power law exponents are the same as provided by the scaling theory. However, the free energy and compression modulus are different because, in contrast to the mean-field theory, the scaling approach accounts for correlations arising due to connectivity of monomers in the chains. The principal advantage of the mean-field approach (as compared with the scaling blob model) is the ability to account explicitly for the combined effect of different interactions that can be tuned independently by varying the external conditions. However, as we demonstrate below, the assumption of uniform and equal stretching of the chains in the brush is oversimplified, and therefore the boxlike model does not provide an adequate picture of the internal brush structure. It also overestimates the brush free energy and the compression modulus.

2.2.3 Self-Consistent Field Theory

A more advanced formalism developed on the basis of the self-consistent mean-field (SCF) approach[25-28] provides a deeper insight into the chain conformations and internal brush structure. In particular, it allows incorporation of the nonuniform polymer density distribution and nonequal and nonuniform stretching of the chains in the brush. In the framework of the so-called strong stretching approximation,[29] tethered chains are envisioned as "trajectories" with specified position of each monomer. If the normalized to unity density profile of the end segments is $c_{end}(z)$, where z is the coordinate perpendicular to the grafting surface (in the lateral direction the brush is assumed to be homogeneous) and the distribution of local stretching in the chain with its free end localized at $z = z'$ is given by $dz/dn = E(z,z')$, then the polymer concentration (the monomer number density) profile is given by

$$c(z) = s^{-1} \int_{z}^{H} \frac{c_{end}(z')dz'}{E(z,z')} \qquad (2.5)$$

The conformational free energy penalty for nonuniform extension of the ensemble of chains characterized by distributions $c_{end}(z)$ and $E(z,z')$ is presented as[29]

$$F_{conf} = \frac{3k_BT}{2a^2}\int_0^H c_{end}(z')dz'\int_0^{z'} E(z,z')dz \tag{2.6}$$

where, in analogy to Equation 2.3, the chains are assumed to exhibit the Gaussian entropic elasticity on an arbitrary small length scale. The central point in the mean-field approximation is an assumption that the density of the free energy of interactions $f\{c(z)\}$ inside the brush is a function of the local monomer concentration $c(z)$,

$$F_{int} = s\int_0^H f\{c(z)\}dz \tag{2.7}$$

In a nonionic brush, only the short-range interactions are relevant, and the free-energy density can be presented in the familiar Flory-Huggins form,

$$\frac{f\{c(z)\}}{k_BT} = \chi a^3 c\ (z)(1-a^3c(z)) + (1-a^3c)\ln(1-a^3c) \tag{2.8}$$

Expansion of Equation 2.8 with respect to polymer concentration $a^3c << 1$ leads to[24]

$$\frac{f\{c(z)\}}{k_BT} = va^3c^2(z) + wa^6c^3(z) \tag{2.9}$$

Functional minimization of the free energy given by Equations 2.6 to 2.9 with the account of two conservation conditions

$$N = \int_0^{z'} \frac{dz}{E(z,z')}, \quad N = s\int_0^H c(z)dz \tag{2.10}$$

provides equilibrium profiles of the polymer density and of the free ends, and distribution of local stretching in the chains. The equilibrium thickness of a nonconfined brush (cutoff for the monomer density profile) H is determined from the condition $\partial F/\partial H = 0$.

In contrast to the boxlike model, a more refined SCF theory predicts that chains in the brush are stretched nonuniformly and nonequally, and individual chains in the brush experience strong fluctuations of extension. The chain ends are distributed throughout the brush, leading to a monotonically decreasing polymer density profile that is determined by a parabolic chemical potential as

$$\delta f\{c(z)\}/\delta c = \mu = \text{const} - 3z^2/8\pi^2 a^2 N^2 \tag{2.11}$$

Brush thickness H obtained in the SCF approach is systematically larger, while the free energy of the brush F is lower, than those following from the boxlike model. Nevertheless, the exponents in asymptotic power dependences of F and H on N, s, and v, obtained in both mean-field boxlike and SCF models, are the same. As we demonstrate below, an important qualitative difference in predictions of both models is found when the combination of attractive and repulsive forces operating in the brush induces internal phase separation in the z-direction in collapsed and extended phases.

After introducing the basic analytical approaches, we proceed with a review of the predictions of theoretical models on conformational transitions and structural rearrangements taking place in the brush when the strength of one or several interaction forces acting in the brush is tuned by varied environmental conditions (external physical or chemical fields).

2.3 Collapse-to-Swelling Transition in Nonionic Brush

In the case of a nonionic brush, the equilibrium extension of grafted chains is governed by the balance of short-range van der Waals interactions between monomers and the conformational entropy of extended chains. In particular, binary (excluded volume) interactions are repulsive under good solvent conditions, but can be tuned to attraction upon decrease in the solvent strength below the theta point. This decrease in the solvent strength corresponding to the decrease in v (increase in χ) can be induced by variation in temperature.

For most organosoluble polymers (e.g., polystyrene in cyclohexane), $\chi(T)$ is a *decreasing* function of temperature. When such polymers are grafted to the surface, tuning of the excluded-volume interactions from repulsion (under good solvent conditions) to attraction (under poor solvent conditions) does not cause macrophase separation (as in the solution), but rather leads to gradual collapse of the brush.[27,30]

Figure 2.2 illustrates the dependence of normalized brush height H/H_θ on temperature via reduced second virial coefficient, $v(s/a^2)^{1/2}$. Figure 2.3a and Figure 2.3b illustrate the corresponding evolution in polymer density profile $c(z)$ and in the ends distribution $c_{\text{end}}(z)$ as predicted by the SCF model.[27] In the limiting cases of good, theta, and poor solvents,

$$c(z) \approx c(0)(1 - z^2/H^2)^{\beta} \tag{2.12}$$

$$c_{\text{end}}(z) \approx \frac{(2\beta+1)z}{H}(1 - z^2/H^2)^{\beta-1/2} \tag{2.13}$$

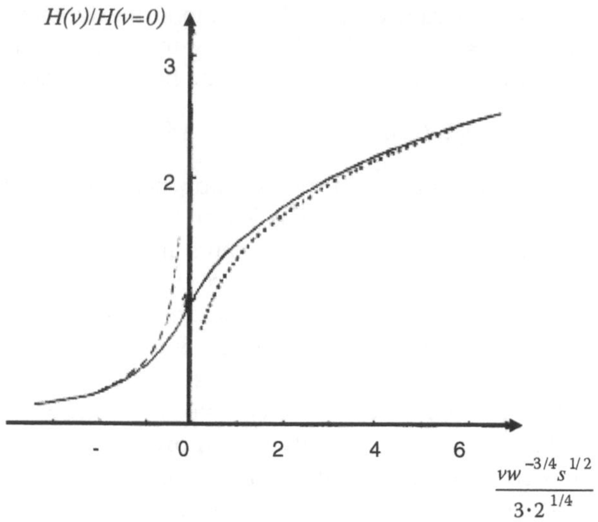

FIGURE 2.2
Dependence of normalized height of the brush $H(v)/H$ ($v = 0$) on reduced second virial coefficient $\approx v(s/a^2)^{1/2}$. Dashed and dotted lines are asymptotes corresponding to extremely poor or good solvent conditions, respectively. (Adapted from Zhulina, E.B. et al., *Macromolecules*, 24, 140, 1991. With permission.)

where $\beta = \{1, 1/2, 0\}$ under good, theta, and poor solvent conditions, respectively. Due to spreading of the chain ends throughout the brush, the force-vs.-compression response is weaker than in the boxlike model. Namely, at weak brush deformation $\Delta H/H << 1$, restoring force π per unit area

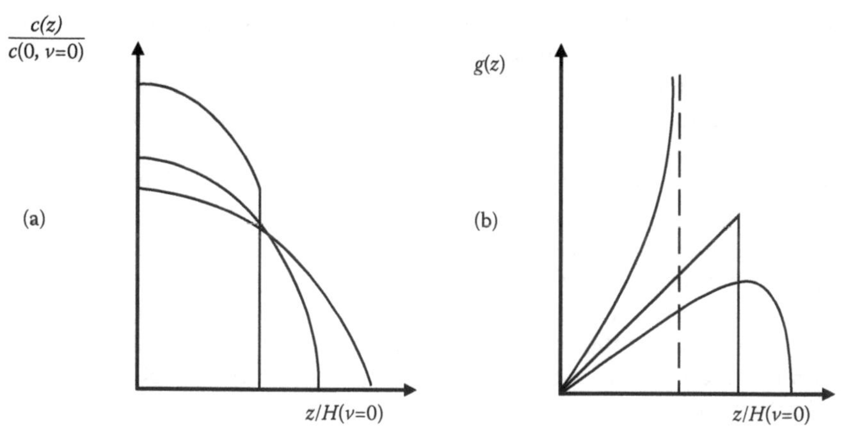

FIGURE 2.3
(a) Polymer density profiles and (b) corresponding free chain-end distributions under good, theta, and poor solvent conditions. (Adapted from Zhulina, E.B. et al., *Macromolecules*, 24, 140, 1991. With permission.)

varies, respectively, as $(\Delta H/H)^{3/2}$ and $(\Delta H/H)^2$ under good and theta-solvent conditions, whereas the boxlike model gives $\pi \approx \Delta H/H$.[28]

Analysis of collapse in an organosoluble neutral brush indicates that

1. In contrast to the second-order coil-to-globule phase transition in individual polymer chains, the collapse of sufficiently densely grafted (laterally homogeneous) brush occurs progressively.
2. As the grafting density increases, the collapse becomes more gradual, and the collapse transition shifts to a larger absolute value of (negative) v.
3. Collapse starts from a more dilute periphery of the brush and continuously propagates to a more dense proximal to the surface regions.
4. The polymer density at the edge of the brush, $z = H$, is equal to that in the collapsed polymer globule (precipitated polymer), $c(H) = |v|/2w$.
5. Within the limit of extremely poor solvent conditions (high polymer density in the brush), the polymer density profile approaches this constant value.
6. The free ends of the chains are distributed throughout the brush under good or theta-solvent conditions, but they become progressively localized at the edge of the collapsed brush under poor solvent conditions.

For neutral water-soluble polymers (NWSPs) the situation is more delicate.[31] Typically $\chi(T)$ is an *increasing* function of temperature, and polymers like poly(ethylene oxide) (PEO), poly(*N*-isopropylacrylamide) (PNIPAAm), and poly(vinylpyrolidon) (PVP) exhibit the so-called lower critical solubility temperature (LCST). The LCST corresponds to inversion in the sign of the second virial coefficient that leads to collapse, aggregation, and macroscopic phase separation (precipitation) of long macromolecules just above the LCST. For many NWSPs the LCST is relatively low (about 100°C for PEO and 28°C for PNIPAAm) and can be reached at normal atmospheric pressure, such polymers are often referred to as "thermosensitive." In addition to reversed temperature dependence, $\chi(T,c)$ often varies with polymer concentration c in aqueous solution. The so-called two-state models[32–35] that incorporate two interconvertible states for a monomer and the n-cluster model introduced by de Gennes[36,37] make it possible to rationalize the $\chi(T,c)$ behavior and the solution phase diagrams for a number of NWSPs. Generalization of the SCF brush theory for an NWSP with an arbitrary $\chi(T,c)$ dependence involves the introduction of an experimentally measurable parameter $\chi'(T,c) = \chi(T,c) - (1 - ca^3)\delta\chi(T,c)/\delta c$ and the formulation of Equation 2.11 in terms of μ and χ'.[31] For NWSPs with decreasing $\chi'(c)$ vs. c dependence (PVP), collapse of the brush is expected to follow the scenario outlined for organosoluble polymers (with the account of reversed $\chi(T)$

dependence). However, for polymers with increasing $\chi'(c)$ vs. c dependence (PEO, PNIPAAm), collapse of a thermosensitive brush can (under certain conditions) involve the so-called vertical phase separation with a jumplike variation in polymer density profile. A dense phase is then found near the surface, and a dilute phase is at the brush periphery, whereas the distribution of chain ends becomes bimodal. A number of experimental observations[38–43] indicate the possibility of this scenario for PNIPAAm brushes. Vertical phase separation in thermosensitive brushes can be also induced by brush compression.[31,44,45] Figure 2.4 illustrates polymer-density profile and chain-end distribution in a vertically segregated polymer brush.[31]

The picture described above refers to the case when, due to high grafting density, the chains remain extended with respect to their Gaussian dimensions, even in the collapsed state. In the case of relatively sparse grafting,

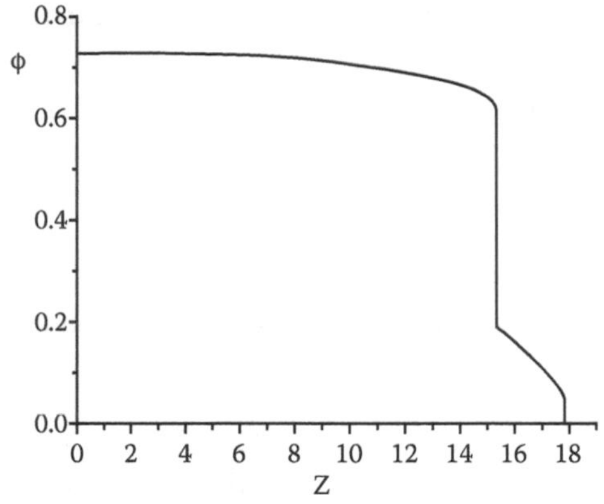

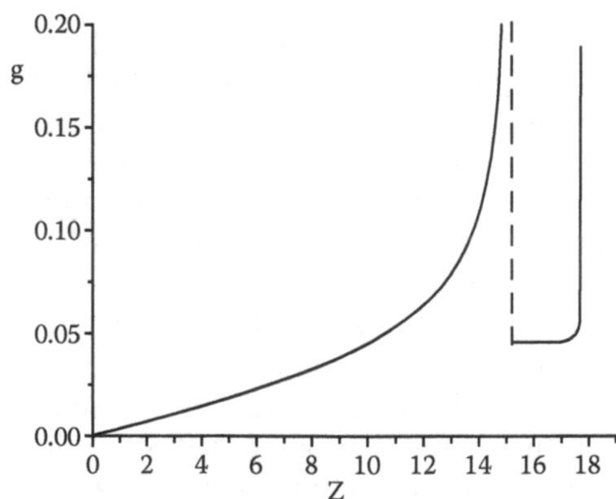

FIGURE 2.4
(a) Polymer density profile and (b) distribution function of free ends in vertically segregated brush. (Adapted from Baulin, V.A. et al., *J. Chem. Phys.*, 119, 10977, 2003. With permission.)

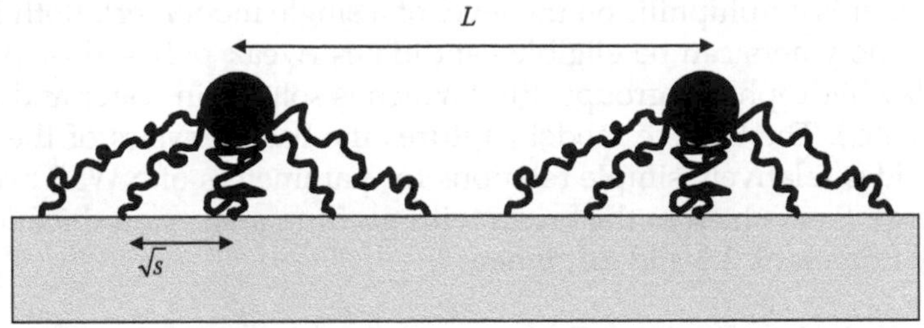

FIGURE 2.5
Lateral aggregation of chains in polymer brush under poor solvent conditions; L = characteristic dimensions of a cluster.

the chains reach their Gaussian dimensions at a certain temperature. Further decrease in the solvent strength results in lateral decomposition of the homogeneous brush into an array of so-called octopus micelles.[46–48] Such a micelle comprises several chains forming a globular core that is connected by extended "legs" to the grafting sites (Figure 2.5).

2.4 Polymer Brush in Solvent Mixture

The properties of a polymer brush can be tuned by varying the composition of mixed solvents.[49–57] Competition between two incompatible solvents A and B (water and oil) for contacts with polymer (selective solvation) can induce sharp conformational transformation in the brush structure (Figure 2.6). A necessary requirement is that the polymer must be soluble in both solvents

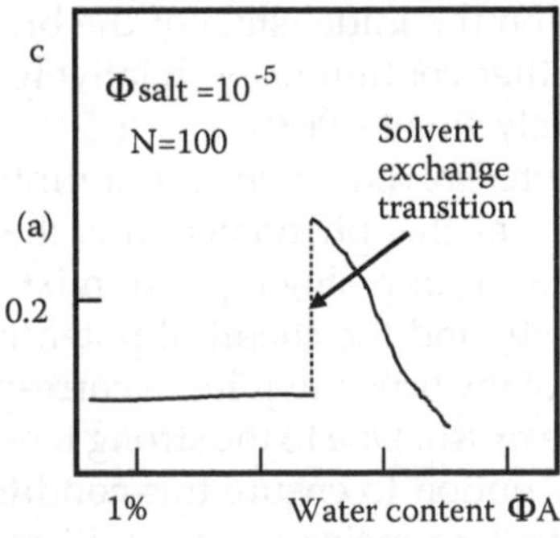

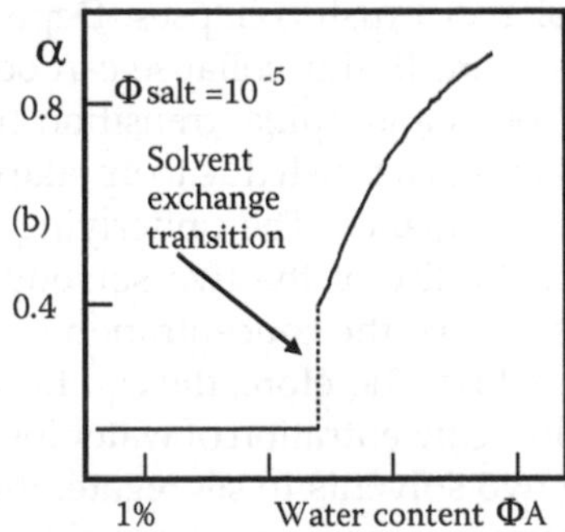

FIGURE 2.6
Polymer brush in a mixed solvent: (a) polymer density profile and (b) degree of chain ionization as functions of water content in water–solvent mixture. (Adapted from Mercurieva, A.A. et al., *Macromolecules*, 35, 4739, 2002. With permission.)

(the polymer is amphiphilic on the level of a single monomer). Both NWSPs and ionic polymers can be eligible candidates (weak polyacid or polybase with bulky hydrophobic groups, PEO which is soluble in water and organic solvents, etc.). The boxlike model captures the basic physics of the system and provides relatively simple relations for parameters of NWSP brushes.[52] Free energy F per chain in the brush with grafting area s and thickness H is given by Equations 2.2 and 2.3, where

$$F_{int}/k_BT = n_A \ln\phi_A + n_B \ln\phi_B + \chi_A n_A(ca^3) + \chi_B n_B(ca^3) + \chi_{AB} n_A \phi_B \quad (2.14)$$

Here, n_A and n_B are the numbers of solvent molecules A and B in volume Hs of the brush, $\phi_i = a^3\, n_i/Hs$ (and $\phi_B + \phi_A + ca^3 = 1$), and χ_B, χ_A, and χ_{AB} are the respective Flory-Huggins parameters of polymer–solvent and solvent–solvent interactions. Under equilibrium conditions, the chemical potentials of solvent molecules in the brush, $\mu_I = \delta F/\delta n_i$, and in the bulk, μ_{bi}, are equal. For a single-phase solvent mixture with concentration $\Phi_A < \Phi_A^*(\chi_{AB})$ of admixed water (here, $\Phi_A^*(\chi_{AB})$ is the binodal concentration corresponding to macrophase segregation of the two solvents), Equation 2.15,

$$\mu_i = \delta F/\delta n_i = \mu_{bi} = k_B T[\ln\Phi_i + \chi_{AB}(1 - \Phi_i)^2],\ (i = A,B) \quad (2.15)$$

allows for examination of the brush rearrangement upon admixture of minor solvent (increase in Φ_A). When $\chi_A < \chi_B$ and water is a better solvent for polymer (but the brush could still swell in pure solvent B), an increase in Φ_A leads to a dramatic transformation in brush structure. An initial increase in water content in bulk mixture does not affect the brush, and average polymer concentration in the brush filled with major solvent B is

$$ca^3(B) \cong (s^2 a^{-4} v_B)^{-1/3} \quad (2.16)$$

Due to subsequent uptake of a better solvent (solvent-exchange transition), the polymer brush collapses. Depending on the initial state of the brush in pure solvent B, this collapse can occur either continuously (relatively large $ca^3(B)$) or as a jumplike transition (relatively small $ca^3(B)$).

Collapse of a brush due to uptake of a better solvent seems counterintuitive at the first glance. The underlying reason for this phenomenon is the high incompatibility of the two solvents ($\chi_{AB} > 2$). In a single-phase mixture of such solvents, the concentration of water Φ_A and the chemical potential μ_{bA} are very low. Therefore, the equilibration of the brush requires a correspondingly low concentration of water inside the brush. Due to the strong tendency of the two solvents to segregate, the only option to ensure this condition is to increase polymer concentration by expelling major solvent (oil) outside of the brush. In the framework of boxlike model, the jump in polymer concentration due to solvent-exchange transition is estimated[52] as

$$\Delta ca^3 = 1 - (v_B/v_A)^{1/2}, \quad v_i = {}^1/_2 - \chi_i \quad (2.17)$$

Further increase in Φ_A (and in chemical potential μ_{bA}) leads to gradual swelling of the brush in almost pure water. In this posttransition regime, polymer concentration ca^3 in the brush is related to water content in the bulk Φ_A as

$$v_A(ca^3)^2\{1 - [ca^3(A)]/(ca^3)^3\} = \ln(\Phi_A{}^*/\Phi_A) \tag{2.18}$$

where

$$ca^3(\mathrm{A}) \cong (s^2a^{-4}v_A)^{-1/3} \tag{2.19}$$

is the equilibrium polymer concentration in a brush swollen in pure water.

A more refined numerical SCF model outlines the possibility of vertical microphase separation instead of jumplike rearrangement in the brush structure and confirms nonmonotonic H vs. Φ_A dependence.[53] It also allows us to address the wetting aspects of the problem.[53,55] As in any vertically segregated state, the dense phase is near the surface, whereas the dilute phase is at the periphery of the brush. Note that the minor solvent (water) is localized in the dense phase, whereas the major solvent (oil) is in a dilute phase. In the posttransition regime, where the brush progressively uptakes minor solvent, the numerical SCF model predicts a peculiar polymer density profile, with a peak at the brush edge reflecting the effect of the A/B solvent boundary.[51,53] To shield unfavorable contacts between solvent molecules, polymer chains localize free ends at the solvent–solvent boundary so that the volume fraction of monomers (ca^3) in the interfacial region is rather large.

For ionic (weakly dissociating) polymers, the salinity and pH in the solution serve as additional stimuli to mediate the brush behavior in mixed solvents. Again, a simple box-type model captures essential physics of the system in certain cases.[57] Generally, the solvent-exchange transition in ionizable brushes is more cooperative due to ionization of the grafted chains in minor solvent (water). Here, the transition pattern is even more anomalous: the brush collapses despite of its ionization. Typically, there must be a sufficient concentration of salt ions to promote water uptake and brush ionization. Similar to the case of NWSPs, a more refined numerical SCF model confirms the basic features of solvent exchange in ionizable brushes and provides additional information[51,57] (possibility of vertical microphase segregation, enrichment of water–oil boundary by polymer segments in the transition and posttransition regimes, etc.).

2.5 Complexation of Grafted Polymers with Amphiphiles

There is abundant experimental evidence supporting the complexation of nonionic water-soluble polymers (PEO, PPO [poly(propylene oxide)], PNIPAAm) with ionic surfactants (e.g., SDS [sodium dodecyl sulfate]) in aqueous solutions.[58,59] Although the mechanism of binding of surfactant molecules to noncharged polymers remains a matter of discussion in the literature, it is generally recognized that

- Surfactants bind to polymer chains cooperatively, i.e., in the form of micellelike nanoaggregates.
- The association of surfactants to the chain starts when the surfactant concentration exceeds a certain critical value, the so-called CAC (critical association concentration), which is lower than the CMC (critical micelle concentration).
- The aggregation number in bound micelles is normally smaller than in "free" micelles formed by surfactants in solution at concentration above the CMC.

Polymer conformation is strongly affected by complexation with surfactants due to (a) "wrapping" of the chain segments around bound micelles and (b) enhancement of intramolecular excluded-volume interactions due to Coulomb repulsion between bound micelles. To describe these effects of coupling between complexation with surfactants, intra- and intermolecular interactions and polymer conformation, a model of "annealed excluded-volume interactions" was proposed.[60,61]

The average monomer concentration in a single polymer coil is relatively low, and therefore binding of surfactants occurs virtually independently of the chain conformation. In contrast, when surfactants make complexes with polymers collapsed under poor solvent conditions, local polymer concentration is high enough, which leads to strong coupling between polymer conformation and complexation. Therefore, the collapse-to-swelling transition occurs upon variation in surfactant concentration or temperature at $T > \text{LCST}$ more cooperatively than conventional temperature-induced coil-to-globule transition (under certain conditions, as the first-order phase transition).[60,61]

Interaction and complexation with surfactants may dramatically affect conformations of grafted chains in polymer brushes (Figure 2.7). The SCF analytical approach coupled to the annealed excluded-volume interaction model provides valuable insight into the brush structural rearrangement induced by complexation with surfactants.[62] The free-energy density in the brush interacting with surfactants depends both on the monomer density profile $c(z)$ and the local degree of loading of the chain by surfactant micelles $\theta(z)$ (fraction of monomer units involved in complexes),

$$\frac{f\{c(z),\theta(z)\}}{k_B T} = v\{\theta(z)\}c^2(z)a^3 + c(z)a^3[u\theta(z) + \theta(z)\ln\theta(z) + (1-\theta(z))\ln(1-\theta(z))] \tag{2.20}$$

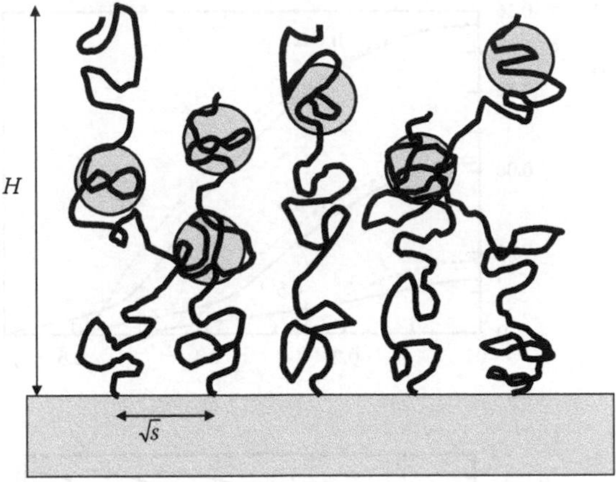

FIGURE 2.7
Schematic presentation of a polymer brush with bound surfactant micelles: s = surface area per grafted chain; H = height of the brush.

Here, the first term accounts for excluded-volume interactions enhanced by bound surfactant micelles, whereas the second and the last two terms account for the free-energy gain upon complexation and the translational entropy of bound micelles along the chain, respectively. Here,

$$v\{\theta(z)\} = v_0(1-\theta^2(z)) + v_1\theta^2(z) + 2v_{01}\theta(z)(1-\theta(z)) \tag{2.21}$$

is the effective parameter of excluded-volume interactions that depends on loading $\theta(z)$, and v_0 refers to the excluded volume of "bare" monomers. Figure 2.8 presents a typical monomer density profile $c(z)$ and the profile of loading of the chain by surfactant micelles $\theta(z)$ in the case of good solvent conditions for polymer $v_0 > 0$ and dominant repulsive interaction between bound micelles $v_1 > 0$. Binding of surfactants to the chains and strong repulsion between bound micelles lead to additional stretching of the chains compared with their conformation in a "bare" (nonloaded) brush. As follows from Figure 2.8, micelles bind to the chains preferably at the sparse periphery of the brush, whereas the region proximal to the grafting surface is depleted of micelles. As a result, additional stretching of the chains is most pronounced at the periphery of the brush. Depletion of micelles from the proximal region of a densely grafted brush results in a nonmonotonic dependence of the total loading amount on the grafting density, as demonstrated by Figure 2.9.

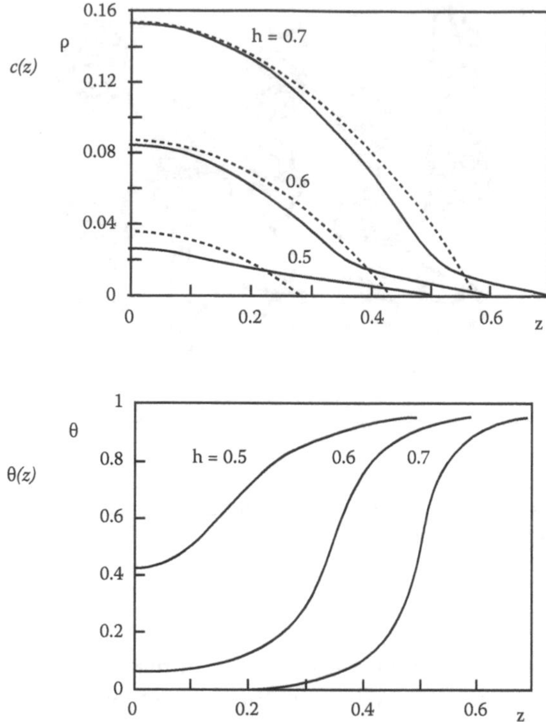

FIGURE 2.8
Profiles of (a) polymer density and (b) degree of loading in the brush complexed with surfactants. Progressively increasing values of $h \equiv H\sqrt{3\pi^2/8}$ correspond to increasing grafting density at constant chemical potential of surfactants. Dotted lines indicate density profiles in bare brush for the same values of grafting density. (Adapted from Currie, E.P.K. et al., *Eur. Physica J. E*, 1, 27, 2000. With permission.)

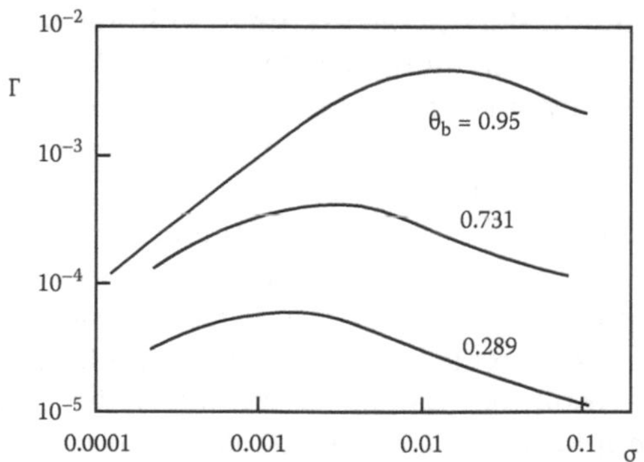

FIGURE 2.9
Amount of surfactants adsorbed in the brush as a function of grafting density. Numbers at the curves indicate equilibrium degree of loading of a chain in the bulk of the solution under the same chemical potential of surfactants. (Adapted from Currie, E.P.K. et al., *Eur. Physica J. E*, 1, 27, 2000. With permission.)

2.6 Polyelectrolyte Brushes

Conformations of end-grafted polyelectrolyte chains can be most efficiently manipulated by a variation of ionic strength or pH in the solution or by an applied external electrical field. In this section, we discuss permanently charged brushes, where the fraction of charged (strongly dissociating) monomer units is independent of external conditions and is predetermined by the chemical sequence of ionic and nonionic (neutral) monomer units. The effects arising due to pH-sensitive properties are considered in Section 2.7.

The basics of polyelectrolyte brush theory have been established in the literature.[63–68] The most essential feature of a polyelectrolyte brush is the trapping of (the majority of) counterions compensating the charge on grafted chains in the interior of the brush, even when the latter is brought in contact with an infinite reservoir of salt-free solution. This is similar to the localization of counterions within the Gouy-Chapmann length near an infinite uniformly charged planar surface. The counterions are retained inside a polyelectrolyte brush by strong Coulomb attraction to grafted polyions or, equivalently, due to excess electrostatic potential inside the brush. As a result, the polyelectrolyte brush swells due to the osmotic pressure of the trapped counterions. The equilibrium degree of swelling follows from the balance of translational entropy of the counterions entrapped in the brush and the conformational entropy penalty for extension of the grafted chains. Using the simple boxlike model, one finds

$$F_{\text{ion}}(H) = \alpha N k_B T[\ln(\alpha c a^3) - 1] \tag{2.22}$$

where $\alpha c = \alpha N/sH$ is the concentration of charged monomer units that is approximately equal to the concentration of counterions in the brush. Minimization of the free energy given by Equations 2.3 and 2.22 leads to a well-known result for the brush height:

$$H_{\text{osm}} = 3^{-1/2}\alpha^{1/2}Na \tag{2.23}$$

which is independent of the grafting density and increases as a square root of the degree of ionization α.

The model described above is known as an osmotic boxlike brush model due to the key role of the osmotic pressure of counterions confined in the brush (which is reflected by subscript "osm" in Equation 2.23). The condition of its applicability implies that the electrostatic screening length in the brush is significantly smaller than the overall thickness of the brush H. It implies that the local compensation of immobilized charge by mobile ions occurs on the length scale smaller than the brush height (local electroneutrality approximation).

Two refinements of the approach have been made:

1. Local electroneutrality approximation coupled to the SCF model for the chains' extension[69]
2. Consistent Poisson-Boltzmann approach allowing independent minimization of the total free energy with respect to chain conformations and distribution of mobile ions[69,70]

The latter is the most quantitative and enables the explicit calculation of the deviation from local (and global) electroneutrality of the brush and all numerical factors and logarithmic (dependent on the grafting density) correction to the brush thickness. However, under good or theta-solvent conditions, all the qualitative predictions that follow from the simple boxlike osmotic model about the dependence of the brush thickness on N, α, and salt concentration remain valid.

Addition of salt in the solution leads to penetration of both counter- and co-ions of salt into the brush (Figure 2.10). Salt ions reduce excess electrostatic potential in the brush and differential osmotic pressure, leading to progressive deswelling of the brush upon an increase in salt concentration. In the boxlike model, excess electrostatic potential $\Delta\Psi$ can be obtained by using the condition of local electroneutrality in the (anionic) brush:

$$\alpha c + \sum_{j-} c_{j-} = \sum_{j+} c_{j+} \tag{2.24}$$

where c_j denotes concentration of ions of type j in the brush, and summations run over all cationic (j+) and anionic (j–) species, together with the Boltzmann

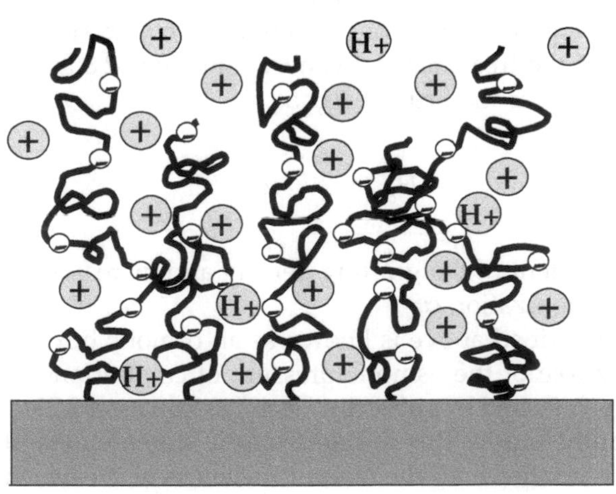

FIGURE 2.10
Schematic presentation of a polyelectrolyte brush.

law for distribution of ions between the interior of the brush and the exterior of the solution (where the concentrations c_{bj} of ions are constant),

$$\frac{c_{j+}}{c_{bj+}} = \frac{c_{bj-}}{c_{j-}} = \exp(-e\Delta\Psi/k_B T) \tag{2.25}$$

By combining Equations 2.24 and 2.25, one obtains for excess electrostatic potential in the anionic brush

$$e\Delta\Psi/k_B T = \ln\left(-\frac{\alpha c}{\Sigma_j c_{bj}} + \sqrt{1+\left(\frac{\alpha c}{\Sigma_j c_{bj}}\right)}\right) \tag{2.26}$$

The contribution of ions to the Gibbs free energy of the brush is

$$f_{ion}\{c_j\}/k_B T = \sum_j c_j(\ln(a^3 c_j)-1) - \sum_j c_{bj} - \sum_j c_j \ln(a^3 c_{bj}) \tag{2.27}$$

where the second term equals to the osmotic pressure, and $k_B T\ln(a^3 c_{bj})$ is the chemical potential of ions of type j in the bulk of the solution.

Minimization of the free energy with respect to polymer concentration leads to the equation for the equilibrium polymer concentration $c = N/sH$ as a function of salt concentration in the bulk of the solution,

$$\left(\frac{c}{c_{osm}}\right)^3 - \frac{c_{osm}}{c} = \frac{2\Sigma_j c_{bj}}{\alpha c_{osm}} \tag{2.28}$$

where $c_{osm} = 3^{1/2}/s\alpha^{1/2}a$ is monomer unit concentration in the salt-free osmotic brush.

As follows from Equation 2.28, the effect of added salt on polymer concentration in the brush is negligible as long as the concentration of salt in the solution is lower than that of counterions trapped inside the brush, $\alpha c >> \Sigma_j c_{bj}$ ("osmotic regime"). Swelling of the brush in this low-salt regime is governed by the osmotic pressure of counterions, $\Delta\Pi/k_B T \approx \alpha c$, and the brush height H obeys Equation 2.23.

In the opposite high-salt limit, $\alpha c << \Sigma_j c_{bj}$ ("salted brush"), an increase in the salt concentration in the solution leads to pronounced deswelling of the brush due to a significant decrease in differential osmotic pressure,

$$\Delta\Pi/k_B T = \sum_j c_j - \sum_j c_{bj} \tag{2.29}$$

Here, the brush height decreases as a function of salt concentration as

$$H \approx 6^{-1/3}(s/a^2)^{-1/3}\alpha^{2/3}\left(\sum_j c_{bj}\right)^{-1/3} Na \tag{2.30}$$

In the salt-dominance regime, the ionic interactions manifest themselves as short-range excluded volume interactions, $\Delta\Pi/k_BT \approx v_{eff}c^2$, with an effective excluded volume parameter

$$v_{eff} = \frac{\alpha^2}{2\sum_j c_{bj}} \tag{2.31}$$

A more refined SCF analysis[71] indicates that salt-induced collapse of the brush leads to progressive transformation of the polymer density profile $c(z)$ from a truncated Gaussian (in the salt-free limit) to a truncated parabolic shape. Recall that the parabolic shape of the density profile is typical for a neutral brush in a good solvent.

The picture of salt-induced collapse described above holds when the short-range excluded volume interactions between both charged and noncharged monomers are repulsive (good solvent conditions). The character of the collapse-to-swelling transition is dramatically different when solvent is poor (like for PSS, poly [styrene sulfonate]) or when its strength can be tuned from good to poor (as in stat-copolymer comprising ionic and thermosensitive monomers, e.g., poly[acrylic acid-co-NIPAAm]).

2.7 Conformational Transitions in pH-Responsive Brushes

End-grafting of a pH-sensitive polyelectrolyte, e.g., a weak polyacid, gives rise to a pH-responsive polyelectrolyte brush.[72,73] Here, the fraction of charged monomers in the chain α is determined by the local value of pH via mass action law as

$$\frac{\alpha}{1-\alpha} = \frac{K}{c_{H^+}} \tag{2.32}$$

where $c_{H+} = 10^{-pH}$ is the local concentration of hydrogen ions, and K is the dissociation constant. For weak polyacids, such as poly(acrylic) or poly(methacrylic) acid, pK ≥ 4.5. By varying the pH in the solution, one can tune the fraction of charged monomers and the strength of Coulomb interactions in the brush. However, in the case of low ionic strength in the solution, the local pH inside the brush may differ significantly from that in the buffer. As a result, ionization of chains in the brush can be quite different from that of the chains in the solution. The local pH in the brush can be found by applying Equations 2.23 and 2.24 if the pH in the solution is known. The degree of ionization of chains in the brush α is related to the degree of ionization α_b of monomers in the bulk of solution as

$$\frac{\alpha}{1-\alpha}\cdot\frac{1-\alpha_b}{\alpha_b}=\exp(e\Delta\Psi/k_BT) \tag{2.33}$$

where excess potential $\Delta\Psi$ is given by Equation 2.25.

Density of the free energy of interactions in such a brush yields

$$\begin{aligned} f_{\text{ion}}\{c_j\}/k_BT &= \sum_j c_j(\ln(a^3c_j)-1)-\sum_j c_{bj}-\sum_j c_j\ln(a^3c_{bj}) \\ &+ca^3[\alpha\ln\alpha+(1-\alpha)\ln(1-\alpha)+\alpha\ln(c_{H^+}a^3)-\alpha\ln(Ka^3)] \end{aligned} \tag{2.34}$$

In addition to terms presented in Equation 2.25, Equation 2.34 comprises (the last line) the contribution describing the free energy of chain ionization. As follows from Equations 2.25 and 2.33, in the limit of high ionic strength in the solution, $\alpha c \ll \Sigma_j c_{bj}$, excess electrostatic potential vanishes, and the brush behaves as that with a constant degree of ionization $\alpha \approx \alpha_b$. The progressive collapse of the brush upon increase in salt concentration follows Equation 2.30.

At low ionic strength in the solution, $\alpha c \gg \Sigma_j c_{bj}$, excess electrostatic potential leads to a significantly lower degree of ionization $\alpha \ll \alpha_b$. Expansion of Equation 2.33 in the low-salt limit provides explicit expression for the degree of ionization of the chains in the brush

$$\alpha \approx \left(\frac{\alpha_b}{1-\alpha_b}\cdot\frac{\Sigma_j c_{bj}}{c}\right)^{1/2} \tag{2.35}$$

As follows from Equation 2.35, α decreases as a function of polymer concentration in the brush c and increases as a function of salt concentration in the bulk of the solution $\Sigma_j c_{bj}$. In this so-called osmotic annealing regime, excess osmotic pressure of counterions inside the brush is $\Delta\Pi/k_BT \approx \alpha(c)\cdot c \approx (\alpha_b c\cdot\Sigma_j c_{bj})^{1/2}$, where $\alpha(c)$ is given by Equation 2.35. This results in an "abnormal" response of a pH-sensitive polyelectrolyte brush to variation in solution salinity: in the high-salt regime, H *decreases* as a function of salt concentration according to Equation 2.30, while in the low-salt regime it *increases* upon an increase in salt concentration as

$$H \approx \left(\frac{1}{2^{1/2}\cdot 3}\right)^{2/3}(s/a^2)^{1/3}\left(\frac{\alpha_b}{1-\alpha_b}\right)^{1/3}\left(\sum_j c_{bj}\right)^{1/3} Na^2 \tag{2.36}$$

This behavior can be rationalized as follows. At high solution salinity, salt-induced reduction in differential osmotic pressure results in collapse of the brush. At low solution salinity, addition of salt ions leads to an increase in local pH due to the ion exchange between the bulk solution and the interior of the brush. As a result, the chain ionization increases, and the brush expands. Remarkably, as follows from Equation 2.36, at low salt concentration, the thickness of a pH-responsive brush H decreases upon an increase in grafting density $1/s$.

The theory[72,73] predicts that in contrast to a strongly dissociating polyelectrolyte brush, the thickness H of a pH-responsive brush exhibits nonmonotonic behavior as a function of salt concentration. For weak polyacids, the nonmonotonic behavior is expected when pH ≤ pK. When pH >> pK, the response of the brush to the salt concentration variation is the same as for strongly dissociating brushes. An increase in α_b (that is, an increase in the pH in the case of polyacid brushes) leads to a progressive increase in brush thickness at any salt concentration (see Equations 2.30 and 2.34). Theoretically predicted nonmonotonic dependence of the pH-sensitive brush thickness on salt concentration was supported by experiments.[74,75]

2.8 Conclusions

The responsive properties of polymer brushes arise due to the ability of the grafted chains to change conformation in response to variations in such environmental conditions as temperature, salinity, pH, solvent composition, etc. These stimuli affect the strength of inter- and intramolecular interactions in the brush, provided that the grafted chains comprise ionic, pH-sensitive, or thermosensitive monomer units. Theoretical models based on different implementations of the mean-field approximation enable us to rationalize properties of both ionic and nonionic polymer brushes in terms of dependences for large-scale and local structural characteristics of the brush on varied environmental conditions. The most complex and sometimes unexpected response of the brush is predicted in the case when internal degrees of freedom, such as ionization or solvation state of monomers, are strongly coupled to the chain conformations, e.g., in pH-sensitive polyelectrolyte brushes. This coupling could lead to nonmonotonic variation of the brush properties, intrabrush microphase separation, etc.

Acknowledgments

OVB gratefully acknowledges support of the Alexander von Humboldt Foundation by the Friedrich Wilhelm Bessel Research Award. This work has been partially supported by the Russian Foundation for Basic Research (grant 05-03-33126) and by a joined program of the Dutch National Science

Foundation (NWO) and the Russian Foundation for Basic Research: "Computational Approaches for Multi-Scale Modelling in Self-Organizing Polymer and Lipid Systems," grant no. 047.016.004/04.01.8900.

References

1. Abu-Lail, N.I. and Camesano, T.A., Role of ionic strength on the relationship of biopolymer conformation, DLVO contributions, and steric interactions to bioadhesion of *Pseudomonas putida* KT2442, *Biomacromolecules*, 4, 1000, 2003.
2. Kumar, S., Yin, X., Trapp, B.D., Hoh, J.H., and Paulaitis, M.E., Relating interactions between neurofilaments to the structure of axonal neurofilament distributions through polymer brush models, *Biophys. J.*, 82, 2360, 2002.
3. Mukhopadhyay, R., Kumar, S., and Hoh, J.H., Molecular mechanisms for organizing the neuronal cytoskeleton, *BioEssays*, 26, 1, 2004.
4. Dean, D., Seog, J., Ortiz, C., and Grodzinsky, A.J., Molecular-level theoretical model for electrostatic interactions within polyelectrolyte brushes: applications to charged glycosaminoglycans, *Langmuir*, 19, 5526, 2003.
5. de Kruif, C.G. and Zhulina, E.B., Kappa-casein as a polyelectrolyte brush on the surface of casein micelles, *Colloid. Surf. A*, 117, 151, 1996.
6. Tuinier, R. and de Kruif, C.G., Stability of casein micelles in milk, *J. Chem. Phys.*, 117, 1290, 2002.
7. Ratner, B.D., Hoffman A.S., and Schoen, F.J., *Biomaterial Science: an Introduction to Materials in Medicine*, Academic Press, London, 2004.
8. Betageri, G.V., Jenkins, S.A., and Parsons, D.L., *Lyposome Drug Delivery Syst.*, CRC Press, London, 1993.
9. Lasic, D. and Martin, F., Eds., *Stealth Lyposomes*, CRC Press, Boca Raton, FL, 1995.
10. Lavasanifar, A., Samuel, J., and Kwon, G.S., Poly(ethylene oxide)-block-poly (l-amino acid) micelles for drug delivery, *Adv. Drug Deliv. Rev.*, 54, 169, 2002.
11. Kabanov, A.V., Batrakova, E.V., and Alakhov, V.Y., Pluronic block copolymers as novel polymer therapeutics for drug and gene delivery, *J. Control. Release*, 82, 189, 2002.
12. Cooper, J.M. and Cass, A.E.G., *Biosensors*, University Press, Oxford, 2004.
13. Halperin, A., Tirrell, M., and Lodge, T.P., Tethered chains in polymer microstructures, *Adv. Polym. Sci.*, 100, 31, 1990.
14. Fleer, G.J., Cohen Stuart, M.A., Scheutjens, J.M.H.M., Cosgrove, T., and Vincent, B., *Polymers at Interfaces*, Chapman and Hall, London, 1993.
15. Milner, S.T., Polymer brushes, *Science*, 251, 905, 1991.
16. Daoud, M. and Cotton, J.P., Star shaped polymers: a model for the conformation and its concentration dependence, *J. Phys. (France)*, 43, 531, 1982.
17. Zhulina, E.B., Phase diagram for semi-rigid macromolecules grafted to a solid sphere, *Polym. Sci. USSR*, 26, 885, 1984.
18. Zhulina, E.B. and Birshtein, T.M., Conformations of block copolymer micelles in selective solvents. (micellar structures), *Polym. Sci. USSR*, 27, 570, 1985.
19. Birshtein, T.M. and Zhulina, E.B., Conformations of star-branched macromolecules, *Polymer*, 25, 1453, 1984.
20. Wang, Z.-G. and Safran, S.A., Size distribution for aggregates of associating polymers, II: linear packing, *J. Chem. Phys.*, 89, 5323, 1988.

21. Harris, J.M., *Poly(ethylene glycol) Chemistry: Biotechnical and Biomedical Applications*, Plenum Press, New York, 1992.
22. Alexander, S., Adsorption of chain molecules with a polar head: a scaling description, *J. Phys. (France)*, 38, 983, 1977.
23. de Gennes, P.-G., Conformations of polymers attached to an interface, *Macromolecules*, 13, 1069, 1980.
24. de Gennes, P.-G., *Scaling Concepts in Polymer Physics*, Cornell University Press, Ithaca, NY, 1979.
25. Skvortsov, A.M., Pavlushkov, I.V., Gorbunov, A.A., Zhulina, E.B., Borisov, O.V., and Pryamitsyn, V.A., Structure of dense-grafted polymer monolayers, *Polym. Sci. USSR*, 30, 1706, 1988.
26. Milner, S.T., Witten, T.A., and Cates, M., Theory of the grafted polymer brush, *Macromolecules*, 21, 2610, 1988.
27. Zhulina, E.B., Borisov, O.V., Pryamitsyn, V.A., and Birshtein, T.M., Coil-globule type transitions in polymers, I: collapse of layers of grafted polymer chains, *Macromolecules*, 24, 140, 1991.
28. Zhulina, E.B., Borisov, O.V., and Pryamitsyn, V.A., Theory of steric stabilization of colloid dispersions by grafted polymers, *J. Colloid Interface Sci.*, 137, 495, 1990.
29. Semenov, A.N., Contribution to the theory of microphase layering in block-copolymer melts, *Sov. Phys. JETP*, 61, 733, 1985.
30. Borisov, O.V., Birshtein, T.M., and Zhulina, E.B., Diagram of states and collapse of grafted chain layers, *Polym. Sci. USSR*, 29, 772, 1988.
31. Baulin, V.A., Zhulina, E.B., and Halperin, A., Self-consistent field theory of neutral water-soluble polymer brushes, *J. Chem. Phys.*, 119, 10977, 2003.
32. Karlstrom, G., A new model for upper and lower critical solution temperatures in poly(ethylene oxide) solutions, *J. Phys. Chem.*, 89, 4962, 1985.
33. Matsuyama, A. and Tanaka, F., Theory of solvation-induced reentrant phase separation in polymer solutions, *Phys. Rev. Lett.*, 65, 341, 1990.
34. Bekiranov, S., Bruinsma, R., and Pincus, P., Solution behavior of polyethylene oxide in water as a function of temperature and pressure, *Phys. Rev. E*, 55, 577, 1997.
35. Dormidontova, E., Role of competitive PEO-water and water-water hydrogen bonding in aqueous solution PEO behavior, *Macromolecules*, 35, 987, 2002.
36. de Gennes, P.-G., A second type of phase separation in polymer solutions, *C.R. Acad. Sci. II (Paris)*, 313, 1117, 1991.
37. Wagner, M., Brochard-Wyart, F., Hervet, H., and de Gennes, P.-G., Collapse of polymer brushes induced by *n*-clusters, *Coll. Polym. Sci.*, 271, 621, 1993.
38. Wu, C., A comparison between the "coil-to-globule" transition of linear chains and the volume phase transition of spherical microgels, *Polymer*, 39, 4609, 1998.
39. Zu, P.W. and Napper, D.H., Experimental observation of coil-to-globule type transitions at interfaces, *J. Colloid Interface Sci.*, 164, 489, 1994.
40. Zu, P.W. and Napper, D.H., Interfacial coil-to-globule transitions: the effects of molecular weight, *Colloid. Surf. A*, 113, 145, 1996.
41. Afroze, F., Nies, E., and Berghmans, H., Phase transitions in the system poly (*N*-isopropylacrylamide)/water and swelling behavior of the corresponding networks, *J. Mol. Struct.*, 554, 55, 2000.
42. Hu, T., You, Y., Pan, C., and Wu, C., The coil-to-globule-to-brush transition of linear thermally sensitive poly(*N*-isopropylacrylamide) chains grafted on a spherical microgel, *J. Phys. Chem. B*, 106, 6659, 2002.

43. Baulin, V.A. and Halperin, A., Signatures of a concentration-dependent Flory χ parameter: swelling and collapse of coils and brushes, *Macromol. Theor. Simul.*, 12, 549, 2003.
44. Halperin, A., Compression induced phase transitions in PEO brushes: the n-cluster model, *Eur. Phys. J. B*, 3, 359, 1998.
45. Sevick, E.M., Compression-induced phase transitions in water-soluble polymer brushes: the n-cluster model, *Macromolecules*, 31, 3361, 1998.
46. Klushin, L.I., private communication.
47. Williams, D.R.M., Grafted polymers in bad solvents: octopus surface micelles, *J. Phys. II (France)*, 3, 1313, 1993.
48. Zhulina, E.B., Birshtein, T.M., Priamitsyn, V.A., and Klushin, L.I., Inhomogeneous structure of collapsed polymer brushes under deformation, *Macromolecules*, 28, 8612, 1995.
49. Lay, P.-I. and Halperin, A., Polymer brushes in mixed solvents: chromatography and collapse, *Macromolecules*, 25, 6693, 1992.
50. Birshtein, T.M. and Lyatskaya, Yu., Theory of the collapse-stretching transition in a polymer brush in a mixed solvent, *Macromolecules*, 27, 1256, 1994.
51. Lyatskaya, Yu. and Balazs, A.C., Phase separation of mixed solvents within polymer brushes, *Macromolecules*, 30, 7588, 1997.
52. Birshtein, T.M., Zhulina, E.B., and Mercurieva, A.A., Amphiphilic polymer brush in a mixture of incompatible liquids, *Macromol. Theor. Simul.*, 9, 47, 2000.
53. Mercurieva, A.A., Leermakers, F.A.M., Birshtein, T.M., Fleer, G.J., and Zhulina, E.B., Amphiphilic polymer brush in a mixture of incompatible liquids: numerical self-consistent-field calculations, *Macromolecules*, 33, 1072, 2000.
54. Leermakers, F.A.M., Mercurieva, A.A., van Male, J., Zhulina, E.B., Besseling, N.A.M., and Birshtein, T.M., Wetting of a polymer brush, a system with pronounced critical wetting, *Langmuir*, 16, 7082, 2000.
55. Leermakers, F.A.M., Zhulina, E.B., van Male, J., Mercurieva, A.A., Fleer, G.J., and Birshtein, T.M., Effect of a polymer brush on capillary condensation, *Langmuir*, 17, 4459, 2001.
56. Birshtein, T.M., Mercurieva, A.A., and Zhulina, E.B., Deformation of polymer brush immersed in a binary solvent, *Macromol. Theor. Simul.*, 10, 719, 2001.
57. Mercurieva, A.A., Birshtein, T.M., Zhulina, E.B., Jakovlev, P., van Male, J., and Leermakers, F.A.M., An annealed polyelectrolyte brush in a polar-nonpolar binary solvent: effect of pH and ionic strength, *Macromolecules*, 35, 4739, 2002.
58. Linse, P., Picullel, L., and Hanson, P., Models of polymer-surfactant complexation, in *Surfactant Science Series*, Kwak, J.C.T., Ed., Marcel Dekker, New York, 1995.
59. Cabane, B., Structure of some polymer-detergent aggregates in water, *J. Phys. Chem.*, 81,1639, 1977.
60. Currie, E.P.K., Cohen Stuart, M.A., and Borisov, O.V., New mechanisms for phase separation in polymer-surfactant mixtures, *Europhys. Lett.*, 49, 438, 2000.
61. Currie, E.P.K., Cohen Stuart, M.A., and Borisov, O.V., Phase separation in polymer solutions with annealed excluded volume interactions, *Macromolecules*, 34, 1018, 2001.
62. Currie, E.P.K., Fleer, G.J., Cohen Stuart, M.A., and Borisov, O.V., Grafted polymers with annealed excluded volume: a model for the surfactant association in brushes, *Eur. Physica J. E*, 1, 27, 2000.

63. Micklavic, S.J. and Marcelia, S.J., Interaction of surfaces carrying grafted polyelectrolytes, *J. Phys. Chem.*, 92, 6718, 1988.
64. Misra, S., Varanasi, S., and Varanasi, P., A polyelectrolyte brush theory, *Macromolecules*, 22, 4173, 1989.
65. Borisov, O.V., Birshtein, T.M., and Zhulina, E.B., Collapse of grafted polyelectrolyte layer, *J. Phys. II*, 1, 521, 1991.
66. Borisov, O.V., Zhulina, E.B., and Birshtein, T.M., Diagram of states of a grafted polyelectrolyte layer, *Macromolecules*, 27, 4795, 1994.
67. Pincus, P., Colloid stabilization with grafted polyelectrolytes, *Macromolecules*, 24, 2912, 1991.
68. Ross, R. and Pincus, P., The polyelectrolyte brush: poor solvent, *Macromolecules*, 25, 2177, 1992.
69. Zhulina, E.B., Borisov, O.V., and Birshtein, T.M., Structure of grafted polyelectrolyte layer, *J. Phys. II*, 2, 63, 1992.
70. Zhulina, E.B. and Borisov, O.V., Structure and interactions of weakly charged polyelectrolyte brushes: self-consistent field theory, *J. Chem. Phys.*, 107, 5952, 1997.
71. Zhulina, E.B., Klein Wolterink, J., and Borisov, O.V., Screening effects in polyelectrolyte brush: self-consistent field theory, *Macromolecules*, 33, 4945, 2000.
72. Zhulina, E.B., Birshtein, T.M., and Borisov, O.V., Theory of ionizable polymer brushes, *Macromolecules*, 28, 1491, 1995.
73. Lyatskaya, Yu.V., Leermakers, F.A.M., Fleer, G.J., Zhulina, E.B., and Birshtein, T.M., Analytical self-consistent-field model of weak polyacid brushes, *Macromolecules*, 28, 3562, 1995.
74. Currie, E.P.K., Sieval, A.B., Fleer, G.J., and Cohen Stuart, M.A., Polyacrylic acid brushes: surface pressure and salt-induced swelling, *Langmuir*, 16, 8324, 2000.
75. Guo, X. and Ballauff, M., Spherical polyelectrolyte brushes: comparison between annealed and quenched brushes, *Phys. Rev. E*, 64, 051406, 2001.

Chapter 3

Conformational Transitions in Cross-Linked Ionic Gels: Theoretical Background, Recent Developments, and Applications

Sergey Starodubtsev, Valentina Vasilevskaya, and Alexei Khokhlov

CONTENTS

Abbreviations

AM	acrylamide
AMPS	2-(acrylamido)-2-methylpropanesulfonate
APTAC	(3-acrylamidopropyl) trimethyl ammonium chloride
BAM	*N,N′*-methylene(bis)acrylamide

DADMAC	*N,N*-diallyl-*N,N*-dimethylammonium chloride
DEAEMA	*N,N*-diethylaminoethylmethacrylate
DMVP	1,2-dimethyl-5-vinylpyridinium methyl sulfate
DTPP	(2,6-diphenyl-4-[2,4,6-triphenylpyridinio]) phenolate
IPN	interpenetrated networks
MAA	methacrylic acid
MADQUAT	2(methacryloyloxy)-ethyltrimethylammonium chloride
MAPTAC	(methacrylamidopropyl)-trimethylammonium chloride
MONT	montmorillonite
NAD+	nicotinamide adenine dinucleotide
NIPAM	*N*-isopropylacrylamide
N-VC	*N*-vinylcaprolactam
N-VI	*N*-vinylimidazole
N-VP	*N*-vinylpyrrolidone
PA	polyampholytes
PAA	poly(acrylic acid)
PAH	poly(allylamine hydrochloride)
PAM	poly(acrylamide)
PDADMAC	poly(*N,N*-diallyl-*N,N*-dimethylammonium chloride)
PDEAAM	poly(*N,N*-diethylacrylamide)
PHEMA	poly(2-hydroxyethylmethacrylate)
PIB	l-polyisobutylene
PMAA	poly(methacrylic acid)
PNIPAM	poly(*N*-isopropylacrylamide)
PSA	poly(sodium acrylate)
PSC	polyelectrolyte gel-surfactant complex
PSMA	poly(sodium methacrylate)
PSS	poly(styrene sulfonate)
PVC	poly(*N*-vinylcaprolactam)
PVME	poly(vinylmethyl ether)
PVP	poly(*N*-vinylpyrrolidone)
PVT	poly(*N*-vinyltriazole)
Py4	1,3,6,8-pyrenetetrasulfonate sodium salt
SA	sodium acrylate
SMA	sodium methacrylate
SSS	sodium styrene sulfonate
VAEE	vinyl 2-aminoethyl ether
VBT	vinyl benzyl trimethylammonium chloride

3.1 Introduction: First Steps

Stimulus-responsive or "smart" polymer gels undergo sharp, often discontinuous conformational transitions under small changes of external conditions, such as pH, ionic strength, the action of light, electric current, mechanical force, etc.[1–3] Smart polymers can be divided into two groups. Most of the gels and elastic polymers can be made "smart" by loading them with stimulus-responsive fillers such as magnetic particles, colloidal semiconductor particles (e.g., TiO_2), etc. Such systems have recently been reviewed in detail, and we will not discuss them here.[4] On the other hand, the gels can also become smart or stimulus responsive due to different physicochemical processes occurring in the swollen networks in the course of absorption or release of different substances by the gel, or as a result of the influence of an external field. The main physicochemical reasons for the response of the gels to changes in the external conditions and the latest experimental works in this field are discussed in Sections 3.2 through 3.5 of this chapter. Section 3.6 introduces some new polymeric systems, including those with an original architecture. Some gel–colloid composites become stimulus responsive due to the ability of colloidal particles to interact with the polymer chains of the networks or with the components of the solution. In this respect, they resemble smart networks, and these are considered in Section 3.7. Smart-gel systems demonstrate interesting properties that can be of significant practical importance. Recent works on practical applications of smart gels and related problems are described in Section 3.8. Note that a list of chemical abbreviations appears at the beginning of this chapter.

The most striking feature of stimulus-responsive gels is their ability to undergo volume phase transition. The existence of volume phase transition was first theoretically predicted by Dusek and Patterson.[5] Their theory was based on the analogy between the coil–globule transition in a single macromolecule and cooperative coil–globule transitions in the cross-linked chains forming a swollen polymer gel.[6] The volume phase transition was first experimentally observed for slightly cross-linked poly(acrylamide) (PAM) gels swollen in water–acetone mixtures by Tanaka in 1979.[7] The next step was the finding that the volume phase transition in PAM gels can be observed only for swollen networks having some amount of charged monomer units of sodium acrylate (SA).[8] Since then, volume phase transitions in gels have been observed for a great number of polymer gels based on networks with different chemical structures, copolymers, interpenetrated polymer networks (IPN), block copolymers, etc. The common feature of all these gels is that they must have a high degree of swelling under good solvent conditions. It was shown in early studies that the factors favorable for the pronounced volume phase transition are: a low concentration of monomers at synthesis, a low cross-linked density, and the presence of some fraction of the charged monomer units in the network.[1–3]

Volume phase transitions were first observed for gels in mixed water–organic solvents. However, further study showed that gels of poly(*N*,*N*-diethylacrylamide) (PDEAAM) and poly(*N*-isopropylacrylamide) (PNIPAM) undergo sharp collapse in water upon an increase of solution temperature.[9,10] Similar behavior was later observed for the gels of several other poly(*N*-alkylacrylamides) and for poly(*N*-vinylcaprolactam) (PVC).[11–13] Sharp gel collapse was also demonstrated for the gels of poly(*N*-vinylpyrrolidone) (PVP) in aqueous media (good solvent) after addition of concentrated aluminum sulfate (salting-out effect).[13]

In the following sections of this chapter we present a theoretical description of the collapse phenomena, review new experimental observations of conformational transitions in cross-linked gels, and attempt to provide a qualitative theoretical interpretation of published results.

3.2 Theory

3.2.1 First Reason for the Appearance of Two Minima in the Free-Energy Profile

The theories of volume phase transitions in gels were developed soon after the phenomenon of gel collapse was observed. Several groups have published their theoretical works on the problem of volume phase transitions in gels.[1–4,14] We begin our discussion by considering the simplest case of a slightly cross-linked neutral gel composed of one type of monomer unit in a one-component solvent of changing quality. The free energy of the gel in infinite solvent depends on two factors: (a) the energy of the short-range interactions of monomer units F_{int} and (b) the elastic free energy of the network F_{el}.

$$F = F_{int} + F_{el} \tag{3.1}$$

Collapse of the network is possible only if there is an attraction between the polymer chains: $F_{int} < 0$. At infinite swelling there is no interaction between monomer units: $F_{int} = 0$. As the degree of the gel swelling decreases, the gain in F_{int} should increase with increasing concentration of monomer units n in the gel. Increase of n is limited only by the concentration of monomer units in a dry network. Thus, qualitatively, the dependence of F_{int} on the volume of the swollen network can be represented by the curve in Figure 3.1a.

The chains in the networks prepared in a polymer melt usually have a Gaussian distribution of conformations. The swelling of such networks leads to an increase of F_{el}, as shown in Figure 3.1b, curve 1. On the other hand, networks prepared by cross-linking of the polymer in a large amount of the solvent at the concentration when the polymer coils are just touching each other behave differently. Their conformations are close to those of the coils in θ-solvent or to the conformations of the coils with excluded volume.[15] In the latter cases, additional swelling and shrinking both will result in an increase of F_{el} (curve 2 in Figure 3.1b). It can be seen that, summing the functions described by the curve in Figure 3.1a and by curve 1 in Figure 3.1b, one will obtain the resulting curve with a single minimum. In contrast, the sum of the functions described by the curve in Figure 3.1a and curve 2 in Figure 3.1b will produce a curve with two minima.

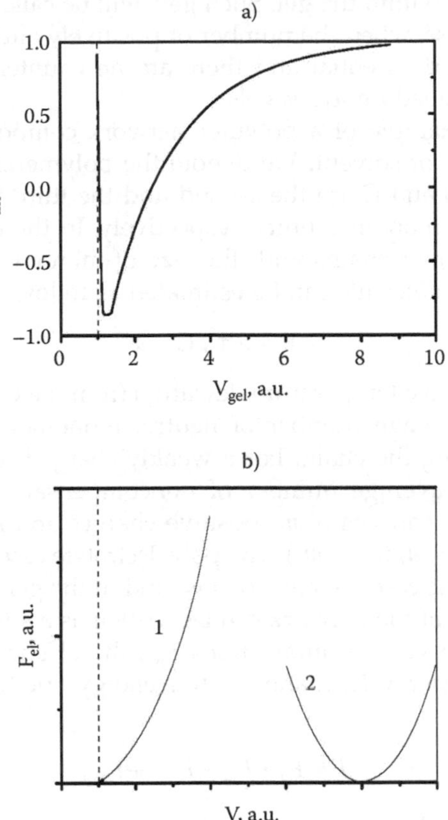

FIGURE 3.1
(a) Schematic representation of the dependence of the free energy of the volume interactions F_{int} on the gel volume. Dotted line corresponds to the volume of the dry gel. (b) Schematic representation of the dependence of the elastic free energy F_{el} on the gel volume for the swollen networks synthesized by cross-linking in the melt (curve 1) and in the presence of a large amount of a good solvent (curve 2).

3.2.2 Theory of Swelling and Collapse of Charged Networks: a General Case

In this section we describe the theory of swelling of a chemically cross-linked charged network in an infinite solvent. The gel network can carry a small fraction of both positively and negatively charged monomer units chemically bound to the chains of the network.[2,16] The net charge of the network cannot be compensated. Suppose the cations of the network are in excess. In this case, due to the electroneutrality condition for the gel, some fraction of negatively charged counterions would be present in the gel so that the total charge is zero. Consider the following two limiting cases. In the first case, the network carries the charges of one sign, while oppositely charged counterions are moving freely within the gel. Such gels will be called *polyelectrolyte* gels. In the opposite case, when the number of positively and negatively charged ions of the network is equal and there are no counterions in the gel, the network will be called *isoelectric* gels.

Consider the example of a polymer network composed of N monomer units in an excess of solvent. We denote the polymerization degree of the subchains as m; B and C are the second and the third virial coefficients of the interaction of monomer units, respectively. In the case of the network with flexible polymer chains with the size of monomer unit a, the second and third virial coefficients can be estimated as follows:

$$B \approx a^3\tau \; ; \; C \approx a^6 \tag{3.2}$$

where τ is the relative temperature deviation from the Θ-point $\tau = (T - \Theta)/T$.

Let σ be the average number of neutral monomer units between two charged units along the chain. For a weakly charged network, $\sigma >> 1$. We denote σ_1 as the average number of noncompensated charges along the chain. If the chain consists of m_+ positive charges and m_- negative charges ($m_+ > m_-$), then $\sigma_1 = m/(m_+ - m_-)$. The polyelectrolyte case corresponds to $\sigma_1 = \sigma$; for the isoelectric case we have $\sigma_1 = \infty$, and in the general case $\sigma < \sigma_1 < \infty$.

The free energy of the network can be written as a sum of four terms: the free energy of the volume interactions F_{int}, the elastic free energy F_{el}, the electrostatic free energy F_{elect}, and the free energy due to the "gas" of counterions $F_{tr\text{-}en}$:

$$F = F_{el} + F_{int} + F_{tr\text{-}en} + F_{elect} \tag{3.3}$$

The free energy of the volume interactions F_{int}, can be written as follows:

$$F_{int} = NT(Bn + Cn^2) \tag{3.4}$$

where n is the concentration of monomer units in the gel, and T is the absolute temperature. Depending on the quality of the solvent (temperature, composition), the second virial coefficient B and the term F_{int} can be negative or positive, corresponding to attraction or repulsion between monomer units of the network.

Elastic free energy describes the increase of the energy of the network due to a loss in conformational entropy under its elongation or compression. The values of F_{el} can be derived from the formula of classic rubber elasticity:

$$F_{el} = 3\frac{N}{m}T\left[\alpha^2 + \frac{1}{\alpha^2}\right] \tag{3.5}$$

where α is the coefficient of swelling. It is very important that the minimum of F_{el} ($\alpha = 1$) corresponds to the reference state, which is usually the state that the network had under the preparation conditions (e.g., cross-linking in melt or in the presence of a large amount of solvent). For the networks prepared by cross-linking in melt, swelling leads to an increase of F_{el}. For the networks prepared in the presence of a large amount of a good solvent, the conditions of the minimum of F_{el} ($\alpha = 1$) correspond to a rather strongly swollen network. Both release and uptake of the solvent by the isotropic gel lead to an increase of F_{el}.

The presence of freely moving counterions in the gel is described by the contribution to the free energy from the translational entropy:

$$\frac{F_{tr\text{-}en}}{T} = \frac{N}{\sigma_1} \ln \frac{n}{\sigma_1} \tag{3.6}$$

Due to the electroneutrality condition, the moving counterions cannot be released from the gel and create additional osmotic pressure in the swollen network. This term increases with increasing concentration of the counterions (e.g., with an increase in the degree of ionization of the network), and in polar media its contribution to the total free energy can be very large.

The electrostatic free energy in the gel with charged monomer units *is always negative*. It *always* leads to some shrinking of the gel. It can be calculated using the Debye-Huckel approximation:

$$\frac{F_{elect}}{T} = -N\left(\frac{u}{\sigma}\right)^{3/2} (na^3)^{1/2} \tag{3.7}$$

where $u = \frac{e^2}{\varepsilon a T}$ is a dimensionless parameter, e is an elementary charge, ε is the dielectric constant of the medium, and T is the absolute temperature. Calculations show that at room temperature (T ≈ 300°K) in aqueous media and for the typical size of monomer unit, $a \approx 1.0$ nm and $u \approx 1$.

Calculations of the total free energy have shown that

1. Under the conditions $\sigma < \sigma_1 < \sigma^2$, the third term in Equation 3.2 dominates and the network is strongly swollen in the polar solvents (water, dimethylsulfoxide). The transition of such networks into the collapsed state occurs as a volume phase transition at the temperature much below the θ point.
2. At $\sigma_1 > \sigma^3$, the network is in the isoelectric point. There are but few counterions and the electrostatic attraction dominates. For strongly charged networks, this attraction usually leads to the collapse of the gel, even in polar media.
3. At an intermediate case $\sigma^2 < \sigma_1 < \sigma^3$, the terms F_{elect} and $F_{tr\text{-}en}$ are approximately equal. In this case, the transition in the collapsed state occurs as a volume phase transition, but at temperature higher than the θ point.

These three regimes of swelling, i.e., the Θ dependences of the parameter α, are shown in Figure 3.2.

The presence of a small fraction of the charged groups in the network composition significantly increases the swelling degree of the gel.[1–3] In the

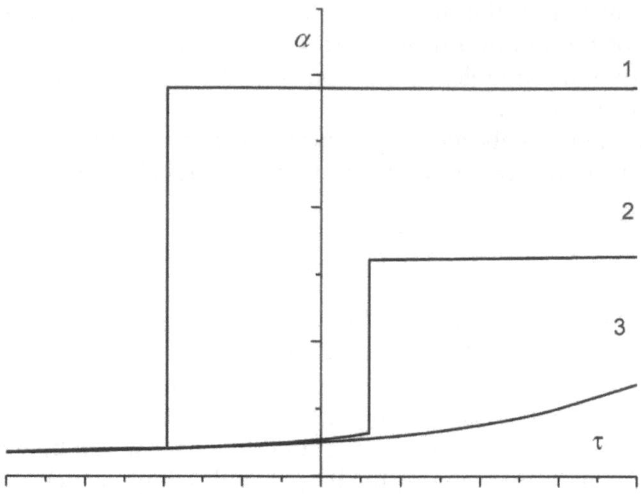

FIGURE 3.2
Dependences of the coefficient of network swelling α on the relative temperature deviation from the Θ-point τ for charged gels: polyelectrolyte (curve 1), intermediate (curve 2), and isoelectric (curve 3) regimes of swelling.

absence of additional salt in polar aqueous media, the volume of the network can be two or even three orders of magnitude higher compared with that of the neutral gel.

The addition of low-molecular-weight salt strongly reduces the swelling degree of the slightly charged polyelectrolyte gels. Detailed studies of the effect of salts with uni- and divalent cations and anions are described in the literature.[1,2] In particular, it was demonstrated that divalent counterions reduce the degree of swelling of the charged gel much more strongly than univalent ions.

All the early experiments are in accordance with the main prediction of the theory. The swelling degree and amplitude of the collapse increase with increasing charge density and, hence, with the concentration of the counterions in the network.[1–3] In water–acetone mixtures, the addition of salt induces a volume phase transition. At fixed acetone content, bivalent counterions initiate collapse of the gels at much smaller concentrations than do univalent ions.[1,17]

3.3 Polyampholytes

Polyampholytes (PA) are the polyions containing both cationic and anionic monomer units. The presence of both signs in the structure of the charges is necessary for the functioning of many biomacromolecules and biological membranes.[18,19] All of the proteins synthesized in the cells are PA. The first synthetic linear polyampholytes, namely copolymers of *N,N*-diethylamino-ethylmethacrylate (DEAEMA) with MAA, have been known since the 1950s.[20,21] Most of the earlier experimental and theoretical works in the field of synthetic PA are reviewed in the literature.[22,23] The theory of PA networks and corresponding experiments are described in our previous review.[2] Nisato and Candau have recently summarized the main achievements in the physics and chemistry of PA up to about 1999.[24] They also tabulate the main monomers that were used for the preparation of the PA gels.

In our first experiments, PAM copolymer gels containing 1.8 mol% of sodium methacrylate (SMA) groups and variable amounts of 1,2-dimethyl-5-vinylpyridinium methyl sulfate (DMVP) groups were studied.[2,25] It was shown that, in salt-free solutions, the addition of cationic groups into the anionic network reduces the swelling ratio of the gel. At equal amounts of positively and negatively charged groups in the network, the swelling ratio reaches its minimum value and then increases. The volume phase transition of the acrylamide (AM)-SMA gel in water–ethanol mixtures transformed to a monotonic decrease of the gel volume for the polyampholyte gels, with rather close content of the charges of both signs. Interesting results were obtained for the gels equilibrated with low-molecular-weight salt. In accordance with theory,[2,16] "antipolyelectrolyte behavior" was observed at the isoelectric point (IEP). The addition of low-molecular-weight salt resulted in the gel swelling. Moreover, if the number of charges of one sign does not strongly exceed that of the charges of the opposite sign, the dependence of the gel volume on the concentration of potassium bromide passes through a minimum.[2,25]

Baker et al. studied the swelling behavior of PA gels based on terpolymers of AM with MAPTAC ([methacrylamidopropyl] trimethylammonium chloride) and sodium styrene sulfonate (SSS), with the charged groups totaling 4.5 mol%.[26] Collapse of the gel close to the isoelectric point was observed. However, closer to the isoelectric point, the hydrogels demonstrated increasingly weak response to ionic strength. Later, the copolymers of PAM containing zwitterionic monomers *N*-(3-sulfopropyl)-*N*-methacrylamidopropyl-*N*-dimethylammonium betaine and *N*-(3-sulfopropyl)-*N*-methacroyloxy-ethyl-*N,N*-dimethylammonium betaine were both, in turn, copolymerized with AM to form ampholytic hydrogels. Antipolyelectrolyte behavior was observed for the ampholytic hydrogels, and hydrogel swelling increased when the concentration of NaCl was increased.[27] Similar results were obtained for PA gels in water, water–salt, and water–organic mixtures.[28–30] Slightly charged PAM gel containing the charged groups of MAPTAC and SA was studied by Katayama et al.[28] Volume phase transitions of the gels under increased temperature were observed. At the IEP, the conditions of the collapse of the PA gel were the same as for a neutral gel, but the amplitude of the transition was lower. The dependence of the swelling ratio on the total charge density of the network had an asymmetric character. Kudaibergenov and Sigitov studied PA gels of copolymers of vinyl 2-aminoethyl ether (VAEE) and SA.[29] In water–acetone mixtures, the collapse of the PA gels was more pronounced near the IEP. English et al. studied PA gels of copolymers of MAPTAC with 2-(acrylamido)-2-methylpropanesulfonate (AMPS).[30] The AMPS was in the form of acid or sodium salt. They observed a new effect in the decrease of the swelling ratio for gels having a slight excess of anionic groups in the network in the range of very low concentrations. This peculiarity was interpreted from the analyses of dissociation equilibrium of the sulfonic groups of the AMPS monomer units. The dependence of the swelling ratio on the total charge density of the network had an asymmetric character.

Nisato et al. studied the swelling and elastic behavior of chemically cross-linked copolymer gels of 2(methacryloyloxy)-ethyltrimethylammonium chloride (MADQUAT) with AMPS in equilibrium with saline solutions.[31] The obtained results were in good agreement with theory[2,16] and with the experimental results obtained previously for weakly charged PA gels based on PAM, containing DMVP and SMA monomer units.[2,25] Due to high total charge density of the network, the observed effects were much more pronounced. The gels containing equivalent amounts of positive and negative charges collapsed at low salt concentration and swelled with rising salt concentration. For the gels with a strong charge imbalance, the degree of swelling decreased with increasing ionic strength of the solution. Weakly unbalanced PA gels showed a characteristic minimum in their swelling curves: increasing the salt concentration resulted in a decrease of the swelling ratio in the region of low salt concentrations and in an increase of the swelling degree at higher ionic strengths. The changes of the elastic modulus of these PA gels equilibrated with salt were quite complex, because the same gel can have the same swelling ratio at two quite different concentrations of low-molecular-weight salt. It was shown that, in the latter case, the gel is characterized by two different values of the shear modulus.

The effect of the presence of low-molecular-weight salt during the synthesis of PA gel was studied by Thanh et al.[32] It was shown that gels synthesized in the presence of salt have a higher degree of swelling compared with gels synthesized in the absence of salt. The observed effect was explained by the fact that the latter have more balanced stoichiometry. Absorption of positively and negatively charged surfactants was also studied. The gels with a large excess of the charges of one sign collapse in the presence of the oppositely charged surfactant. The efficiency of the surfactant absorption depends on the ratio between the charges of both signs.

PA gels were also obtained via radiation polymerization of poly(styrene sulfonate) (PSS) and vinyl benzyl trimethylammonium chloride (VBT).[33] The dynamic swelling of the gels in water showed that they differ by the mechanisms of diffusion. The effects of organic solvents and some neutral organic substances on the collapse behavior of these copolymers is also presented in the same work. Organic additives like urea and glucose and surfactants like Triton-X tend to further swell the polymer matrices, whereas NaCl causes their shrinking.

Interesting new features of the PA gels were observed in a work by Valencia and Pierola.[34] The hydrogels of copolymers of *N*-vinylimidazole (N-VI) and SSS were synthesized by radical cross-linking copolymerization. Swelling measurements in water, at room temperature, revealed an unusual behavior. For some gel compositions, dry hydrogel pellets were swelling up readily, reaching a high degree of swelling and, after that, they spontaneously collapsed. This peculiarity was interpreted in terms of several competing processes in the swelling mechanism: water diffusion inside and outside the gel, chain disentanglement, sodium-proton ion exchange with water, and the approach of chains to allow interaction of sulfonate groups with neighboring protonated imidazole monomer units.

A novel PA gel was prepared from the copolymer of the β-vinyloxyethylamide of acrylic acid with *N*-vinylpyrrolidone (N-VP).[35] The PA character of the gel is associated with the hydrolysis of CO-NH bonds in β-vinyloxyethylamide of acrylic acid units. As was established experimentally, the hydrogels are sensitive to changes in pH, temperature, quality of solvent, and concentration of a low-molecular-weight salt.

Copolymer gels that memorize the positioning of two monomer units–ligands for the bivalent metal ions were recently synthesized.[36] The "holes" for metal ions were formed by an "imprinting method" during the synthesis of the copolymer complexes of the metal with two MA monomer units in the presence of the metal. The complex formation was observed for the gel "fitted" for the metal ion, while the control gel with random distribution of MA groups showed frustration in forming pairs. The effects of imprinting toward specific metal ions on the properties of smart gels are discussed in greater detail in Chapter 7, "Imprinting Using Smart Polymers."

The competing interactions of the charges of the collapsed PA gel with multicharged low-molecular-weight salts (aluminum ions and 1,3,6,8-pyrenetetrasulfonate sodium salt [Py-4]) were studied by Alvarez-Lorenzo et al.[37] The PA network was composed of a copolymer of *N*-isopropylacrylamide (NIPAM) with methacrylic acid (MAA) and MAPTAC taken in equivalent molar ratio. The fraction of each charged monomer in the initial polymerized mixture was 1.33 mol%. In the absence of aluminum, the absorption of Py-4 was suppressed by the gel collapse, presumably because of an increase in the fraction of MAA/MAPTAC ionic pairs that prevented the formation of potential adsorbing centers. Addition of a moderate amount of aluminum salt significantly increases adsorption of Py-4 by the collapsed gel, indicating that aluminum ions compete for binding with MAA and help release vacant MAPTAC centers. At high concentration of aluminum ions, the Py-4 adsorption is again suppressed due to aluminum-mediated effective cross-linking.

3.4 Second Reason for the Appearance of Two Minima in the Profile of the Free Energy: Collapse Due to Ionomer Effect

In this section we discuss the effects arising from the interaction of the counterions with oppositely charged monomer units of a network. In the absence of external fields, the condition of electroneutrality in a swollen polyelectrolyte network is applied. In this case, Coulomb attraction between the oppositely charged ions, described by the term F_{elect} in Equation 3.3, always tends to shrink the gel. Two limiting cases are possible, depending on the polarity of the medium and the charge density. At low charge density in a highly polar medium, and when the network carries the charges of one sign, the effect of Coulomb interactions is negligibly small in comparison with the effect of swelling due to the osmotic pressure of the counterions.[2,16] The charged network is thus in the polyelectrolyte regime of swelling. In the media with low dielectric constant, the situation is the opposite. Eisenberg et al. have shown that ion pairs are formed in media with a low dielectric constant. The ion pairs form multiplets composed of contact ion pairs.[38,39] In particular, the multiplet formation can be observed by small angle X-ray scattering (SAXS) when the characteristic maxima appear in the scattering curves.[38] The free counterions and the multiplets are shown schematically in Figure 3.3a and Figure 3.3c. The charged gels in most cases demonstrate rather complex dual polyelectrolyte–ionomer behavior. During the transition from polyelectrolyte to ionomer regime, the gel shrinks; the osmotic pressure

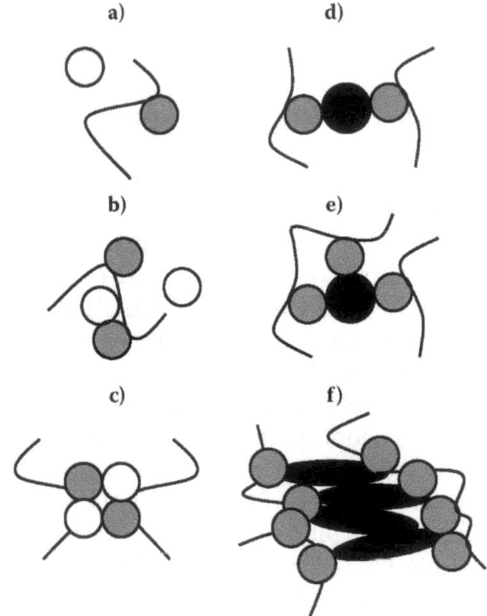

FIGURE 3.3
Schematic representation of different states of the counterions in the charged gel. (a) Free counterions; osmotic pressure is high. (b) Counterions are partly condensed or bounded. (c) Multiplet formation additionally cross-links the network. (d) The formation of strong complex between the ligand-charged groups of the network and the transition metal ion. (e) Interaction of the charges of the network with multivalent ion. (f) Organic ions of dyes associate due to additional nonpolar interactions. Curving lines are polymer chains; open and black circles are univalent and multivalent counterions, respectively; gray circles are the ions of the network; and black ellipses are organic counterions of dyes.

of the counterions becomes smaller, and the multiplets begin to play the role of additional cross-links. The increase of the volume fraction of the polymer in the gel, in turn, usually leads to a further decrease of the dielectric constant. Thus, the process is developing via the positive-feedback mechanism and leads to a strong gel collapse.

The theory of the collapse of the swollen networks due to multiplet formation was developed by Khokhlov and Kramarenko.[40,41] The theory predicted several effects:

1. For slightly charged gel, the collapsed region is expanded and the charged collapsed gel is more shrunken compared with the corresponding neutral gel.
2. The charged collapsed gel can further transform in a "super collapsed," densely packed state, where the multiplets are predominant.
3. The charging of the network due to the change of pH or to the chemical modification of the polymer can induce the gel collapse. For gels with a high charge density, the latter effect is increased due to Manning condensation of the counterions.[23]

Early experiments had already shown that even at a low charge density of the networks, the chemical nature of the univalent counterions has an essential impact on the swelling of slightly charged gels.[42] It was shown that the volume phase transition of slightly charged AAM-SA gels in mixtures of water with organic solvents occurs earlier in the medium with a lower dielectric constant. For highly swollen gel, in the region before the collapse, the swelling ratio of the gel decreases in the following order: methanol ($\varepsilon = 32.6$), ethanol ($\varepsilon = 24.3$), and dioxane ($\varepsilon = 2.2$). These and other similar results indicate that in water–organic media, even at low charge density, some fraction of the counterions form ion pairs with the charges of the network. The collapse induced by ionization of cross-linked gels poly(acrylic acid) (PAA) and poly(methacrylic acid) (PMAA) by univalent sodium methoxide in methanol was observed by Philippova et al.[43] During the neutralization after the initial swelling of the gel, when the threshold degree of ionization has been achieved, the gel collapses sharply. The formation of the ion pairs was demonstrated by an essential increase of electrical conductivity. Similar phenomena were observed for linear PAA in methanol solutions.[44–47] The experimental works in the field of polyelectrolyte–ionomer behavior of gels and linear polymers are summarized in the recent review by Philippova and Khokhlov.[48] The more recent results are discussed below.

Minati and Satoh have studied sodium counterion binding in PA gel in water–organic mixtures using ^{23}Na NMR analysis.[49] They demonstrated that the collapse of the gel is induced by the formation of contact ion pairs. The formation of contact ion pairs was accompanied by precipitation of PAA during its neutralization with tetramethylammonium hydroxide in methanol and was proven by ^{1}H NMR spectroscopy, light scattering, and conductivity methods.[50] It was also observed that the increase of the charge density of PAA leads to a dual effect. When the degree of ionization increased above 0.4, reswelling of the gel occurred.

Generally speaking, the interaction between polyelectrolyte gels with their counterions in a mixed solvent and in salt-containing media is complicated, and many details of this interaction are not yet completely understood, even for the case of univalent ions. New features of the collapse due to ionomer effect were observed recently by Nishiyama and Satoh.[51–53] They showed that, due to ionomer behavior, the gels of poly(sodium acrylate) (PSA) collapse in water–ethanol mixtures. This effect is less pronounced in the presence of 0.1*M* sodium chloride. The position of the swelling-deswelling region depends on the counterion species of the used salt. The presence of Na^+ and K^+ cations is favorable for the collapse of the gel in aqueous ethanol. The cations Li^+ and Cs^+ are both unfavorable for the collapse. Unexpectedly, when both Li^+ and Cs^+ cations are present in the solution, the gel shrinks much more strongly than in the case of a single counterion system. The authors explain the observed effect as a subtle balance between the formation of ion pairs and dipolar aggregation of the ion pairs.

Polyelectrolyte/ionomer behavior of the charged gels can also be observed in aqueous media. In the latter case, the polymers must have a high charge density. The tendency toward formation of charge-transfer complexes between the ions is also favorable for manifestation of the ionomer effect. Kudo et al. showed that the addition of 1,1-salts induces jumpwise collapse of the cationic gels of *N*-*n*-butyl-*N*,*N*-dimethyl-4-vinylbenzylammonium bromide, *N*-*n*-butyl-4-vinylpyridinium bromide, and *N*-*n*-butyl-*N*,*N*-dimethyl-4-vinylanilinium bromide.[54,55] The authors explained the gel collapse by a strong binding of the anions and hydrophobic interactions between the monomer units of the networks. Starodubtsev et al. observed volume phase transition of the cationic gel of poly(*N*,*N*-diallyl-*N*,*N*-dimethylammonium chloride) (PDADMAC) in the presence of sodium iodide.[56] They explained the observed effect by the formation and attraction of ion pairs in the gel. Evidence of the polyelectrolyte/ionomer behavior of the gel is the fact that the decrease of the charge density of the network via copolymerization of *N*,*N*′-diallyl-*N*,*N*-dimethylammonium chloride (DADMAC) with AM leads to disappearance of the collapse transition of the cationic gel. It should be noted that the multiple ionic pairs formed in the phase of the collapsed gel are very stable; therefore, a giant hysteresis is observed when the concentration of iodide salt is decreased. On the other hand, the large local shrinking of the swollen gel is unfavorable for the formation of multiplets, and a long latent period is needed before the phase transition occurs. It was demonstrated that the mechanism of the collapse is nucleus formation, with the initial multiplet nuclei first forming at the edge of the gel; then, like an avalanche, the collapsed region occupies the whole surface of the gel.

Kawaguchi and Satoh studied the swelling behavior of partially quaternized poly(4-vinylpyridine) gels in different water–organic-solvent media.[57] They demonstrated that the counterions I^- or SCN^- induce the collapse of the gels. The authors observed almost no variation of the dielectric constant across different solvent species during the sharp collapse.

3.5 Other Effects Due to Counterions

Complex formation of metal ions with polymer ligands is well known. Ricka and Tanaka observed the collapse of PAM gel containing different fraction of SA monomer units in the presence of Cu(II) ions.[58] The collapse is a result of complex formation between a metal ion and two carboxylate anions. The metal ion plays the role of an additional cross-link and induces the collapse of the gel, even in water. Together with cross-linking, an ionomer effect (see below) is probably responsible for the gel collapse. Complex formation of a polymer ligand gel of poly(*N*-vinyltriazole) (PVT) with Cu(II) ions leads to the appearance of dense blue "skin" on the surface of the gel.[59] SAXS studies have demonstrated the existence of a broad maximum in the scattering curves. Depending on the concentration of Cu(II) ions in the external solution, the positioning of the maxima corresponded to the characteristic size d = 8–18 nm. This result can be explained by the existence of dense aggregates of the complexes analogous to the multiplets formed in water–organic media. Such microheterogeneous structures were observed by SAXS for slightly charged collapsed PAM gels in water–acetone mixtures.[60]

When there is no complex formation between the charged groups of the gel and bivalent metal cations, the gel shrinks, but the true collapse is not necessarily observed, even in the presence of a certain amount of organic solvent. Dembo et al. studied the interaction of PAMPS gel (sodium salt) with copper(II) nitrate in a water–methanol mixture (70/30 by volume).[61] The increased concentration of copper salt resulted in a marked shrinking of the gel. However, this shrinking is much less pronounced compared with the collapse induced by the same salt in the gel of poly(sodium methacrylate) (PSMA). Using electron spin resonance (ESR) technique, the fractions of bound and free counterions were measured for the gels containing different amounts of copper(II) ions.

Organic and metal cations with a valence of three or higher are known to induce collapse of anionic gels and precipitation of linear polyanions. In the case of multivalent cations, even in a polar medium, the electrostatic attraction of univalent monomer units of the charged polymer to the cations with valence 3 or higher is very strong, and $e\psi > kT$. Recently, it was shown that condensed DNA macromolecules can be redissolved at high concentrations of multivalent cations.[62] Similar observations have been reported for PSS: its precipitation was induced by the addition of a multivalent metal salt, followed by redissolution at high salt concentration.[63] Polyelectrolyte gels in aqueous media demonstrate a similar behavior.[64] It was shown that increasing the concentration of cobalt hexamine(III) or iron(III) chlorides (the cations of these salts cannot form complexes with sulfonic groups) induces

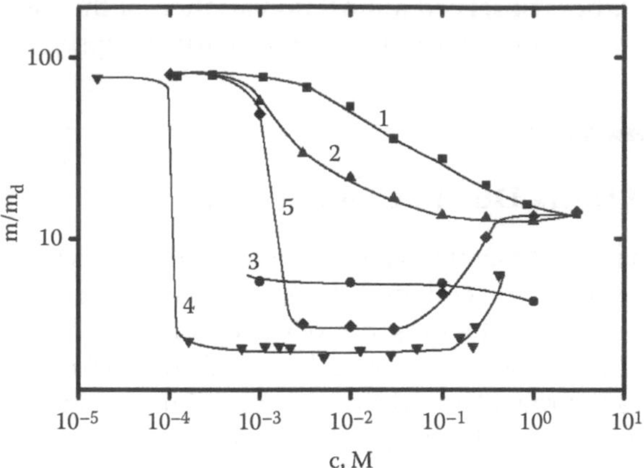

FIGURE 3.4
Dependences of the swelling ratio, m/m_d, of the PAMPS gel on the concentration, c, of NaCl (curve 1), $CaCl_2$ (curve 2), cobalt hexamine trichloride (curve 4), $FeCl_3$ (curve 5), and analogous dependence of the PSA gel on the concentration of copper(II) nitrate (curve 3). Concentration of NaAMPS at synthesis was 0.2 g/ml; m is the mass of the gel at equilibrium, and m_d is the mass of the dried NaAMPS gel. (Reprinted with permission from Lyubimov et al., *J. Phys. Chem.*, 107, 12206, 2003. Copyright 2003, American Chemical Society.)

the collapse of PAMPS gel (see Figure 3.4) at low concentrations. However, at high concentrations of about 0.3*M*, reswelling of the gel is observed. The collapse of the gels was explained by the formation of bridges between the polymer chains due to a very strong electrostatic attraction between the ionic monomer units of the gel and the metal ions. At a very high concentration of metal salts in the solutions and in the gel, these bridges are destroyed due to the competition interactions with the ions of inorganic salt. A SAXS study of the collapsed gels of PAMPS containing Cu(II) and Fe(III) ions did not show any maxima in the scattering curves, indicating the absence of ordered microheterogeneities with the characteristic size of 1 to 100 nm.

Interesting effects are observed if oppositely charged surfactants play the role of counterions in the polyelectrolyte network.[65–67] The formation of polyelectrolyte gel–surfactant complexes (PSC) is accompanied by an aggregation of surfactant ions in the gel phase due to hydrophobic interactions and gel collapse. Despite the irregular structure of the charged networks, their complexes with surfactants often form highly ordered supramolecular structures that provide sharp peaks in the corresponding SAXS profiles.[68,69] The study of PSC has become a large and rather specific branch of polymer science and will not be discussed here. However, one example of a system based on PSC is listed below because other ions, such as charged oligomers and polymers (including proteins) or multi-ions, can be used instead of the ions of surfactants.

It is possible to initiate the swelling of PSCs without a direct contact between the gels.[70] Consider two PSC samples: one formed by a cationic gel and anionic surfactant, and the other formed by an anionic gel and cationic surfactant. In aqueous salt solution, the surfactants then form a common precipitate, while the two gels swell. This phenomenon can be explained by the stability of the surfactant–surfactant complex, which is much higher than that of PSC. If the gel samples are in direct contact, they glue together like oppositely charged polyelectrolyte gels.

Many organic substances have an ionic nature. The interactions of polyelectrolyte networks with ions of organic substances, e.g., dyes, are complex. Rama Rao et al. studied the conformational transitions and the structure of salts of gels of poly(allylamine hydrochloride) (PAH) with organic acids, for instance, with *p*-styrenesulfonate.[71] The gels collapsed when equilibrated with sodium salts of various organic acids. A single broad peak was observed in the SAXS profiles obtained from the gels with organic acid counterions. The strong maximum in the SAXS patterns was attributed to the presence of ordered structures in the collapsed gels. In contrast to this observation, the gels with inorganic counterions did not demonstrate any self-ordering in their structure.

Makhaeva and Nasimova studied the absorption of dyes by oppositely charged polymer gels.[72–74] Anionic dyes (alizarin red S and pyrocatechol violet) formed complexes with cationic PDADMAC gel. The interaction resulted in the collapse of the gel. The UV–Vis spectra of absorbed dyes differed from those of the dyes in aqueous solutions. Alizarin is known to be a specific reagent for aluminum ions. Despite the fact that both the Al^{3+} ion and the gel have a positive charge, metal ions penetrate the gel and form an intracomplex dye–metal-ion compound in the solutions of $Al_2(SO_4)_3$, leading to its additional contraction. A study of the interaction between anionic gels of PMAA with a cationic dye, rhodamine 6G, showed that absorption of organic cationic dye leads to additional contraction of the gel.

Multivalent dyes induce the collapse of the oppositely charged swollen polyelectrolyte network in the same way as other multivalent ions do. Figure 3.5 shows the dependence of the swelling ratio of PDADMAC gel on the concentration c of the anionic dye hydrotype yellow (HY). The sharp collapse occurs in the concentration region between 10^{-5} to $10^{-4}M$. Reswelling of the gel is observed at a high concentration of the dye. The most reasonable explanation for the observed reentrant transition (as in the case of multivalent metal ions) is the destruction of the bridges formed by the dye ions between the polymer chains due to the competition interactions with the other organic and inorganic ions in the concentrate dye solutions. Figure 3.5b shows a SAXS profile obtained from the PDADMAC–HY complex. The scattering curve demonstrates a pronounced maximum corresponding to the characteristic size $d = 2.2$ nm. The strong maximum in the SAXS patterns was attributed to the presence of ordered structures in the collapsed gels.

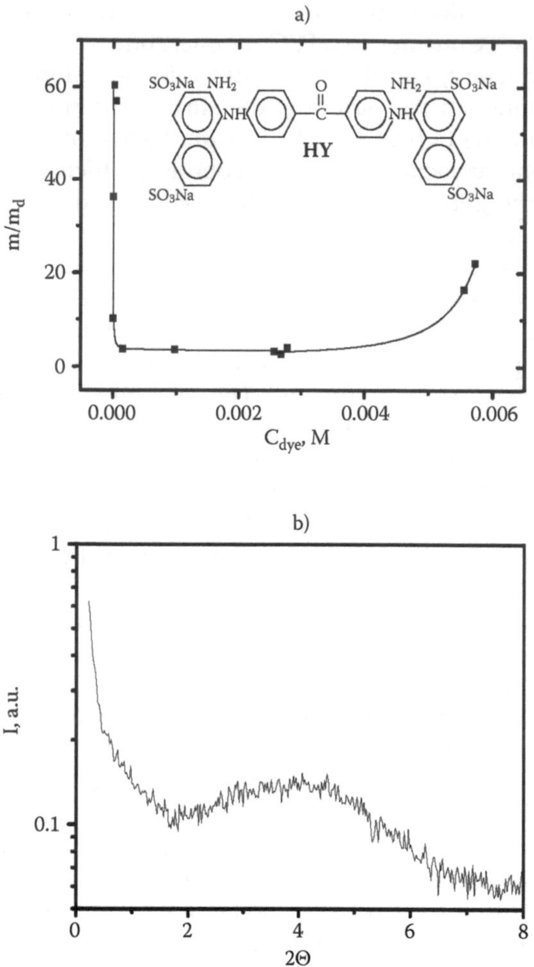

FIGURE 3.5
(a) Dependence of the swelling ratio, m/m_d, of the PDADMAC gel on the concentration, *c*, of hydrotype yellow (HY) in water; *m* is the mass of the gel at equilibrium, and m_d is the mass of the dried PDADMAC gel. (b) SAXS profile obtained from the complex of HY with PDADMAC; concentrations of HY and NaCl were 0.001 and 0.01*M*, respectively. (From Starodubtsev, S. and Dembo, A., unpublished results. With permission.)

3.6 New Polymers

Several new copolymers providing volume phase transition have been synthesized and studied since 2000.[29, 33–35] Ilavsky et al. studied smart gels based on copolymers of N-VP. In particular, the volume phase transitions in the

N-VP–DADMAC and N-VP–*N*-vinylcaprolactam (N-VC) copolymer gels in water and water–acetone mixtures were investigated.[75,76] This group also studied the volume phase transitions in PVC gels and in gels with IPNs of PAM and PVP.[77,78] In the latter case, two transition regions were detected in the dependence of the swelling ratio on acetone concentration.[76] Atta synthesized and studied the gels of copolymers of AMPS with maleate monoester and diester monomers based on poly(ethylene glycol) monomethyl ether.[79] The swelling ratios of the copolymers were measured in water and in solutions of different salts. Zhakupbekova et al. synthesized PA networks of β-vinyloxyethylamide of acrylic acid copolymer with PVP and studied their swelling and collapse behavior.[80] New thermosensitive copolymers were obtained by three-dimensional polymerization of *n*-alkyl methacrylate esters with DEAEMA.[81] The observation of volume phase transition was also reported for cross-linked gels of PVME.[82,83]

One of the modern tendencies in the synthesis of smart gels is the introduction of biopolymers into their structure. New polypeptide polymer networks were synthesized using elastin.[84] The polypeptide was obtained by *in vivo* biosynthesis, utilizing the principles of protein structure and polymer material science. The molecular structure of elastin was determined during the biosynthesis stage. The synthetic polypeptide elastomers and elastomeric networks prepared by this technique were evaluated for chemical and physical properties. Chapter 6, "Protein-Based Smart Polymers," discusses polypeptide elastomers and their networks in greater detail.

A novel class of dextran-maleic anhydride (Dex-MA)/PNIPAM hybrid hydrogels was designed and synthesized by UV photo-cross-linking.[85] The data show that these hybrid hydrogels were responsive to external changes of temperature as well as pH. By changing the composition ratio of the corresponding precursors, the phase transition temperature (lower critical solution temperature, LCST) of the hybrid hydrogels could be adjusted to the body temperature for potential applications in bioengineering and biotechnology.

A new kind of IPN and semi-IPN prepared from chitosan/PNIPAM gels was described in a recent study.[86] It was shown that the properties of the IPN gels (including the cross-linked PNIPAM within it) — the phase transition behavior, the swelling dynamics in aqueous phase, the swelling behavior in ethanol–water mixtures, and even the microstructure — are quite different from those of the semi-IPN, in which PNIPAM was simply embedded. Similar to the semi-IPN chitosan/PNIPAM hydrogels, the IPN gel was also temperature sensitive.

New chiral copolymers of NIPAM with optically active *N*-acryloyl-L-alanine have also been studied. The results showed that the diffusion coefficients of R- and S-[1-(dimethylamino)ethyl]ferrocene are different in the collapsed gel.[87]

PDMAEM-l-polyisobutylene (PIB) amphiphilic co-networks (APCN) were synthesized using methacrylate-telechelic PIB obtained via quasi-living carbocationic polymerization.[88,89] APCNs are new materials composed of covalently bonded and otherwise immiscible hydrophilic and hydrophobic polymer chains. The amphiphilic nature of these new cross-linked polymers is indicated by their swelling ability in both hydrophilic and hydrophobic solvents. New networks, such as poly(2-hydroxyethylmethacrylate) (PHEMA)-PIB, and PMAA-PIB were synthesized. Due to their unique architecture, macrophase separation of the immiscible components is prevented by the chemical bonding in the co-networks. Consequently, phase separation leads to nanophase-separated morphology. The new PMAA-PIB polyelectrolyte APCNs possess smart reversible pH-responsive properties in aqueous media. These and similar block-like structures are shown in Figure 3.6a. The properties of the new emerging materials can lead to numerous new applications, such as smart-material products, sustained drug-release matrices, biomaterials, nanohybrids, nanotemplates, etc. A theoretical model for describing a similar kind of gel was proposed by Dusek et al.[90] They studied the case when the hydrophilic chains that carry ionizable groups are cross-linked by long hydrophobic chains. Jumpwise swelling of such gels after ionization, which is schematically shown in Figure 3.6b, is predicted (Figure 3.6b).

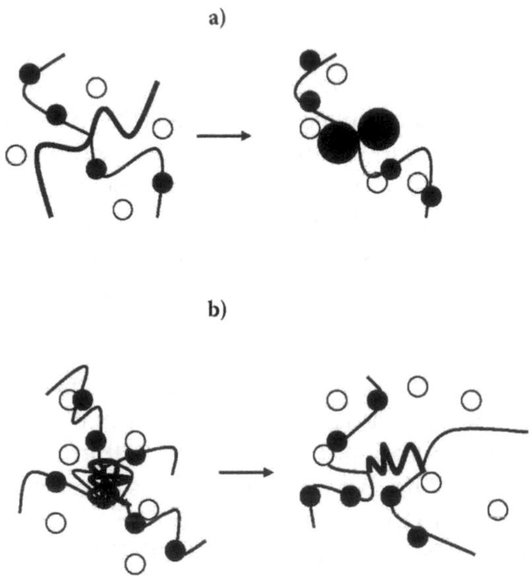

FIGURE 3.6
The modern tendency in the design of the new polymer architectures is to combine polymeric chains with quite different properties in one network. (a) Schema of a fragment of a network combining charged hydrophilic and uncharged hydrophobic polymer subchains in common solvent (left) and under microphase separation (right). (b) The unfolding of a long hydrophobic cross-linking chain in hydrophilic network can be induced by the ionization of the hydrophilic chains. Small black circles are the charges of the hydrophilic chains, open circles are the counterions, and large black circles are the microsegregated hydrophobic chains. Hydrophobic unfolded chains are the fat lines.

The new physical gels composed of hydrophilic charged chains of poly(L-lysine HBr) and hydrophobic chains of poly(L-leucine) have a similar structure.[91] Despite the absence of chemical cross-links, these gels have a very high stability in water. However, the stability decreases with increasing salt concentration. The remarkable properties of these hydrogels make them promising for developing scaffolds for tissue regeneration. The essential feature of most of these networks is that, due to cross-linking, they cannot undergo phase separation, but microsegregation is observed instead.

Swollen networks are interesting new example of gels with cooperative H-bond formation, where the long chains of PMAA are cross-linked with dimethacrylates of polyethylene oxide (PEO) of different chain length.[92] Such gels undergo collapse in the pH range of 4.0 to 6.0 due to cooperative formation of IPC via H-bonds. It was demonstrated that both the position of the collapse on the pH scale and its amplitude are independent of the length of complementary chains if the units involved in complex formation are taken in stoichiometric amounts. For hydrogels with an excessive (up to 77 mol%) content of PEO, the collapse amplitude drops due to (a) the swelling of fragments of PEO chains not involved in complexes and (b) decreases in the length of polyelectrolyte chains and in the contribution of counterions to the swelling pressure of a charged network. Variations in the elastic modulus of hydrogels with changes in pH and ionic strength of solution were studied.

Several known techniques were used for the synthesis of thermosensitive gels with an enhanced rate of swelling-deswelling processes. Copolymers of NIPAM with a very high content of *N,N′*-methylene(bis)acrylamide (BAM) are described Sayil and Okay.[93] It was shown that the obtained copolymers have a macroporous structure with a pore radius ≈0.1 mm. PNIPAM gels with high rates of solvent uptake and release were also obtained by polymerization in mixed water–acetone solvent.[94] The use of a cryogenic technique for the synthesis of highly permeable stimulus-responsive gels shows great potential.[95] It is also possible to increase markedly the rate of swelling–deswelling processes by designing the gel architecture at the molecular level. In particular, the preparation of a gel is described in which the main cross-linked polymer chains bear grafted side chains; the latter create hydrophobic regions, aiding the expulsion of water from the network during collapse.[96] The macroporous gels are discussed in detail in Chapter 6, "Protein-Based Smart Polymers."

3.7 Smart Polymer-Colloid Composite Gels

The properties of hydrogels can be significantly modified through incorporation of colloids in a polymer network. One of the new classes of these composite systems are the gels loaded with dispersed clays. These composite materials combine the elasticity and permeability of gels with the high ability of clays to adsorb different substances.[97,98] Their durable properties make them suitable for use as superadsorbents.[97] On the other hand, immobilization in the cross-linked network prevents coagulation of dispersed clay particles. The montmorillonite (MONT)-type layered clays consist of large (≈30 + nm) and thin (≈1 nm) nanoparticles (platelets) carrying negatively charged groups on their surfaces and the counterions.[99–101] Thus they have a very large surface (up to 800 m^2/g). The structure of the isolated MONT platelet is shown in Figure 3.7a. Figure 3.7b shows that the interaction of the anionic platelets with some substances often leads to the formation of lamellas composed of the alternating platelets and mono- or bilayers of the intercalated substance. The distance between the aluminum silicate platelets in the gel is set by the size of the substrate molecule or ion.[102] This property of the lamellar structures can be used for the preparation of molecular imprinting of the substrates in the gel–clay nanocomposites.[103] A schema of the experiment is shown in Figures 3.7b, c, and d. Immobilization of the obtained complex in the cross-linked gel is fixing the network structure, in particular the relative positioning of the platelets. After removal of the substrate, the gel–clay composite is fitted for the specific adsorption of the same substrate.

Sodium and lithium salts of MONT swell infinitely in water. The high swelling of MONT dispersions in water is explained by a significant increase in the translational entropy of the counterions accompanying the swelling. In fact, the platelets of MONT in water are large 2*d* polyelectrolyte molecules, and their properties in certain aspects are very similar to those of linear polyelectrolytes.

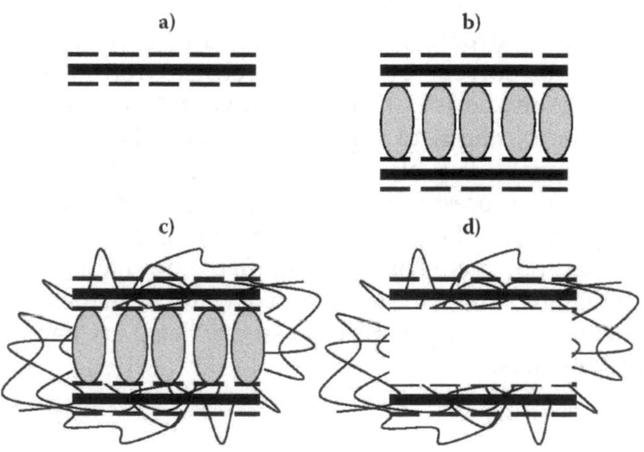

FIGURE 3.7
Schema of the formation of the molecular imprinting in PAM gel-MONT nanocomposite. (a) Isolated MONT platelet (black line) with negative charges on its surface (dashed lines). (b) Fragment of lamellar complex formed between MONT and substrate (gray ellipses) due to H-bond formation or electrostatic interactions. (c) Immobilization of the complex in the cross-linked gel (polymer chains are wavy lines). (d) Removal of the substrate. The vacancies are ready for the adsorption of the substrate.

PAM gels containing MONT and relative clays are sensitive to ionic strength and pH of the media due to ionization of Si-O⁻ groups on their surface.[97] Gel–clay nanocomposites are responsive to the addition of cationic surfactants, neutral linear polymers, polycations, and some drugs.[99,104,105] Stimulus-responsive PAM–clay composite gels can increase their volume several tenfold in the presence of Furacilline, a well-known neutral organic medical product having bacteriostatic action.[105] A gel–clay nanocomposite of copolymer AM with DADMAC has typical polyampholyte properties, e.g., "antipolyelectrolyte swelling," when an increase in the ionic strength of the aqueous solution results in a marked swelling of the gel.[104] Its behavior is similar to that of the polymeric polyampholyte gels described above.[37] If positively charged particles of the linear polymer or the surfactant aggregate are added to the solution, they destroy DADMAC–clay ionic bonds and form their own stronger bonds (the charge density of the linear polymer–guest or of the surfactant aggregate is much higher than that of the AM-DADMAC copolymer). Moreover, binding of the linear polymer or the surfactant aggregate overcharges the surface of the clay platelets. The result is a significant swelling of the polyampholyte gel nanocomposite, as seen in Figure 3.8. On the other hand, if anionic particles (linear polyanions or anionic surfactants) are added to the solution, they destroy DADMAC–clay ionic bonds and form stronger bonds. As in the first case, a significant swelling of the polyampholyte gel nanocomposite is observed.

An interesting gel-colloid composite with regularly packed spherical silica particles is described by Holtz and Asher and by Liu et al.[106,107] After the particles were dissolved in HF, the gel contained regularly arranged voids. Light scattering experiments have demonstrated the cubic symmetry of the

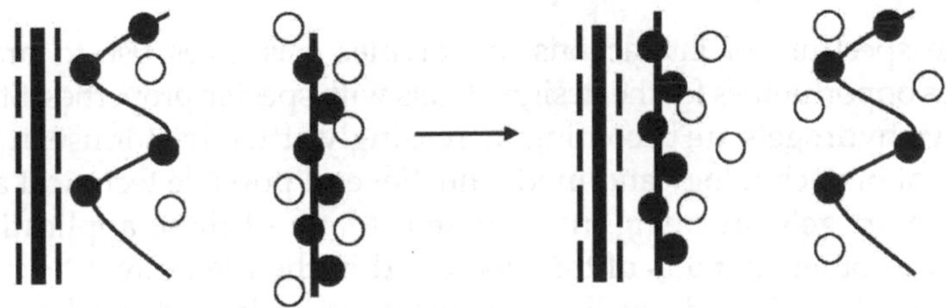

FIGURE 3.8
Schematic presentation of the interaction of the AM-DADMAC–clay nanocomposite copolymer with linear PDADMAC. The PDADMAC polycations–guests have a higher charge density. They destroy DADMAC–clay ionic bonds and form their own stronger bonds instead. Due to the overcharging of the surface of the clay platelets and the appearance of the "free" DADMAC cations of the network, a number of the corresponding mobile counterions create a large osmotic pressure and induce a strong swelling of the gel. (Isolated MONT platelet with negative charges on its surface is the fat black line surrounded by two thin dashed lines; polymer chains of the network are the wavy lines; polymer chains of PDADMAC are the straight lines; black circles are the DADMAC ions of the network and of the linear PDADMAC; open circles are the counterions of the network and of the linear polymer.)

voids obtained in gel crystal. The voids were shown to trap moving linear polymers due to the entropic effect.

A periodically ordered interconnecting porous structure was imprinted in thermosensitive hydrogels by using silica colloidal crystals as mesoscopically sized templates.[108–111] If the silica colloid crystal particles in a similar system form the closest packing, their removal (dissolution in HF) leads to a porous gel. The interconnecting porous structure provides a fast response to changes in temperature for reversible swelling and shrinking of the hydrogels, while the periodically ordered mesoscopical structure endows the gel with a structurally dependent color. The color imprinted in the gel is quickly synchronized with changes in the volume of the gels. The newly invented mesoporous hydrogels show promise as temperature-sensitive smart gels.

Polyelectrolyte gels with embedded isolated voids were recently obtained.[112] The trapping effect in these gels is a result of exclusion of co-ions from the matrix of polymer gel, with the subchains having the charge of the same sign.[113,114] Because of electrostatic forces, the co-ions are concentrated in the voids filled with water. Another new type of polyelectrolyte gel with voids is the PAM-based cationic gel containing a fraction of (3-acrylamidopropyl) trimethyl ammonium chloride (APTAC) obtained by template polymerization.[115] The charge in such gels is localized on the surface of the voids. Such gels show potential as microreactors capable of concentrating reagents having a charge of the sign opposite that of the void surface charge.

3.8 Applications of Smart Gels

3.8.1 Biotechnology-Related Applications of Smart Gels

The wide spectrum of interactions in polymer gels gives rise to practically boundless opportunities for the design of gels with specific properties. Stimulus-responsive hydrogels are becoming increasingly attractive for use in various branches of biotechnology and medicine. Several possible technical applications of smart gels are being investigated. Many of these applications are described in other chapters of this book and in the literature.[116–128]

The rate of swelling–deswelling processes is an important characteristic of many smart gels. Ciszkowska et al. studied the diffusion of some model compounds in thermoresponsive gels of NIPAM-AM copolymers.[129–130] In these studies, a four- to five-fold decrease of the diffusion coefficients was observed after the collapse. The effect was more pronounced for a cationic probe molecule compared with a neutral one.

Permeability of PAM gels and semi-IPNs containing PNIPAM to the dye orange II was studied by Muniz and Geuskens.[131] The permeability of semi-interpenetrated polymer networks increases above 32°C, though a slight shrinking of the gels is observed. This can be explained by the collapse of PNIPAM, resulting in an increase of the pore volume that facilitates diffusion of the dye. The partition of orange II and methylene blue in the same gels was also studied.[132] It was shown that the partition coefficient *K* was higher for the more hydrophobic methylene blue, especially after the collapse of the semi-IPNs.

Binding of different substances by responsive gels plays an essential role in many potential applications. Bianco-Peled and Gryc studied binding of amino acids of different hydrophobicities to PNIPAM microgels.[133] For strongly swollen gel (25°C), no straightforward correlation between the partition coefficient and the hydrophobicity could be found for low hydrophobicity values. At higher hydrophobicities, the partition coefficient increased with increasing hydrophobicity. Correlations were observed for the total hydrophobicity range at 37°C. The mechanisms of binding of the hydrophilic amino acids and the hydrophobic ones were discussed.

Ivanov et al. showed that the conjugation of penicillin acylase to PNIPAM produced a thermosensitive biocatalyst for industrial application in antibiotic synthesis.[134] The kinetics parameters of the reaction of the conjugated and free enzymes were very similar. The diffusion limitations were practically eliminated due to the "smart" nature of PNIPAM.

Tang et al. developed a glucose-sensitive hydrogel membrane.[135] The membrane was synthesized from cross-linked dextran to which Concanavalin A was coupled via a spacer arm. The change of the diffusion rate of model proteins (insulin, lysozyme) through the membrane was chosen as the responsive parameter. Changes in the transport properties of the membrane in response to glucose were explored, and it was found that, while 0.1*M* D-glucose caused a substantial increase in the diffusion rates of both insulin and lysozyme, controls using glycerol or L-glucose (0.1*M*) had no significant effect.

A similar idea was used to design a sensitive hydrogel.[136] Both Cibacron blue (ligand) and lysozyme (receptor) were covalently linked to dextran molecules that were subsequently cross-linked to form a gel membrane. It was demonstrated that the addition of nicotinamide adenine dinucleotide (NAD^+) leads to competitive displacement of lysozyme from lysozyme–Cibacron blue complexes. As a result, the membrane swells and its permeability in respect to proteins (cytochrome C and hemoglobin) becomes higher. This effect was shown to be both specific and reversible.

Liu et al. designed a novel pH-controlled release gel that is based on the inclusion effect of β-cyclodextrin gels.[137] The "host–guest" inclusion effects in such systems can be varied by changing the pH. As an example, the pH-dependent inclusion of methyl orange dye was demonstrated.

3.8.2 Application of Smart Gels in Technology

The novel light-induced shape-memory polymers were recently described by Lendlein et al.[138] The molecular switches in the photoresponsive copolymers were photoresponsive cinnamic acid-type monomers. Upon exposure to UV light with $\lambda > 260$ nm, the polymer becomes cross-linked due to the cycloaddition reaction, and its deformed shape is fixed. When the deformed polymer is irradiated with $\lambda < 260$ nm, the cross-links are cleaved and the polymer reassumes its initial shape. This work shows great promise for the synthesis of photoresponsive smart polymers.

Novel thermochromic gels were prepared using thermoresponsive gel and pH-sensitive dyes.[139] The dyes (2,6-diphenyl-4-[2,4,6-triphenylpyridinio]) phenolate (DTPP) and cresol red exhibited thermochromism when they were embedded in an aqueous polyvinyl alcohol–borax–surfactant gel network. The observed color changes in the isolated hydrogel systems can be explained by a shift of the proton dissociation equilibrium between the phenolate and phenol forms of the dye molecules in the microenvironment of the gel.

The collapse of the gels is usually accompanied by the appearance of turbidity. Zrinyi et al. proposed using this phenomenon for the design of a thermotropic window based on thermoresponsive polymer gels and solutions.[140] The optical properties of the gel layer can also be modified by the Joule heat of audio-frequency ac current. Two types of gel — glasses based on PNIPAM and PVME, respectively — have been developed and investigated. Optical properties, energy consumption, and temperature changes during the switching process were studied in the development of electrically adjustable thermotropic windows.[140–142]

New activity was initiated in the field of mechanical actuators (artificial muscles). The effect is based on the osmotic transport of water through charged gels.[116,143] The medium of an anionic swollen gel can be represented as a system of interpenetrating capillaries with negatively charged walls. After switching the dc current on, the mobile cations move from anode to cathode and involve in this movement most of the water molecules situated between the walls of the capillary. As a result, water is removed from the gel and it shrinks. Our calculations have shown that one elementary charge can move up to 10^4 molecules of water from the gel.[143] The new gels were prepared from an IPN composed of biopolymer, poly(sodium alginate), and PDADMAC.[144] The stimuli response of the IPN hydrogel in the electric fields was also investigated. When swollen IPN hydrogel was placed between a pair of electrodes, the IPN exhibited bending upon the application of an electric field. The effect of the electric field on the bending of charged gels was also studied by Kudaibergenov and Sigitov.[29] Santulli et al. are developing smart variable-stiffness actuators using polymer hydrogels.[145] A number of aspects related to the engineering of gel actuators were studied, including gel selection, response time, as well as modeling and experimentation of constant force and constant displacement behavior. In particular, the actuator was intended for use as a vibration neutralizer.

The idea to design electrically controllable adaptive optical lenses was studied by Paxton et al.[146–147] The effect of changing the swelling degree of the charged gels under the action of electrical current was used in the design of "smart lenses." For instance, reversible changes in optical properties were observed for the gels of PAMPS.

3.9 Conclusion

This chapter has reviewed the state of investigation in the field of polyelectrolyte smart gels, a field that has been intensively developed over the past 30 years. The main features of polyelectrolyte-gel behavior are well understood — based on the synthesis and study of a vast variety of chemically different polyelectrolyte gels — and have been theoretically described. Nevertheless, this remains one of the most attractive fields of study because of its potential applications in science and technology. In particular, the modern tendency is to prepare smart polymer gels that combine in their structure biopolymers and synthetic macromolecules.

References

1. Schibayama, M. and Tanaka, T., Phase transition and related phenomena of polymer gels, in *Responsive Gels: Volume Transitions I*, Disek, K., Ed., Springer-Verlag, Berlin, 1993, p. 1.
2. Khokhlov, A.R., Starodubtsev, S.G., and Vasilevskaya, V.V., Conformational transitions in polymer gels: theory and experiment, in *Responsive Gels: Volume Transitions I*, Disek, K., Ed., Springer-Verlag, Berlin, 1993, p. 123.
3. Ilavsky, M., Effect of phase transition on swelling and mechanical behavior of synthetic hydrogels, in *Responsive Gels: Volume Transitions I*, Disek, K., Ed., Springer-Verlag, Berlin, 1993, p. 173.
4. Zrinyi, M. et al., Electrical and magnetic field-sensitive smart polymer gels, in *Polymer Gels and Networks*, Osada, Y. and Khokhlov, A.R., Eds., Marcel Dekker, New York, 2002, p. 309.
5. Duzek, K. and Patterson, D., Transition in swollen polymer networks induced by intramolecular condensation, *J. Polym. Sci.*, A2, 1209, 1968.
6. Lifshitz, I.M., Grosberg, A.Yu., and Khokhlov, A.R., Some problems of the statistical physics of polymer chains with volume interactions, *Rev. Modern. Phys.*, 50, 683, 1978.
7. Tanaka, T., Phase transitions in gels and a single polymer, *Polymer*, 20, 1404, 1979.
8. Steiskai, J., Gordon, M., and Torkington, J.A., Collapse of polyacrylamide gels, *Polym. Bull.*, 3, 621, 1980.
9. Ilavsky, M., Hrouz, J., and Ulbrich, K., Phase transitions in swollen gels, 3: the temperature collapse and mechanical behavior of poly(*N*,*N*-diethylacrylamide) networks in water, *Polym. Bull.*, 7, 107, 1982.
10. Hirokawa, Y. and Tanaka, T., Volume phase transition in a nonionic gel, *J. Chem. Phys.*, 81, 6379, 1984.
11. Saito, S., Konno, M., and Inomata, H., Volume phase transition of *N*-alkylacrylamide gels, in *Responsive Gels: Volume Transitions I*, Disek, K., Ed., Springer-Verlag, Berlin, 1993, p. 207.
12. Machaeva, E.E. et al., Thermoshrinking behavior of poly(vinylcaprolactam) gels in aqueous solution, *Macromol. Chem. Phys.*, 197, 1973, 1996.

13. Starodubtsev, S.G. and Ryabina, V.R., Collapse of polymer gels: concentrated, microheterogeneous and neutral networks, *Vysokomolekul. Soed. (Polym. Sci. USSR)*, 29B, 224, 1987.
14. Onuki, A., Theory of phase transition in polymer gels, in *Responsive Gels: Volume Transitions I*, Disek, K., Ed., Springer-Verlag, Berlin, 1993, p. 63.
15. de Gennes, P.G., *Scaling Concepts in Polymer Physics*, Cornell University Press, Ithaca, NY, 1979.
16. Vasilevskaya, V.V. and Khokhlov, A.R., Theory of charged polymer networks, in *Mathematical Methods for the Investigations of Polymers*, Lifshiz, I.P. and Molchanov, A.M., Eds., Nanka, Puschino, Russia, 1982.
17. Ohmine, I. and Tanaka, T., Salt effect on the phase transitions of ionic gels, *J. Chem. Phys.*, 77, 5725, 1984.
18. Lehninger, A.L., *Principles of Biochemistry*, 4th ed., Nelson, L. and Cox, M., Eds., W.H. Freeman, New York, 2004.
19. Voet, D., Voet, J.G., and Pratt, C.W., *Fundamentals of Biochemistry*, John Wiley & Sons, New York, 1999.
20. Alfray, J.T. et al., Synthetic electrical analog of proteins, *JACS*, 72, 1864, 1950.
21. Alfray, J.T. et al., Amphoteric polyelectrolytes, II: copolymers of methacrylic acid and diethylaminoethyl methacrylate, *JACS*, 74, 438, 1952.
22. Katchalsky, K. et al., Polyelectrolyte solutions, in *Chemical Physics of Ionic Solutions*, Conway, B.E. and Barrada, R.G., Eds., John Wiley, New York, 1966.
23. Dautzenberg, H. et al., *Polyelectrolytes: Formation, Characterization and Application*, Hanser, Munich, 1994.
24. Nisato, G. and Candau, S.J., Structure and properties of polyampholyte gels, in *Polymer Gels and Networks*, Osada, Y. and Khokhlov, A., Eds., Marcel Dekker, New York, 2002, p. 131.
25. Starodubtsev, S.G. and Ryabina, V.R., Swelling and collapse of polyampholyte networks of acrylamide with methacrylic acid and 1,2-dimethyl-5-vinylpyridinium sulphate, *Visokomolekul. Soed.* (*Polym. Sci. USSR*), 29A, 2281, 1987.
26. Baker, J.P. et al., Swelling equilibria for acrylamide-based polyampholyte hydrogels, *Macromolecules*, 25, 1955, 1992.
27. Baker, J.P., Blanch, H.W., and Prausnitz, J.M., Swelling properties of acrylamide-based ampholytic hydrogels: comparison of experiments with theory, *Polymer*, 36, 1061, 1995.
28. Katayama, S., Mioga, A., and Akahori, Y., Swelling behavior of amphoteric gel and the volume phase transition, *J. Phys. Chem.*, 96, 4698, 1992.
29. Kudaibergenov, S.E. and Sigitov, V.B., Swelling, shrinking, deformation, and oscillation of polyampholyte gels based on vinyl 2-aminoethyl ether and sodium acrylate, *Langmuir*, 15, 4230, 1999.
30. English, A.E. et al. Equilibrium swelling properties of polyampholyte hydrogels, *J. Chem. Phys.*, 104, 8713, 1996.
31. Nisato, G., Munch, J.P., and Candau, S.J., Swelling, structure, and elasticity of polyampholyte hydrogels, *Langmuir*, 15, 4236, 1999.
32. Thanh, L.T.M., Makhaeva, E.E., and Khokhlov, A.R., Polyampholyte gels: swelling, collapse and interaction with ionic surfactants, *Polym. Gels Netw.*, 5, 357, 1997.
33. Bhardwaj, Y.K., Kumar, V., and Sabharval, S., Swelling behavior of radiation polymerized polyampholytic two-component gels: dynamic and equilibrium swelling kinetics, *J. Appl. Polym. Sci.*, 88, 730, 2003.
34. Valencia, J. and Pierola, I.F., Swelling kinetics of poly(*N*-vinylimidazole-*co*-sodium styrenesulfonate) hydrogels, *J. Appl. Polym. Sci.*, 83, 191, 2002.

35. Zhakupbekova, E.Zh. et al., Swelling and collapse of polyampholytic networks of β-vinyloxyethylamide of acrylic acid copolymer with *N*-vinylpyrrolidone, *Polymer Sci. B*, 47, 104, 2005.
36. Alvarez-Lorenzo, C. et al., Polymer gels that memorize elements of molecular conformation, *Macromolecules*, 33, 8693, 2000.
37. Alvarez-Lorenzo, C. et al., Simultaneous multiple point adsorption of aluminum ions and charge molecules by a polyampholyte thermoreversible gel: controlling frustractions in a heteropolymer gel, *Langmuir*, 17, 3616, 2001.
38. Moore, R.B. et al., Heterogeneities in random ionomers: a small-angle X-ray scattering investigation of alkylated poly(styrene) based materials, *Macromolecules*, 25, 5769, 1992.
39. Eisenberg, A. and Kim, J.S., *Introduction to Ionomers*, Wiley, New York, 1998.
40. Khokhlov, A.R. and Kramarenko, E.Yu., Polyelectrolyte/ionomer behavior in polymer gel collapse, *Macromol. Theory Simul.*, 3, 45, 1994.
41. Khokhlov, A.R. and Kramarenko, E.Yu., Weakly charged polyelectrolytes: collapse induced by extra ionization, *Macromolecules*, 29, 681, 1996.
42. Starodubtsev, S.G., Vasilevskaya, V.V., and Khokhlov, A.R., Collapse of poly(acrylamide) gels: effect of mechanical deformation and the type of a solvent, *DAN SSSR (Doklady Phys. Chem.)*, 282, 392, 1985.
43. Philippova, O.E., Mixed polyelectrolyte/ionomer behavior of poly(methacrylic acid) gel upon titration, *Macromolecules*, 29, 4642, 1996.
44. Klooster, N., van der Touw, F., and Mandel, M., Solvent effects in polyelectrolyte solutions, 1: potentiometric and viscosimetric titration of poly(acrylic acid) in methanol and counterion specificity, *Macromolecules*, 17, 2070, 1984.
45. Klooster, N., van der Touw, F., and Mandel, M., Solvent effects in polyelectrolyte solutions, 2: osmotic, elastic light scattering, and conductometric measurements on (partially) neutralized poly(acrylic acid) in methanol, *Macromolecules*, 17, 2078, 1984.
46. Klooster, N., van der Touw, F., and Mandel, M., Solvent effects in polyelectrolyte systems, 3: spectrophotometric results with (partially) neutralized poly(acrylic acid) in methanol and general conclusions regarding these systems, *Macromolecules*, 17, 2087, 1984.
47. Morawetz, H. and Wang, Y., Titration of poly(acrylic acid) and poly(methacrylic acid) in methanol, *Macromolecules*, 20, 194, 1987.
48. Philippova, O.E. and Khokhlov, A.R., Polyelectrolyte/ionomer behavior of polymer gels, in *Polymer Gels and Networks*, Osada, Y. and Khokhlov, A., Eds., Marcel Dekker, New York, 2002, p. 163.
49. Minati, T. and Satoh, M., Counterion binding in poly(acrylate) gel: an ^{23}Na-NMR study, *J. Polym. Sci. B: Polym. Phys.*, 42, 4412, 2004.
50. Volkov, E.V., Philippova, O.E., and Khokhlov, A.R., Dual polyelectrolyte-ionomer behavior of poly(acrylic acid) in methanol, 1: salt free solutions, *Colloid J.*, 66, 663, 2004.
51. Nishiyama, Y. and Satoh, M., Swelling behavior of poly(acrylic acid) gels in aqueous ethanol: effect of counterion species and ionic strength, *Macromol. Rapid Commun.*, 21, 174, 2000.
52. Nishiyama, Y. and Satoh, M., Solvent and counterion specific swelling behavior of poly(acrylic acid) gels, *J. Polym. Sci.*, 38, 2791, 2001.
53. Nishiyama, Y. and Satoh, M., Collapse of a charged polymer gel induced by mixing "unfavorable" counterions, *Polymer*, 42, 3919, 2001.

54. Kudo, S. et al., Volume-phase transitions of cationic polyelectrolyte gels, *Polymer*, 33, 5040, 1992.
55. Kudo, S., Konno, M., and Saito, S., Swelling equilibria of cationic polyelectrolyte gels in aqueous solutions of various electrolytes, *Polymer*, 34, 2370, 1993.
56. Starodubtsev, S.G. et al., Evidence for polyelectrolyte/ionomer behavior in the collapse of polycationic gel, *Macromolecules*, 28, 3930, 1995.
57. Kawaguchi, D. and Satoh, M., Swelling behaviour of partially quaternized poly(4-vinylpyridine) gels in water/organic solvent media, *Macromolecules*, 32, 7828, 1999.
58. Ricka, J. and Tanaka, T., Phase transition in ionic gels induced by copper complexation, *Macromolecules*, 18, 83, 1985.
59. Polynsky, A.S. et al., The study of binding of copper(II) ions by the gel of poly(*N*-vinyltriazole) by small-angle Roentgen scattering, *Vysokomolecul. Soed. (Polym. Sci. USSR)*, 25B, 283, 1983.
60. Khokhlov, A.R. et al., Supramolecular structures in polyelectrolyte gel, *Faraday Discuss.*, 101, 125, 1995.
61. Dembo, A.T. et al., Structure of binary and ternary complexes formed by sodium poly(2-acrylamide-2-methyl-1-propane-sulfonate) gel in the presence of copper(II) nitrate and cetylpyridinium chloride, *Langmuir*, 19, 7845, 2003.
62. Delsanti, M. et al., Macro-ion characterization from dilute solutions to complex fluids, in *ACS Symposium Series*, Vol. 548, Schmitz, K.S., Ed., American Chemical Society, Washington, DC, 1994, p. 195.
63. Olvera de la Cruz, M. et al., Precipitation of highly charged polyelectrolyte solutions in the presence of multivalent salts, *J. Chem. Phys.*, 103, 5781, 1995.
64. Starodubtsev, S.G., Lyubimov, A.A., and Khokhlov, A.R., Interaction of sodium poly(2-acrylamide-2-methyl-1-propane-sulfonate) linear polymer and gel with metal salts, *J. Phys. Chem. B*, 107, 12206, 2003.
65. Starodubtsev, S.G., Ryabina, V.R., and Khokhlov, A.R., Interaction of polyelectrolyte networks with oppositely charged micelle forming surfactants, *Visokomolekul. Soed.* (*Polymer Sci. USSR*), 32A, 969, 1990.
66. Khokhlov, A.R. et al., Collapse of polyelectrolyte networks induced by their interaction with oppositely charged surfactants, *Macromolecules*, 25, 4779, 1992.
67. Khokhlov, A.R. et al., Collapse of polyelectrolyte networks induced by their interaction with oppositely charged surfactants, *Macromol. Theory Simul.*, 1, 105, 1992.
68. Khandurina, Yu.V. et al., On the structure of polyacrylate–surfactant complexes, *J. Phys. II*, 5, 337, 1995.
69. Okuzaki, H. and Osada, Y., Ordered aggregate formation by surfactant-charged gel interaction, *Macromolecules*, 28, 380, 1995.
70. Dembo, A.T. and Starodubtsev, S.G., Interaction between polyelectrolyte gel-surfactant complexes with oppositely charged polymer and surfactant components, *Macromolecules*, 34, 2635, 2001.
71. Rama Rao, G.V., Konishi, T., and Ise, I., Ordering in poly(allylamine hydrochloride) gels, *Macromolecules*, 32, 7582, 1999.
72. Jeon, C.H., Makhaeva, E.E., and Khokhlov, A.R., Complexes of polyelectrolyte hydrogels with organic dyes: effect of charge density on the complex stability and intragel dye aggregation, *J. Polym. Sci. B: Polym. Phys.*, 37, 1209, 1999.
73. Nasimova, I.R., Makhaeva, E.E., and Khokhlov, A.R., Polymer gel/organic dye complexes in aqueous salt solutions, *Macromol. Symp.*, 146, 199, 1999.

74. Nasimova, I.R., Makhaeva, E.E., and Khokhlov, A.R., Interaction of poly(diallyldimethylammonium chloride) gel with oppositely charged organic dyes: behavior in salt solutions, *Polym. Sci. A*, 42, 319, 2000.
75. Mamytbekov, G. et al., Phase transition in swollen gels, 25: effect of the anionic comonomer concentration on the first-order phase transition of poly(1-vinyl-2-pyrrolidone) hydrogels, *Eur. Polym. J.*, 35, 451, 1999.
76. Ilavsky, M. et al., Phase transition in swollen gels, 28: swelling and mechanical behavior of poly(1-vinyl-2-pyrrolidone-co-*N*-vinylcaprolactam) gels in water/acetone mixtures, *Eur. Polym. J.*, 33, 214, 2001.
77. Mamytbekov, G., Bouchal, K., and Ilavsky, M., Phase transition in swollen gels, 26: effect of charge concentration on temperature dependence of swelling and mechanical behaviour of poly(*N*-vinylcaprolactam) gels, *Eur. Polym. J.*, 35, 1925, 1999.
78. Ilavsky, M. et al., Phase transition in swollen gels, 31: swelling and mechanical behaviour of interpenetrating networks composed of poly(1-vinyl-2-pyrrolidone) and polyacrylamide in water/acetone mixtures, *Eur. Polym. J.*, 38, 875, 2005.
79. Atta, A.M., Swelling behaviors of polyelectrolyte hydrogels containing sulfonate groups, *Polym. Adv. Technol.*, 13, 567, 2002.
80. Zhakupbekova, E.Zh. et al., Swelling and collapse of polyampholytic networks of a β-vinyloxyethylamide of acrylic acid copolymer with N-vinylpyrrolidone, *Polym. Sci. B*, 47, 104, 2005.
81. Siegel, R.A. and Firestone, B.A., pH-dependent equilibrium swelling properties of hydrophobic polyelectrolyte copolymer gels, *Macromolecules*, 21, 3254, 1988.
82. Moerkerke, R. et al., Phase transitions in swollen networks, 3: swelling behavior of radiation cross-linked poly(vinyl methyl ether) in water, *Macromolecules*, 31, 2223, 1998.
83. Reichelt, R. et al., Structural characterization of temperature-sensitive hydrogels by field emission scanning electron microscopy (FESEM), *Macromol. Symp.*, 210, 501, 2004.
84. McMillan, R.A., Wright, E.R., and Conticello, V.P., Elastin-mimetic protein polymers: biologically derived smart materials, *Proc. SPIE, Int. Soc. Optical Eng.*, 4332, 439, 2001.
85. Zhang, X., Wu, D., and Chu, C.-C., Synthesis and characterization of partially biodegradable, temperature and pH sensitive Dex-MA/PNIPAAm hydrogels, *Biomaterials New*, 25, 4719, 2004.
86. Wang, M., Fang, Y., and Hu, D., Preparation and properties of chitosan-poly(*N*-isopropylacrylamide) full-IPN hydrogels, *Reactive Functional Polym.*, 48, 215, 2001.
87. Nakayama, D., Nishio, Y., and Watanabe, M., Inter/intramolecular interaction and chiral recognition of water-soluble copolymers and their hydrogels containing optically active groups, *Langmuir*, 19, 8542, 2003.
88. Suvegh, K. et al., Free volume and swelling dynamics of the poly(2-dimethylamino)ethyl methacrylate)-I-polyisobutylene amphiphilic network by positron annihilation investigations, *Macromolecules*, 31, 7770, 1998.
89. Ivan, B. et al., New nanophase separated intelligent amphiphilic conetworks and gels, *Macromol. Symp.*, 227, 265, 2005.
90. Dusek, K. et al., Swelling pressure induced phase-volume transition in hybrid biopolymer gels caused by unfolding of folded crosslinks: a model, *Biomacromolecules*, 4, 1818, 2003.

91. Nowak, A.R. et al., Unusual salt stability in highly charged diblock co-polypeptide hydrogels, *JACS*, 125, 15666, 2003.
92. Lagutina, M.A. et al., New network polymers composed of poly(ethyleneoxide) and poly(methacrylic acid) chains, *Polymer Sci. B*, 44, 1295, 2002.
93. Sayil, C. and Okay, O., Macroporous poly(*N*-isopropylacrylamide) networks, *Polym. Bull.*, 48, 499, 2002.
94. Zhang, X., Zhuo, R., and Yang, Y., Using mixed solvent to synthesize temperature sensitive poly(*N*-isopropylacrylamide) gel with rapid dynamics properties, *Biomaterials*, 23, 1313, 2002.
95. Lozinsky, V.I. et al., Study of cryostructuring of polymer systems, 25: the influence of surfactants on the properties and structure of gas-filled (foamed) poly(vinyl alcohol) cryogels, *Colloid J.*, 67, 589, 2005.
96. Yoshida, R. et al., Comb-type grafted hydrogels with rapid de-swelling response to temperature changes, *Nature*, 374, 240, 1995.
97. Gao, D. and Heimann, R.B., Structure and mechanical properties of superabsorbent poly(acrylamide)-montmorillonite composite hydrogels, *Polym. Gels Netw.*, 1, 225, 1993.
98. Starodubtsev, S.G., Churochkina, N.A., and Khokhlov, A.R., Swelling and collapse of the gel composites based on neutral and slightly charged poly(acrylamide) gels containing Na-montmorillonite, *Polym. Gels Netw.*, 6, 205, 1998.
99. Theng, B.K.G., *Formation and Properties of Clay–Polymer Complexes*, Elsevier, Amsterdam, 1979, Chap. 1.
100. Grim, R.E., *Clay Mineralogy*, McGraw-Hill Series in Geology, New York, 1953.
101. Hoffmann, U., On the chemistry of clays, *Angew. Chem. Int. Ed.*, 7, 681, 1968.
102. Starodubtsev, S.G. et al., Smectic arrangement of bentonite platelets incorporated in gels of poly(acrylamide) induced by interaction with cationic surfactants, *Langmuir*, 19, 10739, 2003.
103. Starodubtsev, S.G., Complexes of clay-gel nanocomposites with cationic surfactants and linear polycations, *European Polymer Congress, Extended Abstracts, Section 9, Macromolecules in Solution, Polymer Gels, o-9-2-1, 3617*, Moscow, 1995.
104. Starodubtsev, S.G., Churochkina, N.A., and Khokhlov, A.R., Hydrogel composites of neutral and slightly charged poly(acrylamide) gels with incorporated clays: interaction with salt and ionic surfactants, *Langmuir*, 16, 1529, 2000.
105. Evsikova, O.V., Starodubtsev, S.G., and Khokhlov, A.R., Synthesis, swelling and adsorption properties of the composites based on poly(acrylamide) gel and sodium bentonite, *Polym. Sci. A*, 44, 1, 2002.
106. Holtz, J.H. and Asher, S.A., Polymerized colloidal crystal hydrogel films as intelligent chemical-sensing materials, *Nature*, 389, 829, 1997.
107. Liu, L., Li, P., and Asher, S.A., Entropic trapping of macromolecules by mesoscopic periodic voids in a polymer hydrogel, *Nature*, 397, 141, 1999.
108. Takeoka, Y. and Watanabe, M., Polymer gels that memorize structures of mesoscopically sized templates: dynamic and optical nature of periodic ordered mesoporous chemical gels, *Langmuir*, 18, 5977, 2002.
109. Takeoka, Y. and Watanabe, M., Tuning structural color changes of porous thermosensitive gels through quantitative adjustment of the cross-linker in pre-gel solutions, *Langmuir*, 19, 9104, 2003.
110. Takeoka, Y. and Watanabe, M., Template synthesis and optical properties of chameleonic poly(*N*-isopropylacrylamide) gels using closest-packed self-assembled colloidal silica crystals, *Adv. Mater.*, 15, 199, 2003.

111. Takeoka, Y. and Watanabe, M., Controlled multistructural color of a gel membrane, *Langmuir*, 19, 9554, 2003.
112. Komarova, G.A., Starodubtsev, S.G., and Khokhlov, A.R., Synthesis and properties of polyelectrolyte gels with embedded voids, *Macromol. Chem. Phys.*, 206, 1752, 2005.
113. Vasilevskaya, V.V. and Khokhlov, A.R., Swelling and collapse of Swiss-cheese polyelectrolyte gels in salt solutions, *Macromol. Theory Simul.*, 11, 623, 2002.
114. Vasilevskaya, V.V., Aerov, A.A., and Khokhlov, A.R., "Swiss-cheese" polyelectrolyte gels as media with extremely inhomogeneous distribution of charged species, *J. Chem. Phys.*, 120, 18, 2004.
115. Starodubtsev, S.G. and Khokhlov, A.R., Synthesis of polyelectrolyte gels with embedded voids having charged walls, *Macromolecules*, 2004, 37, 2004.
116. Osada, Y. and Gong, J.P., Electrical behavior and mechanical responses of polyelectrolyte gel, in *Polymer Gels and Networks*, Osada, Y. and Khokhlov, A., Eds., Marcel Dekker, New York, 2002, p. 177.
117. Stevin, H.G., Synthesis, equilibrium swelling, kinetics, permeability and applications of environmentally responsive gels, in *Responsive Gels: Volume Transitions II*, Disek, K., Ed., Springer-Verlag, Berlin, 1993, p. 81.
118. Wang, K.L., Burban, J.H., and Cussler, E.L., Hydrogels as separation agent, in *Responsive Gels: Volume Transitions II*, Disek, K., Ed., Springer-Verlag, Berlin, 1993, p. 67.
119. Okano, T., Molecular design of temperature-responsive polymers as intelligent materials, in *Responsive Gels: Volume Transitions II*, Disek, K., Ed., Springer-Verlag, Berlin, 1993, p. 179.
120. Suzuki, M. and Hiraza, O., An approach to artificial muscle using polymer gels formed by micro-phase separation, in *Responsive Gels: Volume Transitions II*, Disek, K., Ed., Springer-Verlag, Berlin, 1993, p. 241.
121. Hirotsu, S., Coexistence of phases and the nature of first-order phase transition in poly-N-isopropylacrylamide gels, in *Responsive Gels: Volume Transitions II*, Disek, K., Ed., Springer-Verlag, Berlin, 1993, p. 1.
122. Kokufuta, E., Novel applications for stimulus-sensitive polymer gels in the preparation of functional immobilized biocatalysts, in *Responsive Gels: Volume Transitions II*, Disek, K., Ed., Springer-Verlag, Berlin, 1993, p. 157.
123. Irie, M., Stimuli-responsive poly(*N*-isopropylacrylamide): photo- and chemical-induced phase transitions, in *Responsive Gels: Volume Transitions II*, Disek, K., Ed., Springer-Verlag, Berlin, 1993, p. 49.
124. Verdugo, P., Polymer gel phase transition in condensation–decondensation of secretory product, in *Responsive Gels: Volume Transitions II*, Disek, K., Ed., Springer-Verlag, Berlin, 1993, p. 145.
125. Shiga, T., Deformation and viscoelastic behavior of polymer gels in electric fields, *Adv. Polym. Sci.*, 134, 130, 1997.
126. Snowden, M.J., Murray, M.J., and Chowdry, B.Z., Some like it hot: thermosensitive polymers, *Chemistry and Industry (London)*, 14, 531, 1996.
127. Galaev, I.Y. and Mattiasson, B., Smart polymers and what they could do in biotechnology and medicine, *Trends Biotechnol.*, 336, 17, 1999.
128. Yu, Y. and Ikeda, T., Photodeformable polymers: a new kind of promising smart material for micro- and nano-applications, *Macromol. Chem. Phys. Rev.*, 206, 1705, 2005.

129. Ma, C., Zhang, W., and Ciszkowska, M., Transport of ions and electrostatic interactions in thermoresponsive poly(*N*-isopropylacrylamide-*co*-acrylic acid) gels: electroanalytical studies, *J. Phys. Chem. B*, 105, 10446, 2001.
130. Xhang, W., Gaberman, I., and Ciszkowska, M., Effect of volume phase transition on diffusion and concentration of molecular species in temperature responsive gels: electroanalytical studies, *Elecroanalyses*, 15, 409, 2003.
131. Muniz, E.C. and Geuskens, G., Influence of temperature on the permeability of polyacrylamide hydrogels and semi-IPNs with poly(*N*-isopropylacrylamide), *J. Membr. Sci.*, 172, 287, 2000.
132. Guilherme, M.R. et al., Hydrogels based on PAM network with PNIPAM included: hydrophobic-hydrophilic transition measured by the partition of orange II and methylene blue in water, *Polymer*, 44, 4213, 2003.
133. Bianco-Peled, H. and Gryc, S., Binding of amino acids to "smart" sorbents: where does hydrophobicity come into play? *Langmuir*, 20, 169, 2004.
134. Ivanov, A.E. et al., Conjugation of penicillin acylase with the reactive copolymer of *N*-isopropylacrylamide: a step toward a thermosensitive industrial biocatalyst, *Biotech. Progress*, 19, 1167, 2003.
135. Tang, M. et al., A reversible hydrogel membrane for controlling the delivery of macromolecules, *Biotech. Bioeng.*, 82, 47, 2003.
136. Tang, M. et al., NAD-sensitive hydrogel for the release of macromolecules, *Biotech. Bioeng.*, 87, 791, 2004.
137. Liu, Y.-Y. et al., A cyclodextrin microgel for controlled release driven by inclusion effects, *Macromol. Rapid Commun.*, 25, 1912, 2004.
138. Lendlein, A. et al., Light induced shape memory polymers, *Nature*, 434, 879, 2005.
139. Seeboth, A., Kriwanek, J., and Vetter, R., The first example of thermochromism of dyes embedded in transparent polymer gel networks, *J. Mater. Chem.*, 9, 2277, 1999.
140. Zrinyi, M. et al., Smart gel-glass based on the responsive properties of polymer gels, *Polym. Adv. Technol.*, 12, 501, 2001.
141. Gyenes, T. et al., Electrically adjustable thermotropic windows based on polymer gels: electrically adjustable thermotropic windows based on polymer gels, *Polym. Adv. Technol.*, 14, 757, 2003.
142. Szilagyi, A. et al., Thermotropic polymer gels: smart gel glass, *Macromol. Symp.*, 227, 357, 2005.
143. Starodubtsev, S.G., Khokhlov, A.R., and Makhaeva, E.E., Electroosmotic transport of water in polyelectrolyte networks, *Polym. Bull.*, 25, 373, 1991.
144. Kim, S.J. et al., Electrical characterizations of smart hydrogel based on alginate/poly(diallyldimethylammonium chloride) in solutions, *Proc. SPIE*, 5116 II, 756, 2003.
145. Santulli, C. et al., Development of smart variable stiffness actuators using polymer hydrogels, *Smart Mater. Struct.*, 14, 434, 2005.
146. Paxton, R.A. et al., Gel-type changeable focal length lens, *Proc. SPIE*, 5051, 504, 2003.
147. Paxton, R.A., Al-Jumaily, A.M., and Easteal, A.J., An experimental investigation on the development of hydrogels for optical applications, *Polym. Test.*, 22, 371, 2003.

Chapter 4

Thermally Responsive Polymers with Amphiphilic Grafts: Smart Polymers by Macromonomer Technique

Antti Laukkanen and Heikki Tenhu

CONTENTS

4.1 Polymers with Temperature-Dependent Solubility in Water

Several nonionic water-soluble polymers, such as poly(*N*-isopropylacrylamide) (PNIPAM),[1] poly(*N*-vinylcaprolactam) (PVCL),[2] and poly(vinyl methyl ether) (PVME),[3] show temperature-dependent solubility in water. The structure of these polymers typically shows a delicate balance between hydrophilic and hydrophobic moieties, and therefore even small changes in temperature can trigger drastic changes in the hydration layer.[4] This can lead to chain contraction and, eventually, to phase separation. In the case of PNIPAM, it has been possible to monitor the conformational change of a polymer chain from a hydrated coil to a collapsed globule while the quality of the solvent is gradually decreased by heating the aqueous solution.[5–7]

Poly(*N*-vinylcaprolactam) has a repeating unit consisting of a cyclic amide, where the amide nitrogen is directly connected to the hydrophobic polymer backbone, as seen in Figure 4.1. Because of the position of the amide linkage in the caprolactam ring, PVCL does not produce low-molecular-weight amines upon hydrolysis, unlike thermosensitive poly(*N*-alkylacrylamides).

The first report on the temperature-dependent solubility of PVCL was published in 1968 by Solomon et al.[8] Up to the 1990s, the great majority of the studies on PVCL were made in the former Soviet Union. Those early studies are extensively reviewed in Kirsh's book on poly(*N*-vinylamides).[2] The interest in PVCL and its solution properties has recently expanded, and now the phase transition of aqueous PVCL solutions has been investigated over the entire water–polymer composition range.

The temperature-dependent solubility of these types of polymers has fascinated scientists in both academia and industry since the first observations of the thermosensitivity of PNIPAM[9] and PVCL.[8] Based on the controllable change of polymer conformation, various "smart" structures have been created that are sensitive to external stimulus, in this case to temperature. For example, a thermally responsive polymer can be cross-linked to form a responsive hydrogel, or a porous surface can be grafted with smart polymers to create controllable porosity and be used as a thermally responsive filter unit.[10] Phase separation of PNIPAM and PVCL takes place at approximately 32°C, close to physiological temperature. Therefore, these polymers are considered to be suitable materials for novel biotechnological applications.[11] Based on their thermosensitivity, several applications are proposed in drug delivery, bioseparation, diagnostics, etc.[12–15] PNIPAM has been used, for example, in drug targeting for solid tumors with local hyperthermia,[16,17] in thermosensitive coatings or micelles for controlled release of drugs,[18,19] and as a cell attachment/detachment surface.[20,21] The use of PVCL instead of

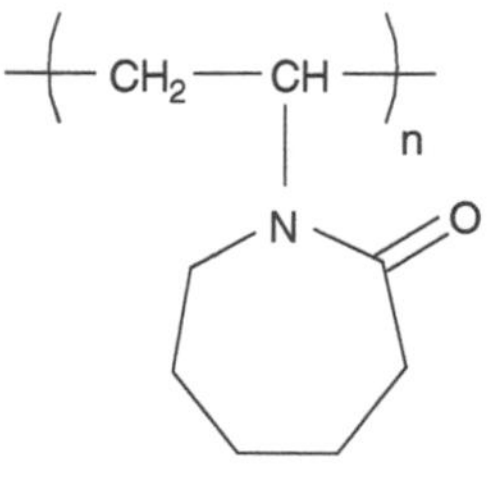

$$\left[CH_2 - CH \right]_n$$ (CH attached to N of a seven-membered caprolactam ring, C=O adjacent to N)

FIGURE 4.1
Poly(*N*-vinylcaprolactam).

PNIPAM is, however, considered advantageous because of the assumed lower toxicity of PVCL. In a recent cytotoxicity study of various thermally responsive polymers, Vihola et al.[22] have observed that PVCL was well tolerated below and above the phase transition temperature.

The general phase behavior of these polymers is now well understood. The balance between the hydrophilic and hydrophobic groups along the polymer chain determines the lower critical solution temperature (LCST), and therefore it is easily shifted to higher or lower temperatures by increasing the fraction of hydrophilic or hydrophobic groups, respectively.[23] Tailoring the polymer chain with hydrophilic or hydrophobic groups can also solve problems that have been observed in certain applications. The unavoidable increase in hydrophobicity at elevated temperatures can cause serious problems in many applications. For example, protein adsorption on hydrophobic surfaces is enhanced,[24] and the colloidal stability of thermosensitive nanoparticles is drastically lowered at elevated temperatures. The problem can be solved by grafting polymers with *hydrophilic* chains, e.g., with poly(ethylene oxide) (PEO) to provide stealth properties that enhance the circulation time in the bloodstream.[25] Introduction of *hydrophobic* groups, on the other hand, can provide interesting properties as well. Hydrophobic groups have been used to achieve self-assembled structures or hydrophobic pockets, which could be utilized to solubilize, for example, sparingly water-soluble drugs, etc.

Polymeric nanoparticles and micelles are considered as potential materials for various pharmaceutical applications.[26,27] Introduction of their thermally responsive character further broadens the possibilities of smart applications, for example, in diagnostics[28] and pulsed drug release.[29] A common problem with nanosized particles is their stability against flocculation, especially if the particle contains hydrophobic substances. The characteristic feature of the thermally responsive polymers is their increased hydrophobicity at elevated temperatures above the LCST. This feature can be a serious problem in some applications because it can lead to the coagulation of the colloidal dispersion. Therefore, the particle surfaces must be modified, either by the use of various amphiphilic additives or by a careful chemical modification of the surface. The stability of aqueous polymer dispersions can be enhanced by steric stabilization. Steric stabilization can be achieved by grafting the particle surface with, e.g., PEO or with some other highly water-soluble polymer. The use of PEO is often considered advantageous because PEO also considerably prevents the adsorption of proteins onto polymer surfaces and thus increases the biocompatibility of the polymer.[30]

One of the most popular methods of preparing either hydrophilically or hydrophobically grafted polymers is to use macromonomers of desired solubility as co-monomers. Several studies have appeared where NIPAM has been copolymerized with PEO macromonomers end-capped with methacrylates.[31] Acrylamides with long alkyl chains have also been copolymerized with NIPAM to obtain associative thermally responsive polymers.[32] In this chapter, we describe how to use *amphiphilic* PEO macromonomers to produce

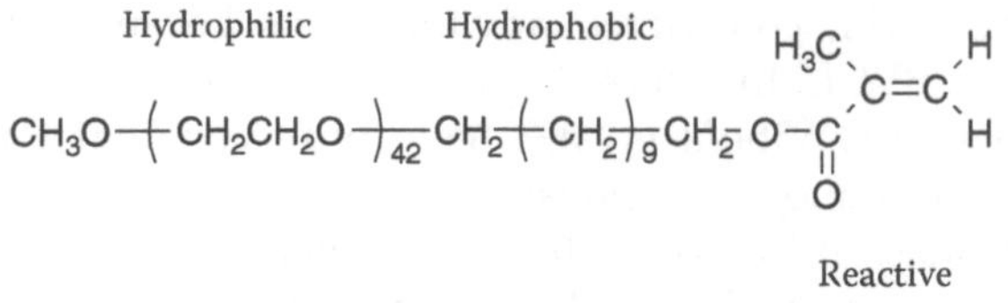

Hydrophilic Hydrophobic

$$CH_3O\left(CH_2CH_2O\right)_{42}CH_2\left(CH_2\right)_9CH_2{-}O{-}\overset{\displaystyle O}{\underset{}{C}}{-}C(CH_3){=}CH_2$$

Reactive

FIGURE 4.2
Amphiphilic PEO-alkyl macromonomer $MAC_{11}EO_{42}$.

thermally responsive grafted polymers that show enhanced stability against aggregation at elevated temperatures and, in certain cases, associative properties at lower temperatures. Our interest is focused on a methacrylate functionalized PEO macromonomer that has a short alkyl segment between the reactive unit and the hydrophilic PEO part, as seen in Figure 4.2.

Examples are provided to illustrate how to use this amphiphilic macromonomer as a reactive surfactant in emulsion polymerization in the synthesis of grafted PVCL microgels. In addition, several examples are also given on how to prepare linear PVCL with amphiphilic grafts.

4.2 Polymer Design by Amphiphilic Macromonomers

4.2.1 Amphiphilic Macromonomers

Over the last 15 years or so, reactive surfactants have received much attention in the stabilization of latex particles. The stabilization mechanism is based on covalently bound surfactant molecules that serve as either electrostatic or steric stabilizers on the latex, depending on the nature of the polymerizable surfactant. Compared with conventional surfactants, the surface-bound stabilizers have many advantages. Conventional surfactants are held on the particle surface by physical forces; thus, adsorption/desorption equilibrium always exists, which may not be desirable and, consequently, the surfactants can be easily desorbed when subjected to high shear or to a freeze/thaw cycle. Surfactant migration in the end product, e.g., in a film, is also prevented, and the quality of the product is increased when using reactive surfactants. Another advantage is that the consumption of polymerizable surfactants in emulsion polymerization is lower compared with conventional surfactants due to more effective stabilization by the surface-bound compounds.

In 1958, Freedman et al.[33] reported the first synthesis of vinyl monomers, which also functioned as emulsifying agents. Since then, a great variety of reactive surfactants have been developed. In recent reviews, both ionic[34–36] and nonionic[37,38] reactive surfactants have been well described, as well as their use in heterogeneous polymerization processes. The general features of the polymerization mechanisms in emulsion and dispersion polymerization have been reviewed by Asua et al.[39] Some nonionic reactive surfactants are also called amphiphilic macromonomers, since the molecular weights are in the range of 10^3 to 10^4. Typically, the hydrophilic part is composed of poly(ethylene oxide) and the hydrophobicity is achieved by an alkyl chain, although other types of macromonomers have been synthesized as well.[40] The main application of these monomers is in the heterogeneous polymerizations; the surface-active macromonomers can be used to replace conventional surfactants in emulsion and dispersion polymerization.

One of the major problems that limits the use of the macromonomer technique is the lack of commercially available monomers; usually the work has to be started with monomer synthesis. The most straightforward way to obtain amphiphilic PEO-containing macromonomers is to utilize commercial PEO surfactants such as various types of Brij® or Triton® as starting material. The problem in this, however, might be the position of the functional hydroxy end group; it is located at the hydrophilic end of the molecule. Thus, simple esterification, for example with methacryloyl chloride, leads to a structure where the polymerizable group is in the hydrophilic end of the amphiphilic molecule. To obtain structures as in Figure 4.2, where the reactive methacrylate is at the hydrophobic end, multistep synthesis has to be conducted. A typical four-step synthesis route is well described in the work of Liu et al.,[41] who synthesized PEO alkyl methacrylates with different lengths of the PEO part.

Detailed information on the association properties of the amphiphilic macromonomer in water is needed to evaluate its potential in heterogeneous polymerization where the surface activity is essential. The self-association property of the macromonomer $MAC_{11}EO_{42}$ (see Figure 4.2) has been studied with various methods, and the critical micellization concentration (CMC) has been determined by light scattering, by fluorescent probe experiments, and with isothermal titration calorimetry measurement.

A light scattering instrument has been used to follow the intensity of light scattered from aqueous macromonomer solutions. The intensity increases sharply when the amphiphilic macromonomers start to form micelles, as seen in Figure 4.3. The hydrodynamic size distribution of the micelle is narrow, and the mean hydrodynamic diameter is approximately 8 nm.[42,43]

The association of the macromonomers has also been evaluated with fluorescence measurements using the pyrene, Py, probe method.[44] Pyrene is a hydrophobic probe that reports changes in its microenvironment by changes in the fine structure of its emission band.[45] In the presence of micelles, Py is preferentially solubilized in the hydrophobic core of the micellar assemblies. The change in the microenvironment of pyrene, from hydrophilic to hydrophobic, can be detected by monitoring the ratio, I_1/I_3, of the intensities of the [0,0] band ($\lambda = 376$ nm) and the [0,3] band ($\lambda = 387$ nm). The ratio decreases, from a value of ≈ 1.60 recorded in water to a value of ≈ 0.60 in a hydrocarbon medium.[45] As the macromonomer concentration exceeds

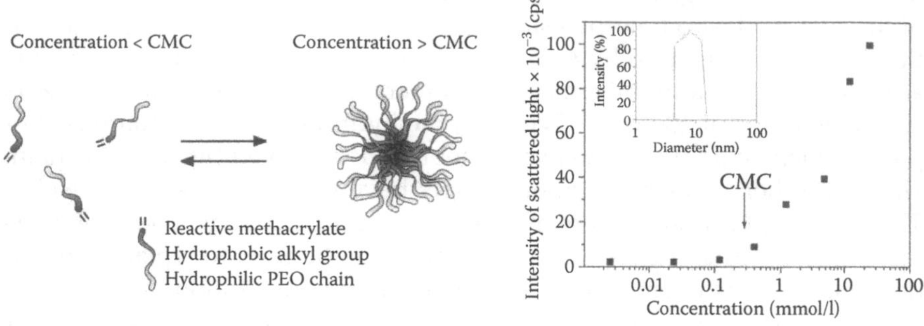

FIGURE 4.3
Left: schematic drawing of the micellization of the macromonomer. Right: dependence of the scattered light intensity on the concentration of the macromonomer. The intensity is given in counts of photons per second (cps). Size distribution of the $MAC_{11}EO_{42}$ micelle is presented in the inset (c = 5 mmol/l).

a concentration of ≈0.5 mmol/l, the I_1/I_3 ratio decreases, an indication that micellization of the macromonomers is taking place in the solution. The ratio I_1/I_3 reaches a constant low value in macromonomer solutions with concentration higher than ≈3 mmol/l. The lowest I_1/I_3 value (≈1.20) recorded in solutions of $[MAC_{11}EO_{12}]$ > 3 mmol/l is similar to the value 1.18 measured for pyrene in micellar solution of PEO surfactants, such as Brij 35 ($C_{12}EO_{23}$).[45]

It was further observed that as the macromonomer concentration increases, the total fluorescence intensity decreases, a trend opposite to the commonly observed *increase* in the emission intensity as Py passes from hydrophilic to hydrophobic environments. The decrease in Py emission intensity may reflect the fact that pyrene solubilized in the micellar core is located in close proximity to the methacrylate end groups of the macromonomer, which are known to act as quenchers of Py fluorescence.[46] A similar observation was made earlier for styrene functionalized reactive surfactants.[47] The information on the locus of the reactive unit of a polymerizable surfactant is crucial when the molecule is intended to be used in emulsion polymerization. Since the reactive methacrylate is located in the interior of the micelles (Figure 4.3), the unfavorable aggregation between the propagating micelles in emulsion polymerization is not likely.

The micellization of $MAC_{11}EO_{42}$ was also monitored by isothermal titration calorimetry, adding concentrated solution of the macromonomer (8.62mM > CMC) into water. Each injection produced an exothermic signal that, with the number of injections, remained constant and then decreased. When the macromonomer concentration in the cell exceeds 0.95mM, the heat evolved decreases, signaling that an increasing fraction of the macromonomer micelles injected per aliquot remains in the associated form rather than disintegrating into isolated macromonomers. The transition point noted in the enthalpogram corresponds to the value recorded by the fluorescence probe experiment for the onset of micellization.[44]

In all the heterogeneous polymerizations reported here with the macromonomer as an emulsifier, the macromonomer concentration was well above the critical. Hence, the hydrophobic monomer *N*-vinylcaprolactam most probably diffuses into the hydrophobic cores of the micelles either before or during the polymerization.

4.2.2 Emulsion Polymerization

Although amphiphilic macromonomers are widely used in heterogeneous polymerizations to obtain latex particles, there are not many reports on the synthesis of grafted microgels, i.e., submicron-sized hydrogels. In our recent studies, the amphiphilic PEO-alkyl macromonomer was used as a reactive surfactant in the emulsion polymerization of *N*-vinylcaprolactam to synthesize grafted PVCL microgels.[42,48] A schematic representation of the mechanism of emulsion polymerization of VCL with the aid of the amphiphilic macromonomer is presented in Figure 4.4. It was assumed that the polymerization of VCL follows the general mechanism of emulsion polymerization of hydrophobic monomers. The main ingredients in the reaction mixture include monomers, water, surfactants (the macromonomer), and initiators. When stirring the mixture, the amphiphilic macromonomers assemble into micelles, and their hydrophobic cores get more or less swollen by the monomer. However, the majority of the monomers are located in the larger-sized monomer droplets, which act as monomer reservoirs during the polymerization. The water-soluble radical initiator is likely to initiate the polymerization in the aqueous phase, and the particle nucleation takes place after the oligomeric radicals become insoluble (T >> LCST) and are stabilized by the reactive emulsifiers. When the monomers have been consumed, monodisperse cross-linked PVCL particles have been formed. The surfaces of the particles are covered with

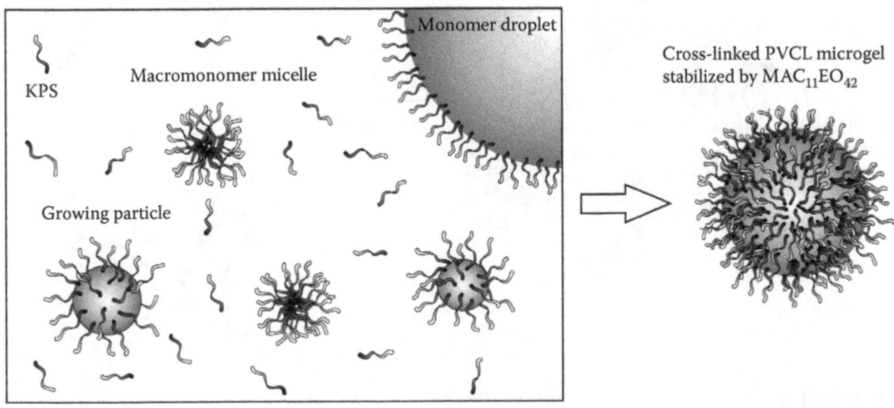

FIGURE 4.4
Schematic drawing of the emulsion polymerization of *N*-vinylcaprolactam using amphiphilic macromonomers.

covalently bound surfactants. At the polymerization temperature, the particle is in a collapsed state, and the polymer chains dissolved upon cooling without the use of a cross-linker. For comparison, we also prepared several nongrafted microgel particles using conventional anionic surfactants, such as sodium dodecyl sulfate (SDS).

4.2.3 Microemulsion Polymerization

Polymerization of reactive surfactants in other structured media, such as in micelles, vesicles, lyotropic liquid crystals, and microemulsions, were described in a recent review.[49] However, the number of studies where reactive surfactants have been used in microemulsion polymerization is still scarce. Amphiphilic PEO macromonomers are very suitable for microemulsion polymerization, since it is common to use a nonreactive PEO derivative as a cosurfactant with anionic emulgators.[38,50] By changing the ratio between reactive and nonreactive surfactant, it is possible to achieve graft copolymers with different degrees of grafting (Figure 4.5).

Eisele and Burchard[51] have synthesized poly(*N*-vinylcaprolactam) in microemulsions by using anionic surfactant sodium Bis (2-ethylhexylsulfosuccinate) (AOT) as an emulsifying agent. In our experiments, AOT has been partially replaced with the amphiphilic macromonomer $MAC_{11}EO_{42}$. After removal of the unbound surfactant by anion exchange resin, the amphiphilically grafted polymers were obtained.[43]

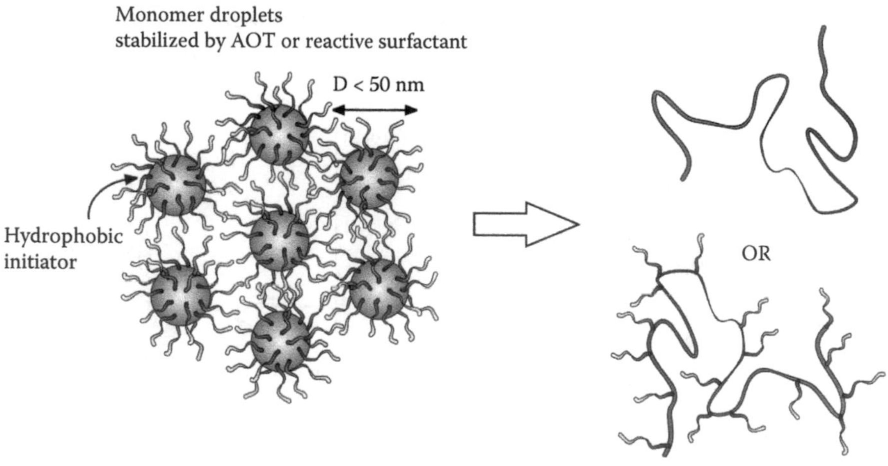

FIGURE 4.5
Schematic presentation of microemulsion polymerization of *N*-vinylcaprolactam with or without a reactive surfactant. After purification, either a linear polymer (polymerization in the absence of $MAC_{11}EO_{42}$) or a graft copolymer is achieved (AOT has been partially replaced with $MAC_{11}EO_{42}$).

4.2.4 Solution Copolymerization

Well-defined graft copolymers have also been synthesized by copolymerizing macromonomers in solution. Utilizing the macromonomer technique, a great variety of polymeric architectures, such as comblike, brushlike, and starlike copolymers, have been synthesized.[40,52,53] Especially interesting structures are achieved when a hydrophobic monomer is copolymerized with a hydrophilic macromonomer, and vice versa. The synthesized graft copolymers show a wide range of solution properties in water due to the varying hydrophilic/hydrophobic balance. For example, hydrophobically modified water-soluble macromolecules associate in water, forming hydrophobic microdomains.[54] These assemblies give both dilute and semidilute solutions unique properties with numerous industrial applications.[55–57]

Schulz et al. have used methacrylate functionalized amphiphilic macromonomers composed of an alkyl segment and a PEO chain as co-monomers with acrylamide.[58] The structure of the macromonomer employed in this study was such that the polymerizable group was attached to the hydrophilic end of the amphiphile. As a consequence, in the graft copolymer, the hydrophobic anchor was separated from the polymer backbone by a hydrophilic PEO spacer. The resulting graft copolymers predominantly formed interpolymeric associates in aqueous solutions, with the polymer concentration higher than the overlap concentration C^*. The presence of such associates results in an enhancement of the solution viscosity, compared with that of the unsubstituted polymer. In dilute solutions, however, the graft copolymer adopts a contracted conformation via intramolecular interactions and the amphiphilic grafts form micellar structures within a polymer chain. Thermosensitive PNIPAM has also been grafted with hydrophobically modified PEO chains.[59] In that study, Berlinova et al. showed that the copolymers of NIPAM and the macromonomers consisting of PEO with a terminal perfluorooctyl group form intra- or intermolecular associates in water, depending on the grafting density and the concentration of the solution.

In our recent study, an amphiphilic PEO–alkyl methacrylate was used as a comonomer of *N*-vinylcaprolactam in solution polymerization to prepare *amphiphilically* grafted poly(*N*-vinylcaprolactam).[44] The graft copolymers were obtained by conventional free-radical polymerization of VCL and $MAC_{11}EO_{42}$ in benzene using 2, 2′-azo-bis-isobutyronitrile (AIBN) as a radical initiator. Copolymers with different degrees of grafting were obtained by altering the monomer ratio in the synthesis.

Contrary to the previous studies with amphiphilically grafted associative polymers,[58,59] the macromonomer $MAC_{11}EO_{42}$ is designed such that the polymerizable group is linked to the hydrophobic end of the amphiphile (see Figure 4.6) to favor the intramolecular association of the alkyl segments. On the other hand, it is known that hydrophilic PEO chains stabilize colloidal particles formed by thermally responsive polymers at elevated temperatures.[60,61] It was therefore expected that the introduction of amphiphilic grafts on a polymer chain would modify the structure of the heat-induced aggregates and possibly stabilize them against flocculation, as in the case of sterically stabilized latexes.

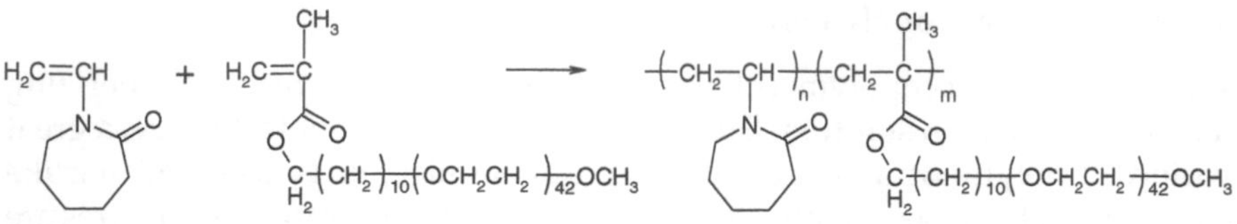

FIGURE 4.6
PVCL grafted with $MAC_{11}EO_{42}$.

4.3 Grafted PVCL Microgels

The size distributions of the cross-linked PVCL microgel particles synthesized by a batch emulsion polymerization were monomodal and reasonably narrow.[42,48] This was the case regardless of the choice of the emulgator, either SDS or the macromonomer. In the syntheses, particular attention was paid to the surface properties of the microgels. Nonionic microgel particles were prepared in emulsions where both the initiator and the surfactant (macromonomer) were electrically neutral. Negatively charged particles were obtained by using an ionic initiator, potassium persulfate (KPS), which upon decomposition forms sulfate anion radicals that covalently bind to the growing polymer. Thus, it has been possible to compare the effectiveness of the stabilization in each case and, further, to study the effects of electric charges and nonionic polymeric grafts on the thermal properties of the polymers.

4.3.1 Thermosensitive Properties of the Microgels

The thermal collapse of the various PVCL particles turned out to be a more or less continuous process, regardless of the particle charge and the presence or absence of amphiphilic grafts. The PEO derivatives bound to the particle surfaces had only a minor effect on the transition temperature. Even though the sizes of the particles at lower temperatures varied, in every case the hydrodynamic radius decreased to approximately half of its original value upon heating above the LCST. The transition is reversible. One example is given in Figure 4.7 (curve b).

Molecular-level information on the conformation of surface-bound $MAC_{11}EO_{42}$ chains is needed to evaluate the efficiency of the macromonomers in sterically shielding the particles against coagulation. The mobility of the amphiphilic graft was followed with relaxation-time measurements over a broad temperature range using ^{1}H-NMR spectroscopy. The relaxation times T_1 of the CH_2 protons of the PEO component of the grafts were measured for a micellar solution of $MAC_{11}EO_{42}$ and for various $MAC_{11}EO_{42}$-grafted microgels. The measurements were conducted at various temperatures, i.e., below and above the volume transition temperature of the microgels.

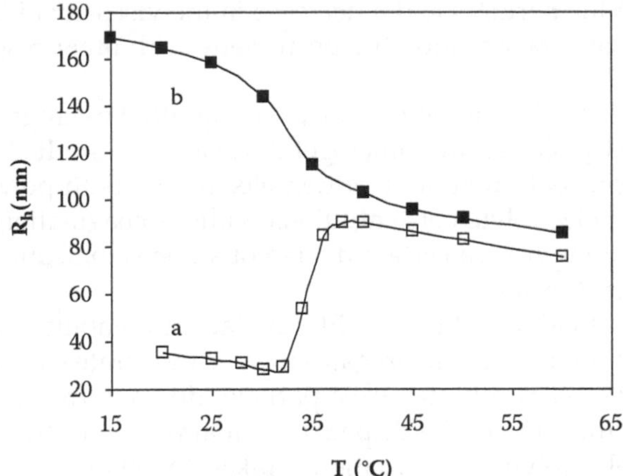

FIGURE 4.7
(a) Radius of a linear grafted PVCL single chain at low temperatures and the heat-induced particle formation above the LCST. (b) Temperature dependence of the hydrodynamic radius of a grafted PVCL microgel.

The relaxation time of all the samples increased with increasing temperature. The relaxation profiles were found to fit well to a single exponential function. Different samples behaved quite similarly, and no significant change in the relaxation behavior could be observed in any of the samples at the transition temperature around 35°C. All measured relaxation times fall into the same straight line, within an experimental error.

It can be concluded that the dynamics of the PEO segment is not affected by the collapse of the PVCL particle. This implies that the grafts are in a brush conformation, as in the case of the macromonomer micelle (Figure 4.3), at temperatures below and above the critical.

4.3.2 Colloidal Stability of the Microgels

Colloidal stability of the studied microgels arises from different origins. $MAC_{11}EO_{42}$-grafted microgels are expected to be sterically stabilized against coagulation, while the stability of the nongrafted particles has basically an electrostatic origin. Different charge densities were observed on the surface of the particles, depending on the choice of the initiator and the surfactant in the emulsion polymerization. When either an ionic initiator or surfactant was used in the synthesis, a particle with a slightly negative zeta potential was obtained. A particle with the highest negative potential was obtained by the use of an ionic initiator together with an ionic surfactant. As expected, the particles synthesized using a nonionic initiator and surfactant (the macromonomer) had zero charges. The electrophoretic mobility of the negatively charged particles increased with increasing temperature, both as a result of the collapse of the particles and a decrease in the viscosity of the electrolyte solution. The same observation has been made with other responsive particles as well.[62,63]

The colloidal stability of the responsive nongrafted microgel and the electroneutral $MAC_{11}EO_{42}$-grafted microgel particles was studied as a function of increasing concentration of an added electrolyte. Both polymer particles can stand high electrolyte concentrations without coagulation at 25°C, this obviously resulting from a certain degree of steric stabilization of dangling PVCL chains on the surface.

At 45°C the situation is different. Electrostatically stabilized particles start to coagulate and form large aggregates in very moderate electrolyte concentration. However, sterically stabilized particles do not form such large aggregates, and the turbidity of the dispersion remains practically constant, even at very high electrolyte concentrations. Taken together, the grafting of the responsive PVCL particles with PEO derivatives increased their stability toward added electrolytes considerably, the effect being especially pronounced at high temperatures, where the PVCL particles were shrunken. Without the stabilizing grafts, the repulsive forces between the negatively charged PVCL particles decreased sharply with the addition of an electrolyte, as was expected according to the DLVO (Derjaguin-Landau-Verwey-Overbeek) theory. It can be concluded that the macromonomer technique is a method of choice for the synthesis of thermally responsive PVCL microgels that do not coagulate or precipitate in solutions of high ionic strength.

4.4 Amphiphilic Graft Copolymers

Amphiphilic macromonomers are commonly used as reactive surfactants in heterogeneous polymerizations, whereas copolymerizations with other monomers by conventional solution polymerization are rarely reported. It has been noticed that PEO-methacrylate macromonomers can lead to branching and cross-linking during the polymerization reaction.[64,65] This finding may have reduced the interest for such a direct method to produce graft copolymers. However, direct solution copolymerization of macromonomers provides an easy and straightforward method to produce well-defined graft copolymers, assuming that the chain-transfer reactions can be controlled. Interesting solution properties can be expected for the graft copolymers, especially when an amphiphilic macromonomer such as $MAC_{11}EO_{42}$, is used.

In our recent studies, various copolymers of *N*-vinylcaprolactam and the amphiphilic macromonomer were synthesized by solution polymerization in benzene[44] and by microemulsion polymerization.[44] In the solution polymerizations, the macromonomer content varied from 0 to 34 wt%, and the molecular weights of the products were approximately 300,000 g/mol. Graft copolymers with molecular weights of over 1 million were obtained from microemulsion polymerizations. The size distributions were monomodal, and no evidence of cross-linking by chain-transfer reactions was found. The long alkyl chain between the reactive methacrylate group and the PEO segment may be the reason for the better copolymerization behavior compared with ordinary PEO-methacrylates. The reactivity of VCL is known to be quite low compared with other monomers, for example to methacrylates,[66] and this might result in a nonrandom distribution of the grafts in the chain. However, the distribution along the polymer backbone is still expected to be random because the larger size of the macromonomer balances out the difference in the reactivity between the macromonomer and VCL.[67]

4.4.1 Self-Association below the LCST

The association of the graft copolymers in water at room temperature was studied with dynamic light scattering.[44] The hydrodynamic radii of the various copolymers in dilute aqueous solutions at 20°C, well below the cloud point (see Chapter 1, Section 1.4.1), were determined for solutions ranging in concentration from 1 to 10 g/l. Similar measurements were also conducted for solutions of an unmodified PVCL of similar molecular weight. The homopolymer has a hydrodynamic radius of the same order of magnitude (5 to 30 nm), whether it is dissolved in water or in tetrahydrofuran (THF), both good solvents for PVCL at this temperature. Thus, the polymer is dissolved as single chains in water at 20°C, and this is the situation over the entire concentration range covered.

Two different patterns emerged from the dynamic light scattering (DLS) studies of the copolymers, depending on their grafting density. The hydrodynamic radius distribution recorded with solutions of the copolymers of high grafting density is monomodal (c = 1 to 10 g/l), with no indication of the presence of larger objects. Thus, in aqueous solutions, the graft copolymer, which contains 34 wt% of $MAC_{11}EO_{42}$, does not form interpolymeric associates, i.e., the amount of amphiphilic grafts is high enough to force the polymer chain to adopt a stable intrapolymeric structure. A similar observation has previously been reported in a study of PNIPAM grafted with amphiphilic PEO macromonomer bearing terminal perfluorooctyl groups,[59] another type of amphiphilically modified polymers that tend to form unimers rather than interpolymeric associates.

In contrast, the hydrodynamic radius distribution recorded for aqueous solutions of copolymers with a low grafting density is bimodal, with a contribution of small entities of hydrodynamic radius R_h = 5 to 30 nm being assigned to single polymer chains and another contribution of larger particles of R_h = 80 to 150 nm. The relative abundance of the two populations depends on the copolymer concentration, with the relative amount of the larger particles increasing with increasing copolymer concentration.

The self-association at 20°C was also studied with fluorescence probe measurements.[44] The measurements were conducted using solutions with increasing copolymer concentration under experimental conditions identical to those used to monitor the macromonomer micellization, as described in the preceding section. For all copolymer solutions, the ratio I_1/I_3 decreased with increasing polymer concentration from its value in water (≈1.65) and eventually leveled off for solutions with copolymer concentration in excess of 0.3 to 1.0 g/l, depending on the grafting level. The onset of the decrease in I_1/I_3, which is often taken as an indication of the onset of micellization, depends on the grafting level of the copolymer: the higher the grafting level, the lower the onset of micellization. The fact that the minimum $C_{11}EO_{42}$ concentration required for copolymer micellization is lower by more than two orders of magnitude than the CMC of the macromonomer is also in agreement with previous observations.

The fluorescence-probe experiments yield information on the minimum copolymer concentration needed for association and on the micropolarity of the assemblies, but they do not allow one to assess whether the hydrophobic assemblies are formed intra- or intermolecularly. Dynamic light scattering measurements performed on copolymer solutions in the same concentration region indicate that either intra- or intermolecular association can take place or, in some cases, even both. Thus, in the studied polymer concentration range, the pyrene molecules are solubilized by hydrophobic domains formed by single polymer chains, by the interpolymeric associates, or by both. If the concentration is increased well above the overlap concentration, very strong interpolymeric association is likely to take place.

4.4.2 Heat Induced-Phase Transition

When an aqueous solution of PVCL is heated above its LCST, phase separation takes place, a phenomenon accompanied by the loss of the hydration layer surrounding the polymer chains in a cold solution. The dehydration is an endothermic process and is therefore easily followed by microcalorimetry. The endotherms recorded for the solutions of PVCL homopolymer and the grafted copolymers are broad and markedly asymmetric, with a sharp increase in heat capacity on the low-temperature side (onset of the transition, T_{onset}) and a gradual decrease of the heat capacity at temperatures higher than the maximum temperature. T_{onset} is closely related to the onset of aggregation, and thus the measured cloud-point temperatures were nearly identical to the T_{onset} values. Similar endotherms can also be measured for PVCL homopolymers with varying molecular weights[68] as well as for PVCL grafted with PEO.[61]

Introduction of amphiphilic grafts onto PVCL increases the onset temperature of the endothermic transition compared with the unmodified PVCL. Surprisingly, the degree of grafting does not affect the T_{onset} very much, which is the same for all copolymers. In contrast, the heat of transition, ΔH, depends on the grafts content of the copolymers, corresponding closely to the concentration of VCL units. The heat of transition is not dependent on the molecular weight of the PVCL chain, either: $\Delta H = 4.4 \pm 0.4$ kJ·mol^{-1} in each case.

A negative change in partial heat capacity (ΔC_p) due to the transition can be detected for each graft copolymer. For all polymers, ΔC_p was −70 ±

20 J·mol^{-1}K^{-1}. A similar observation was also made for the cooling scans (100 to 10°C), although the sign of ΔC_p was opposite. A similar negative change in C_p upon heating was also observed for PNIPAM ($\Delta C_p \approx -63$ J·mol^{-1}K^{-1})[69] as well as for the phase transitions of various pluronic-type block copolymers.[70] A negative heat-capacity change during the phase transition can be taken as an indication of diminished interaction between water molecules and polymer chains.

4.4.3 Formation of Stable Mesoglobules

Heating of an aqueous solution of a thermally responsive polymer above the LCST increases the attraction between the hydrophobic parts of the polymer. This leads to a conformational change of the polymer, and the swollen polymer chain starts to contract. If the concentration is very low, a transition from a coil to a globule can occur, as demonstrated by Wu et al. with PNIPAM.[5–7] Most often, however, the contraction of the polymer chains does not lead to the formation of a single-chain globule but, rather, to interpolymeric aggregation followed by precipitation.

It has recently been observed that heating of aqueous PNIPAM solution does not necessarily lead to the formation of single-chain globules or to macroscopic phase separation. If the polymer concentration is 0.1 to 1.0 g/l, the polymer chains tend to form stable aggregates upon heating.[71,72] These aggregates are called *mesoglobules* to separate the concept of collapsed aggregate particles from single-chain globules. The recent observations of colloidally stable mesoglobules are based on PNIPAM samples that were prepared using ionic radical initiators, and therefore the observed, somewhat surprising, stability of the particles may have an electrostatic origin.[71]

Formation of colloidally stable mesoglobules seems to be a more general phenomenon for thermally responsive polymers and not a specific feature of PNIPAM. Recently, it has been observed that PVCL[68,73] and PVME[73] form stable mesoglobules in dilute solutions at elevated temperatures. The mesoglobules are also very stable against further coagulation, although the particles are entirely nonionic.[68,73]

The formation of stable particles of PNIPAM[60] or PVCL[61] grafted with PEO has been detected earlier, and it is expected that the introduction of the amphiphilic PEO-alkyl chains on PVCL will stabilize the PVCL aggregates upon heating.

The effect of temperature on the size of $MAC_{11}EO_{42}$-grafted PVCL[43] is presented in Figure 4.7 (curve a). The polymer (14 wt% $C_{11}EO_{42}$, M_w 1.3 million g/mol) is, at $T <$ LCST, fully hydrated and dissolved as single chains. Upon heating, the hydrodynamic size of the polymer slightly decreases, and at the cloud point the polymer starts to aggregate, causing an increase in the observed size. At $T >$ LCST the size of the formed particle slightly decreases upon further heating. For comparison, the temperature-dependent shrinking of a grafted PVCL microgel particle is presented in the same figure. The grafted microgel experiences a continuous volume decrease upon heating and finally forms a collapsed particle at temperatures above 45°C (Figure 4.7, curve b). At temperatures well above the LCST, the sizes of these two particles are actually very close to each other.

The same phenomenon, i.e., the transition from a coil (to a globule) to an aggregate, was observed for the other amphiphilically grafted PVCLs: stable nanoparticles are formed, ranging in size from ≈50 to 120 nm, depending on the concentration of the solution.[44] The size of the particles is only slightly dependent on the grafting density of the polymer, and the size distributions of the particles are rather narrow and monomodal. The particles remained colloidally stable, with no interparticle flocculation over periods of several days, as judged by DLS monitoring of aqueous polymer samples kept at 50°C.

The density of particles ($\rho = M_w^{agg}/(N_A(4/3)\pi R_h^3)$) indicates that particles are partially hydrated, as in the case of collapsed poly(*N*-isopropylacrylamide) (ρ = 0.34 g/cm^3).[7] A particle formed by a PVCL with 18 wt% of $MAC_{11}EO_{42}$ has the aggregation number of approximately 1600 ($M_w^{agg} = 4.8 \times 10^8$ g/mol) and the density ρ = 0.40 g/cm^3. The shape of the particles was estimated from the ratio of the radius of gyration (R_g) to the hydrodynamic radius, R_g/R_h, which varies from 0.77 for hard spheres to 1.5 for random coils. In the present experiments, the ratio ranges from 0.72 to 0.75, indicating that the particles have a homogeneous spherical geometry, even after being subjected to dilution. Qualitatively similar observations were also made for the mesoglobules formed by the homopolymer PVCL.[73]

In the mesoglobular phase, at temperatures above the LCST, the polymers are entangled with each other and stuck together via hydrophobic forces. Consequently, the particles disintegrate immediately upon cooling. The physical network that is present at elevated temperatures can, however, be strengthened by adding phenols into the particles. Phenols are known to complex with PVCL via hydrogen bond formation between the carbonyl groups of the lactam ring and the hydroxyl units of the phenol. These specific interactions have been used to prepare hydrogels via physical cross-linking with polyphenols, such as pyrocatechol and phloroglucinol, of linear PVCL in concentrated aqueous solutions below and above the LCST.[2,74,75] The same cross-linking mechanism was tested with the colloidal PVCL particles in an attempt to fix permanently the mesoglobule structure, so that the particles preserve their integrity upon cooling.

The effect of added phenols on the size of heat-induced particles was studied via DLS in the case of PVCL and PVCL-g-18 particles. Cross-linking was performed by injecting a hot aqueous 1,2-benzenediol solution into a colloidal polymer dispersion placed in the light scattering cell holder at 50°C. Addition of the phenol at 50°C did not affect the size of the PVCL particles, but it triggered a slight decrease in the size of the PVCL-g-18 particles, from R_h = 77 nm to 64 nm. The decrease in the size of the *graft copolymer* particles may be a result of the cross-linking of a less dense structure of the PEO-containing particle compared with that of a PVCL mesoglobule. The treated solutions were allowed to cool down to 20°C. Their hydrodynamic radii did not change, which indicates that physical cross-linking occurred within the particles. In the absence of the cross-linker, the particles disintegrated immediately as the temperature was lowered below 32°C. Cross-linked particles were monitored at 20°C for several weeks by DLS, and no flocculation was detected.

4.5 Conclusions

Utilization of a synthesized amphiphilic macromonomer in emulsion polymerization turned out to be successful, and grafted thermally responsive PVCL microgels could be prepared. Microgels prepared by emulsion polymerization with the macromonomer proved to be colloidally very stable, even in concentrated electrolyte solutions, which indicates an effective steric stabilization generated by the amphiphilic macromonomer. The synthesized microgels underwent a reversible volume change from expanded to shrunken state when the temperature was raised, as seen in Figure 4.8.

The high surface activity of the macromonomer was also exploited in microemulsion polymerization. This method was especially effective in producing very high-molecular-weight copolymers of VCL and $MAC_{11}EO_{42}$. Due to the general properties of the microemulsion polymerization, namely the environmentally friendly process and the high polymerization rate, it could be the method of choice to produce well-defined graft copolymers also on an industrial scale. Various graft copolymers of VCL and $MAC_{11}EO_{42}$ were also synthesized in organic solvents using conventional solution polymerization. Below the LCST, the graft copolymers form intra- and interpolymeric associates in water, depending on the concentration and the grafting density. The associates solubilize hydrophobic substances, such as pyrene, inside their nonpolar domains, which are composed of self-assembled $C_{11}EO_{42}$ grafts. Upon heating, the PVCL backbone collapses, triggering a change in

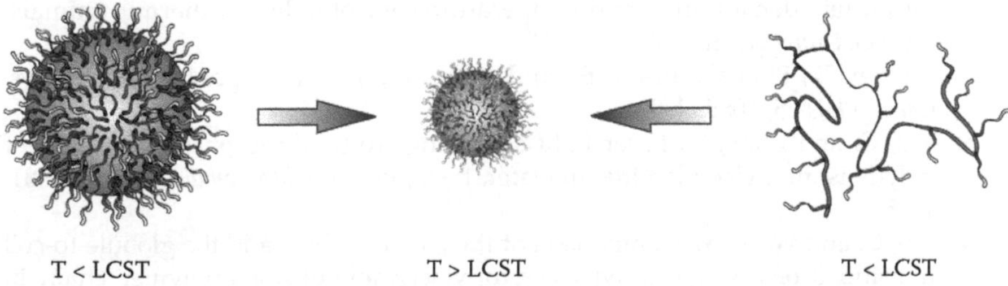

FIGURE 4.8
A scheme on the thermal response of different polymeric structures based on PVCL and amphiphilic macromonomer $MAC_{11}EO_{42}$. Left: shrinking of the grafted PVCL microgel. Right: heat-induced aggregation of the graft-copolymer and the formation of a mesoglobule.

the hydration of the chain and the release of the polymer-bound water molecules. Thermally induced aggregation leads to the formation of colloidally stable mesoglobules, as illustrated in Figure 4.8. The graft copolymers form less-dense particles than the PVCL homopolymers. Each particle is composed of several thousands of collapsed polymer chains, although the particles still contain much water at 50°C. The size of the mesoglobules can be altered by changing the polymer concentration or the composition of the graft copolymer.

4.5.1 Potential Applications in Biotechnology and Biomedicine

The studied polymer, i.e., PVCL grafted with amphiphilic poly(ethylene oxide) chains, shows no sign of cytotoxicity in standard viability (MTT) or cellular damage (LDH) tests.[22] Thermosensitive PVCL particles are also designed to be stable in various environments[48] and most probably have so-called stealth properties in the bloodstream. Intravenous applications are hardly suitable, however, mainly because the nanoparticles are nonbiodegradable. The harmful accumulation of PVCL might possibly be prevented, however, by using lower-molecular-weight PVCL chains as the building blocks of the nanoparticles and conducting the cross-linking with degradable or pH-sensitive molecules. With the above-mentioned technique, the well-tolerable PVCL chains could be removed from the body after the therapeutic act.

References

1. Schild, G.H., Poly(*N*-isopropylacrylamide): experiment, theory and application, *Prog. Polym. Sci.*, 17, 163, 1992.
2. Kirsh, Y.E., *Water Soluble Poly-N-Vinylamides*, John Wiley & Sons, Chichester, U.K., 1998.
3. Horne, R.A. et al., Macromolecule hydration and the effect of solutes on the cloud point of aqueous solutions of poly(vinyl methyl ether): possible model for protein denaturation and temperature control in homeothermic animals, *J. Colloid Interface Sci.*, 35, 77, 1971.
4. Widom, B., Bhimalapuram, B., and Koga, K., The hydrophobic effect, *Phys. Chem. Phys.*, 5, 3085, 2003.
5. Wu, C. and Zhou, S., Laser light scattering study of the phase transition of poly(*N*-isopropylacrylamide) in water, 1: single chain, *Macromolecules*, 28, 8381, 1995.
6. Wu, C. and Wang, X., Comparison of the coil-to-globule and the globule-to-coil transitions of a single poly(*N*-isopropylacrylamide) homopolymer chain in water, *Phys. Rev. Lett.*, 80, 4092, 1998.
7. Wang, X., Qiu, X., and Wu, C. Comparison of the coil-to-globule and the globule-to-coil transitions of a single poly(*N*-isopropylacrylamide) homopolymer chain in water, *Macromolecules*, 31, 2972, 1998.

8. Solomon, O.F. et al., Properties of solutions of poly(*N*-vinylcaprolactam), *J. Appl. Polym. Sci.*, 12, 1835, 1968.
9. Heskins, M. and Guillet, J.E., Solution properties of poly(*N*-isopropylacrylamide), *Macromol. Sci. Chem. A2*, 2, 1441, 1968.
10. Shtanko, N.I. et al., Preparation of permeability controlled track membranes on the basis of "smart" polymers, *J. Membrane Sci.*, 179, 155, 2000.
11. Galaev, I.Y. and Mattiasson, B., Thermoreactive water-soluble polymers, nonionic surfactants, and hydrogels as reagents in biotechnology, *Enzyme Microb. Technol.*, 15, 354, 1993.
12. Vihola, H. et al., Binding and release of drugs into and from thermosensitive poly(*N*-vinyl caprolactam) nanoparticles, *Eur. J. Pharm. Sci.*, 16, 69, 2002.
13. Hoffman, A.S. and Stayton, P.S., Bioconjugates of smart polymers and proteins: synthesis and applications, *Macromol. Symp.*, 207, 139, 2004.
14. Piskin, E., Molecularly designed water soluble, intelligent, nanosize polymeric carriers, *Int. J. Pharm.*, 277, 105, 2004.
15. Jeong, B. and Gutowska, A., Lessons from nature: stimuli-responsive polymers and their biomedical applications, *Trends Biotech.*, 20, 305, 2002.
16. Meyer, D.E. et al., Drug targeting using thermally responsive polymers and local hyperthermia, *J. Control Release*, 74, 213, 2001.
17. Chilkoti, A. et al., Targeted drug delivery by thermally responsive polymers, *Adv. Drug Delivery Rev.*, 54, 613, 2002.
18. Gutowska, A. et al., Heparin release from thermosensitive polymer coatings: *in vivo* studies, *J. Biomed. Mater. Res.*, 29, 811, 1995.
19. Chung, J.E. et al., Thermo-responsive drug delivery from polymeric micelles constructed using block copolymers of poly(*N*-isopropylacrylamide) and poly(butylmethacrylate), *J. Control Release*, 62, 115, 1999.
20. Okano, T. et al., Mechanism of cell detachment from temperature-modulated, hydrophilic–hydrophobic polymer surfaces, *Biomaterials*, 16, 297, 1995.
21. Ebara, M. et al., Temperature-responsive cell culture surfaces enable "on-off" affinity control between cell integrins and RGDS ligands, *Biomacromolecules*, 5, 505, 2004.
22. Vihola, H. et al., Cytotoxicity of thermosensitive polymers poly(*N*-isopropylacrylamide), poly(*N*-vinylcaprolactam) and amphiphilically modified poly(*N*-vinylcaprolactam), *Biomaterials*, 26, 3055, 2005.
23. Taylor, L.D. and Cerankowski, L.D., Preparation of films exhibiting a balanced temperature dependence to permeation by aqueous solution: lower consolute behavior, *J. Polym. Sci., Polym. Chem. Ed.*, 13, 2551, 1975.
24. Lee, J.H., Li, T., and Park, K., Solvation interactions for protein adsorption to biomaterial surfaces, in *Water in Biomaterials Surface Science*, Morra, M., Ed., Wiley, New York, 2001.
25. Otsuka, H., Nagasaki, Y., and Kataoka, K., PEGylated nanoparticles for biological and pharmaceutical applications, *Adv. Drug Deliv. Rev.*, 55, 403, 2003.
26. Yokoyama, M., Novel passive targetable drug delivery with polymeric micelles, in *Biorelated Polymers and Gels: Controlled Release and Applications in Biomedical Engineering*, Okano, T., Ed., Academic Press, San Diego, 1998, chap. 6.
27. Francis, M.F., Cristea, M., and Winnik, F.M., Polymeric micelles for oral drug delivery: why and how, *Pure Appl. Chem.*, 76, 1321, 2004.
28. Pichot, C. et al., Functionalized thermosensitive latex particles: useful tools for diagnostics, *J. Disp. Sci. Tech.*, 24, 423, 2003.

29. Kikuchi, A. and Okano, T., Pulsatile drug release control using hydrogels, *Adv. Drug Deliv. Rev.*, 54, 53, 2002.
30. Sofia, S.J. and Merrill, E.W., Protein adsorption on poly(ethylene oxide)-grafted silicon surfaces, in *Poly(ethylene glycol) Chemistry and Biological Applications, ACS Symposium Series 680*, Harris, J.M. and Zaplisky, S., Eds., American Chemical Society, Washington, DC, 1997, chap. 22.
31. Qiu, X. and Wu, C., Study of the core-shell nanoparticle formed through the "coil-to-globule" transition of poly(*N*-isopropylacrylamide) grafted with poly(ethylene oxide), *Macromolecules*, 30, 7921, 1997; Chen, H. et al., Folding and unfolding of individual PNIPAM-g-PEO copolymer chains in dilute aqueous solutions, *Macromolecules*, 38, 4403, 2005; Chen, H. et al., Formation of mesoglobular phase of PNIPAM-g-PEO copolymer with a high PEO content in dilute solutions, *Macromolecules*, 38, 8045, 2005.
32. Ringsdorf, H., Venzmer, J., and Winnik, F.M., Fluorescence studies of hydrophobically modified poly(*N*-isopropylacrylamides), *Macromolecules*, 24, 1678, 1991; Ringsdorf, H., Simon, J., and Winnik, F.M., Hydrophobically modified poly(*N*-isopropylacrylamides) in water: probing of the microdomain composition by nonradiative energy transfer, *Macromolecules*, 25, 5353, 1992.
33. Freedman, H.H., Mason, J.P., and Medalia, A.I., Polysoaps, II: preparation of vinyl soaps, *J. Org. Chem.*, 23, 76, 1958.
34. Guyot, A., Recent progress in reactive surfactants in emulsion polymerization, *Macromol. Symp.*, 179, 105, 2002.
35. Guyot, A. et al., Reactive surfactants in heterophase polymerization, *Acta Polym.*, 50, 57, 1999.
36. Paleos, C.M., *Polymerization in organized media*, Gordon and Breach Science Publishers, Paris, 1992.
37. Capek, I., Surface active properties of polyoxyethylene macromonomers and their role in radical polymerization in disperse systems, *Adv. Colloid Interface Sci.*, 88, 295, 2000.
38. Capek, I., Radical polymerization of polyoxyethylene macromonomers in disperse systems, *Adv. Polym. Sci.*, 145, 1, 1999.
39. Asua, J.M. and Schoonbrood, H.A.S., Reactive surfactants in heterophase polymerization, *Acta Polym.*, 49, 671, 1998.
40. Ito, K. and Kawaguchi, S., Poly(macromonomers): homo- and copolymerization, *Adv. Polym. Sci.*, 142, 129, 1999.
41. Liu, J., Chew, C.H., and Gan, L.M., Synthesis and polymerization of a nonionic surfactant: poly(ethylene oxide) macromonomer, *J. Macromol. Sci., Pure Appl. Chem.*, A33(3), 337, 1996.
42. Laukkanen, A. et al., Poly(*N*-vinylcaprolactam) microgel particles grafted with amphiphilic chains, *Macromolecules*, 33, 8703, 2000.
43. Laukkanen, A., *Thermally responsive polymers based on N-vinylcaprolactam and an amphiphilic macromonomer*, Yliopistopaino, Helsinki, 2005; available on-line at http://ethesis.helsinki.fi.
44. Laukkanen, A. et al., Thermosensitive graft copolymers of an amphiphilic macromonomer and *N*-vinylcaprolactam: synthesis and solution properties in dilute aqueous solutions below and above the LCST, *Polymer*, 46, 7055, 2005.
45. Kalyanasundaram, K. and Thomas, J.K., Environmental effects on vibronic band intensities in pyrene monomer fluorescence and their application in studies of micellar systems, *J. Am. Chem. Soc.*, 99, 2039, 1977.

46. Encinas, M.V., Guzman, E., and Lissi, E.A., Intramicellar aromatic hydrocarbon fluorescence quenching by olefins, *J. Phys. Chem.*, 87, 4770, 1983.
47. Kawaguchi, S. et al., Fluorescence probe study of micelle formation of poly (ethylene oxide) macromonomers in water, *J. Phys. Chem.*, 98, 7891, 1994.
48. Laukkanen, A. et al., Stability and thermosensitive properties of various poly(*N*-vinylcaprolactam) microgels, *Colloid Polym. Sci.*, 280, 65, 2002.
49. Summers, M. and Eastoe, J., Applications of polymerizable surfactants, *Adv. Colloid Interface Sci.*, 100, 137, 2003.
50. Antonietti, M. and Bremser, W., Microgels: model polymers for the crosslinked state, *Macromolecules*, 23, 3796, 1990.
51. Eisele, M. and Burchard, W., Hydrophobic water-soluble polymers, 1: dilute solution properties of poly(1-vinyl-2-piperidone) and poly(*N*-vinylcaprolactam), *Makromol. Chem.*, 191, 169, 1990.
52. Velichkova, R.S. and Christova, D.C., Amphiphilic polymers from macromonomers and telechelics, *Prog. Polym. Sci.*, 20, 819, 1995.
53. Ito, K., Polymeric design by macromonomer technique, *Prog. Polym. Sci.*, 23, 581, 1998.
54. Varadaraj, R. et al., Analysis of hydrophobically associating copolymers utilizing spectroscopic probes and labels, in *Macromolecular Complexes in Chemistry and Biology*, Dubin, P. et al., Eds., Springer-Verlag, Berlin, 1994.
55. Taylor, K.C. and Nasr-El-Din, H.A., Water-soluble hydrophobically associating polymers for improved oil recovery, *J. Petrol. Sci. Eng.*, 19, 265, 1998.
56. Schulz, D.N. and Glass, J.E., *Polymers as Rheology Modifiers, ACS Symposium Series 462*, American Chemical Society, Washington DC, 1991.
57. Kastner, U., The impact of rheological modifiers on water-borne coatings, *Coll. Surf. A*, , 183, 805, 2001.
58. Schulz, D.N. et al., Copolymers of acrylamide and surfactant macromonomers: synthesis and solution properties, *Polymer*, 28, 2110, 1987.
59. Berlinova, I.V. et al., Associative graft copolymers comprising a poly(*N*-isopropylacrylamide) backbone and end-functionalized polyoxyethylene side chains. Synthesis and aqueous solution properties, *Polymer*, 42, 5963, 2001.
60. Virtanen, J. and Tenhu, H., Thermal properties of poly(*N*-isopropylacrylamide)-g-poly(ethylene oxide) in aqueous solutions: influence of the number and distribution of the grafts, *Macromolecules*, 33, 5970, 2000.
61. Verbrugghe, S. et al., Light scattering and microcalorimetry studies on aqueous solutions of thermo-responsive PVCL-g-PEO copolymers, *Polymer*, 44, 6807, 2003.
62. Daly, E. and Saunders, B.R., A study of the effect of electrolyte on the swelling and stability of poly(*N*-isopropylacrylamide) microgel dispersions, *Langmuir*, 13, 5546, 2000.
63. Duracher, D., Elaissari, A., and Pichot, C., Characterization of cross-linked poly(*N*-isopropylmethacrylamide) microgel latexes, *Colloid Polym. Sci.*, 277, 905, 1999.
64. Bo, G., Wesslén, B., and Wesslén, K.B., Amphiphilic comb-shaped polymers from poly(ethylene glycol) macromonomers, *J. Polym. Sci.: Part A: Polym. Chem.*, 30, 1799, 1992.
65. Jannasch, P. and Wesslén, B., Synthesis of poly(styrene-graft-ethylene oxide) by ethoxylation of amide group-containing styrene copolymers, *J. Polym. Sci.: Part A: Polym. Chem.*, 31, 1519, 1993.
66. Okhapkin, I.M. et al., Effect of complexation of monomer units on pH- and temperature-sensitive properties of poly(*N*-vinylcaprolactam-co-methacrylic acid), *Macromolecules*, 36, 8130, 2003.

67. Yanul, N.A. et al., Thermoresponsive properties of poly(*N*-vinylcaprolactam)-poly(ethylene oxide) aqueous systems: solutions and block copolymer networks, *Macromol. Chem. Phys.*, 202, 1700, 2001.
68. Laukkanen, A. et al., Formation of colloidally stable phase separated poly(*N*-vinylcaprolactam) in water: a study by dynamic light scattering, microcalorimetry, and pressure perturbation calorimetry, *Macromolecules*, 37, 2268, 2004.
69. Kujawa, P. and Winnik, F., Volumetric studies of aqueous polymer solutions using pressure perturbation calorimetry: a new look at the temperature-induced phase transition of poly(*N*-isopropylacrylamide) in water and D_2O, *Macromolecules*, 34, 4130, 2001.
70. Beezer, A.E. et al., An investigation of dilute aqueous solution behavior of poly(oxyethylene) + poly(oxypropylene) + poly(oxyethylene) block copolymers, *Langmuir*, 10, 4001, 1994.
71. Chan, K., Pelton, R., and Zhang, J., On the formation of colloidally dispersed phase-separated poly(*N*-isopropylacrylamide), *Langmuir*, 15, 4018, 1999.
72. Gorelov, A.V., Du Chesne, A., and Dawson, K.A., Phase separation in dilute solutions of poly(*N*-isopropylacrylamide), *Physica*, A240, 443, 1997.
73. Aseyev, V. et al., Mesoglobules of thermoresponsive polymers in dilute aqueous solutions above the LCST, *Polymer*, 46, 7118, 2005.
74. Kuz'kina, I.F., Pashkin, I.I., Markvicheva, E.A., Kirsh, Y.E., Bakeeva, I.V., and Zubov, V.P., Hydrogel granules of poly(*N*-vinyl-cuprolactam). Preparation, properties, and uses. *Khim-Farm. Zh.*, 1, 39, 1996.
75. Markvicheva, E.A., Kuz'kina, E.F., Pashkin, I.I., Plechko, T.N., Kirsh, Y.E., and Zubov, V.P., Anovel technique for entrapment of hybridoma cells in synthetic thermally reversible polymers, *Biotechnol. Tech.*, 5, 223, 1991.

Chapter 5

Microgels from Smart Polymers

Nighat Kausar, Babur Z. Chowdhry, and Martin J. Snowden

CONTENTS

5.1 Introduction

The term "microgel" is often used to describe discrete intermolecular cross-linked polymers or products formed by heterogeneous polymerization techniques (emulsion or precipitation polymerization). Microgels are also referred to as intramolecular cross-linked macromolecules or nanogels.[1] Microgels have also been described as superabsorbent polymers because they consist of a network of flexible cross-linked molecules that exhibits an intriguing combination of the physicochemical properties of bulk solids and liquids.[2] These terms are not always restricted to cross-linked structures and cause confusion with respect to the exact nature of the system under study. In contrast to true single-phase solutions, microgels are dispersions of particles because of their typical characteristic colloidal behavior, e.g., turbidity. Colloidal microgels can be defined as a disperse phase of discrete polymeric gel particles, which are typically in the size range of 1 nm to 1 µm and uniformly dispersed in a continuous solvent medium, swollen by a good solvent (Figure 5.1).[3]

The synthesis of poly(*N*-isopropylacrylamide) (poly[NIPAM]) microgel particles by surfactant-free emulsion polymerization (SFEP) of *N*-isopropylacrylamide (NIPAM) in water at 70°C in the presence of a cross-linker, *N,N*-methylenebis-acrylamide (BA) was first reported in 1986. Since then, poly(NIPAM) microgel particles have attracted a significant amount of

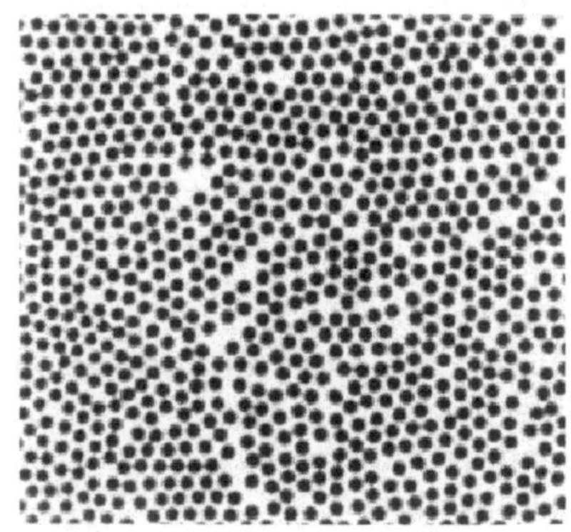

FIGURE 5.1
Transmission electron micrograph of poly(NIPAM) microgel particles with a diameter of 450 nm.

cross-disciplinary scientific and technological attention.[4] A common misconception is the use of the terms "hydrogels" and "microgels." They have similar polymer chemistry, but their physical molecular arrangements are different. Microgels, as compared with hydrogels, are differentiated as discrete gel-like particles (see Figure 5.2).[2,5]

Any material that can respond to subtle changes in environmental conditions (e.g., a change in temperature, pH, ionic strength, solvency, light intensity, or the influence of an applied electric field) could be considered a "smart" or "intelligent" material. Colloidal microgels behave like microsponges, absorbing solvated materials into the particles under one set of conditions, and then undergoing rapid conformational changes and releasing them again following environmental changes (such as temperature, pH, electric field, or ionic strength; e.g., rapid thermal and solution response).[6–8] Microgels possess many interesting physicochemical properties, including much lower viscosities than macrogels and very high surface areas. Their properties make them useful in a variety of fields, including the paint industry, in ink-jet printing, cements, enzyme immobilization, oil recovery, molecular separation, and environmentally sensitive display devices. Microgels also have potential uses as drug-delivery devices in the biomedical field and have been used as on/off molecular switches, responding to small changes in physiological conditions. The development of microgels responding simultaneously to temperature and pH stimuli is a broad and interesting area of research for the development of specific drug carriers, mainly for oral delivery.[5]

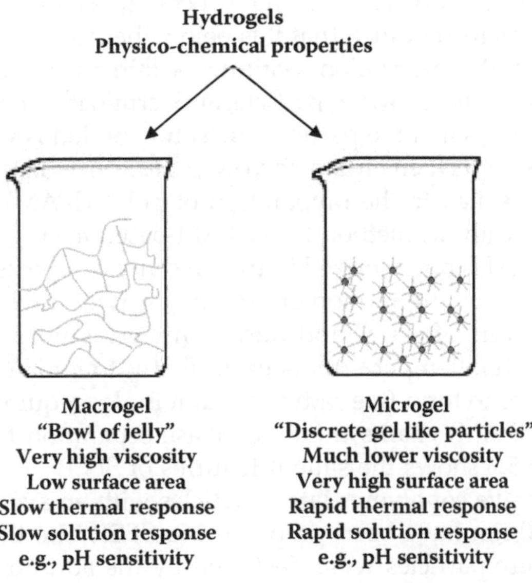

FIGURE 5.2
Illustration of the fundamental differences between macrogels and microgels.

5.2 Synthesis

Microgel particles are synthesized by the following most commonly used methods: emulsion polymerization,[9,10] inverse emulsion polymerization,[11] living free-radical polymerization,[12,13] or via methods involving the use of radiation.[14,15]

5.2.1 Emulsion Polymerization

Emulsion polymerization is a versatile technique that yields narrow particle size distributions. Emulsion polymerization can be performed in the presence of added surfactant (conventional emulsion polymerization [EP]) or in the absence of added surfactant (SFEP), also called precipitation polymerization.[8,16] EP enables preparation of small microgel particles with diameters of less than 150 nm. A problem associated with this synthetic methodology is the difficulty of completely removing residual surfactant. Obviously, SFEP does not suffer from residual surfactant contamination.

EP and SFEP typically yield microgel particles with diameters between 100 and 1000 nm. The continuous phase must have a high dielectric constant (e.g., water) and ionic initiators (e.g., $K_2S_2O_8$) in the SFEP method. Thermal decomposition of the ionic initiator $S_2O_8^{2-}$ initiates free-radical polymerization. The oligomers produced are surface active and form nuclei when the length of the oligomers exceeds the solubility limit of the solvent. The nuclei then undergo limited aggregation, thereby increasing the surface charge until electrostatic stabilization is achieved. Further particle growth occurs through absorption of monomer or oligomeric chains. This process results in a decrease in the concentration of oligomers such that it is below the critical value required for particle formation. Polymerization continues within the particles until another radical species enters the growing particle, and termination occurs. The important feature of SFEP is that the particle nucleation period is very short (on the order of minutes), which ensures a narrow particle size distribution. SFEP is, therefore, ideally suited to the preparation of poly(NIPAM). This method is now a standard synthesis method for NIPAM-based microgels.

Typically, NIPAM is polymerized in the presence of a potassium persulfate initiator and cross-linking monomer containing two vinyl groups (usually BA) at a temperature of ≈70°C and under an inert atmosphere of nitrogen. The elevated reaction temperature is used, firstly, to initiate the decomposition of the initiator to form free radicals, but it is also required to ensure that growing poly(NIPAM) chains undergo phase separation to form colloidal particles. Figure 5.3 shows the salient features of SFEP.[8]

Thermoresponsive copolymer latex particles with an average diameter of about 200 to 500 nm have been prepared via SFEP. The thermoresponsive properties of these particles were designed by the addition of hydrophilic monomers (acrylic acid [AA] and sodium acrylate [SA]) to copolymerize with NIPAM. It has been observed that the addition of hydrophilic AA or SA affected the mechanism and kinetics of polymerization.[17]

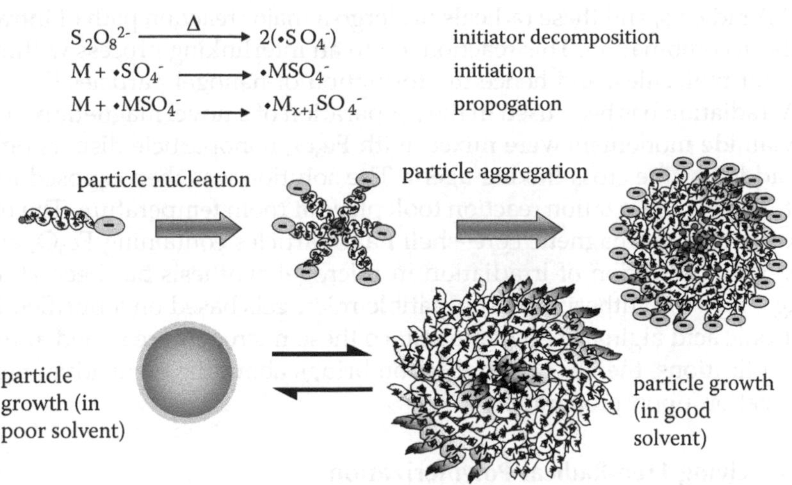

FIGURE 5.3
Mechanism for the preparation of microgel particles by SFEP. (M represents a vinyl monomer.)

5.2.2 Inverse Emulsion Polymerization

Neyret and Vincent (see Reference 6) developed a new approach, inverse microemulsion polymerization for the formation of microgel particles. The oil phase consisted of anionic 2-acrylamido-2-methylpropanesulfonate and cationic 2-meth-acryloyloxy-ethyl-trimethylammonium monomers in addition to the cross-linker BA.

The copolymerization reaction was initiated using UV irradiation and the product isolated and redispersed in aqueous electrolyte solution to yield polyampholyte microgel particles. The particles swelled in the presence of high electrolyte concentrations as a result of screening of the attractive electrostatic interactions between neighboring chains.[8] An inverse emulsion polymerization is one of the most convenient ways to synthesize water-swellable microgels, particularly for the polymerization of a monomer that dissolves only in aqueous media. A novel inverse microemulsion polymerization method using a nonionic surfactant has been developed recently to synthesize poly(dimethylacrylamide-*co*-2-acrylamide-2-methyl-1-propane-sulfonic acid) cross-linked microgel particles.[18]

5.2.3 Radiation Polymerization

Radiation techniques are very suitable tools for the synthesis of gels due to additive-free initiation and easy process control.[14] Radiation has been employed for the synthesis of various microgel dispersions. An aqueous solution of linear poly(acrylic acid) (PAA) was exposed to pulse irradiation produced by fast electrons. The irradiation energy facilitates the formation of PAA radicals, and these radicals undergo a major reaction path of intramolecular recombination. This reaction led to an interlinking process within the polymer molecules, and hence the formation of nanogel particles.[13]

UV radiation has been used in the preparation of a novel magnetic nanogel. Acrylamide monomers were mixed with Fe_3O_4 nanoparticle dispersion. BA was added as the cross-linking agent. The solution was then exposed to UV light. The polymerization reaction took place at room temperature. The result took the form of magnetic core–shell nanoparticles containing Fe_3O_4 as the core. The application of irradiation in microgel synthesis has been studied using γ rays to synthesize biocompatible microgels based on a purified high guluronic acid alginate copolymer. Since these microgels are aimed at medical applications, the use of γ irradiation brings about the extra advantage of sterilization upon preparation.[19,20]

5.2.4 Living Free-Radical Polymerization

Conventional free-radical polymerization does not present problems associated with the controlled polymerization methodologies, and reactions can be carried out in a single step, using a wide range of monomers and under a wide range of experimental conditions. The ability to form starlike structures using a conventional free-radical polymerization method has potential for economically viable industrial-scale microgel products.[21] The network formation of cross-linked polymer microgels made via free-radical polymerization mechanism is significantly influenced by the polymerization conditions (e.g., monomer concentration, ionic strength, and pH).[22] It has been found that living free-radical polymerization provides much better control over the formation of statistical microgels than traditional free-radical polymerization and can be used successfully for the synthesis of statistical and star microgels using divinyl monomer, which previously posed the problem of gelation during traditional polymerization. Moreover, this method is reported to allow better control over the molecular-weight properties of the polymers and has potential for the synthesis of star microgels with "tunable" physicochemical properties[15] (Figure 5.4).

Star microgels have been produced using a living linear polymer as the arms of the microgel structure, which was prepared first. The living polymer

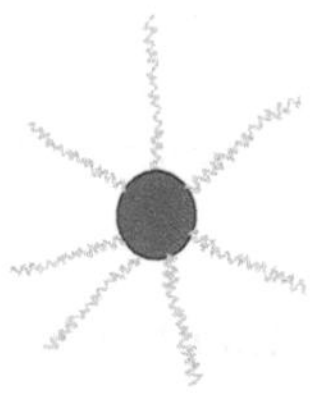

FIGURE 5.4
Schematic illustrating the structural differences between statistical and star microgels.

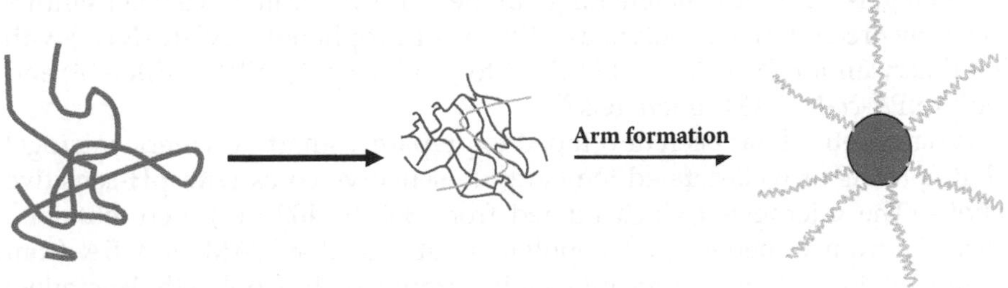

FIGURE 5.5
Schematic illustration of the formation of the starlike microgels.

is then reacted with a divinyl cross-linker to form a star microgel consisting of a central core surrounded by linear polymeric arms. Experimental details are reported elsewhere[4] (Figure 5.5).

Magnetic nanogels with amino groups, or poly(acrylamide-vinyl amine) magnetic nanogels, possessing core–shell diameters in the range of 25 to 180 nm (with narrow size distributions) were prepared by a photochemical method in a surfactant-free aqueous system. The system included the use of superparamagnetic Fe_3O_4 nanoparticles. Several methods have been developed to prepare magnetic microgels and nanogels, such as inverse microemulsion polymerization and EP. It is not easy to obtain magnetic nanogels with controllable particle size below 200 nm using conventional methods, which are desirable for *in vivo* systemic circulation applications. Moreover, the thermochemical polymerization procedures usually involve surfactants and initiators, giving rise to problems of toxic residual additives in the final microgel products used as drug carriers and in other biomedical applications.[19,21]

A one-pot free-radical polymerization process has been used to prepare methyl acrylate/ethyleneglycol-dimethacrylate and methyl methacrylate/ethyleneglycol-dimethacrylate polymers. In this work, a simple and cost-effective synthetic route for the production of polymers with high molecular weight and low viscosity, with considerable potential for industrial-scale processing, was highlighted.[21,22]

Cylindrical poly(NIPAM)-based microgels have been synthesized successfully by a novel strategy in which template-guided synthesis and photochemical polymerization were combined. Cylindrical poly(NIPAM)-based microgels of relatively uniform sizes were easily obtained corresponding to the pore size of the template phosphatidylcholine membranes.[23]

Novel molecularly imprinted microgels incorporating arginine or tyrosine side chains as functional monomers have been designed and synthesized with percentages of cross-linker ranging from 70 to 90%. Kinetic characterization of the catalytic activity of the different preparations indicated that a critical monomer concentration (C_m) and percentage of cross-linker play an important role in determining the catalytic efficiency of the different preparations.[24]

Microgels whose diameters range in the nanosize scale have been synthesized by precipitation copolymerization of 4-nitrophenol acrylate (NPA) with methacrylamide (MeAM) and NIPAM to produce poly(NPA-*co*-MeAM) and poly(NPA-*co*-NIPAM) microgels.[25]

A new method has been developed to prepare a smart polymeric microgel that consists of well-defined temperature-sensitive cores with pH-sensitive shells. The microgels (which ranged from 300 to 400 nm) were obtained directly from aqueous graft copolymerizations of NIPAM and BA from water-soluble polymers containing amino groups such as poly(ethyleneimine) and chitosan. The unique core–shell nanostructures, which had narrow size distributions, exhibited tunable responses to pH and temperature.[26]

Novel pH-responsive microgels (250 to 700 nm) have been synthesized by EP of 2-(diethylamino)ethyl methacrylate (DEA) in the presence of a bifunctional cross-linker at pH 8 to 9. Both batch and semicontinuous synthesis were explored using thermal and redox initiators. The PDEA-based microgels were compared with poly(2-[di-isopropylamino]ethyl methacrylate) (PDPA) microgels prepared with identical macromonomer stabilizers that showed a lower critical swelling pH at pH 5.0 to 5.5. The kinetics of swelling for the PDPA microgels was slower than that observed for PDEA microgels.[27]

The following synthetic challenges remain[28]:

Removal of sols, surfactants, and other impurities

Obtaining colloidally stable particles of less than 50 nm in diameter

Synthesis of concentrated, colloidally stable microgel dispersions

Accurate measurement of water content

Control of microgel particle morphology

Spherical silver-poly(4-vinylpyridine) hybrid microgels have been successfully synthesized in a single step by γ irradiation and SFEP. This hybrid microgel dispersion exhibited a peculiar photoluminescence phenomenon that might arise from the monomer emission of the pyridine group enhanced by Ag nanoparticles.[29]

5.3 Properties of Microgels

Temperature-sensitive polymers exhibit a critical solution temperature (CST) behavior where phase separation is induced by surpassing a certain temperature. Polymers of this type undergo a thermally induced, reversible phase transition; they are soluble in a solvent (water) at low temperatures but become insoluble as the temperature rises above the CST.[30]

The liquid–liquid phase diagram at constant pressure of binary polymer solutions is usually determined by plotting the temperature of incipient phase separation as a function of the overall polymer concentration. Although the solution is homogeneous at low temperature, a macroscopic phase separation appears when the temperature exceeds the CST or the *cloud point* of the mixture. The minimum in the phase diagram (known also as cloud-point curve) is called the precipitation threshold, or lower critical solution temperature (LCST), since it denotes the extreme temperature at which phase separation can occur at all.

Figure 5.6 shows a typical curve of cloud point vs. composition that one might find for a thermoresponsive system. The right-hand branch of the experimental phase diagram defines the composition of the polymer that precipitates at various temperatures. It can also be viewed as the equilibrium swelling ratio of the polymer solvent system as a function of the temperature. In a CST system, the right-hand branch of the curve is characterized by a positive slope, indicating that the polymer (or gel) will precipitate (collapses) as the temperature increases.

A phase transition from the liquid to the crystalline and finally to the glassy state is observed with increasing volume fraction (φ_{eff}), but phase transitions in poly(NIPAM) microgels occur at very different volume fractions compared with model hard-sphere suspensions. The shift of the volume fraction when crystallization occurs toward higher volume fractions is related to the softness of the microgel particles[31,32] (Figure 5.7).

5.3.1 Swelling Behavior of Microgels

Microgels are particularly temperature-sensitive if most of the polymer in the microgel network displays temperature-sensitive phase behavior in the

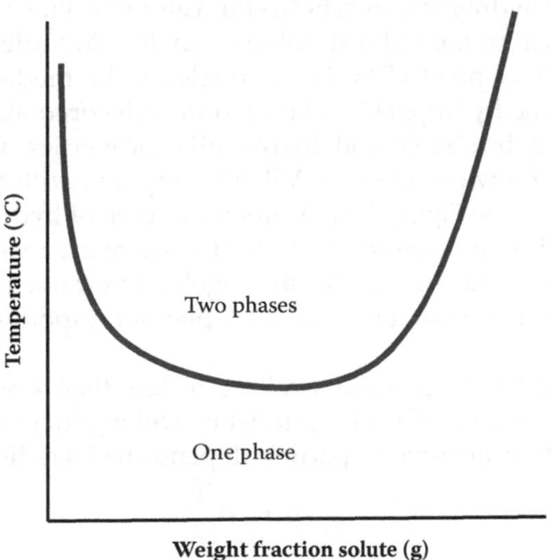

FIGURE 5.6
Phase diagram of a system exhibiting a LCST.

FIGURE 5.7
Phase behavior of colloidal poly(NIPAM) microgel particles. At $\varphi_{eff} < 0.59$, the microgel particles are disordered in the fluid phase (left); crystallization starts at $\varphi_{eff} < 0.59$ (center); and at high volume fractions the glassy state is reached (right).

swelling solvent. Poly(NIPAM) gels contain temperature-sensitive sponge-like particles that can swell/deswell in response to changes in temperature and undergo a large-magnitude volume change near the volume phase transition temperature (VPTT). The mechanism of the VPT of the poly(NIPAM) microgels is generally attributed to the reversible formation and breakage of hydrogen bonds between water molecules and hydrophilic amide groups within the NIPAM polymer chains. According to this mechanism, the VPTT of poly(NIPAM) microgels can be modulated by tuning the hydrophilic and hydrophobic balance through incorporation of monomers with different hydrophobic and hydrophilic properties. Copolymer-type microgels exhibit a wider range of VPTTs compared with their homopolymer counterparts (see Figure 5.8). A greater degree of hydrophobicity does not always result in a decrease of the VPTT. There are many other factors, e.g., the network structure of the microgels, determined by the relative reactivity of the monomer, etc., that also play an important role in determining the VPTT.[16]

Microgels achieve steady-state swelling in less than a second when the temperature is changed. The characteristic swelling/deswelling relaxation time (t_c) for a spherical microgel particle depends on its radius (R) as follows:

$$t_c \propto \frac{R^2}{D}$$

where D is the diffusion coefficient of the polymer chain in the gel network.[33]

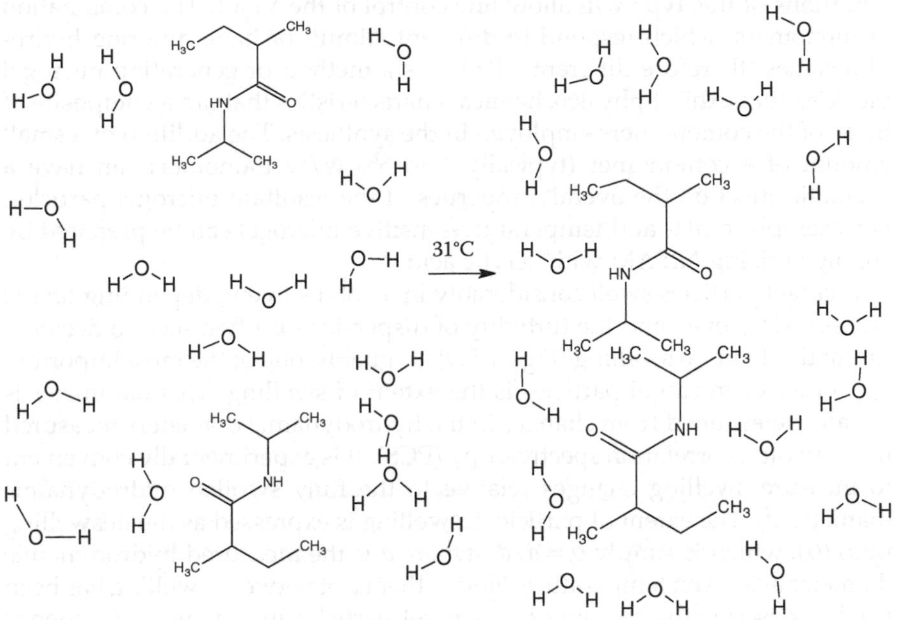

FIGURE 5.8
Volume phase transition in poly(NIPAM) microgels.

5.3.2 Rheological Properties

The rheological behavior of homogeneous poly(NIPAM) microgels as a function of temperature, shear rate, and concentration has been investigated. The elastic modulus decreases by an order of magnitude between 28 and 50°C, reflecting a decrease in the effective volume fraction of the dispersed gel phase.[11]

5.3.3 Electrical Properties

Poly(NIPAM) microgel dispersions show interesting electrical properties due to the presence of covalently bonded electrically charged groups originating from the initiator. Poly(NIPAM) microgels provide an extensive database of electrophoretic mobility values as functions of temperature, electrolyte, and surfactant concentration.[11]

5.3.4 Effect of Monomer on Microgel Properties

There is considerable scope for variation of microgel properties by changes in the structure of the monomers. For example, *N*-vinyl-*n*-butyramide is isomorphous with NIPAM and is a good candidate for microgel preparation.

Variations of this type will allow fine control of the VPTT. The combination of monomers, which respond to different stimuli or have differing hydrophobicities (therefore different VPTTs), is a method of generating microgel particles that exhibit physicochemical characteristics that are a composite of those of the comonomers employed in the synthesis. The addition of a small amount of a comonomer (typically 1 to 5% w/w monomer) can have a dramatic effect on the overall properties of the resultant microgel particles. For example, a pH- and temperature-sensitive microgel can be prepared by copolymerizing NIPAM with acrylic acid.[34]

Microgel particles swell considerably in a good solvent, depending on the cross-linking level, and the turbidity of dispersions is often seen to decrease dramatically upon swelling (Figure 5.9). Arguably, one of the most important properties of microgel particles is the extent of swelling. This parameter is usually determined from changes in the hydrodynamic diameters measured using photon correlation spectroscopy (PCS). It is experimentally convenient to measure swelling changes relative to the fully swollen hydrodynamic diameter, d_0. The extent of particle deswelling is expressed as the deswelling ratio (α), which is simply $\alpha = d/d_0$, where d is the measured hydrodynamic diameter at a given temperature. Flory's theory of network swelling has been used to describe the swelling of microgel particles in organic and aqueous systems. A polymer network immersed in a good solvent imbibes solvent to balance the solvent chemical potential inside and outside the gel network; the presence of cross-links restricts the extent of swelling. Thus, swelling continues until the sum of the elastic forces between cross-links is equal to the osmotic force. The extent of network swelling is usually described by the polymer volume fraction obtained at equilibrium. The electrophoretic mobility is an important property of microgel particles that is frequently reported in the literature. If the frictional coefficient of the segments in the swollen microgel particles is known, then the space charge density can be derived.[5]

A new spherically shaped poly(*N*-isopropylacrylamide)/poly(ethylene-glycol) microgel was used as an additive during the polymerization and gelation process of poly(NIPAM) hydrogels at room temperature for the

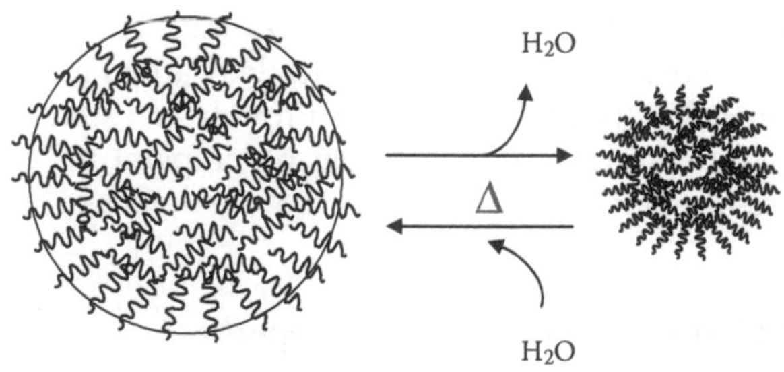

FIGURE 5.9
A microgel particle dispersed in (a) good solvent and (b) poor solvent.

purpose of improving their thermosensitivity without any significant alteration to their mechanical properties. The microgel-impregnated poly(NIPAM) hydrogels exhibited a tighter and more constrained porous network than pure poly(NIPAM) hydrogels without microgel additives (Figure 5.10). The pore size of the microgel-impregnated poly(NIPAM) hydrogels was reduced with an increase in the amounts of impregnated microgel. The mechanical property of microgel-impregnated poly(NIPAM) hydrogels was significantly higher than pure poly(NIPAM) hydrogels. Compared with a normal poly(NIPAM) hydrogel, VPTTs of the microgel-impregnated poly(NIPAM) hydrogels did not change because of the same chemical nature shared by the microgels and their surrounding poly(NIPAM) matrix. The microgel additives greatly improved the thermosensitive properties of poly(NIPAM) hydrogels in terms of swelling ratio, shrinking rate, swelling rate, as well as the oscillatory shrinking-swelling kinetics upon temperature changes around the VPTT; the level of improvement depended on the amounts of the microgel impregnated. Such improvements in mechanical and thermosensitive properties in the microgel-impregnated poly(NIPAM) hydrogels may find applications in the biomedical and biotechnology fields.

This faster and larger magnitude of oscillating responses from microgel-impregnated poly(NIPAM) hydrogel is advantageous for practical applications in fields such as bioengineering and biotechnology because the faster response kinetics of the oscillating shrinking-swelling property of hydrogels to small temperature cycles (e.g., cycled around the physiological temperature) should be useful.[35]

5.3.5 pH Sensitivity

The mechanism controlling the swelling of microgel particles containing pH-responsive groups is governed by the internal osmotic pressure attributed to the mobile counterions contained within the particles, which balance the internal electrostatic repulsion. The swelling of microgels is determined by

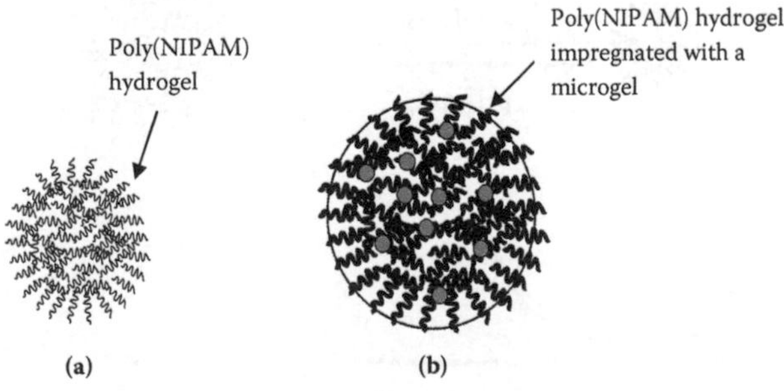

FIGURE 5.10
The difference in network structures of (a) the normal poly(NIPAM) hydrogel and (b) modified poly(NIPAM) hydrogel with a microgel additive.

the balance between the osmotic pressure inside the microgels and the osmotic pressure outside, as described by Equation 5.1.

$$\Pi_{in} + \Pi_{el} = \Pi_{out} \tag{5.1}$$

where Π_{in} and Π_{out} are, respectively, the osmotic pressure of the mobile ions inside the microgels and bulk solution, and Π_{el} is the elastic pressure of the polymeric network.

The behavior of microgel particles containing a weak acid or base, i.e., pH-dependent groups, is more complex. The ionization of the sample is governed by the pK_a or pK_b of the groups concerned, but these parameters are functions of the local charge group chemistry (a higher charge density suppresses ionization) and background ionic strength (screens local electrostatic repulsion). Irrespective of the addition of any inert electrolyte, adjustment of pH inevitably leads to changes in background ionic strength.

Incorporation of co-monomers containing acidic or basic functionalities into poly(NIPAM) microgels yields particles with pH-driven swelling. The preparation of poly(NIPAM) microgel particles containing acrylic acid groups has been reported. It was found that the hydrodynamic diameter (volume change) of the microgel particles increased with a corresponding rise in pH ($pH > pK_a$). At a pH value above that of the pK_a, the polymer chains of the microgel become ionized, forcing the microgel to adopt a more extended conformation as a result of intramolecular charge repulsion. At a pH below the pK_a, the microgel particles adopt a compact structure.[36,37] Figure 5.11 shows the shrinking and swelling of a poly(NIPAM/acrylic acid) microgel.

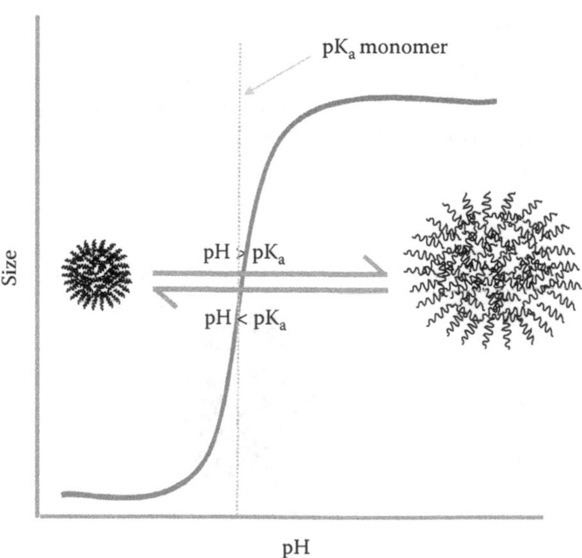

FIGURE 5.11
The shrinking and swelling behavior of a poly(NIPAM/acrylic acid) microgel.

5.3.6 Polyelectrolyte Microgels

Polyelectrolyte microgels are soft particles due to their ability to deswell osmotically. The flow properties of suspensions and pastes are universal. It is the osmotic pressure associated with the high translational entropy of the small ions that is the origin of the softness of the individual particles in polyelectrolyte microgels. This softness controls most of the flow properties of microgel dispersions. It is possible to obtain the desired macroscopic behavior by tuning the microscopic structure of the individual microgel particles.

5.3.7 Swelling and Elasticity of Polyelectrolyte Microgels

It is a challenge to describe the swelling of polyelectrolyte systems, since the equilibrium swelling arises from a delicate balance of several factors. The swelling of polyelectrolyte gels mainly results from the combined action of three contributions:

1. The mixing entropy of the polymer
2. The osmotic pressure exerted by the counterions trapped in the network against the ions in the solution
3. The configurational elasticity of the network

Suspensions of polyelectrolyte microgels share several features common to other pastes and slurries: elasticity, yielding behavior, and aging.[2,38] The macroscopic flow properties exhibit remarkable universality. The cross-linked polymers and copolymers possess high and fast swelling properties in aqueous media.[39,40]

A study of the internal structure of poly(NIPAM/xBA) (where x is the wt% of BA = 1 to 10) microgel particles has been carried out. In this study, affine swelling (i.e., the macroscopic hydrodynamic diameter of the microgel particles and the submicroscopic mesh size of the network are linearly related) for poly(NIPAM/xBA) microgels was shown to apply for the first time. The data suggest that the swollen particles have a mostly homogeneous structure, although evidence for a thin, low segment density shell is presented. It has been confirmed by this study that poly(NIPAM/xBA) microgel particles have a core–shell structure.[41]

The effects of L-phenylalanine (L-Phe) on the synthesis of poly(*N,N'*-methylenebisacrylamide-*co*-4-vinylpyridine) (poly[bis-*co*-4-VP]) microgels by γ-ray irradiation has been studied. The addition of L-Phe not only decreased the gelation dose (D_g) of the synthesis, but also transformed the morphology of the copolymer from a microgel to a gel. In addition, the swelling ability of the (micro)gels was affected by the presence of L-Phe. The decrease of D_g was ascribed to the effect of pH, while the transformation of the morphology was ascribed to the effect of L-Phe on the stability of the poly(bis-*co*-4-VP) microgel. Such an effect was further confirmed by comparison with the effects of L-alanine, L-glutamic acid, L-arginine, sulfuric acid, and aqueous ammonia.[42]

5.4 Characterization

A variety of techniques for characterizing microgels have been reported in the literature (Table 5.1). The most commonly used methods include dynamic light scattering (DLS), photon correlation spectroscopy (PCS), differential scanning calorimetry (DSC), small-angle neutron scattering (SANS), transmission electron microscopy (TEM), isothermal titration calorimetry, rheology, and nuclear magnetic resonance (NMR). Some of the most recently used methods for microgel characterization are described in the following subsections. Surfactant and microgel interactions have been investigated using SANS, DLS, NMR, and fluorescence probe studies.[4,25,26] A good tool to obtain information about the size of microgel particles and the polydispersity of microgel is scanning electron microscopy (SEM). The structure of nanogels has been analyzed by using field emission scanning electron microscopy.[1,6]

5.4.1 Static Light Scattering

Static light scattering (SLS) measures and analyzes the time-averaged scattering intensities. The method is often used to determine microscopic properties of particles, such as the z-average radius of gyration (R_g), the weight-average molecular weight ($\bar{M}_w$), and the second virial coefficient (A_2):

$$\frac{Kc}{R_\theta} = \frac{1}{M_w}\left[1 + 16\pi^2 n^2 \left\langle R_g{}^2 \right\rangle \sin^2\left(\frac{\theta}{2}\right) \Big/ 3\lambda^2\right] + 2A_2 c$$

where the Rayleigh ratio

$$R_\theta = \left(I_s r^2 \Big/ I_t \sin\theta \right)$$

$$K = 4\pi^2 n^2 \left(\frac{dn}{dc}\right)^2 \Big/ (N_A \lambda^4)$$

c is the concentration of the polymer solution; n is the refractive index of the solvent; θ is the angle of measurement; λ is the wavelength of laser light; and N_A is Avogadro's constant.

TABLE 5.1

Commonly Used Characterization Techniques for Microgels

Experimental Techniques	Measurement	References
Transmission electron microscopy (TEM)/FF-TEM	particle size, shape, diameter	1–6
Scanning electron microscopy (SEM)	particle size, shape	7–10
Atomic force microscopy (AFM)	surface morphology	11–13
Ultraviolet-visible spectroscopy (UV-Vis)	stability, kinetics, concentration	14–24
Quasi-elastic light scattering (QELS)	average particle diameter, particle size distribution	25
Fluorescence resonance energy transfer (FRET)	swelling behavior	26–27
Static light scattering (SLS)	particle molecular weight	28–30
Photon correlation spectroscopy/ dynamic light scattering (PCS/DLS)	hydrodynamic size	31–32
Raman spectroscopy	structural/quantitative analysis of composition	33
Small-angle X-ray scattering (SAXS)	internal structure	34
Small-angle neutron scattering (SANS)	internal structure	35–36
Turbidimetric methods	stability of microgel dispersions	40–44
Thermogravimetric analysis (TGA)	thermal properties	33
High-sensitivity differential scanning calorimetry (HSDSC)	thermodynamic properties	34
Differential scanning calorimetry (DSC)	thermodynamic properties	35–36
Ultracentrifugation	weight-average molecular weight	37
Conductometric and potentiometric titration	surface charge	38
Isothermal titration calorimetry (ITC)	thermodynamic properties	39
Rheometer	rheology	40–41
^{1}H-NMR (nuclear magnetic resonance)	structural analysis	42–44
Gel permeation chromatography (GPC)	weight/number-average molecular weight, polydispersity	45–47

References (Table 5.1):

1. Eke, I., A new, highly stable cationic-thermosensitive microgel: uniform isopropylacrylamide-dimethylaminopropylmethacrylamide copolymer particles, *Colloid. Surf. A*, 279, 247, 2006.
2. Kim, K.S. et al., Colloidal behavior of polystyrene-based copolymer microgels composed of a hydrophilic hydrogel layer, *J. Indust. Eng. Chem.*, 12, 91, 2006.
3. Kim, K.S. et al., pH and temperature-sensitive behavior of poly(*N*-isopropylacrylamide-*co*-methacrylic acid) microgels, *J. Indust. Eng. Chem.*, 11, 736, 2005.
4. Ma, X.M., Influence of ethyl methacrylate content on the volume-phase transition of temperature-sensitive poly[(*N*-isopropylacrylamide)-*co*-(ethyl methacrylate)] microgels, *Polym. Int.*, 54, 83, 2005.
5. Perez, L., Synthesis and characterization of reactive copolymeric microgels, *Polym. Int.*, 54, 963, 2005.
6. Kaneda, I., Water-swellable polyelectrolyte microgels polymerized in an inverse microemulsion using a nonionic surfactant, *J. Colloid Interface Sci.*, 275, 450, 2004.

7. Zhang, X.Z., Fabrication and characterization of microgel-impregnated, thermosensitive PNIPAAm hydrogels *Polymer*, 46, 9664, 2005.
8. Rehim, H.A. et al., Swelling of radiation crosslinked acrylamide microgels and their potential applications, *Radiat. Phys. Chem.*, 74, 111, 2005.
9. Li, X. et al., Preparation and characterization of narrowly distributed nanogels with temperature-responsive core and pH-responsive shell, *Macromolecules*, 37, 10042, 2004.
10. Zhang, Y., Preparation of spherical nanostructured poly(methacrylic acid)/PbS composites by a microgel template method, *J. Colloid Interface Sci.*, 272, 321, 2004.
11. Gotzamanis, G.T. et al., Cationic telechelic polyelectrolytes: synthesis by group transfer polymerization and self-organization in aqueous media, *Macromolecules*, 39, 678, 2006.
12. Gattas-Afura, K.M., Cinnamate-functionalized gelatin: synthesis and "smart" hydrogel formation via photo-cross-linking, *Biomacromolecules*, 6, 1503, 2005.
13. Serre, C. et al., Atomic force microscopy, a powerful tool to study blend morphologies based on polyester resins, *J. Mater. Sci.*, 36, 113, 2001.
14. Ma, X.M., Influence of ethyl methacrylate content on the volume-phase transition of temperature-sensitive poly[(*N*-isopropylacrylamide)-*co*-(ethyl methacrylate)] microgels, *Polym. Int.*, 54, 83, 2005.
15. Xiao, Y., Study on the interaction of poly (*N*-isopropylacrylamide-*co*-acrylic acid) microgel with terbium (III), *Acta Polymerica Sinica*, 3, 458, 2005.
16. Perez, L., Synthesis and characterization of reactive copolymeric microgels, *Polym. Int.*, 54, 963, 2005.
17. Jones, C.D. et al., Characterization of cyanine dye-labeled poly(*N*-isopropylacrylamide) core/shell microgels using fluorescence resonance energy transfer, *J. Phys. Chem. B*, 108, 12652, 2004.
18. Boyko, V., Thermo-sensitive poly(*N*-vinylcaprolactam-*co*-acetoacetoxyethyl methacrylate) microgels, 1: synthesis and characterization, *Polymer*, 44, 7821, 2003.
19. Tan, B.H., Osmotic compressibility of soft colloidal systems, *Langmuir*, 21, 4283, 2005.
20. Pyett, S., Structures and dynamics of thermosensitive microgel suspensions studied with three-dimensional cross-correlated light scattering, *J. Chem. Phys.*, 122, 34709, 2005.
21. Tan, B.H., Dynamics and microstructure of charged soft nano-colloidal particles, *Polymer*, 45, 5515, 2004.
22. Lopez, V.C. et al., The use of colloidal microgels as a (trans) dermal drug delivery system, *Int. J. Pharm.*, 292, 137, 2005.
23. Kim, K.S. et al., pH and temperature-sensitive behavior of poly(*N*-isopropylacrylamide-*co*-methacrylic acid) microgels, *J. Indust. Eng. Chem.*, 11, 736, 2005.
24. Pyett, S., Structures and dynamics of thermosensitive microgel suspensions studied with three-dimensional cross-correlated light scattering, *J. Chem. Phys.* 122, 34709, 2005.
25. Yan, H. et al., Template-guided synthesis and individual characterization of poly(*N*-isopropylacrylamide)-based microgels, *Langmuir*, 21, 7076, 2005.
26. Baek, K.Y., Star-shaped polymers by Ru(II)-catalyzed living radical polymerization, 2: effective reaction conditions and characterization by multi-angle laser light-scattering/size-exclusion chromatography and small-angle X-ray scattering, *J. Pol. Sci. Part A, Pol. Chem.*, 40, 2245, 2002.
27. Seelenmeyer, S. et al., Small-angle X-ray and neutron scattering studies of the volume phase transition in thermosensitive core-shell colloids, *J. Chem. Phys.*, 114, 10471, 2001.
28. Cabarcos, E.L., Small angle neutron scattering study of the structural modifications induced by the entrapped glucose oxidase in polyacrylamide microgels, *Physica A*, 344, 417, 2004.
29. Stieger, M., Structure formation in thermoresponsive microgel suspensions under shear flow, *J. Phys. Cond. Mater.*, 16, 38, 2004.
30. Saunders, B.R. et al., On the structure of poly(*N*-isopropylacrylamide) microgel particles, *Langmuir*, 20, 3925, 2004.
31. Tan, B.H., Osmotic compressibility of soft colloidal systems, *Langmuir*, 21, 4283, 2005.
32. Amalvy, J.I. et al., Synthesis and characterization of novel pH-responsive microgels based on tertiary amine methacrylates, *Langmuir*, 20, 8992, 2004.

33. Perez, L. et al., Synthesis and characterization of reactive copolymeric microgels, *Polym. Int.*, 54, 963 2005.
34. Woodward, N.C. et al., Small angle neutron scattering study of the structural modifications induced by the entrapped glucose oxidase in polyacrylamide microgels, *Eur. Polym. J.*, 36, 1355, 1999.
35. Ma, X.M., Deswelling comparison of temperature-sensitive poly(*N*-isopropylacrylamide) microgels containing functional-OH groups with different hydrophilic long side-chains, *J. Polym. Sci. Part B, Polym. Phys.*, 43, 3575, 2005.
36. Perez, L., Synthesis and characterization of reactive copolymeric microgels, *Polym. Int.*, 54, 963, 2005.
37. Kuckling, D. et al., Preparation and characterization of photo-cross-linked thermosensitive PNIPAAm nanogels, *Macromolecules*, 39, 1585, 2006.
38. Kratz, K. et al., Influence of charge density on the swelling of colloidal poly(*N*-isopropylacrylamide-*co*-acrylic acid) microgels, *Colloid. Surf. A*, 170, 137, 2000.
39. Seidel, J. et al., Isothermal titration calorimetric studies of the acid–base properties of poly(*N*-isopropylacrylamide-*co*-4-vinylpyridine) cationic polyelectrolyte colloidal microgels, *Thermochimica Acta*, 414, 47, 2004.
40. Kaneda, I., Water-swellable polyelectrolyte microgels polymerized in an inverse microemulsion using a nonionic surfactant, *J. Colloid Interface Sci.*, 275, 450, 2004.
41. Tan, B.H., Osmotic compressibility of soft colloidal systems, *Langmuir*, 21, 4283, 2005.
42. Camerlynk, S., Control of branching vs. cross-linking in conventional free radical copolymerization of MMA and EGDMA using CoBF as a catalytic chain transfer agent, *J. Macromol. Sci. Phys.*, B44, 881, 2005.
43. Martin, B.S., Microscopic signature of a microgel volume phase transition, *Macromolecules*, 38, 10782, 2005.
44. Amalvy, J.I., Synthesis and characterization of novel pH-responsive microgels based on tertiary amine methacrylates, *Langmuir*, 20, 8992, 2004.
45. Ho, A.K. et al., Synthesis and characterization of star-like microgels by one-pot free radical polymerization *Polymer*, 46, 6727, 2005.
46. Gurr, P.A. et al., Synthesis, characterization, and direct observation of star microgels, *Macromolecules*, 36, 5650, 2003.
47. Biffis, A. et al., Efficient aerobic oxidation of alcohols in water catalysed by microgel-stabilised metal nanoclusters, *J. Catal.*, 236, 405, 2005.

5.4.2 Dynamic Light Scattering

Dynamic light scattering (DLS) provides information regarding the overall size and shape of mesoscopic particles in solution. DLS measures the intensity fluctuations with time and correlates these fluctuations to the properties of the scattering objects. In general, the terms of correlation functions of dynamic variables are always used to describe the response of the scattering molecules to the incident light.[43] The translational diffusion coefficient (*D*) can be calculated from the following expression:

$$\Gamma = Dq^2$$

where Γ is the decay rate, which is the inverse of the relaxation time (τ), and q is the scattering vector.

If the Stokes-Einstein equation is used, the apparent hydrodynamic radius (R_h) can be calculated using the relation

$$R_h = \frac{kT}{6\pi\eta_s D}$$

where k is the Boltzmann constant, T is the absolute temperature, and η_s is the solvent viscosity. The hydrodynamic radius of the microgel particles can be determined using DLS. The value of R_h is calculated from the decay constant, which is obtained by a second-order fit of the cumulant analysis.[33] DLS has been used to monitor the temperature-induced swelling/deswelling process and polydispersity of microgels; however, swollen particle diameters often approach 1000 nm, which is outside the sensitive range for DLS. Furthermore, swelling changes are usually computed as volume changes, so the DLS diameters must be cubed, which limits the accuracy of the volumetric data.

DLS measurements give particle size, whereas the concentrations of polymer and water in the gel are usually the required quantities. However, if the average mass of polymer per gel particle is known, then the diameters are easily converted to absolute measures of swelling. The first DLS measurements for poly(NIPAM) microgels as functions of temperature and electrolyte concentration were carried out by Pelton.[9]

The influence of the cross-linker type on the swelling behavior and the local structure of colloidal poly(NIPAM) microgel particles has been investigated by DLS and SANS. The cross-linkers ethyleneglycol-dimethacrylate and triethyleneglycol-dimethacrylate are found to form particles with larger swelling/deswelling capacities compared with BA cross-linked particles at identical cross-linker concentration.[38]

A novel series of hydrophobically modified poly(NIPAM) microgels was prepared by using isopropyl-methacrylate (iPMA) as a comonomer. The deswelling behavior of the microgels was investigated by DLS technique and turbidimetric method. Experimental results show that the VPTT of the poly(NIPAM) microgels can be tuned successfully by incorporating different amounts of iPMA[16] (see Figure 5.12).

5.4.3 NMR Spectroscopy

Nuclear magnetic resonance is another useful technique for the characterization of microgel particles. The technique provides information regarding the internal environment and thermoresponsive behavior of a microgel particle.[21] The dynamics of the solvent inside swelling microgel particles has been studied using ^{1}H-NMR. NMR measurements indicate that solvent dynamics is controlled only by the degree of swelling of the microgel particles. In the collapsed state, water molecules still exist inside the microgel particles, being strongly confined within the polymer network. The variation of solvent relaxation is a consequence of the change in particle size, indicating

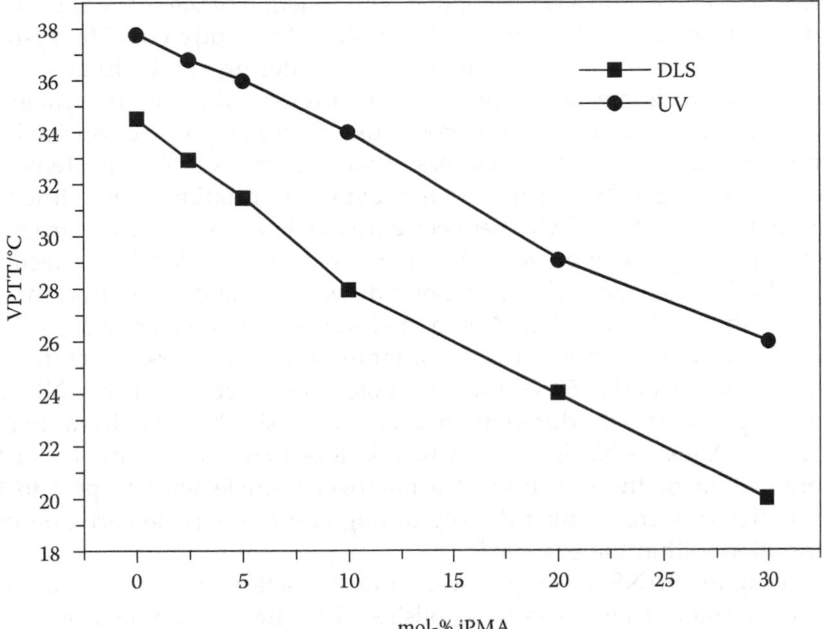

FIGURE 5.12
Dependence of VPTT on the mol % of iPMA in polymer microgels.

that the VPT and the microscopic dynamics of solvent are directly related. At the volume transition, the microgel causes a change in the confinement state of water bounded at the polymer network due to the increasing polymer–polymer interactions above the VPTT.

The flocculation behavior of triethyleneglycol-methacrylate-modified temperature-sensitive poly(NIPAM) microgels in NaCl solutions has been investigated. The critical flocculation temperature of the microgels shifts to higher temperature with more incorporation of polar $-(-OCH_2CH_2-)_3-OH$ chains. The flocculation rate of the microgels at 45°C increases with the increase of chains of $-(-OCH_2CH_2-)_3-OH$, which is verified by variable-temperature ^1H-NMR spectroscopy.[44] It was concluded that the nature of the cross-linker played a negligible role in the polymerization kinetics, and overall conversion (particles plus water-soluble polymer) was determined to be 80% by NMR. The cross-linker was found to have a marginal influence on the VPTT. It was also observed that the nature of the cross-linker dramatically affects the final hydrodynamic particle size and the electrophoretic mobility behavior.[45]

5.4.4 Small-Angle Neutron Scattering

SANS has been used to study the internal structure of microgels by a number of research groups.[46,47] It involves a measurement of the scattered intensity, I, of a neutron beam with a wavelength, λ, as a function of the scattering vector, Q. It is a structural probe, commonly employed to study colloidal systems with particle sizes of 1 to 500 nm. Neutron scattering results from a short-range repulsive interaction between the neutrons and nuclei of a material. As a result of the presence of the cross-links, microgel particles do not have a uniform density at the length scales of the neutron wavelength. Temperature-dependent SANS measurements provide information about structural changes across the VPT. SANS has been employed to investigate the structural changes taking place in poly(NIPAM) particles across the VPTT as a result of thermally induced deswelling, osmotic deswelling, and co-nonsolvency. It has been shown that poly(NIPAM) particles subjected to osmotic deswelling and co-nonsolvency appear to have a more diffuse network structure than pure particles heated to 50°C. The morphology of the colloidal poly(NIPAM) microgel particles with different cross-linker density has also been investigated by SANS. SANS is also a valuable tool that can be applied to the determination of the structure of a microgel particle with respect to the distribution of a cross-linker density throughout the particle and monomer distribution within the particle.[48]

By using the SANS technique, structural characterization of polymer systems on the nanoscale can be accomplished. This method can provide important information about morphological differences between systems on different length scales. The effects of surfactant addition, monocaprin concentration, and pH on the properties of the microgel-forming polymer Carbopol 974P have been evaluated by using rheological methods and SANS. Swelling properties of temperature-sensitive core–shell microgels consisting of two thermosensitive polymers with VPTT at, respectively, 34°C in the core and 44°C in the shell have been investigated by SANS. A core–shell form factor has been employed to evaluate the structure, and the real-space particle structure is expressed by radial density profiles. Higher shell/core mass ratios lead to an increased expansion of the core at temperatures between the LCSTs, whereas a variation of cross-linker in the shell mainly affects the dimensions of the shell.[49,50]

The internal structure of thermoreversible poly(*N*′-vinylcaprolactam-*co*-*N*′-vinylpyrrolidone) microgels has been investigated by SLS/DLS and SANS. A dramatic decrease in the radius of gyration (R_g) and the hydrodynamic radius (R_h) is observed during the phase transition near 30°C.[51]

The volume transition and structure changes of the poly(NIPAM) microgel particles as a function of temperature have been studied using LS and SANS, with the objective of determining the small-particle internal structure and particle–particle interactions. It has been discovered that these particles have a uniform radial cross-linker density on either side of the transition temperature. The particle interactions change across the transition temperature. Comparison of the osmotic second virial coefficient determined by SLS at low concentrations with similar values determined from SANS at 250-times greater concentration suggests a strong concentration dependence of the interaction potential.[52]

The temperature-induced structural changes and thermodynamics of ionic microgels based on poly(acrylic acid) networks bonded with poly(ethylene oxide)-poly(propylene oxide)-poly(ethylene oxide) (PEO-PPO-PEO) (pluronic) copolymers have been studied by SANS, ultra-small-angle neutron scattering, DSC, and equilibrium swelling techniques. The neutron-scattering results indicate the formation of micelle-like aggregates within the pluronic F127-based microgel particles, while the L92-based microgels formed fractal structures of dense nanoparticles. The microgels exhibit thermodynamically favorable VPTT within certain temperature ranges due to reversible aggregation of the PPO chains, which occurs because of hydrophobic associations.[53]

Shear-induced structures of concentrated temperature-sensitive poly(NIPAM) microgel suspensions have been studied employing rheo-SANS. The interaction potential of swollen poly(NIPAM) microgels could be varied from repulsive at temperatures below the VPTT to attractive at temperatures above the VPTT. In contrast to the case for suspensions of rigid spheres, the effective volume fraction could be changed by means of temperature while the mass concentration and particle number density were kept constant. The shear-induced particle arrangements strongly depended on the particle–particle interaction potential (Figure 5.13).[54]

5.4.5 Turbidimetric Analysis

Turbidimetric analysis is a valuable technique for determining the temperature dependence of the conformational transition of the microgel dispersions and is performed using UV-visible spectroscopy. It is also used to monitor the effect of pH, concentration, and the hydrophobic and hydrophilic content of the microgel on the turbidity of the dispersion. The turbidity of microgels is largely dominated by the amount of water contained within the interstitial regions between cross-linked polymer chains, as this controls the difference in refractive index between the microgel–water macrocomplex and the bulk water. As temperature increases up to a certain point, water contained in the microgels is expelled from the microgels due to (a) the disruption of H-bonding between water and the hydrophilic amide groups and (b) the hydrophobic association interactions between isopropyl groups, leading to an increase of the refractive difference between solvent (water) and microgels.[55,56]

Poly(NIPAM) microgel particles labeled with 3-(2-propenyl)-9-(4-*N*,*N*-dimethylaminophenyl)phenanthrene (VDP) as an intramolecular fluorescent probe have been prepared by EP. The thermoresponsive behavior of the VDP-labeled poly(NIPAM) microgel particles dispersed in water has been studied by turbidimetric and fluorescence analyses. It has been suggested that the gradual shrinking of microgel particles occurs with increasing temperature, and the subsequent dramatic shrinking results in an increase in turbidity. In these three systems, the microenvironments around the fluorescent probes above the transition temperatures became more hydrophobic than those below the transition temperature, and the estimated values of microenvironmental polarity around the VDP units on their collapsed states were almost the same.[57]

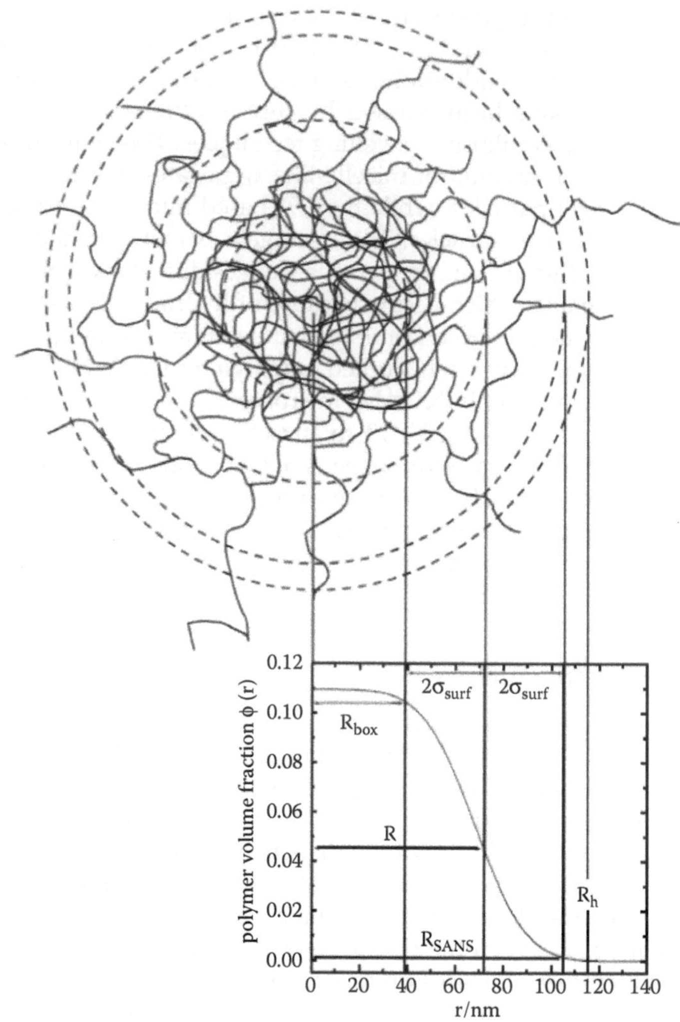

FIGURE 5.13
The overall size of poly(NIPAM) microgels obtained from rheo-SANS, which is obtained from $R_{SANS} = R + \sigma_{surf}$ (where σ_{surf} is the width of the smeared particle surface).

5.4.6 Transmission Electron Microscopy

The morphology and particle size of the poly(methacrylic acid) or poly(NIPAM) microgel particles have been characterized using a combination of TEM and PCS. The electrophoretic mobility and zeta potential for the microgel particles were found to be negative, and a marked maximum negative value occurred at around pH 5. The colloidal stability of the microgel particles was very sensitive to the concentration of the KCl solution and to the temperature.[58–60]

The shape, size dispersity, and VPT behavior of the temperature-sensitive poly([*N*-isopropylacrylamide]-*co*-[ethylmethacrylate]) (poly[NIPAM-*co*-EMA]) microgels have been investigated by TEM, UV-visible spectroscopy, DLS, and DSC. The TEM micrographs and DLS results showed that microgels with narrow size distributions were prepared. Measurements of the UV-Vis, DLS, and DSC indicate that the VPTT of the poly(NIPAM-*co*-EMA) microgels decrease with increasing incorporation of EMA, but the temperature sensitivity was impaired, causing the VPT of the microgels to become more continuous.[61]

The molecular sizes of four star-shaped microgels, synthesized by atom-transfer radical polymerization of poly(methyl methacrylate) with ethyleneglycol-dimethacrylate as the cross-linker, have been investigated by LS methods and TEM. Gel permeation chromatography right-angle laser light scattering performed with a triple detector system and DLS together confirm that individual particles of one star microgel in tetrahydrofuran and styrene solutions possess mean diameters of 31 and 18 nm, respectively. TEM reveals that another star microgel appears as discrete particles measuring 18 to 30 nm in diameter when added to styrene monomer subsequently polymerized to form a solid matrix. Additional TEM examination of a star microgel deposited by dilute-solution casting provides direct evidence for the existence of individual microgel molecules with a mean (dry) diameter of ≈15 nm.[62]

A new family of microgel particles that contains methylmethacrylate, ethylacrylate, acrylic acid (AA), glycerol propoxytriacrylate, and Ernulsogen (Ern) has been investigated. TEM and PCS data reveal that the extent of microgel swelling originates from a pH-independent contribution (due to Ern) as well as a pH-dependent contribution (due to AA). The major contribution to swelling comes from pH-independent swelling.[63]

5.4.7 Scanning Electron Microscopy

The influence of the additive poly(*N*-isopropylacrylamide)/poly(ethylene glycol)-diacrylate (poly[NIPAM/PEGDA]) on the properties of poly(NIPAM) hydrogels has been investigated and characterized by SEM and DSC. The study of interior morphology by SEM revealed that microgels impregnated with poly(NIPAM/PEGDA) have tighter and more constrained porous network structures, although large cavities (of 30 to 40 μm in diameter) occupied by the microgels were sporadically distributed in this constrained network. DSC studies did not show an apparent difference in VPTT between normal and microgel-impregnated poly(NIPAM) hydrogels.[35]

The structural changes in irradiated cross-linked polyacrylamide microgels have been investigated using FT-IR (Fourier transform infrared spectroscopy) and SEM. The ability of the microgels to absorb and retain large amounts of solutions suggests their possible uses in horticulture and in hygienic products such as disposable diapers.[2]

5.4.8 Other Techniques

A number of other interesting techniques have been used to investigate macroscopic gels and may have a role in future microgel characterization. Positron annihilation lifetime spectroscopy has been used to measure the free volume in macroscopic poly(NIPAM-*co*-AM) gels. Raman microscopy has been used to image the pore structure in poly(NIPAM) macrogels; however, the resolution of this technique seems to be too low to be of interest in thermosensitive microgels based on the assumption that bound water does not freeze, and hence does not contribute to the heat of fusion.[9]

5.5 Theoretical Aspects

A theoretical understanding of the properties and behavior of microgels from a *detailed* chemical and physical viewpoint has, to date, proved elusive. One of the major reasons is that microgel particles display properties of both a liquid and a soft solid state (i.e., hard spheres covered with a layer of polyelectrolytes), which makes modeling complex and difficult. Ohshima's theory for soft particles has only been partially successful in accounting for the electrophoretic mobility of microgel particles. The foregoing has proved to be difficult because the surface of a microgel particle is diffuse; therefore, the location of a Stern plane, and hence the plane of shear, is difficult to pinpoint. This renders inappropriate the typical equations used to convert an electrophoretic mobility to a classical zeta potential (e.g., the Smoluchowski equation). Nieves and coworkers have nevertheless achieved a considerable degree of agreement between experimental and theoretical parameters for the mobility of core–shell microgels by combining Ohshima's model with (1) particle size changes and their influence on charge density and (2) the Heaviside model for the drag coefficient.[64] The more compact the conformation of the microgel (i.e., the more hard spherelike the structure), the more accurately these models describe the electrokinetic behavior of microgel particles.

5.6 Applications of Microgels

Well-defined polymer particles, particularly those in the submicrometer size range, are offering useful and suitable colloidal supports in a large number of biotechnological, pharmaceutical, and medical applications such as diagnostics (assays), biological analysis and research, bioseparation, NMR imaging, microsystems (microarrays, labs on a chip), cosmetics, drug-delivery systems (DDS) and targeting, etc. (see Figure 5.14). Stimuli-responsive microgels

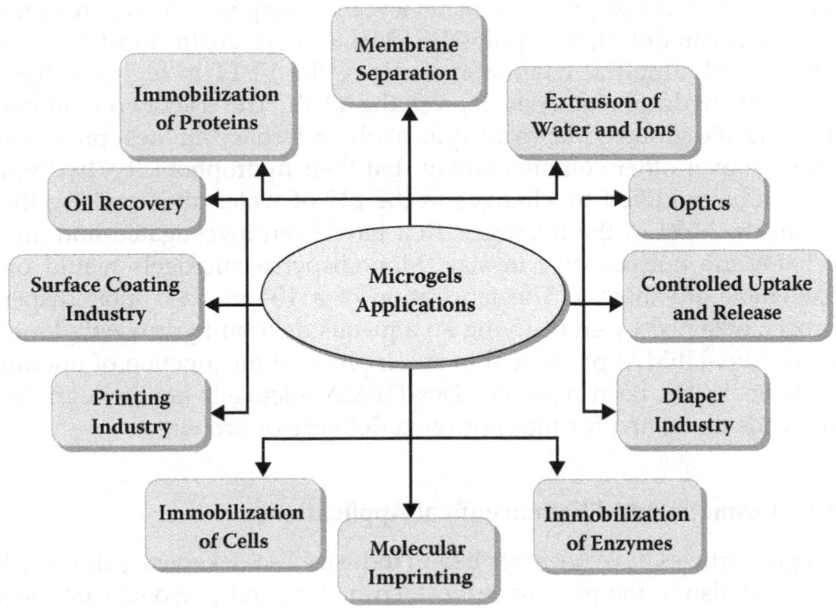

FIGURE 5.14
Applications of microgels.

have shown a great potential for applications, especially in medicine and biological fields, mostly due to the friendly environment they offer for the immobilization or interaction of biomolecules.[65] A fast response to potential stimuli is a prerequisite for most of the applications. Microgels are good candidates for gaining insights into the physical nature of the volume phase transition because of their fast kinetic response.[66] Intelligent colloidal microgels prepared from NIPAM are the subject of an ever-growing number of publications. NIPAM microgel particles find extensive application, since they exhibit dramatic changes in volume with external stimuli such as changes in temperature, pH, and ionic strength.

In biological systems, it is well known that iron nitrogen and iron sulfur reactions are common, hence the binding of iron(II/III) protoporphyrin centers (hemes) to microgels presents an important model for mammalian biology. Novel copolymer microgels were prepared incorporating specific functional groups to potentially enhance the binding of iron(III) protoporphyrin IX hydroxide (hematin). The incorporated functional groups contain either nitrogen or sulfur ligands to increase the affinity of the heme complex to the microgel. The heme–microgel complex produced may potentially find use in the reversible binding of oxygen and carbon monoxide and in catalysis.[67]

Precise control and tuning of the VPTT is a requirement for different applications of these polymeric materials. Adding solvents, salts, surfactants, and polyelectrolytes can change the CST of poly(NIPAM) solutions.

From a materials development point of view, the incorporation of co-monomers and chemical modification of poly(NIPAM) backbone are the most important alternative. Hydrophilic co-monomers drive the VPTT to higher temperatures, while hydrophobic ones depress the VPTT. The use of co-monomers with ionizable groups, like carboxylic acids or tertiary amines, present the advantage over other comonomers in that their hydrophobicity/hydrophilicity can be modified by changes in the pH of water, thus making them pH-tunable. Most of the microgels that have been investigated and are in use today are polydisperse in size. Monodisperse microgels would offer considerable advantages. The fabrication of a 10-μm-sized monodisperse microgel, obtained by emulsifying an aqueous dextran-hydroxyethyl methacrylate (dex-HEMA) phase within an oil phase at the junction of microfluidic channels, has been reported. Dex-HEMA microgels are biodegradable and are ideally suited for the controlled delivery of proteins.[68]

5.6.1 Cosmetic and Pharmaceutical Applications

Microgel particles have been applied in industry as thickeners, primarily for paints, but also in the pharmaceutical, cosmetics, and personal care industries. For example, Corbopol is one of the most popular microgel thickeners for use in aqueous systems in the cosmetic and pharmaceutical industries. Microgel particles can also be used as carrier particles for small molecules, such as drugs. Alternative carrier systems, such as surfactant micelles, have been proposed and tested. However, their size (a few nm) is too small for general application, and they break down easily. Block copolymer micelles are not only somewhat larger, but are also more robust and, therefore, are more effective. However, microgel particles offer an even more robust and stable environment. If microgel particles are to be used under physiological conditions, e.g., in pharmaceutical or cosmetic applications, they need to be stable in aqueous media over the necessary ranges of pH and ionic strength. Many microgel systems for use in aqueous media are charge-stabilized by incorporating acrylic acid groups into the particles. However, such microgel particles can also become destabilized at low pH or at high electrolyte concentration. An alternative strategy is to use sterically stabilized particles. There have been several reports in the literature relating to the preparation of polymer latex particles with terminally grafted PEO chains by copolymerizing the main monomer with a PEO-terminated macromonomer, using a dispersion polymerization route. The resultant polymer particles have a core–shell structure, where the shell consists of hydrated PEO chains and the core is the hydrophobic polymer. The particles are expected to be stable under high ionic strength and low-pH conditions.[33]

5.6.2 Drug Delivery Systems (DDS)

Recent research work has been focused on the development of devices that can control the release of rapidly metabolized drugs or have the ability to protect sensitive drugs. Microgel particles are expected to fulfill these requirements. They have been used to achieve both zero-order and pulsatile release patterns.[3] These materials in various physical formats have shown intelligent loading and release capabilities for drugs, proteins, nanoparticles, and DNA as a result of temperature modulation, ionic strength, pH values, solvents, and even light. Poly(NIPAM), a typical paradigm of thermosensitive polymer, is known to exhibit a LCST in aqueous media, below which the polymer is soluble and above which it is in a collapsed phase or water insoluble.[69]

Skin toxicity is a major problem in transdermal drug delivery systems. Drug molecules are required that are innocuous, creating neither irritation nor allergenicity, so in some circumstances using a larger patch area can alleviate the problem. Because of this issue, and taking into consideration one of the possible practical applications of "smart materials" in wound treatment, the idea of using pH/temperature-sensitive microgels as transdermal drug-delivery systems was considered. The microgels have potential in drug release to the skin. They may be of particular importance where the skin barrier is compromised, as in a diseased state or in wound management. In this situation, controlled delivery to the skin of active materials can provide therapeutic levels where required and minimize systemic uptake. The microgels could be used at higher temperatures and release more material, as can be anticipated within wound tissue. They appear to be pH insensitive in the release of the compounds, and therefore any pH effects in the wound would be negligible.[3]

Although much progress has been made in the field of pharmaceuticals, delivery of protein drugs is still quite challenging. Biomacromolecules such as proteins are sensitive to their environment and easily degraded; therefore, topical parenteral delivery systems are the main approach used today. A principal advantage of entrapment of these biomacromolecules is that microparticles enable administration via injection, and fabrication from a biodegradable polymer eliminates the need for surgical removal. The commonly used materials are poly(L-lactic acid) and its copolymers with D-lactic acid or glycolic acid, which provide a wide range of degradation periods from weeks to years. Compared with hydrophobic biomaterials such as poly(D,L-lactide-coglycolide), microgels more closely resemble natural living tissues due to their high water contents and soft and rubbery consistency, thus minimizing irritation to surrounding tissues. Furthermore, the drug loading and release can be controlled by environmental stimuli. Required properties for efficient drug release, e.g., intelligence, injectability, degradability, and biocompatibility, are combined in smart microgel particles. The loading strategy of proteins is also of great importance.

The microgel and associated postfabrication encapsulation is presented schematically in Figure 5.15. In this case, the "intelligent" aspect of microgel behavior is its negative temperature sensitivity, i.e., the gel swells at low temperatures and contracts at temperatures above the VPTT of the microgel. The FT-IR spectra of the microgels absorbing hemoglobin show the vibrational

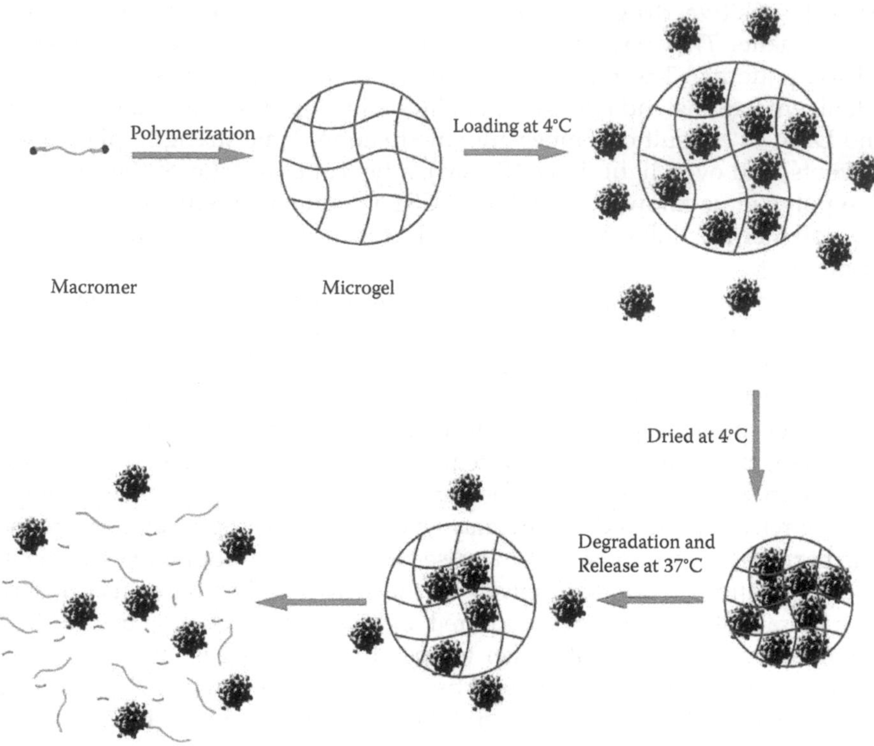

FIGURE 5.15
Schematic representation of a postfabrication encapsulation strategy of protein drugs based upon an "intelligent" and biodegradable microgel.

bands of blank microgel plus hemoglobin (Figure 5.16). These results indicate that the hemoglobin did penetrate into the microgels and the structure was not destroyed.[20,70]

Immobilization of enzymes on the electrode surface is a key factor in fabricating a biosensor, and various immobilization methods have been employed in preparation of reagentless amperometric enzyme electrodes. One method consists of encapsulation within polymers.[71] Microgels are used as matrices to immobilize enzymes for application in biosensor development. Polyacrylamide microgels are synthesized with entrapped glucose oxidase to be used for fabricating a glucose sensor. It is found that some physical properties of the PA, such as the glass transition temperature, changed after the incorporation of the enzyme.[72] It is reported that these materials have the ability to behave as controlled uptake and release agents, where species such as dextran and heavy metal ions are absorbed by the microgels at temperatures below the conformational transition temperature and expelled following contraction of the microgels at elevated temperatures. The addition of acrylic acid to the reaction medium, during microgel preparation, modifies the physical properties

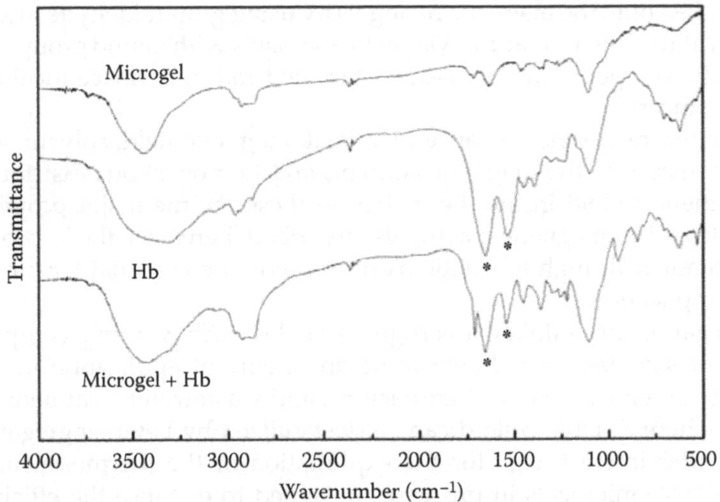

FIGURE 5.16
FT-IR spectra of microgels, native hemoglobin, and microgels plus hemoglobin.

of the dispersions in such a way that the copolymer produced is responsive to changes in temperature, ionic strength, and pH. The dispersion behavior and physicochemical properties of the poly(NIPAM) microgel have also been reported to be modified by the presence of surfactants.[73]

In recent years, research efforts have been made in developing targeted drug carriers with magnetic nanogels. Magnetic nanogels are nanometer-scale particles of inorganic/polymer core–shell composites. Magnetic nanogels of common interest are ferromagnetic magnetite (Fe_3O_4) coated with cross-linked polymer nanogel. The Fe_3O_4 core, which possesses strong magnetic properties and super-paramagnetic behavior, is of relatively low toxicity to the human body when encapsulated in the protective polymer shell, which consists of cross-linked polymer hydrogels. The shell prevents the Fe_3O_4 core from oxidation and aggregation. With good hydrophilicity and biocompatibility, magnetic nanogels with a hydrogel shell can be desirable for biomedical applications such as drug-delivery systems. Modified magnetic nanoparticles have gained considerable attention because of their potential in medical applications such as protein and enzyme immobilization, bioseparation, immunoassays, biosensors, and improved magnetic resonance imaging (MRI) contrast agents for cancer diagnosis. This technique is also expected to be useful in removal of toxic elements from industrial wastes.

Magnetic nanogels with an average diameter of 78 nm and a polydispersity index of 0.217, along with amino groups, can be applied as a radiopharmaceutical carrier of L-histidine labeled with one of the most effective radioisotopes (^{188}Re) for radiotherapy of cancers. The labeled L-histidine could be easily linked onto the magnetic nanogels by using glutaraldehyde in aqueous solution at room temperature. Magnetic nanogels with amino groups labeled with radioisotopes could be used as targeted radiopharmaceutical carriers against cancers.[19]

Rapid and reversible response to a weak magnetic field, solvent swelling without release of the magnetic content, and homogeneous distribution of the magnetic colloid inside the matrix — these are the major properties of the hydrophilic magnetic microgels described here. All these properties, concomitant with high magnetic oxide content, are essential for biotechnological applications.[74]

Aggregation of colloidal microgel particles with varying composition, charge, or size has been shown to be an important application in mineral recovery, ceramic materials manufacture, and wastewater treatment. Applications where a stable colloid can be destabilized by heteroaggregation are of particular interest, e.g., for the exploitation of the thermosensitivity of poly(NIPAM) microgels in processes designed to enhance the efficiency of crude oil recovery. The process relies on the charged microgel particles forming stable dispersions below their VPTT. When injected into the ground, the microgels flocculate in response to the higher temperatures prevalent in oil-bearing rock (above the VPTT) and the presence of electrolytes in the pumping material, e.g., seawater. The colloidal aggregates block off channels of high permeability, enabling the oil in the less permeable areas to be mobilized. The process relies on the strength of the aggregation, i.e., its ability to endure large shear forces resulting from high pressures. Mixed-charge colloidal dispersions can exhibit stronger aggregation characteristics than simple van der Waals forces when electrostatic attractive forces also contribute to the robustness of the resulting aggregates. However, the microgels undergo aggregation when heated above their VPTT. This behavior is attributed to the microgels being swollen with solvent at room temperature. The heteroaggregation characteristics of mixed colloidal dispersions of anionic poly(NIPAM) and cationic poly(NIPAM-*co*-4-VP) have been examined in relation to their relative concentrations. It has been shown that heteroaggregation in systems of variable charge and charge density is influenced by a complex balance of forces, both electrostatic and steric, and is affected significantly by the environment of the particles. Aggregation under various conditions of pH, electrolyte, temperature, charge density, and relative concentration can be induced by the careful manipulation of any one or more of the environmental conditions.[75]

In a new approach, a group of scientists used a synthetic microgel system for delivering protein antigens that incorporates responsiveness to the biological pH gradient found in the phagosome of antigen-presenting cells. The microgels have a well-controlled, spherical morphology with sizes ranging from 200 to 500 nm. At the center of the microgel approach is the ability to incorporate cross-linking chemistries that introduce pH-dependent hydrolytic degradation rates. The gel was designed to break down to nontoxic substances while releasing the protein antigens. The components of the system have received FDA approval. As with the biodegradable polymeric microspheres, it should also be possible to load multiple antigens and adjuvants. The microgel-delivered model antigen was shown to significantly enhance MHC class I presentation and cytotoxic T lymphocyte activation when compared with the free antigen.[76] In particular, microgels can be applied as rheology control agents due to their unique rheological properties. For example, it is reported that agar microgel dispersions are good viscosity thickeners for cosmetics. As an interesting fact, the texture of such products is strongly affected by the rheological properties of the applied microgels.[18]

In replacing amalgam, resin composites and glass ionomer cements are used to restore cavities in the primary and permanent dentition with tooth-colored materials, and a variety of substances are used for the bonding process. However, some components of the composites and bonding materials may become segregated in an aqueous environment during implantation and even after polymerization.[77–79] Here they may exert adverse effects on the organism, e.g., allergic reactions such as urticaria and contact dermatitis, systemic toxicity, cytotoxicity, estrogenicity, and mutagenicity.[80]

Noble metal nanoclusters stabilized by *N,N*-dimethylacrylamide-based soluble cross-linked microgel have been prepared and tested as catalysts in the selective oxidation of secondary alcohols to the corresponding carbonyl compounds with molecular oxygen in water.[81] Novel microgels composed of cross-linked copolymers of poly(acrylic acid) and pluronics were evaluated as possible permeation enhancers for doxorubicin transport using Caco-2 cell monolayers as a gastrointestinal model.[82]

5.6.3 Surface-Coatings Industry

The main applications involving microgel particles have been in the surface-coatings industry. Microgel particle dispersions are shear thinning and provide a rheological control for automotive surface coatings. The particles also have good film-forming properties and favor the alignment of added metallic flakes parallel to the substrate surface. The original motive for employing microgel particles in surface coatings arose from U.S. Environment Protection Agency regulations that required a decrease in the volatile components of surface-coating formulations. This was achieved by increasing the total solids content and decreasing the molecular weight of the linear polymer resin; however, this led to an unacceptably low viscosity. The microgel particles had the added effect of imparting a yield stress to the dispersion.[5]

5.6.4 Molecular Imprinting

Molecular imprinting can be generally defined as a synthetic approach by which a molecular receptor is assembled via template-guided synthesis. Molecular-imprinted microgel spheres can be easily synthesized using a novel precipitation method.[83] Microgels are valuable materials to be used with the molecular-imprinting technology. Their main advantage is stability, and therefore the possibility of using them for a variety of different applications compared with their insoluble counterparts.[24,84]

5.6.5 The Diaper Industry

The most important property in a commercial superabsorbent used in personal-care applications is the extent of swelling. There has been increasing interest in the synthesis and applications of superabsorbent polymers or hydrogels. Acrylamide hydrogels have found extensive commercial applications as sorbents in personal care and for agriculture purposes. The swelling behavior of the hydrogel is very sensitive to the network microstructure and its interaction with water. The promising results reveal that such prepared materials may be of great interest in horticulture and industrial applications.[2]

5.6.6 Optics

Photoactivated template-based synthesis of fluorescent Ag nanoclusters in the interior of microgel particles has been demonstrated. Photoluminescence intensity and emission wavelength varies with the time of photoactivated synthesis. Optical properties of the photogenerated Ag nanoclusters strongly depend on the acidity of the medium. The resulting hybrid microgels show stable photoluminescence and strong response to external stimuli.[85]

5.7 Future Areas of Research

There are numerous areas that should be investigated to ensure that the full scientific potential of microgels is realized. The foregoing include (but are not limited to) the following. There is wide scope in terms of new methodologies and materials used for the synthesis of microgels. In particular, the use of naturally occurring or nontoxic (FDA approved) synthetic monomers/cross-linkers, etc., is to be encouraged. Such research could pave the way for novel uses of microgels in terms of, for example, drug delivery and other clinical as well as biotechnological uses. Obviously, further research relating to the physicochemical properties of both a fundamental and applied nature is required pertaining to different types of microgels. Potential applications of microgels in the field of solid-state chemistry need to be explored further, thereby making use of their large surface-area to volume ratio (in terms of their mass). Similarly, possible future applications of microgels in areas such as catalysis, immunoassays, and for the storage and use of fragile biomolecules such as peptides/proteins and DNA need to be examined further in relation to their possible advantages and disadvantages compared with currently used methodologies. Certainly our current ability to model the properties of microgels, at the molecular level, is in its infancy.

References

1. Schmidt, T. et al., Pulsed electron beam irradiation of dilute aqueous poly(vinyl methyl ether) solutions, *Polymer*, 46, 9908, 2005.
2. Rehim, H.A. et al., Swelling of radiation cross-linked acrylamide-based microgels and their potential applications, *Radiat. Phys. Chem.*, 74, 111, 2005.
3. Lopez, V.C. et al., The use of colloidal microgels as a transdermal drug delivery system, *Int. J. Pharm.*, 292, 137, 2005.
4. Matsumura, Y. et al., Synthesis and thermo-responsive behaviour of fluorescent labeled microgel particles based on poly(*N*-isopropylacrylamide) and its related polymers, *Polymer*, 46, 10027, 2005.
5. Saunders, B.R. and Vincent, B., Microgel particles as model colloids: theory, properties and applications, *Adv. Colloid Interface Sci.*, 80, 1, 1999.
6. Gracia, L.H. and Snowden, M.J., Preparation, properties and applications of colloidal microgels, *Handbook of Industrial Water Soluble Polymer*, Williams, P., Ed., Blackwell, Oxford, U.K., 2006.
7. Martin, B.S. et al., Solvent relaxation of swelling poly(NIPAM) microgels by NMR, *Colloid. Surf. A*, 270–271, 296, 2005.
8. Steiger, M. et al., Small-angle neutron scattering study of structural changes in temperature sensitive microgel colloids, *J. Chem. Phys.*, 13, 6197, 2004.
9. Pelton, R., Temperature-sensitive aqueous microgels, *Adv. Colloid Interface Sci.*, 85, 1, 2000.
10. Wu, X. et al., The kinetics of poly(*N*-isopropylacrylamide) microgel latex formation, *Colloid Polym. Sci.*, 272, 467, 1994.
11. Kaneda, I. et al., Water-swellable polyelectrolyte microgels polymerized in an inverse microemulsion using a nonionic surfactant, *J. Colloid Interface Sci.*, 275, 450, 2004.
12. Connal, L.A. et al., From well defined star-microgels to highly ordered honeycomb films, *J. Mater. Chem.*, 15, 1286, 2005.
13. Abrol, S., Studies on microgels, 5: synthesis of microgels via living free radical polymerisation, *Polymer*, 42, 5987, 2001.
14. Rosiak, J.M. et al., Nano-, micro- and macroscopic hydrogels synthesized by radiation technique, *Nucl. Instrum. Meth. Phys. Res. Sect. B*, 208, 325, 2003.
15. Ulanski, P., Synthesis of poly(acrylic acid) nanogels by preparative pulse radiolysis, *Radiat. Phys. Chem.*, 63, 533, 2002.
16. Ma, X. et al., Novel hydrophobically modified temperature-sensitive microgels with tunable volume-phase transition temperature, *Mater. Lett.*, 58, 3400, 2004.
17. Lin, C.L. et al., Preparation, morphology, and thermoresponsive properties of poly(*N'*-isopropylacrylamide) based copolymer microgels, *J. Polym. Sci. A*, 44, 356, 2006.
18. Kaneda, I. et al., Rheological properties of water swellable microgel polymerized in a confined space, *Colloid. Surf. A*, 270–271, 163, 2005.

19. Sun, H. et al., Novel core–shell magnetic nanogels synthesized in an emulsion-free aqueous system under UV irradiation for targeted radiopharmaceutical applications, *J. Magnetism Magn. Mater.*, 294, 273, 2005.
20. Giovannaon, G. et al., Microgels of poly(aspartamide) and poly(ethylene glycol) derivatives obtained by γ-irradiation, *Radiat. Phys. Chem.*, 65, 159, 2002.
21. Ho, A.K. et al., Synthesis and characterization of star-like microgels by one-pot free radical polymerization, *Polymer*, 46, 6727, 2005.
22. Elliott, J.E. et al., Structure and swelling of poly(acrylic acid) hydrogels: effect of pH, ionic strength, and dilution on the crosslinked polymer structure, *Polymer*, 45, 1503, 2004.
23. Yan, H. et al., Template-guided synthesis and individual characterization of poly(*N*′-isopropylacrylamide)-based microgels, *Langmuir*, 21, 7076, 2005.
24. Pasetto, P. et al., Synthesis and characterisation of molecularly imprinted catalytic microgels for carbonate hydrolysis, *Anal. Chim. Acta*, 542, 66, 2005.
25. Perez, L. et al., Synthesis and characterization of reactive copolymeric microgels, *Polym. Int.*, 54, 963, 2005.
26. Leung, M.F. et al., New route to smart core–shell polymeric microgels: synthesis and properties, *Macromol. Rap. Comm.*, 25, 1819, 2004.
27. Amalvy, J.I. et al., Synthesis and characterization of novel pH-responsive microgels based on tertiary amine methacrylates, *Langmuir*, 20, 8992, 2004.
28. Pelton, R., Unresolved issues in the preparation and characterization of thermoresponsive microgels, *Macromol. Symp.*, 207, 57, 2004.
29. Chen, Q. et al., One-step synthesis of silver-poly(4-vinylpyridine) hybrid microgels by gamma-irradiation and surfactant-free emulsion polymerisation: the photoluminescence characteristics, *Colloid. Surf. A*, 275, 45, 2006.
30. Taylor, L.D. and Cerankowski, L.D., Preparation of films exhibiting a balanced temperature-dependence to permeation by aqueous solutions study of lower consolute behaviour, *J. Polym. Sci., Polym. Chem. Ed.*, 13, 2551, 1975.
31. Senff, H. and Richtering, W., Influence of cross-link density on rheological properties of temperature-sensitive microgel suspensions, *J. Chem. Phys.*, 111, 1705, 1999.
32. Nayak, S.P., Design, Synthesis and Characterization of Multiresponsive Microgels, Ph.D. thesis, Georgia Institute of Technology, Atlanta, 2004.
33. Kaneda, I. and Vincent, B., Swelling behavior of PMMA-g-PEO microgel particles by organic solvents, *J. Colloid Interface Sci.*, 274, 49, 2004.
34. Morris, G.E., Vincent, B., and Snowden, M.J., Adsorption of lead ions onto *N*′-isopropylacrylamide and acrylic acid copolymer microgels, *J. Colloid Interface Sci.*, 190, 198, 1997.
35. Zhang, X.Z. et al., Fabrication and characterization of microgel-impregnated, thermosensitive poly(NIPAM) hydrogels, *Polymer*, 46, 9664, 2005.
36. Snowden, M.J. et al., Colloidal copolymer microgels of *N*′-isopropylacrylamide and acrylic acid: pH, ionic strength and temperature effects, *J. Chem. Soc. (Faraday Trans.)*, 92, 5103, 1996.
37. Loxley, A. and Vincent, B., Equilibrium and kinetic aspects of the pH-dependent swelling of poly(2-vinylpyridine-*co*-styrene) microgels, *Colloid Polym. Sci.*, 275, 1108, 1997.
38. Tan, B.H. et al., Microstructure and rheological properties of pH-responsive core–shell particles, *Polymer*, 46, 10066, 2005.

39. Kratz, K. et al., Volume transition and structure of triethyleneglycol dimethacrylate, ethyleneglycol dimethacrylate, and *N,N'*-methylene bis-acrylamide cross-linked poly(*N*-isopropyl acrylamide) microgels: a small angle neutron and dynamic light scattering study, *Colloid. Surf. A*, 197, 55, 2002.
40. Boyko, V. et al., Thermo-sensitive poly(*N*-vinylcaprolactam-*co*-acetoacetoxyethyl methacrylate) microgels, 1: synthesis and characterization, *Polymer*, 44, 7821, 2003.
41. Saunders, B.R., On the structure of poly(*N*-isopropylacrylamide) microgel particles, *Langmuir*, 20, 3925, 2004.
42. Chen, Q.D. and Shen, X.H., Effects of L-phenylalanine on the radiation synthesis of poly(*N,N'*-methylenebisacrylamide-*co*-4-vinylpyridine)(micro)gels, *Chin. J. Polym. Sci.*, 23, 643, 2005.
43. Tan, B.H. et al., Microstructure and rheology of stimuli-responsive microgel systems: effect of cross-linked density, *Adv. Colloid Interface Sci.*, 113, 111, 2005.
44. Ma, X. et al., Flocculation behavior of temperature-sensitive poly(*N*-isopropylacrylamide) microgels containing polar side chains with -OH groups, *J. Colloid Interface Sci.*, 299, 217, 2006.
45. Hazot, P. et al., Poly(*N*-ethylmethacrylamide) thermally sensitive microgel latexes: effect of the nature of the crosslinker on the polymerization kinetics and physicochemical properties, *C. R. Chimie*, 6, 1417, 2003.
46. Crowther, H.M. et al., Poly(NIPAM) microgel particles deswelling: a light scattering and small angle neutron scattering study, *Colloid. Surf. A.*, 152, 327, 1998.
47. Steiger, M. et al., Small-angle neutron scattering study of structural changes in temperature sensitive microgel colloids, *J. Chem. Phys.*, 120, 6197, 2004.
48. Koh, A.Y.C. and Saunders, B.R., Small-angle neutron scattering study of temperature-induced emulsion gelation: the role of sticky microgel particles, *Langmuir*, 21, 6734, 2005.
49. Berndt, I. et al., Influence of shell thickness and cross-link density on the structure of temperature-sensitive poly-*N*-isopropylacrylamide-poly-*N*-isopropylmethacrylamide core-shell microgels investigated by small-angle neutron scattering, *Langmuir*, 22, 459, 2006.
50. Mason, T.G., Lin, M.Y. et al., Density profiles of temperature sensitive microgel particles, *Phys. Rev. E*, 71, 4, 2005.
51. Boyko, V. and Richter, S., Structure of thermosensitive poly(*N*-vinylcaprolactam-*co*-*N*-vinylpyrrolidone) microgels, *Macromolecules*, 38, 5266, 2005.
52. Arleth, L. and Xia, X.H., Volume transition and internal structures of small poly(*N*-isopropylacrylamide) microgels, *J. Polym. Sci. B*, 43, 849, 2005.
53. Bromberg, L. et al., Thermodynamics of temperature-sensitive polyether-modified poly(acrylic acid) microgels, *Langmuir*, 20, 5683, 2004.
54. Stieger, M. et al., Structure formation in thermoresponsive microgel suspensions under shear flow, *J. Phys. Cond. Mater.*, 16, S3861, 2004.
55. Ma, X. et al., Novel hydrophobically modified temperature-sensitive microgels with tunable volume-phase transition temperature, *Mater. Lett.*, 58, 3400, 2004.
56. Guo, Z. et al., Evidence for spatial heterogeneities observed by frequency dependent dielectric and mechanical measurements in vinyl/dimethacrylate systems, *Polymer*, 46, 12452, 2005.

57. Matsumura, Y. and Iwai, K., Synthesis and thermo-responsive behavior of fluorescent labeled microgel particles based on poly(*N*-isopropylacrylamide) and its related polymers, *Polymer*, 46, 10027, 2005.
58. Kim, K.S. et al., Colloidal behavior of polystyrene-based copolymer microgels composed of a hydrophilic hydrogel layer, *J. Indust. Eng. Chem.*, 12, 91, 2006.
59. Kim, K.S. et al., pH and temperature-sensitive behaviour of poly(*N*-isopropylacrylamide-*co*-methacrylic acid) microgels, *J. Indust. Eng. Chem.*, 11, 736, 2005.
60. Kim, K.S. and Vincent, B., pH and temperature-sensitive behaviours of poly(4-vinylpyridine-*co*-*N*-isopropyl acrylamide) microgels, *Polym. J.*, 37, 565, 2005.
61. Ma, X.M. et al., Influence of ethylmethacrylate content on the volume-phase transition of temperature sensitive poly([*N*-isopropylacrylamide]-*co*-[ethyl methacrylate]) microgels, *Polym. Int.*, 54, 83, 2005.
62. Gurr, P.A. et al., Synthesis, characterization, and direct observation of star microgels, *Macromolecules*, 36, 5650, 2003.
63. Kaggwa, G.B. et al., A new family of water-swellable microgel particles, *J. Colloid Interface Sci.*, 257, 392, 2003.
64. Garcia-Salinas, M.J., Romero-Cano, M.S., and de las Nieves, F.J., Electrokinetic characterization of poly(*N*-isopropylacrylamide) microgel particles: effect of electrolyte concentration and temperature, *J. Colloid Interface Sci.*, 241, 280, 2001.
65. Pichot, C. et al., Surface-functionalized latexes for biotechnological applications, *Curr. Opinion Colloid Interface Sci.*, 9, 213, 2004.
66. Kratz, K. et al., Structural changes in poly(NIPAM) microgel particles as seen by SANS, DLS, and EM techniques, *Polymer*, 42, 6631, 2001.
67. Cornelius, V.J. et al., A study of the binding of the biologically important hematin molecule to a novel imidazole containing poly(*N*-isopropylacrylamide) microgel, *React. Funct. Polym.*, 58, 165, 2004.
68. De Geest, B.G. et al., Synthesis of monodisperse biodegradable microgels in microfluidic devices, *Langmuir*, 21, 10275, 2005.
69. Vihola, H. et al., Binding and release of drugs into and from thermosensitive poly(*N*-vinyl caprolactam) nanoparticles, *Eur. J. Pharm. Sci.*, 16, 69, 2002.
70. Zhang, Y. et al., A novel microgel and associated post-fabrication encapsulation technique of proteins, *J. Control. Release*, 105, 260, 2005.
71. Retama, J.R. et al., High stability amperometric biosensor based on enzyme entrapment in microgels, *Talanta*, 68, 99, 2005.
72. Cabarcos, E.L. et al., Small angle neutron scattering study of the structural modifications induced by the entrapped glucose oxidase in polyacrylamide microgels, *Physica A*, 344, 417, 2004.
73. Woodward, N.C. et al., The interaction of sodium dodecyl sulphate with colloidal microgel particles, *Eur. Polym. J.*, 36, 1355, 2000.
74. Ménager, C. et al., Preparation and swelling of hydrophilic magnetic microgels, *Polymer*, 45, 2475, 2004.
75. Hall, R.J. et al., Heteroaggregation studies of mixed cationic co-polymer/anionic homopolymer microgel dispersions, *Colloid. Surf. A*, 233, 25, 2004.
76. Murthy, N. et al., A macromolecular delivery vehicle for protein-based vaccines: acid-degradable protein-loaded microgels, *Trends Biotechnol.*, 21, 4995, 2003.
77. Geurtsen, W. et al., Substances released from dental resin composites and glass ionomer cements, *Eur. J. Oral Sci.*, 106, 687, 1998.

78. Spahl, W. et al., Determination of leachable components from four commercial dental composites by gas and liquid chromatography/mass spectrometry, *J. Dent.*, 26, 137, 1998.
79. Ortengren, U. et al., Water sorption and solubility of dental composites and identification of monomers released in an aqueous environment, *Oral Rehab.*, 12, 1106, 2001.
80. Geurtsen, W. et al., Biocompatibility of resin modified filling materials, *Crit. Rev. Oral Biol. Med.*, 11, 333, 2000.
81. Biffis, A. et al., Efficient aerobic oxidation of alcohols in water catalysed by microgel-stabilised metal nanoclusters, *J. Catalysis*, 236, 405, 2005.
82. Bromberg, L. et al., Effects of polyether-modified poly(acrylic acid) microgels on doxorubicin transport in human intestinal epithelial Caco-2 cell layers, *J. Control. Release*, 88, 11, 2003.
83. Ye, L. et al., Molecular imprinting on microgel spheres, *Anal. Chim. Acta*, 435, 187, 2001.
84. Zhu, W. et al., Preparation of a thermosensitive and biodegradable microgel via polymerization of macromonomers based on diacrylated pluronic/oligoester copolymers, *Eur. Polym. J.*, 41, 2161, 2005.
85. Zhang, J.G., Xu, S.Q. et al., Photogeneration of fluorescent silver nanoclusters in polymer microgels, *Adv. Mater.*, 17, 2336, 2005.

Chapter 6

Protein-Based Smart Polymers

J. Carlos Rodríguez-Cabello, Javier Reguera, Susana Prieto, and Matilde Alonso

CONTENTS

6.1 Introduction

Nature provides an unbelievable amount of protein materials that exhibit an ample range of functional applications. These proteins must have properties that fulfill the most diverse functions and purposes that the living organisms need to live. Materials science is learning from nature to obtain polymers that mimic the superior properties of the natural macromolecules. Protein-based polymers (PBPs) are polymers that include sequences of the natural proteins and other sequences designed *de novo* using the knowledge obtained from the observation of the natural proteins and synthetic polymers.

Elastinlike polymers (ELPs) are a very interesting kind of protein-based polymers. These polymers show the mechanical properties of the natural elastin,[1,2] are highly biocompatible,[3] show self-assembly properties, and can be tailored using DNA recombinant technologies.[4] However, one of the more interesting properties of the ELPs is their smart behavior.[5] They show an inverse temperature transition (ITT) that makes them smart polymers with responsiveness to changes in temperature. Furthermore, this ITT can be modified in some ELPs as a response to external stimuli, thereby obtaining smart-protein polymers with a wide range of sensitivity.[5]

This chapter provides some examples of smart behavior, including pH-sensitive polymers and photosensitive polymers. In addition, some applications for ELPs are presented, in particular their use in biomedicine.

6.2 Genetically Engineered Protein-Based Polymers

Natural biopolymers provide an impressive example of how nearly all of the properties displayed by biological materials and systems are almost exclusively determined by the physical–chemical properties of their monomers and the sequence in which they are arranged. Here, competing interactions, structural flexibility, and functional properties are tailored by the succession of monomeric units taken from a rather limited set. Materials science is nowadays taking advantage of this fact, especially because the new techniques of molecular biology and genetic engineering allow the creation of almost any DNA duplex coding any amino-acid sequence at will. Using these techniques, called DNA recombinant technologies, it is also possible to introduce a synthetic gene into the genetic content of a microorganism, plant, or other organisms and induce the production of its encoded protein-based polymer (PBP) as a recombinant protein.[4,6]

This approach can provide many advantages. First, genetically engineered protein-based polymers (GEPBPs)[7] will, in principle, be able to show the same simple or complex properties present in natural proteins. In this sense, this method offers an opportunity to exploit the huge functional resources that have been hoarded and refined to the extreme by biology during the long process of natural selection. GEPBPs can easily make use of the vast functionality present in hundreds of thousands of different proteins that amply exist in living organisms, from the smallest prions or viruses to the higher animals.

Second, because we can construct a coding gene, base by base, by following our own original designs and without being restricted to gene fragments found in living organisms, we can design and produce GEPBPs to obtain materials, systems, and devices exhibiting functions of particular technological interest that are not displayed in living organisms.

Third, from the point of view of a polymer chemist, the degree of control and complexity attained by genetic engineering is clearly superior to that achieved by more conventional synthesis technologies. GEPBPs are characterized as being strictly monodisperse and can be obtained from a few hundred daltons to more than 200 kDa, and this upper limit is continuously increasing.[8] Among other things, this has provided the opportunity to study, in a simple and highly precise manner, the dependence of different material properties on the molecular weight (MW).[9] This knowledge opens the possibility of fine-tuning the properties while designing the materials. Furthermore, in contrast to the polymers obtained by chemical synthesis, GEPBPs are produced independently of the complexity, and the cost, of course, is not related to this complexity.

Fourth, the number of different combinations attainable by combining the 20 natural amino acids is practically infinite. In a simple calculation, if we consider the number of possible different combinations to obtain a small protein consisting of, for example, 100 amino acids (their modest MW would range between 5.7 and 18.6 kDa), the figure is as high as 1.3×10^{130} possibilities, and the matter necessary to produce just a single copy of them would be 55 orders of magnitude higher than the whole dark and luminous matter of the universe.[10]

Genetic engineering of protein-based polymers is still in its early infancy. The radically different approach in the methodology used to produce such polymers has limited the number of research groups and companies committed to develop this field of study. Among these pioneer groups, the main interest has been concentrated mainly in two major polymer families: spider-silk-like polymers and elastinlike polymers, although some other interesting protein polymers have also been researched, including coiled-coil motifs and their related leucine zippers,[11–14] β-sheet-forming polymers, poly(allylglycine), and homopolypeptides such as poly(glutamic acid).[15] Some of the representative examples of GEPBPs produced to date are summarized in the literature.[7] The different strategies and methodologies for gene construction and the associated iterative, random, and recursive ligations have also been summarized recently.[16–18]

6.3 Elastinlike Polymers (ELPs)

ELPs are a promising class of protein-based polymers. The basic structure of ELPs is a repeating sequence identified in the repeating sequences found in mammalian elastic protein, in elastin, or in some modification of those sequences. The most striking and longest sequence between cross-links in pig and cow is the undecapentapeptide $(VPGVG)_{11}$, where G stands for glycine, V for L-valine, and P for L-proline.[19,20] Along with this repeating sequence, others can be pointed out, such as $(VPGG)_n$ and $(APGVGV)_n$, where A stands for L-alanine.[21] From the bovine elastin, the most repetitive sequence is $(VPGVG)_n$. Initially, the monomers, oligomers, and high polymers of these repeats, or their modifications, were chemically synthesized and conformationally characterized.[22–24] More recently, with the development of molecular biology, these and other complex ELPs have been bioengineered, allowing the production of more-complex and well-defined polymers.[9,16,24–26] These polymers are important because they show a versatile and ample range of properties that are difficult to find in other materials and that go beyond their simple mechanical performance.

6.4 The Fundaments of the Inverse Temperature Transition (ITT): The Thermal Response

The polymer poly(VPGVG) is considered as a model for the ELPs. Most of the ELPs are based on the pentapeptide VPGVG (or its permutations), with amino acid side chains, excluding glycine, comprising simple aliphatic and mainly hydrophobic chains without further functionalization. A wide variety of ELPs have been (bio)synthesized based on the model ELP with the general formula (VPGXG), where X represents any natural or modified amino acid excepting proline.[5,22,24] All of the polymers with this general formula that can be found in the literature are functional, i.e., all show a sharp smart behavior.

All functional ELPs exhibit a reversible phase transition in response to changes in temperature,[5] i.e., they show an acute thermoresponsive behavior associated with the ITT. In aqueous solution below a certain transition temperature T_t, the free polymer chains remain disordered, random coils in solution[27] that are fully hydrated, mainly by hydrophobic hydration. This hydration is characterized by ordered clathratelike water structures surrounding the apolar moieties of the polymer,[28–30] structures somewhat similar to that described for crystalline gas hydrates,[30,31] although more heterogeneous and of varying perfection and stability.[29] However, above T_t, the chain folds hydrophobically and assembles to form a phase-separated state — 63% water and 37% polymer in weight[32] — in which the polymer chains adopt a dynamic, regular, nonrandom structure, called a β-spiral, involving type II β-turns as the main secondary feature and stabilized by intraspiral inter-turn and interspiral hydrophobic contacts.[5]

This behavior is the result of the ITT. In their folded and associated state, the chains essentially lose all of the ordered water structures of hydrophobic hydration.[28] During the initial stages of polymer dehydration, hydrophobic association of β-spirals takes on fibrillar form. This process starts from the formation of filaments composed of three-stranded dynamic polypeptide β-spirals that grow to several hundred nanometers before settling into a visible phase-separated state.[5,33] The process of the ITT is completely reversible, and it goes back to the first state when the temperature is lowered below T_t.[5] It has been shown that one repeat unit of VPGVG is sufficient to allow the transition from random coil to an ordered β-spiral structure,[34] although higher-molecular-weight polymers are required for useful materials properties.

6.5 Measuring the ITT

The T_t of the ITT can be measured by different techniques. Turbidity measurements, also called the "cloud point," are the most commonly used measurement, followed by calorimetric methods that measure the heat flow during the transition. The first method is characterized by a turbidity profile showing a sharp step. This increase in turbidity is caused by the formation of aggregates. T_t can be taken as the temperature at 10 to 50% change in the relative turbidity (Figure 6.1). In contrast, the differential scanning calorimetry (DSC) measurements are always characterized by a broad peak (Figure 6.1) extending over 20°C or more. In this case, T_t can be taken as either the onset or the peak temperature; furthermore, with this technique, it is possible to obtain the enthalpy of the process as the area of the peak.

The T_t values obtained by these two methods usually differ due to two different factors. The first is the thermal lag caused by the dynamic nature of DSC measurements. This can be eliminated by extrapolating T_t values to zero heating rate.[35] The second factor is due to the different concentrations used for

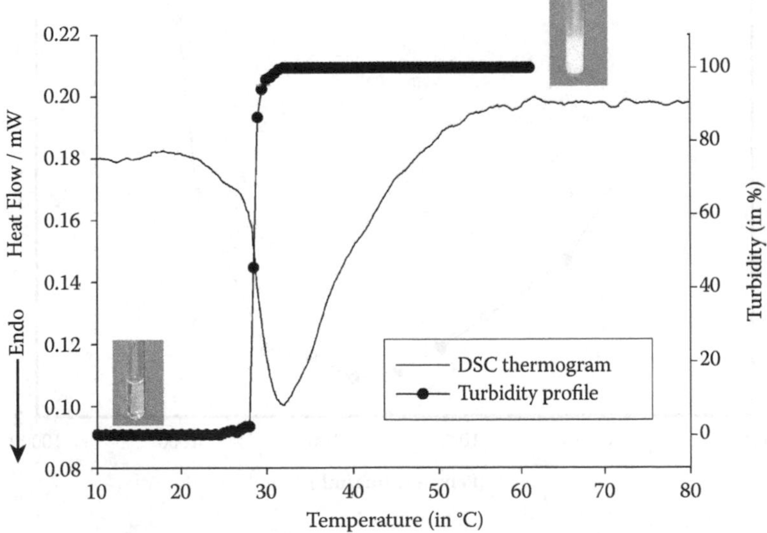

FIGURE 6.1
Turbidity profile as a function of temperature for a poly(VPGVG) 5 mg/l in water. The two photographs are below (5°C) and above (40°C) its T_t. Also, a DSC thermogram is shown of a 50-mg/l aqueous solution of the same polymer (heating rate 5°C/min).

the two methods. As has been reported, a change in concentration can cause a change in both the T_t and the enthalpy of the process (ΔH).[5,29] Typical concentrations for turbidity experiments are in the range of 1 to 5 mg/ml, while concentrations for DSC are usually in the range of 50 to 150 mg/ml. The dependence of T_t and ΔH on polymer concentration can be seen in Figure 6.2, which shows an increase in T_t and a decrease in ΔH for high and low polymer concentrations, respectively. At higher polymer concentration, the deficiency of water reduces the formation of some water clathrates among the different populations of clathrate-like water structures. Only the most solidly attached structures remain, and this results in an increment in T_t and a decrease in ΔH, as the enthalpy is related to the disruption of these structures. In contrast, the effect at lower concentrations can be caused by diminution of the interchain cooperation, making it more difficult to achieve hydrophobic hydration, and the hydrophobic hydration that does occur is achieved with less aggregation.

T_t depends on (a) the composition of the polymer being higher when it contains fewer apolar amino acids and (b) the polymer chain length, i.e., on the molecular weight (MW): T_t decreases as MW increases.[2,9,36] Furthermore, the presence of ions, such as those of the buffer and dissolved molecules, can also affect the T_t value. Hence, a well-defined polymer and well-defined conditions are important to establish the desired T_t. Furthermore, all of these factors complicate the comparison of T_t values among different techniques and different authors.

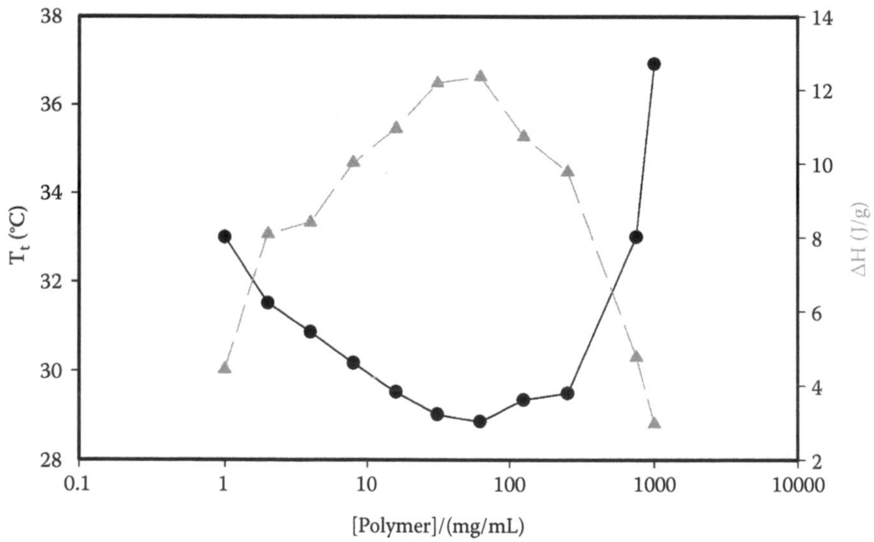

FIGURE 6.2
Transition temperature (solid line) and enthalpy (dashed line) for a water solution of $(GVGVP)_{251}$ as a function of the polymer concentration.

6.6 Designing the ELP with a Desired T_t

A series of experiments have been performed by Urry et al.[37–39] for the study of designed molecules with the general formula (f_v[GVGVP], f_x[GXGVP]), where f_v and f_x are mole fractions with $f_v + f_x = 1$ and X can be any one of the 20 naturally occurring amino acid residues or a biologically relevant chemical modification thereof. To avoid the chain-length and concentration dependence in those studies, the experiments were performed with about 100-kDa molecular weight and a concentration of 40 mg/ml, as 40 mg/ml occurs at a broad minimum in the concentration-vs.-T_t curve. The T_t(X) and ΔH_t(X) of the polymer (GXGVP) can be then obtained by linearly extrapolating the molar fraction to 1, as these parameters form a linear function with f_x. The extrapolated T_t can be seen in Figure 6.3a; hence, with a chosen amino acid and f_x, it is possible to obtain a desired T_t that fits a specific application.

These data have been used not only to design a polymer with a correct transition temperature, but also to give a reliable measure of the hydrophobicity of the amino acids. A hydrophobicity scale has been obtained as a function of a fundamental quantity, the change in Gibbs free energy of hydrophobic association (ΔG_{HA}), that takes place when a hydrophobic side chain goes from being fully surrounded by water to being hydrophobically associated with loss of essentially all of its hydrophobic hydration. Hence, the hydrophobicity scale is based on ΔG^0_{HA}, which is the change in ΔG_{HA} upon substitution of a reference amino acid residue by any other amino acid residue (Figure 6.3b). ΔH_t(GGGVP) is defined as the reference state because the hydrogen atom side chain for the glycine is neutral, neither hydrophobic nor hydrophilic. The change in Gibbs free energy at $T_t = T_t$(GXGVP) can be written as $\Delta G^0_{HA} = -\delta \Delta H_t(X) = \Delta H_t(GGGVP) - \Delta H_t(GXGVP)$.[37,38]

As can be seen in Figure 6.3, there is more than a 10-kcal/mol change in the free energy from the most hydrophobic residue (W, Trp) to the least hydrophobic residue (E^-, Glu^-), or, for example, there is a change in ΔG^0_{HA} due to ionization of the carboxylate of the glutamic acid residue, ($\Delta G^0_{HA}[GE^-GVP] - \Delta G^0_{HA}[GE^0GVP]$), that represents a change of +5.2 kcal/mol (GEGVP) toward destabilization of hydrophobic association. (Recall that these substantial ΔG^0_{HA} values have been diminished by the oppositely signed van der Waals interaction energies.[40]) Furthermore, these calculations of the ΔG^0_{HA} are applicable to the variables that change the ITT, such as the addition of salts to the medium, organic solutes, change in pressure, etc. Thus $\Delta G^0_{HA}(\chi) = [\Delta H_t(\mathrm{ref}) - \Delta H_t(\chi)]$.

The scale based on the fundamental quantity ΔG^0_{HA} of hydrophobicity can be useful for understanding protein folding and function as well as for the improved design of more-complex and better characterized ELPs.

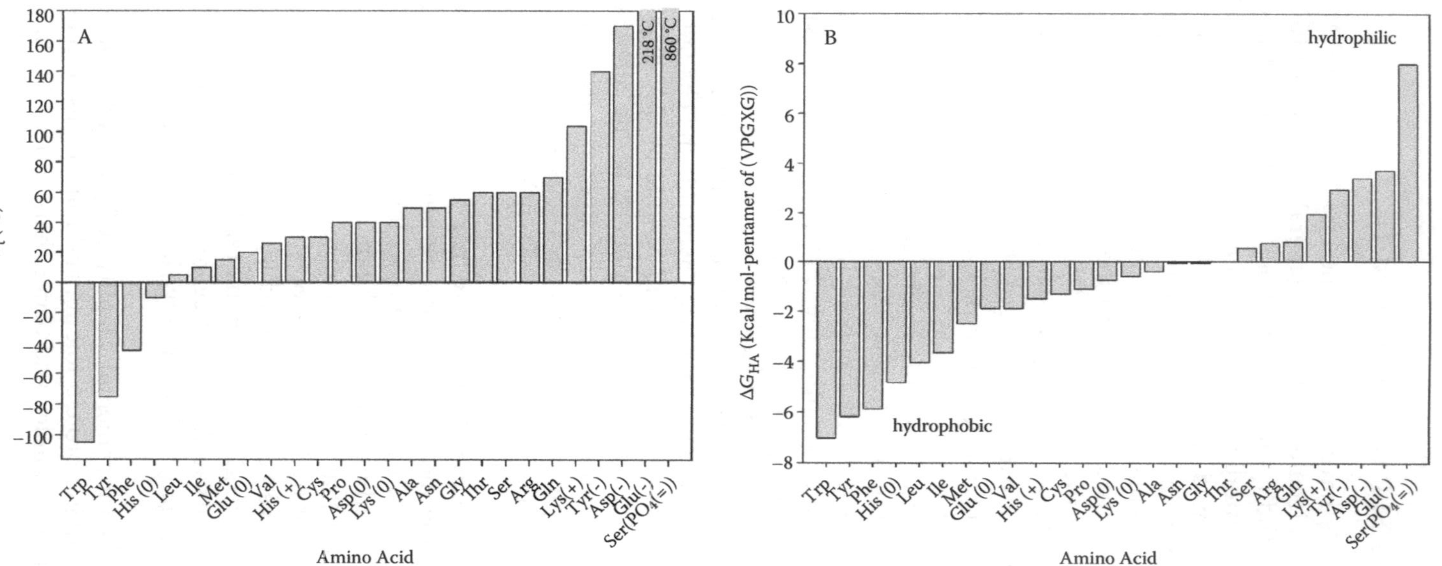

FIGURE 6.3
(A) T_t of the polymer poly(GXGVP) for X, one of the 20 natural amino acids. (B) Hydrophobicity scale of 20 natural amino acids in their different polarized states. (Adapted from data taken from Urry, D.W., *Chem. Phys. Lett.*, 399, 177, 2004; *Protein-Based Nanotechnology*, Renugopalakrishnan, V. and Lewis, R., Eds., Kluwer Academic Publishers, Dordrecht, 2006, p. 141. With permission.)

6.7 Separating Components in the ITT

The endothermic peak found in a DSC heating run is in fact the net result of a complex process reflecting different thermal contributions. Once an ELP solution reaches its T_t, a destruction of the ordered hydrophobic hydration structures surrounding the polymer chain takes place. This is further accompanied by an ordering of the polymer chain into the β-spiral structure. In their turn, these β-spirals further establish interchain hydrophobic contacts (van der Waals cohesive interactions) that cause the formation of nano- and microaggregates that separate from the solution. The first process must be endothermic, while the second one must be exothermic. Although the events take place simultaneously, they are very different in nature. In particular, it is reasonable to expect that the two phenomena occur with different kinetics. In effect, previous kinetic studies made on poly(VPGVG) showed that the phase-separation process is faster than redissolution and that both are multistep processes.[41] This difference offers an opportunity to separate the different contributions to the ITT. This has recently been achieved for the first time using temperature-modulated DSC (TMDSC).[40] In principle, TMDSC provides a clear split of two overlapping processes when, under the particular dynamic conditions, one is reversible and the other is not. The processes can be split by finding a frequency for the periodic component low enough for the faster process to follow the oscillating temperature changes (reversing) but high enough to impede alternating behavior of the slower (nonreversing) one. This approach has been used to study the ITT of three ELPs: chemically synthesized poly(VPGVG), recombinant $(VPGVG)_{251}$, and recombinant $(IPGVG)_{320}$, where I stands for L-isoleucine.[40] Figure 6.4a shows an example of the TMDSC thermogram found for $(VPGVG)_{251}$, while Figure 6.4b shows the results of its analysis. Under the experimental conditions, the

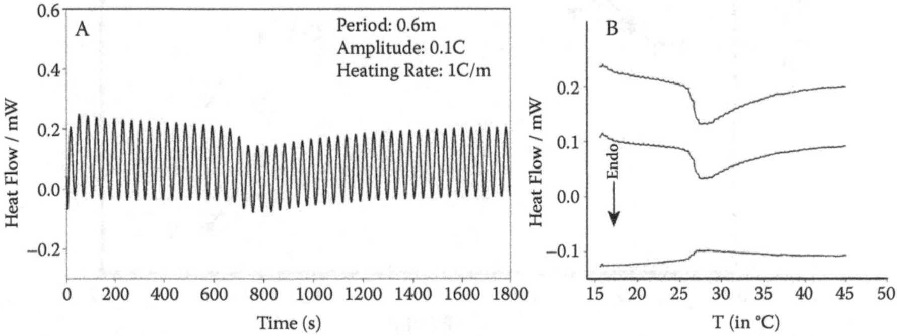

FIGURE 6.4
(A) Heat flow vs. time in TMDSC analysis of a 125-mg/ml aqueous solution of $(VPGVG)_{251}$. (B) Reversing, nonreversing, and total thermograms. (Reproduced with permission from Elsevier, *Chem. Phys. Lett.*, 36, 8470, 2004).

endothermic total curve ($\Delta H_{tot} = -10.40$ J·g^{-1}, $T_t = 27.72$°C) is composed of a nonreversing endothermic component ($\Delta H_{nonrev} = -13.98$ J·g^{-1}, $T_t = 27.63$°C) and a reversing exotherm ($\Delta H_{rev} = 3.33$ J g^{-1}, $T_t = 27.30$°C).

A detailed analysis has been carried out on reversing and nonreversing components as functions of the heating rate r, the amplitude A, and the period P of the alternating component. For the total contribution, the changes in r (0.5 to 1.5°C/min), A (0.1 to 1°C), and P (0.1 to 1.0 min) did not significantly affect the total enthalpy and T_t values, which are similar to those obtained by DSC. Also the reversing and nonreversing components were not affected by changes in r and A. However, P exerts a strong influence on the enthalpy values of both components. ΔH_{rev} has been plotted in Figure 6.5 as a function of P for the three polymers. In all cases, at low frequencies (high P), the reversing component shows an endothermic peak with enthalpy comparable with that of the endothermic peak of the nonreversing component. Thus, at these high P values, the chain folding and dehydration contributions are not well separated. However, as P decreases, ΔH_{rev} undergoes a substantial increase. At P about 0.8 to 1 min, the reversing component turns into a positive exothermic peak, which reaches a maximum P_M at 0.5 to 0.6 min, and simultaneously, ΔH_{nonrev} suffers an equivalent decrease. Therefore, as P decreases, the reversing component is enriched in the exothermic component (chain folding), while the nonreversing component is enriched in the endothermic contribution (dehydration).

The ΔH_{rev}, ΔH_{nonrev}, and ΔH_{tot} values found at P_M can be seen in Table 6.1. Further decrease in P results in a progressive reduction in ΔH_{rev} to zero and an increase in ΔH_{nonrev} to the total enthalpy as a result of the complete overlap

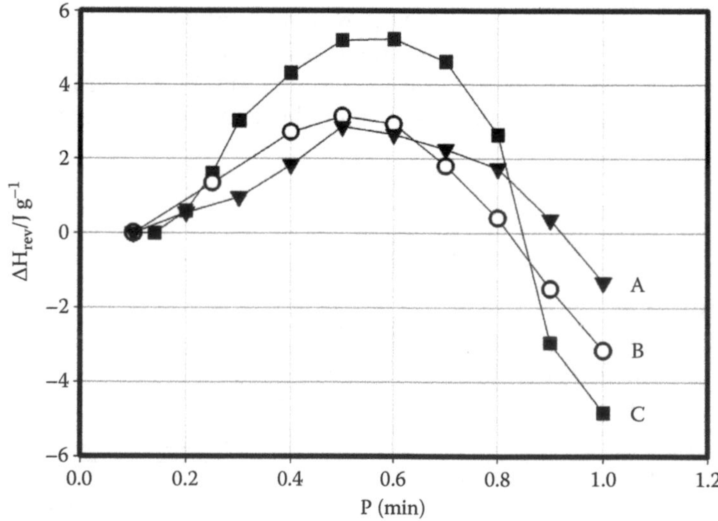

FIGURE 6.5
ΔH_{rev} as a function of P for 125-mg/ml aqueous solution of (A) synthesized poly(VPGVG); (B) recombinant $(VPGVG)_{251}$; and (C) recombinant $(IPGVG)_{320}$ (r = 1°C/min, A = 0.1°C). (Reproduced with permission from Elsevier, *Chem. Phys. Lett.*, 36, 8470, 2004.)

TABLE 6.1
Enthalpy Values of the Reversing, Nonreversing, and Total Components Found at P_M

Polymer	ΔH_{rev} (J·g^{-1})	ΔH_{nonrev} (J·g^{-1})	ΔH_{tot} (J·g^{-1})	P_M (min)
IPGVG)$_{320}$	5.61	−22.82	−17.21	0.6
(VPGVG)$_{251}$	3.14	−11.34	−7.50	0.5
poly(VPGVG)	2.96	−11.11	−8.79	0.5

Source: Reproduced with permission from Elsevier, *Chem. Phys. Lett.*, 36, 8470, 2004.

of both processes in the nonreversing component. The maximum splitting was found at approximately the same P_M, regardless of the polymer. Additionally, comparison of the data for (VPGVG)$_{251}$ and (IPGVG)$_{320}$ indicates that the reversing component at the maximum is higher for (IPGVG)$_{320}$. Given the higher hydrophobicity of the isoleucine (I) as compared with the valine (V), its chain folding has to show a higher exothermic ΔH_{rev} (see Table 6.1). Additionally, the increased hydrophobicity of (GVGIP)$_{320}$ would also induce a greater extent of hydrophobic hydration; hence its higher endothermic ΔH_{nonrev} is also reasonable.

There are no significant differences in data from (VPGVG)$_{251}$ and poly(VPGVG) (see Table 6.1). Since the only difference between these polymers is their MW dispersity, the TMDSC results are practically the same, implying that the reversing and nonreversing TMDSC components depend mainly on the mean hydrophobicity of the monomer.

Therefore, TMDSC has been demonstrated to be an effective method of resolving the overlapping kinetic processes implicated in the ITT of elastic protein-based polymers. By tuning the frequency of the periodic component, a maximum split can be achieved that shows an exothermic contribution arising from the van der Waals contacts attending chain folding and assembly, and an endothermic contribution associated with loss of hydrophobic hydration, the former being about one fourth of the latter. To the best of our knowledge, TMDSC is the only method currently available to separate the two contributions. Accordingly, its utilization in future research to evaluate hydrophobicity of the full complement of naturally occurring amino acids (and relevant modifications thereof) is clear, as is its relevance to hydrophobic folding of polymers and natural proteins.

6.8 (VPAVG)$_n$, an Unconventional ELP

The polymers based in the pentapeptide VPAVG (V – L-valine, P – L-proline, A – L-alanine, G – glycine) are ELPs that exhibit an acute smart nature; however, they show a hysteresis in its heating–cooling cycle and a different mechanical behavior. This fact opens new possibilities for materials that not only exhibit a smart response under different stimuli, but also that can remain in their states after the stimulus has finished. In this way, a polymer with memory could be triggered by different stimuli between two different states and remain in that state unless an opposite stimulus acts, thereby working as a "set-reset" system.

As we have seen before, most of the ELPs are based in the general formula (VPGXG), where X represents any natural or modified amino acid, except proline. All the polymers with this general formula that can be found in the literature are functional, i.e., all show a sharp smart behavior. However, the achievement of functional ELPs by the substitution of any of the other amino acids in the pentamer is not so straightforward. For example, the proline cannot be substituted, and the first glycine cannot be substituted by any natural amino acid other than L-alanine.[5] This is because the type II β-turn per pentamer involves this glycine together with the proline in the folded state of the polymer.[5] The presence of bulky moieties in amino acids with L chirality impedes the formation of the β-turn, and the resulting polymer is not functional. Thus, the substitution by alanine is the only possibility reported that still leads to a functional polymer, though even in this case, the resulting polymer shows significantly different and "out of trend" mechanical and thermal properties. This fact is clearly seen in the thermoresponsive behavior of this polymer. Figure 6.6 shows the DSC thermograms of poly(VPGVG) and poly(VPAVG) water solutions. (See Reguera et al.[41] for details on the synthesis and characterization of both polymers.)

The DSC patterns of both polymers share common features. During the first heating run, the presence of an endothermic transition is evident for both polymers. This endotherm is caused by the characteristic process of chain folding, which is accompanied by the destruction of the ordered shells of hydrophobic hydration. Obviously, this last endothermic contribution dominates this calorimetric run. T_t can be identified as the peak temperature. In contrast, the subsequent cooling stage shows a clear exotherm for both polymers. This reflects the reverse process, i.e., the unfolding of the polymer chain and its concurrent hydrophobic hydration. However, although the main features are common for both polymers, there are significant differences pointing out their divergent behavior. For poly(VPGVG), heating T_t (T_{tH}) and cooling T_t (T_{tC}) show only marginal differences that can be attributed to the inherent thermal lags of the DSC experiment (see Figure 6.6a). However, for poly(VPAVG), the difference $T_{tH} - T_{tC}$ scores 25.6°C for this cycle run at 8°C min^{-1} (T_{tH} = 30.7°C and T_{tC} = 6.1°C) (see Figure 6.6a). This indicates that a clear hysteresis behavior exists for poly(VPAVG); the polymer chain folds at 30.7°C but does not unfold until the temperature is undercooled down to 6.1°C. Additionally, the difference $T_{tH} - T_{tC}$, i.e., the degree of undercooling, for poly(VPAVG) is found to be, to a significant extent, heating/cooling-rate dependent. This is especially true for T_{tC}. This difference increases with the heating/cooling rate,[41] being still greater than 15°C at rates as low as 2°C min^{-1}.

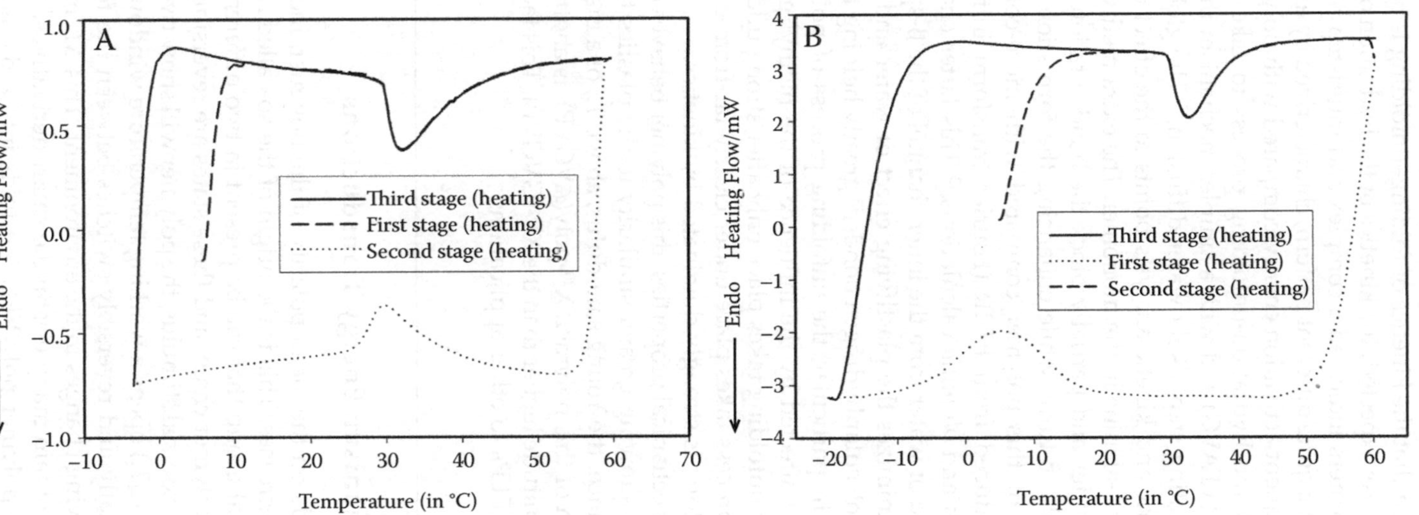

FIGURE 6.6
Typical DSC run of 125-mg/ml samples standing a heating–cooling–heating cyclic temperature program: (A) poly(Val-Pro-Gly-Val-Gly); (B) poly(Val-Pro-Ala-Val-Gly). A heating rate of 8°C min^{-1} and −8°C min^{-1} cooling were used in this example. (Reproduced with permission from the American Chemical Society, *Macromolecules*, 36, 8470, 2003.)

The use of a model-free kinetic analysis, the Friedman's isoconversional method,[42] also revealed that the kinetics of folding–unfolding is quite different for both polymers. As expected, the kinetics of the hydration of nonpolar moieties for both polymers indicated a complex and multistep process, but poly(VPAVG) showed a quite different pattern, characterized by a significant decrease in the subsequent activation energy compared with poly(VPGVG). This decrease caused the hydration-unfolding process to take place at a higher speed for poly(VPAVG) and with a simpler mechanism that is dominated by just one limiting step, likely water diffusion.[41] The global picture taken from calorimetric and kinetic analysis points to the effect of the substitution of glycine by L-alanine in the monomer. The extra methyl group of L-alanine seems to hinder and partially block the bond rotation needed to establish or destroy the β-turns while enhancing the formation of a more perfect folding state of this polymer compared with more-conventional poly(VPGVG), as deduced from FT-IR (Fourier transform infrared) and Raman spectroscopies and *ab initio* calculations.[43] This latter characteristic enhances the exclusion of water from the inner channel of the β-spiral in the folded chain. This minimizes the plastifying effect of water and its role in disrupting the intramolecular hydrogen bonds,[43] greatly helping to stabilize the folded structure. In particular, the unfolding process of poly(VPAVG) seems to be especially affected by the influence of the methyl group of L-alanine, such that this unfolding takes place only after strong undercooling. This means that the process takes place under clearly far-from-equilibrium conditions and, therefore, is strongly dominated by kinetics.

With regard to the mechanical properties, this polymer behaves differently than the others ELPs, showing greater similarity to the plastics than to the elastomers.[44] For instance, the Young's modulus, above T_t, of a matrix cross-linked by γ-irradiation of this polymer, X^{20}-poly(AVGVP), is about 2.0×10^{10} Pa, two orders of magnitude higher than the X^{20}-(GVGVP). This fact expands the possible utility of ELPs to other applications.

6.9 The ΔT_t Mechanism: Energy Transductions

In all ELPs, T_t depends on the mean polarity of the polymer, increasing as the hydrophobicity decreases. This is the origin of the so-called ΔT_t mechanism;[5] i.e., if a chemical group that can be present in two different states of polarity exists in the polymer chain, and these states are reversibly convertible by the action of an external stimulus, the polymer will show two different T_t values. This T_t shift (ΔT_t) opens a working temperature window in which the polymer isothermally and reversibly switches between the folded and unfolded states, following changes in the environment. This ΔT_t mechanism has been exploited to obtain many elastinlike smart derivatives.[5,45–47]

During the ITT, the polymer poly(VPGVG) folds into a dynamic helical structure optimizing the hydrophobic contacts; this action constitutes a molecular machine or a transducer between thermal energy and mechanical work. Those deductions at the molecular level have been tested at the macroscopic scale by cross-linking the polymer in an elastomeric matrix. For example, a matrix of poly(GVGVP) can be formed by γ-irradiation.[5] This matrix swells in water below 20°C, and it contracts less than one-half its extended length upon raising the temperature to 40°C, performing in this way useful mechanical work with an increase in temperature. This is thermally driven contraction or, in other words, a thermomechanical transduction, but it is not always the desired transduction. For example, the proteins in living organisms function as chemically driven molecular machines; they work at physiological temperatures and remain folded or unfolded, depending on whether the transition temperature is above or below the physiological temperature.

As the T_t can change under different types of stimuli, the ELP can have a smart nature under these stimuli, obtaining different kinds of energy transduction. This energy transduction can be chemomechanical, e.g., proton-driven contraction or salt-driven contraction. Proton-driven contraction is obtained in polymers that contain amino acids that have two states as a function of the pH.[9] An increase in the salt concentration, in most cases, decreases the T_t, thereby obtaining a ΔT_t between two different solution salt concentrations.[5,48] The energy transduction can also be an electromechanical transduction with the oxidation–reduction of some groups that change the T_t. There are other examples, such as photomechanical transduction, in which the polymer must contain a photochromic group, or baromechanical transduction,[49,50] where a change in the pressure causes a change in T_t. In general, any stimulus that can change the T_t can give rise to some type of energy transduction.

6.10 pH-Responsive ELPs

The mechanism of the ΔT_t is exploited in the model pH-responsive polymer ([VPGVG]$_2$-VPGEG-[VPGVG]$_2$)$_n$, where E stands for L-glutamic acid. In the present ELP, the γ-carboxylic function of the glutamic acid suffers strong polarity changes between its protonated and deprotonated states as a consequence of pH changes around its effective pK_a. Figure 6.7 shows the folded chain content as a function of T and pH for a genetically engineered polymer with the above general formula ($n = 45$). At pH = 3.5, in the protonated state, the T_t shown by the polymer is 30°C. Below this temperature, the polymer is unfolded and dissolved, while above the T_t the polymer folds and segregates from the solution. However, at pH = 8.0, the rise in the polarity of the γ-carboxyl groups, as they lose their protons and become carboxylate, is enough to increase T_t at values above 85°C, opening a working temperature window wider than 50°C. At temperatures above 30°C, the polymer would fold at low pH and unfold at neutral or basic conditions. In addition, this reveals the extraordinary efficiency of ELPs as compared with other pH-responsive

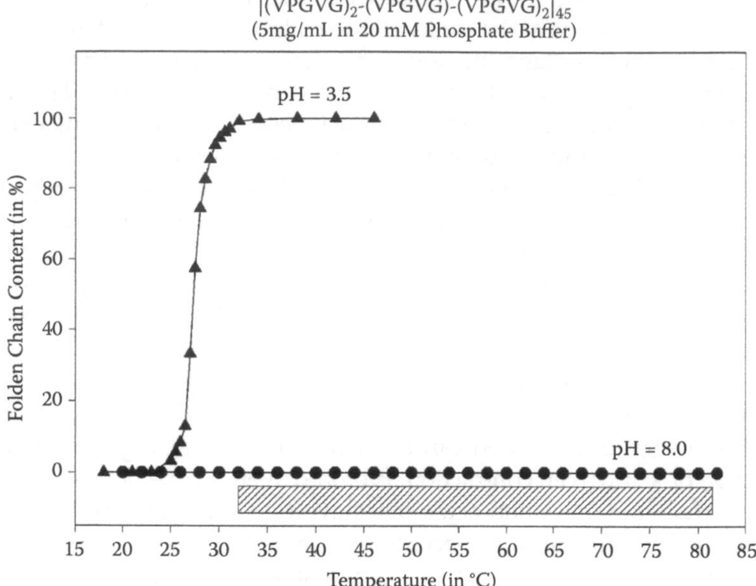

FIGURE 6.7
Turbidity temperature profiles of a model genetically engineered pH-responding ELP. (For details on bioproduction of this polymer, see Girotti, A. et al., *Macromolecules*, 37, 3396, 2004, and Urry, D.W., *What Sustains Life? Consilient Mechanisms for Protein-Based Machines and Materials*, Springer-Verlag, New York, 2005.) Box at bottom: window of working temperatures. Experimental conditions are given in plot. (Reproduced with permission from Elsevier, *Prog. Polym. Sci.*, 30, 1119, 2005.)

polymers, since this huge ΔT_t is achieved with just four E residues per 100 amino acids in the polymer backbone. This high efficiency is of practical importance in the use of these polymers in the design of molecular machines and nanodevices, such as nanopumps or nanovalves, because just a limited number of protons are needed to trigger the two states of the system. The influence of a charge over a side chain in a given ELP due to acid–basic equilibrium has been considered in the literature as a highly efficient way to achieve high ΔT_t. In a number of ELPs designed and studied to date, the capability of the free carboxyl or amino groups of aspartic acid, glutamic acid, or lysine to drive those T_t shifts is only overcome by the ΔT_t caused by the phosphorylation of serine.[5]

6.11 Photoresponsive ELPs

With a further chemical modification of an ELP, it is possible to obtain polymers that exploit the ΔT_t mechanism under different stimuli, increasing in this way the range of applicable stimuli for an energy transduction. It is possible, for instance, to modify some amino acid side chains to achieve systems with extended properties. A good example of this is the photoresponsive ELPs, which bear photochromic side chains either coupled to functionalized side chains in the previously formed polymer (chemically or genetically engineered) or by using nonnatural amino acids that are already photochromic prior to chemical polymerization.

The first example corresponds to this latter configuration. The polymer is an azobenzene derivative of poly(VPGVG), the copolymer poly(f_V[VPGVG], f_X[VPGXG]), where X is L-*p*-(phenylazo) phenylalanine, and f_V and f_X are mole fractions. The *p*-phenylazobenzene group suffers a photoinduced *cis–trans* isomerization. Dark adaptation or irradiation with visible light around 420 nm induces the presence of the *trans* isomer, the most apolar. On the other hand, UV irradiation (at around 348 nm) causes the appearance of high quantities of the *cis* isomer, which is a little more polar than the *trans*. Although the polarity change is not high, it is enough to obtain functional polymers due to the sensitivity and efficiency of ELPs. Figure 6.8 shows the photoresponse of one of these polymers with $f_X = 0.15$, which represents only 3 L-*p*-(phenylazo)phenylalanine groups per 100 amino acid residues in the polymer chain. In spite of the low polarity change and the exiguous presence of chromophores, it is evident (see Figure 6.8a) that there is a working temperature window at around 13°C (see Figure 6.8b).[47]

In another example, a different chromophore, a spiropyrane derivative, is attached over the free γ-carboxyl group of a Glu-containing ELP either chemically synthesized or genetically engineered. Figure 6.9 represents the photochromic reaction for this polymer.[46] As compared with *p*-phenylazo benzenes, spiropyrane compounds have photoreactions that can be driven by natural cycles of sunlight–darkness without the employment of UV sources, although UV irradiation causes the same effect as darkness and at a higher rate.[46]

Again, the difference in polarity between the spiro- and merocyanine forms (see Figure 6.9) is enough to cause a significant T_t shift. Figure 6.10 shows the turbidity profiles of the polymer in the different illumination regimens (Figure 6.10a) and the photomodulation of the polymer folding and unfolding (Figure 6.10b and c). Again, the efficiency of the polymer is outstanding, since just 2.3 spiropyran chromophores per 100 amino acid residues in the polymer backbone was sufficient to render the clear photomodulation shown in Figure 6.9 and Figure 6.10.

It is possible to increase and further control the smart behavior of ELPs without increasing the number of sensitive moieties. This is possible if this moiety has one state that is able to interact with a different compound, and this interaction further increases the difference in polarity between both states. This is the basis of the so-called amplified ΔT_t mechanism and has been proved for a *p*-phenylazobenzene-containing polymer of the kind shown above, poly(0.8[VPGVG], 0.2[VPGXG]), in the presence of α-cyclodextrin (αCD).[45] αCD is able to form inclusion compounds with the *trans* isomer of the *p*-phenylazobenzene group and not with the *cis* due to steric

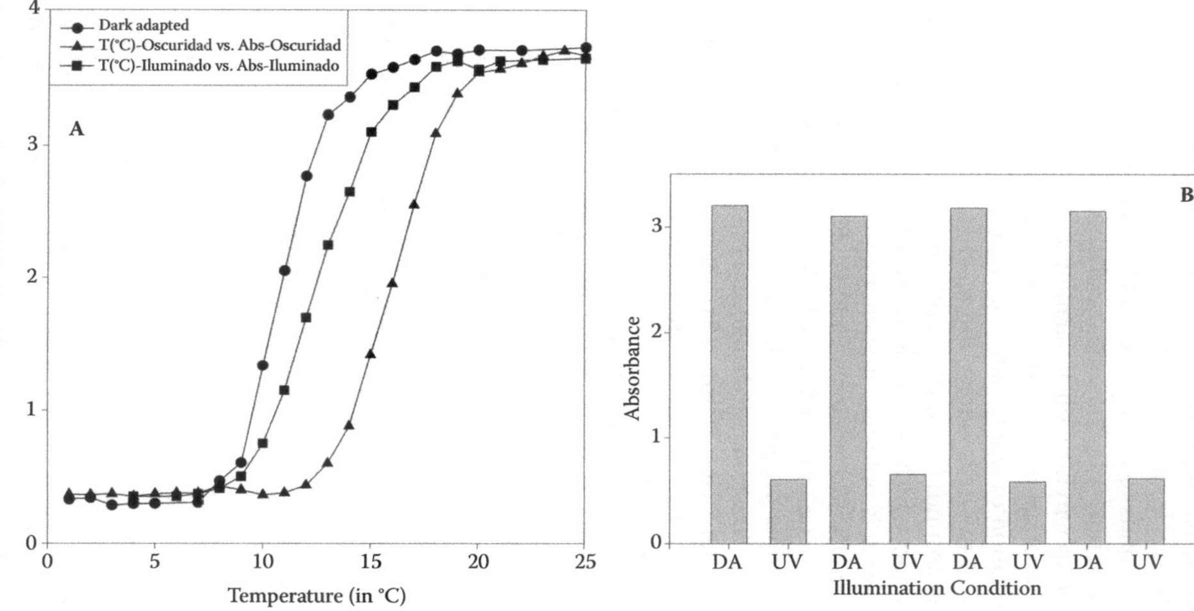

FIGURE 6.8
(A), Temperature profiles of aggregation of 10-mg·ml^{-1} water solutions of the photoresponsive poly(0.85[VPGVG], 0.15[VPGXG]) (X ≡ L-*p*-[phenylazo]phenylalanine) under different illumination regimens. The correspondence between each profile and its illumination condition is indicated in the plot. (Details on polymer synthesis and illumination conditions can be found in Tanford, C., *The Hydrophobic Effect: Formation of Micelles and Biological Membranes*, Wiley, New York, 1973.) (B), Photomodulation of phase separation of 10-mg·ml^{-1} aqueous samples of poly(0.85[VPGVG], 0.15[VPGXG]) at 13°C. The illumination conditions prior to measurement are indicated in the horizontal axis: DA, dark adaptation; UV, UV irradiation. (Reproduced with permission from the American Chemical Society, *Macromolecules*, 34, 8072, 2001.)

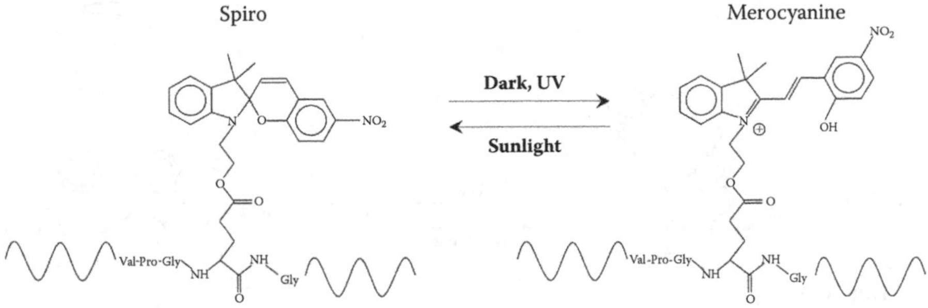

FIGURE 6.9
Photochemical reaction responsible for the photochromic behavior of the ELP spiropyran derivative. (Reproduced with permission from the American Chemical Society, *Macromolecules*, 33, 9480, 2000.)

hindrance[45] (see Figure 6.11) as a consequence of the relatively high polarity of the αCD outer shell, which, of course, is quite more polar than the *p*-phenylazobenzene moiety in both the *trans* or *cis* states. The change in polarity between the dark-adapted sample (*trans* isomer buried inside the αCD) and the UV-irradiated one (*cis* isomer unable to form inclusion compounds) led to an enhanced ΔT_t (see Figure 6.12).

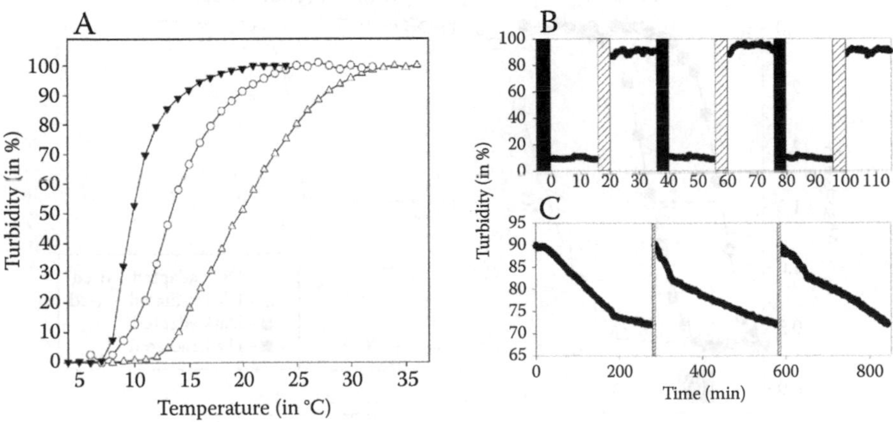

FIGURE 6.10
(A) Temperature profiles of aggregation of 20-mg·ml^{-1} phosphate-buffered (0.01*N*, pH 3.5) water solutions of the photoresponsive polymer under different illumination regimens. The correspondence between each profile and its illumination condition is indicated in the plot. Turbidity was calculated from the absorbance values obtained at 600 nm on a Cary 50 UV-Vis spectrophotometer equipped with a thermostatized sample chamber. (B and C) Photomodulation of phase separation of 5-mg·ml^{-1} aqueous samples of the photochromic polymer (T = 14°C, 0.01*N* phosphate buffer at pH = 3.5). (B) UV–sunlight cycles; boxes in the subplot represent periods of irradiation: black boxes = UV, white boxes = sunlight. (C) Darkness–sunlight cycles; boxes in the subplot represent periods of sunlight irradiation. (Reproduced with permission from the American Chemical Society, *Macromolecules*, 33, 9480, 2000.)

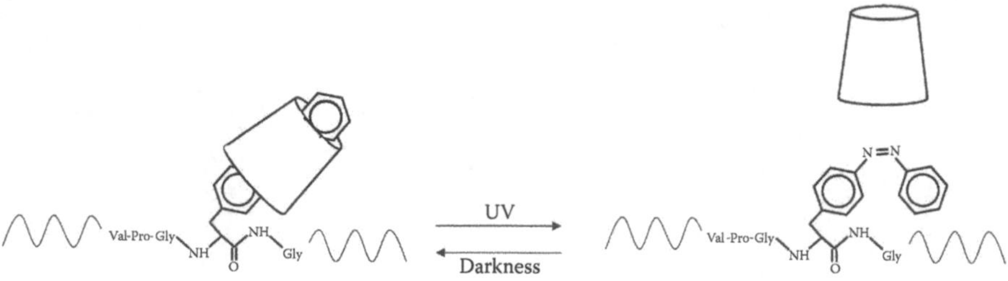

FIGURE 6.11
Schematic diagram of the proposed molecular mechanism on the interaction between the *p*-phenylazobenzene pendant group and the αCD. (Reproduced with permission from Wiley, *Adv. Mater.*, 14, 1151, 2002.)

In conclusion, in the coupled αCD/poly(0.8[VPGVG], 0.2[VPGXG]) photoresponsive system, αCD acts like an amplifier on an electronic circuit. αCD promoted a tunable offset, gain, and inversion of the photoresponse of the polymer (Figure 6.12). In this way, the polymer photoresponsiveness could be shifted to room or body temperature and with a wider range of working temperatures. Furthermore, the amplified ΔT_t mechanism is not

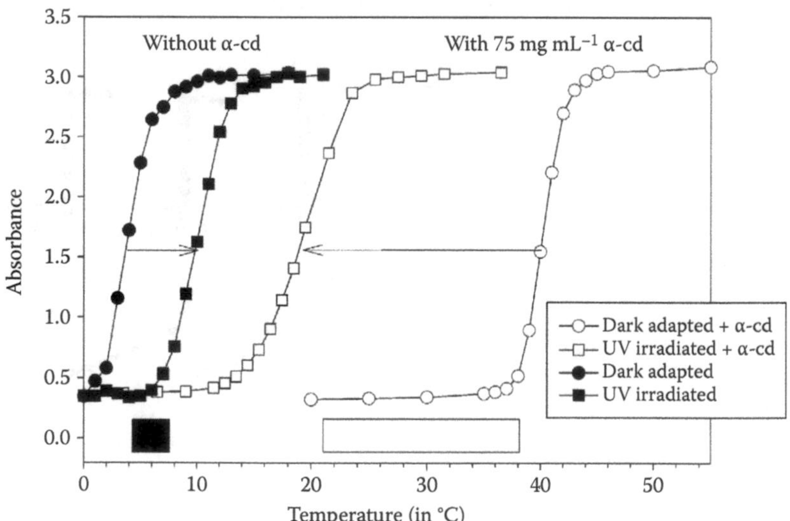

FIGURE 6.12
Temperature profiles of aggregation of 10-mg·ml^{-1} water solutions of the photoresponsive ELP in the absence and presence (75 mg·ml^{-1}) of αCD under both illumination regimens. Circles represent dark-adapted samples, and squares represent UV-irradiated samples; hollow symbols represent the presence of αCD, and filled symbols represent the absence of αCD; arrows represent the sense of displacement of the turbidity profile caused by UV irradiation of the corresponding dark-adapted sample. Boxes at the bottom indicate the window of working temperatures open when the system is in absence (filled box) and presence (hollow box) of αCD. (Reproduced with permission from Wiley, *Adv. Mater.*, 14, 1151, 2002.)

restricted to photoresponsive ELPs and could be exploited in some other smart ELPs responding to stimuli of a different nature. It also adds a further possibility of control, since the ability of CDs to form inclusion compounds can be controlled by different stimuli in some modified CDs.[51–53]

6.12 Self-Assembling Behavior

Polymer self-assembly is one of the most active and promising fields of materials science due to its potential in nanotechnological and nanobiotechnological applications. This is especially so with the tunable properties of the self-assembling block copolymers.[54] Their applicability rests on their capability of producing structures at the nanometer scale whose molecular order can be tuned to have a certain periodicity and hierarchy. These structures can be employed in many different areas of technological interest, including nanolithography, nanostructured membranes, templates for nanoparticle synthesis, enhanced and controlled catalysis, photonic crystals, high-density information-storage media, nanodevices for drug delivery, and a multitude of other applications.

Nature uses self-assembling materials for nanostructures, such as the components for living cells, thereby demonstrating an impressive example of the bottom-up approach.[55] Furthermore, most biological systems exhibit hierarchy, with different features at different length scales at the molecular, nanoscopic, microscopic, and macroscopic levels. For example, natural elastin undergoes a self-aggregation process in its natural environment. Elastin is produced from a water-soluble precursor, tropoelastin, which spontaneously aggregates to yield fibrillar structures that are finally stabilized by enzymatic interchain cross-links. This produces the well-known insoluble and elastic elastin fibers that can be found in abundance in the skin, lungs, arteries, and, in general, those parts of the body undergoing repeated stress–strain cycles.

The self-assembling ability of elastin seems to reside in certain relatively short amino acid sequences, as has recently been probed by Yang et al.[56] working in recombinant ELPs. Some of these polypeptides have shown that, above their T_t, they are able to form nanofibrils that further organize into hexagonally close-packed arrangements when the polymer is deposited onto a hydrophobic graphite substrate.[56]

However, in ELPs, this tendency to self-assemble in nanofibers can be expanded to other topologies and nanostructured features.[57–59] Taking advantage of (a) the opportunities and potential given by genetic engineering in designing new polymers, (b) the possibility of obtaining them monodisperse with an absolute control on the primary structure, (c) the growing understanding of the molecular behavior of ELPs, and (d) the wealth of experimental and theoretical experience gained in recent years on the self-assembling characteristics of different types of block copolymers, different

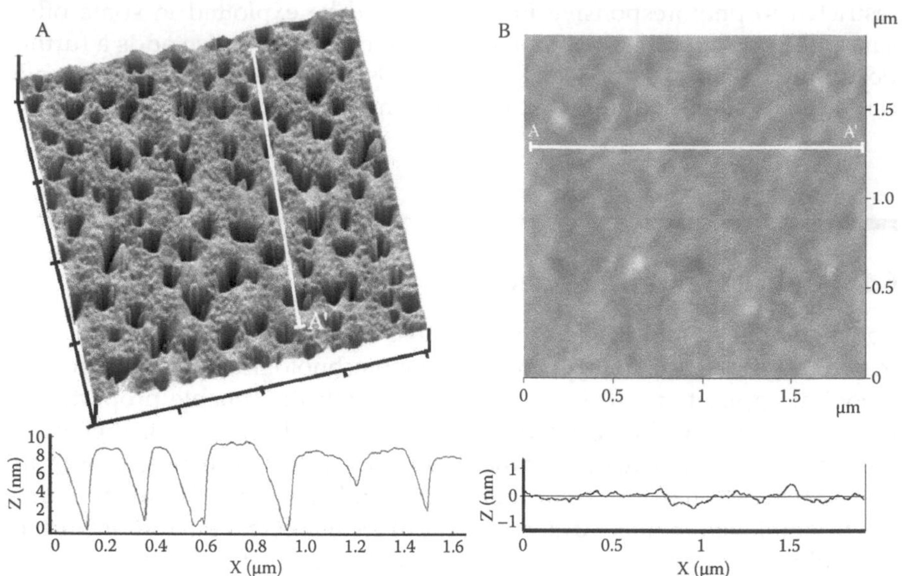

FIGURE 6.13
Tapping-mode AFM image of ([VPGVG]$_2$-[VPGEG]-[VPGVG]$_2$)$_{15}$ deposited from a water solution on a Si hydrophobic substrate. Sample conditions: (A) 10 mg·ml^{-1} in 0.02*M* NaOH–water solution (basic solution) and (B) 10 mg·ml^{-1} in 0.02*M* HCl–water solution (acid solution). (Reproduced with permission from the American Chemical Society, *J. Am. Chem. Soc.*, 126, 13212, 2004.)

self-assembling properties are starting to be unveiled within the ELP family. For example, it has been shown that the ELP previously shown as a pH-responding polymer, ([VPGVG]$_2$[VPGEG][VPGVG]$_2$)$_{15}$, is able to form polymer sheets showing self-assembled nanopores[59] (Figure 6.13).

When this Glu-containing polymer is deposited by spin coating from a basic medium on a Si hydrophobic substrate, the formation of a quite homogeneous and aperiodic nanopore distribution is a noteworthy topological feature. Figure 6.13a shows a three-dimensional AFM image of the nanopores. The nanopores are approximately 70 nm in width, at least 8 nm in depth, and they are separated by approximately 150 nm. Figure 6.13b shows the result of the spin coating of an acidic solution of the same polymer. Under these acidic conditions, the polymer shows just a flat surface without any topologically remarkable features.

This different behavior as a function of pH has been explained in terms of the different polarity state exhibited by the free γ-carboxyl groups of glutamic acid. When dissolved in basic media, the negatively charged carboxylate becomes the point of highest polarity along the chain, contrasting with the quite hydrophobic surroundings composed by the rest of polymer, which is predominantly apolar and is a hydrophobic substrate. These domains, due to their charge, impede in their vicinities both hydrophobic contacts between chains and hydrophobic interaction with the substrate.

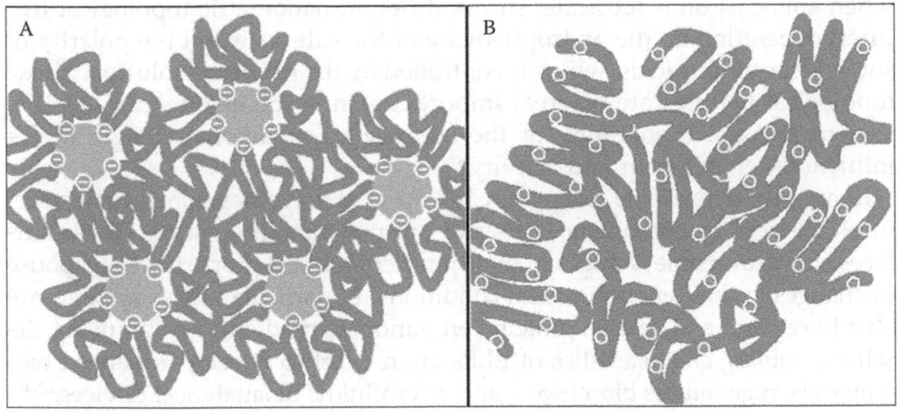

FIGURE 6.14
Schematic representation of polymer distribution on hydrophobic substrate: (A) in a basic medium and (B) in an acid medium. Counterions have been not drawn for clarity. (Reproduced with permission from the American Chemical Society, *J. Am. Chem. Soc.*, 126, 13212, 2004.)

Since the position of the polar domains is well controlled to be equally spaced within nanometer distances along the polymer chain, these domains and their hydration water are finally segregated from the predominantly hydrophobic environment, giving rise to the formation of pores of nanometer dimensions (Figure 6.14). However, in the acidic solution, the free γ-carboxylic group is protonated, and in this state this group shows a low polarity, yielding a quite homogeneous polymer in polarity. Then, in this protonated state, spin coating of the solution over a hydrophobic substrate yields a flat and homogeneous deposition with no particular topographical features (Figure 6.14).

The controlled amphiphilicity that has been shown here can be exploited in polymers designed for other purposes, e.g., for tissue engineering. Prieto et al.[60] have shown that the polymer containing regularly spaced REDV (Arg Glu Asp Val) and VPGKG (where K stands for L-lysine) sequences (Figure 6.15),

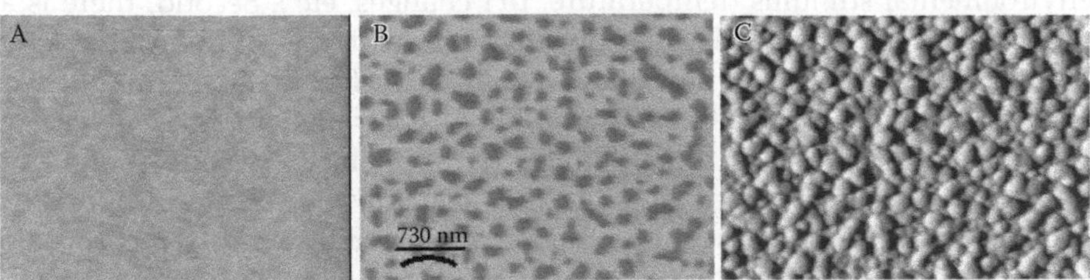

FIGURE 6.15
AFM images of the deposition of a polymer for tissue engineering with different conditions: (A) polymer solution at pH 7 deposited on hydrophobic silicon substrate; (B) polymer solution at pH 12 deposited on hydrophobic silicon substrate; (C) polymer solution at pH 12 deposited on hydrophilic mica. (Adapted from Prieto, S. et al., *Nano Lett.*, submitted. With permission.)

when spin-cast on a substrate, shows different nanometric topological features, depending on the hydrophobicity of the substrate and the polarity of some of the amino acids, which is controlled by the pH of the solution. Those topological features are of great importance in tissue engineering, as it is known that the topography of the substratum at different length scales influences cell behavior and activity.[61]

In another remarkable example, Chilkoti et al. have developed nanostructured surfaces by combining ELPs and dip-pen nanolithography. These surfaces show reversible changes in their physicochemical properties in response to changes in their environmental conditions. In particular, these systems are able to capture and release proteins on nanopatterned surfaces by using the self-assembling characteristics of ELPs in an effort to develop advanced biomaterials, regenerable biosensors, and microfluidic bioanalytical devices.[62–64]

There are already noteworthy examples of diblock and triblock copolymers of ELPs. For example, Wright et al.[57] have used BAB triblock copolymers in which the endblock domains undergo a self-assembly in aggregates when the temperature is raised above T_t, whereas the central blocks act as virtual cross-links between aggregates, giving rise to the formation of a gel state. However, despite the extraordinary potential of ELPs in producing self-assembling block copolymers, exploitation of this potential remains poor. It is possible to use genetic engineering to produce ELPs with blocks of extremely controlled size and composition. These ELPs show different physicochemical properties that, in an adequate environment, segregate in various immiscible phases and, with the right combination of adequate external fields, are able to self-organize into a variety of highly interesting nanostructures. One of the physicochemical properties that can be used to trigger phase segregation among the different blocks is their hydrophobic–hydrophilic nature. This opens an interesting possibility for using ELP blocks to construct self-assembling block copolymers.

The tendency of these ELPs to show controlled hydrophobic association can be exploited to obtain advanced multiblock copolymers with three main advantages. First, the hydrophobic association of ELPs can be externally controlled, since it is associated with the ITT, and this is triggered by an environmental stimulus (temperature, pH changes, etc.). Second, there is a deep and quantitative body of knowledge on the degree of hydrophobicity of the different amino acid side chains. Third is the fact that the unparalleled ability to achieve complex and completely controlled PBPs is given by genetic engineering. The block length, hydrophobicity, composition, and position can be engineered at will with absolute precision. Additionally, genetically engineered elastinlike block copolymers can easily incorporate any other structural feature of interest for self-assembly and function, such as β-sheet-forming domains, leucine zippers, binding of domains to different substrates, and any biofunctionality imparted by bioactive peptides (cell attachment sequences, etc.). All three of these characteristics will certainly open new ways of creating advanced multiblock copolymers with applications spreading to a wide variety of technological fields.

6.13 ELPs for Biomedical Applications: Introducing Tailored Biofunctionality

One field in which the ELPs are being used with incredible success is in biomedical applications. The suitability of ELPs as polymers for biomedical applications is based on three important pillars.[10] The first is the existence of the ITT, which is the base of the ELPs' remarkable smart and self-assembling properties. A second pillar in the development of such extraordinary materials is the apparent power of genetic engineering in obtaining complex and well-defined polymers with controlled and multiple (bio)functionality. Additionally, ELPs show a third property, which is highly relevant when planning the use of these polymers in the most advanced biomedical applications, such as tissue engineering and controlled drug release. This third pillar is the tremendous biocompatibility shown by ELPs.

The complete series of the ASTM-recommended generic biological tests for materials and devices in contact with tissues and tissue fluids and blood demonstrate an unmatched biocompatibility.[3] Despite the polypeptide nature of these polymers, it has not been possible to obtain monoclonal antibodies against most of them. Apparently, the immune system ignores these polymers because it cannot distinguish them from natural elastin. It has been argued that the dynamic nature of segments in the β-spirals of the phase-separated (hydrophobically associated) state at physiological temperatures constitutes a barrier to interaction that would otherwise be required for identification as a foreign material by the host.[65] In addition, the secondary products of their bioabsorption are just simple and natural amino acids.

With this set of properties, it is not surprising that the biomedical uses of ELPs seem to be the first area where ELPs will enter the market. This is especially true considering that the biomedical (and cosmetic) market shows a clear disposition to quickly adopt new developments that show superior performance. Additionally, this sector is not so conditioned by the costs associated with the materials used in their devices and developments, in contrast to commodity manufacturing and other applications. Thus the companies producing ELPs will find the biomedical sector a good option for amortizing the costs incurred in developing the know-how and technology required to produce ELPs.

Two main fields of interest for ELPs have been targeted in biomedical applications. The first application of ELPs is for drug-delivery systems, in which a polymer that can be in a matrix, aggregate, or device releases a necessary drug gradually or when a certain stimulus triggers it. The second major application is the use of ELPs for tissue engineering in which the polymer, normally in the form of a matrix, works as a temporary scaffold that is gradually replaced by endogenous growing tissue. Hence, the artificial tissue must possess the mechanical properties of the natural tissue for which it is intended as a substitute while also favoring cell proliferation and gradual replacement by new tissue.

Different versions of ELPs have been designed in an effort to develop polymers for drug delivery. However, these efforts do not share a common basic strategy on design. Indeed, ELPs display many different properties that can be useful for drug-delivery purposes, including smart behavior (sensitivity to certain stimuli), self-assembly, biocompatibility, etc., so the number of diverse strategies is understandable.

The firsts ELP-based drug-delivery systems were reported by Urry.[66] They were quite simple devices in which γ-radiated cross-linked poly(VPGVG) hydrogels of different shapes were loaded with a model water-soluble drug, and this drug was then released by diffusion. From that simple system, using only the property of high biocompatibility, other systems have been developed. In a further design, the basic VPGVG pentapeptide was functionalized by including some glutamic acid to provide free carboxyl groups for cross-linking purposes. The cross-linkers, of a type that forms carboxyamides, were selected because of their ability to hydrolyze at a controlled rate, thus releasing the polymer chains and, concurrently, any drug entrapped within the hydrogel.[67] This was an apparently simple and conventional degradation-based drug-delivery system. However, due to the characteristics of ELPs, the displayed behavior was a little more complex and efficient. While the cross-links were intact, the carboxyl groups were amidated and, consequently, uncharged. This state of lower polarity yielded a cross-linked ELP material with a T_t below body temperature. At that temperature the chains were folded, the material contracted and deswelled, and the polymer chains were essentially insoluble, entrapping the loaded drug quite efficiently in the models studied. When hydrolysis took place on the outer surface of the hydrogel, charged carboxylates appeared, which strongly increased the T_t (well above body temperature) in this zone. The skin of the slab swelled, and the loaded drug readily escaped from the outer layer of the pellet. Additionally, since the fully released chains were completely soluble, they soon diffused away and were reabsorbed. This ensured the presence of an ever-fresh surface on the slab hydrogel, ready to release more drug.[67] Accordingly, the kinetics of drug release was almost zero order and, hence, the performance of the system was superior to others based on equivalent polymers but lacking the ΔT_t mechanism.

In a different example, the trend to form stable, drug-loaded nano- and microparticles by some ELPs, especially those based on the pentapeptide VPAPG (thanks to its hysteresis cycle), has facilitated the development of injectable systems for controlled release.[68]

These examples are based on simple polymer formulations that still fall far short of the full potential of ELPs for drug-delivery systems. Their smart and self-assembly properties, as well as deeper knowledge of the molecular basis of the ITT, have only been marginally exploited. However, new systems are beginning to appear in the literature that show a more decided bent toward exploiting the very special characteristics of ELPs and the powerful way they can be produced by genetic engineering. For example, Chilkoti's group[62–64] has produced nice examples of ELPs specially designed for targeting and intracellular drug delivery. They exploited the soluble–insoluble transition of the ELPs to target a solid tumor by local hyperthermia, and then, in the most sophisticated versions, an additional pH response of these ELPs was used to mimic the membrane-disruptive properties of viruses and toxins to cause effective intracellular drug delivery. Among the most evident advantages of this kind of advanced drug-delivery system is the more efficient dosage of antitumor drugs. However, these polymers could also serve as alternatives to fusogenic peptides in gene-therapy formulations and to enhance the intracellular delivery of protein therapeutics that function in the cytoplasm.[17,69,70]

On the other hand, recent progress in understanding the molecular characteristics of the ITT has allowed development of advanced systems for more-general drug release that have achieved practically ideal zero-order release kinetics without the concerns caused by previous designs. The first examples are based on Glu-containing ELPs mentioned previously, in which the highly hydrophobic microenvironment of the γ-carboxyl groups is maintained by positioning Phe residues in the polymer sequence. In neutral or basic conditions (including physiological pH), those carboxylate moieties show a strong propensity to neutralize their charge by ion-coupling contacts with positively charged drugs if such coupling causes an effective decrease of the polarity of the carboxyl microenvironment. As a result, these polymers can form insoluble aggregates at neutral or basic conditions and in the presence of a drug of opposite sign. These materials are characterized by a high rate of drug loading and, upon implantation, release the drug slowly as it is leached from its coupling on the outer surface of the aggregate. The release rate can be tuned by modifying the hydrophobic environment of the carboxyl with a properly chosen amino acid sequence in the polymer.[38,39] Once the drug is released and the polymer drug interaction is lost (as a consequence of the charged state of the carboxyl group [carboxylate]), the polymer unfolds and finally dissolves. At the same time, the interface between the remaining insoluble, still-loaded aggregate and the body fluids is continuously renewing without changing the physical–chemical properties for practically all of the functional period of the system. This behavior results in near-ideal zero-order release.[38,39] Others versions of that system have been formed with a lysine-containing ELP in which the γ-amino group of this lysine groups is maintained highly hydrophobic by positioning Phe residues in the polymer sequence. As a consequence, in neutral or acidic conditions (including physiological pH), those amino moieties show a strong propensity to neutralize negatively charged drugs.[65]

Another important field in materials science for biomedical application is the field of tissue engineering (or the currently preferred term, "regenerative medicine"). In this discipline, a polymer matrix has to substitute, at least transiently, the natural extracellular matrix (ECM). However, the ECM is a complex system with certain mechanical properties and includes many proteins (fibronectin, collagen, elastin, etc.) containing in their sequences a huge number of bioactive peptides that are of crucial importance in the natural processes of wound healing. A material that fulfills some important characteristics of the ECM must be a complex material. In this regard, the GEPBPs are well situated, since it is possible to include bioactive sequences as building blocks in a complex polymer.

Some ELPs have been designed for tissue engineering. These polymers have good properties that make them excellent candidates for such applications, including their high biocompatibility, their mechanical properties in the range of the natural elastin, the fact that they are GEPBPs and thus easily include bioactive peptides sequences, and the advantage that they are smart materials that can give different responses from the simple purification of the polymer to the more-complex systems that respond to changes in their environment.

The starting point in the current development of ELP for tissue engineering is poly(VPGVG) and its cross-linked matrices. Surprisingly, the cross-linked matrices of poly(VPGVG)s, when tested for cell adhesion, showed that cells do not adhere to this matrix and that no fibrous capsule forms around it when implanted.[71] This matrix and other states of the material have a potential use in the prevention of postoperative, posttrauma adhesions.[71] However, most directed polymers for tissue engineering include other sequences, e.g., cross-linking or bioactive sequences, such as the cell-adhesion peptide RGD[72] (where R stands for L-Grginine, G for glycine, and D for aspartate) or REDV, which is specific to endothelial cells.[73,74] Figure 6.16 shows an example of these complex polymers for tissue engineering.[74] The designed polymer has a monomer that is 87 amino acids long, comprising four different building blocks. Such polymers are based on an ELP sequence (VPGIG) that donates its properties to the polymer; the second domain is another ELP sequence that contains a lysine for cross-linking proposes; the third block is the CS5 human fibronectin domain, which contains the well-known endothelial cell attachment sequence, REDV; and the fourth block is the hexapeptide VGVAPG, which is an elastase target sequence to favor its bioprocessability by natural routes.

These ELPs for tissue engineering use some of the advantages of the GEPBPs. They can be produced at low cost; they have a strict control in the primary composition; and they can be designed with a high level of complexity. However, huge possibilities await a better understanding of cell-materials interactions, protein folding, and the ITT. The ultimate goal is the design of advanced systems that maximize the potential of smart materials to respond to changes in their environment.

(VPGIG)$_2$ (VPGKG) (VPGIG)$_2$ (EEIQGHIPREDVDYHLYP) **(VPGIG)$_2$ (VPGKG) (VPGIG)$_2$** (VGVAPG)$_3$

FIGURE 6.16
Schematic composition of monomer used in ELP design described in text. The scheme shows the different functional domains of monomer, which can be easily identified by their corresponding peptide sequences. (Reproduced with permission from Elsevier, *Prog. Polym. Sci.*, 30, 1119, 2005.)

6.14 Outlook

ELPs are being intensely studied because they can shed light on the processes of protein folding and energy conversions in living cells. Moreover, their exceptional biocompatibility should provide important clues as to why some systems avoid detection by the immunologic system.

The ΔT_t mechanism, which has proved to be an efficient system for energy transductions, opens a vast number of applications in the technological world. ELPs can be exploited, for example, as molecular switches, molecular motors, molecular sensors, and nanopumps or nanovalves that are activated by external or internal stimuli. They can also be used in microfluidic devices, in labs-on-a-chip, or in smart surfaces or membranes. Furthermore, using self-assembly properties, they can be designed to obtain systems with smart behavior at a high length scale using nature's bottom-up approach for construction of nanoscale components in living cells.

Given their outstanding biocompatibility, ELPs are most likely to find success in the biomedical field. Some new systems have appeared for drug delivery, and some of these take advantage of the smart nature of ELPs for drug release. The smart behavior of ELPs opens the possibility for development of more-complex systems in which the ΔT_t can be triggered by a stimulus caused by a certain disease, e.g., systems that are triggered by the local hyperthermia of tumors. Such targeted drug-delivery systems could also include zero-order release systems in which ELPs are gradually dissolved or deswelled as the drug is released.

ELPs have also been used in tissue engineering. These ELPs are produced by genetically engineered synthesis to obtain extremely complex polymers that maintain the properties of the ELPs and include new sequences with other functionalities mimicking some of the characteristics of the ECM or other characteristics that favor tissue regeneration. New protein functionalities continue to be discovered, and these will lead to the development of more sequences with a better design. Also, as these ELPs for tissue engineering are smart polymers, they could be satisfactorily used in advanced systems that respond to a certain stimulus or that release a drug under certain circumstances.

ELPs show interesting and unconventional properties, and they have great potential for use in many different fields. Furthermore, as they can be tailored with unlimited complexity, they can be designed to suit a wide variety of applications.

References

1. Ayad, S. et al., *The Extracellular Matrix Facts Book*, Academic Press, San Diego, 1994.
2. Urry, D.W. et al., Protein-based materials with a profound range of properties and applications: the elastin ΔT_t hydrophobic paradigm, in *Proteins and Modified*

Proteins as Polymeric Materials, McGrath, K. and Kaplan, D., Eds., Birkhäuser, Boston, 1997, pp. 133–177.

3. Urry, D.W. et al., Biocompatibility of the bioelastic materials, Poly(GVGVP) and its gamma-irradiation cross-linked matrix: summary of generic biological test results, *J. Bioact. Compat. Polym.*, 6, 263, 1991.
4. Cappello, J., Genetic production of synthetic protein polymers, *MRS Bull.*, 17, 48, 1992.
5. Urry, D.W., Molecular machines: how motion and other functions of living organisms can result from reversible chemical changes, *Angew. Chem., Int. Ed.*, 32, 819, 1993.
6. McGrath, K. and Kaplan, D.S., *Protein-Based Materials*, Birkhäuser, Boston, 1997.
7. Rodríguez-Cabello, J.C. et al., Genetic engineering of protein-based polymers: the example of elastin-like polymers, *Adv. Polym. Sci.*, 200, 119, 2006.
8. Lee, J., Macosko, C.W., and Urry, D.W., Elastomeric polypentapeptides cross-linked into matrixes and fibers, *Biomacromolecules*, 2, 170, 2001.
9. Girotti, A. et al., Influence of the molecular weight on the inverse temperature transition of a model genetically engineered elastin-like pH-responsive polymer, *Macromolecules*, 37, 3396, 2004.
10. Rodriguez-Cabello, J.C. et al., Developing functionality in elastin-like polymers by increasing their molecular complexity: the power of the genetic engineering approach, *Prog. Polym. Sci.*, 30, 1119, 2005.
11. Yu, Y.B., Coiled-coils: stability, specificity, and drug delivery potential, *Adv. Drug Delivery Rev.*, 54, 1113, 2002.
12. Bilgicer, B., Fichera, A., and Kumar, K., A coiled coil with a fluorous core, *J. Am. Chem. Soc.*, 123, 4393, 2001.
13. Tang, Y. et al., Fluorinated coiled-coil proteins prepared *in vivo* display enhanced thermal and chemical stability, *Angew. Chem., Int. Ed.*, 40, 1494, 2001.
14. Potekhin, S.A. et al., Synthesis and properties of the peptide corresponding to the mutant form of the leucine-zipper of the transcriptional activator gcn4 from yeast, *Protein Eng.*, 7, 1097, 1994.
15. Zhang, G.H. et al., Biological synthesis of monodisperse derivatives of poly(alpha,L-glutamic acid): model rodlike polymers, *Macromolecules*, 25, 3601, 1992.
16. Meyer, D.E. and Chilkoti, A., Genetically encoded synthesis of protein-based polymers with precisely specified molecular weight and sequence by recursive directional ligation: examples from the elastin-like polypeptide system, *Biomacromolecules*, 3, 357, 2002.
17. Chilkoti, A., Dreher, M.R., and Meyer, D.E., Design of thermally responsive, recombinant polypeptide carriers for targeted drug delivery, *Adv. Drug Delivery Rev.*, 54, 1093, 2002.
18. Haider, M., Megeed, Z., and Ghandehari, H., Genetically engineered polymers: status and prospects for controlled release, *J. Controlled Release*, 95, 1, 2004.
19. Sandberg, L.B. et al., Elastin covalent structure as determined by solid-phase amino-acid sequencing, *Pathol. Biol.*, 33, 266, 1985.
20. Yeh, H. et al., Sequence variation of bovine elastin messenger-RNA due to alternative splicing, *Coll. Relat. Res.*, 7, 235, 1987.
21. Sandberg, L.B., Soskel, N.T., and Leslie, J.G., Elastin structure, biosynthesis, and relation to disease states, *N. Engl. J. Med.*, 304, 566, 1981.
22. Gowda, D.C. et al., Synthesis, characterization and medical applications of bioelastic materials, in *Peptides: Design, Synthesis and Biological Activity*, Basava, C. and Anantharamaiah, G.M., Eds., Birkhäuser, Boston, 1994, p. 81.

23. Kurkova, D. et al., Structure and dynamics of two elastin-like polypentapeptides studied by NMR spectroscopy, *Biomacromolecules*, 4, 589, 2003.
24. Martino, M., Perri, T., and Tamburro, A.M., Biopolymers and biomaterials based on elastomeric proteins, *Macromol. Biosci.*, 2, 319, 2002.
25. McPherson, D.T., Xu, J., and Urry, D.W., Product purification by reversible phase transition following *Escherichia coli* expression of genes encoding up to 251 repeats of the elastomeric pentapeptide GVGVP, *Protein Expr. Purif.*, 7, 51, 1996.
26. Welsh, E.R. and Tirrell, D.A., Engineering the extracellular matrix: a novel approach to polymeric biomaterials, I: control of the physical properties of artificial protein matrices designed to support adhesion of vascular endothelial cells, *Biomacromolecules*, 1, 23, 2000.
27. San Biagio, P.L. et al., The overlap of elastomeric polypeptide coils in solution required for single-phase initiation of elastogenesis, *Chem. Phys. Lett.*, 145, 571, 1988.
28. Urry, D.W., Physical chemistry of biological free energy transduction as demonstrated by elastic protein-based polymers, *J. Phys. Chem. B*, 101, 11007, 1997.
29. Rodríguez-Cabello, J.C. et al., Differential scanning calorimetry study of the hydrophobic hydration of the elastin-based polypentapeptide, poly(VPGVG), from deficiency to excess of water, *Biopolymers*, 54, 282, 2000.
30. Tanford, C., *The Hydrophobic Effect: Formation of Micelles and Biological Membranes*, Wiley, New York, 1973.
31. Pauling, L. and Marsh, R.E., The structure of chlorine hydrate, *Proc. Nat. Acad. Sci. U.S.A.*, 38, 112, 1952.
32. Urry, D.W., Trapane, T.L., and Prasad, K.U., Phase-structure transitions of the elastin polypentapeptide-water system within the framework of composition-temperature studies, *Biopolymers*, 24, 2345, 1985.
33. Manno, M. et al., Interaction of processes on different length scales in a bioelastomer capable of performing energy conversion, *Biopolymers*, 59, 51, 2001.
34. Reiersen, H., Clarke, A.R., and Rees, A.R., Short elastin-like peptides exhibit the same temperature-induced structural transitions as elastin polymers: implications for protein engineering, *J. Mol. Biol.*, 283, 255, 1998.
35. Alonso, M. et al., Effect of alpha-, beta- and gamma-cyclodextrins on the inverse temperature transition of the bioelastic thermo-responsive polymer poly(VPGVG), *Macromol. Chem. Phys.*, 202, 3027, 2001.
36. Meyer, D.E. and Chilkoti, A., Quantification of the effects of chain length and concentration on the thermal behavior of elastin-like polypeptides, *Biomacromolecules*, 5, 846, 2004.
37. Urry, D.W., The change in Gibbs free energy for hydrophobic association: derivation and evaluation by means of inverse temperature transitions, *Chem. Phys. Lett.*, 399, 177, 2004.
38. Urry, D.W., Deciphering engineering principles for the design of protein-based nanomachines, in *Protein-Based Nanotechnology*, Renugopalakrishnan, V. and Lewis, R., Eds., Kluwer Academic Publishers, Dordrecht, 2006, p. 141.
39. Urry, D.W., *What Sustains Life? Consilient Mechanisms for Protein-Based Machines and Materials*, Springer-Verlag, New York, 2005.
40. Rodríguez-Cabello, J.C. et al., Endothermic and exothermic components of an inverse temperature transition for hydrophobic association by TMDSC, *Chem. Phys. Lett.*, 388, 127, 2004.
41. Reguera, J. et al., Thermal behavior and kinetic analysis of the chain unfolding and refolding and of the concomitant nonpolar solvation and desolvation of two elastin-like polymers, *Macromolecules*, 36, 8470, 2003.

42. Friedman, H.L., Kinetics of thermal degradation of char-forming plastics from thermogravimetry: application to phenolic plastic, *J. Polym. Sci., Part C*, 183, 1964.
43. Schmidt, P. et al., Role of water in structural changes of poly(AVGVP) and poly(GVGVP) studied by FTIR and Raman spectroscopy and *ab initio* calculations, *Biomacromolecules*, 6, 697, 2005.
44. Luan, C.-H. and Urry, D.W., Elastic, plastic, and hydrogel protein-based polymers, in *Polymer Data Handbook*, Mark, J.E., Ed., Oxford University Press, New York, 1999, p. 78.
45. Rodríguez-Cabello, J.C. et al., Amplified photoresponse of a *p*-phenylazobenzene derivative of an elastin-like polymer by alpha-cyclodextrin: the amplified Delta T-t mechanism, *Adv. Mater.*, 14, 1151, 2002.
46. Alonso, M. et al., Spiropyran derivative of an elastin-like bioelastic polymer: photoresponsive molecular machine to convert sunlight into mechanical work, *Macromolecules*, 33, 9480, 2000.
47. Alonso, M. et al., Novel photoresponsive *p*-phenylazobenzene derivative of an elastin-like polymer with enhanced control of azobenzene content and without pH sensitiveness, *Macromolecules*, 34, 8072, 2001.
48. Luan, C.H. et al., Differential scanning calorimetry studies of NaCl effect on the inverse temperature transition of some elastin-based polytetrapeptides, polypentapeptides, and polynonapeptides, *Biopolymers*, 31, 465, 1991.
49. Urry, D.W. et al., Pressure effect on inverse temperature transitions: biological implications, *Chem. Phys. Lett.*, 182, 101, 1991.
50. Tamura, T. et al., Effects of temperature and pressure on the aggregation properties of an engineered elastin model polypeptide in aqueous solution, *Biomacromolecules*, 1, 552, 2000.
51. Kuwabara, T. et al., Inclusion complexes and guest-induced color changes of pH-indicator-modified beta-cyclodextrins, *J. Phys. Chem.*, 98, 6297, 1994.
52. Chokchainarong, S., Fennema, O.R., and Connors, K.A., Binding constants for complexes of alpha-cyclodextrin with L-phenylalanine and some related substrates, *Carbohydr. Res.*, 232, 161, 1992.
53. Reguera, J. et al., Effect of modified alpha, beta and gamma-cyclodextrins on the thermo-responsive behavior of the elastin-like polymer, poly(VPGVG), *Carbohydr. Polym.*, 57, 293, 2004.
54. Park, C., Yoon, J., and Thomas, E.L., Enabling nanotechnology with self-assembled block copolymer patterns, *Polymer*, 44, 6725, 2003.
55. Hyde, S. et al., *The Language of Shape*, Elsevier, Amsterdam, Netherlands, 1997.
56. Yang, G.C., Woodhouse, K.A., and Yip, C.M., Substrate-facilitated assembly of elastin-like peptides: studies by variable-temperature *in situ* atomic force microscopy, *J. Am. Chem. Soc.*, 124, 10648, 2002.
57. Wright, E.R. and Conticello, V.P., Self-assembly of block copolymers derived from elastin-mimetic polypeptide sequences, *Adv. Drug Delivery Rev.*, 54, 1057, 2002.
58. Lee, T.A.T. et al., Thermo-reversible self-assembly of nanoparticles derived from elastin-mimetic polypeptides, *Adv. Mater.*, 12, 1105, 2000.
59. Reguera, J. et al., Nanopore formation by self-assembly of the model genetically engineered elastin-like polymer [(VPGVG)(2)(VPGEG)(VPGVG)(2)](15), *J. Am. Chem. Soc.*, 126, 13212, 2004.

60. Prieto, S. et al., Self-assembly and nanostructure formation in genetically engineered protein-based polymers: study of pH effect and substrate interaction, *Nano Lett.*, submitted.
61. Abrams, G.A. et al., Effects of substratum topography on cell behavior, in *Biomimetic Materials and Design*, Lowman, A.K.D. and Lowman, A.M., Eds., Marcel Dekker, New York, 2002, p. 91.
62. Nath, N. and Chilkoti, A., Interfacial phase transition of an environmentally responsive elastin biopolymer adsorbed on functionalized gold nanoparticles studied by colloidal surface plasmon resonance, *J. Am. Chem. Soc.*, 123, 8197, 2001.
63. Nath, N. and Chilkoti, A., Creating "smart" surfaces using stimuli responsive polymers, *Adv. Mater.*, 14, 1243, 2002.
64. Nath, N. and Chilkoti, A., Fabrication of a reversible protein array directly from cell lysate using a stimuli-responsive polypeptide, *Anal. Chem.*, 75, 709, 2003.
65. Urry, D.W. et al., Elastic protein-based biomaterials: elements of basic science, controlled release and biocompatibility, in *Biomaterials Handbook: Advanced Applications of Basic Sciences and Bioengineering*, Wise, D.L. et al., Eds., Marcel Dekker, New York, 2003.
66. Urry, D.W. et al., Bioelastic materials and the ΔTt-mechanism in drug delivery, in *Polymeric Drugs and Drug Administration*, Ottenbrite, R.M., Ed., ACS, Washington, DC, 1994, p. 15.
67. Urry, D.W., Preprogrammed drug delivery systems using chemical triggers for drug release by mechanochemical coupling, *Abstr. Pap. Am. Chem. Soc.*, 200, 74, 1990.
68. Herrero-Vanrell, R. et al., Self-assembled particles of an elastin-like polymer as vehicles for controlled drug release, *J. Controlled Release*, 102, 113, 2005.
69. Stayton, P.S. et al., Molecular engineering of proteins and polymers for targeting and intracellular delivery of therapeutics, *J. Controlled Release*, 65, 203, 2000.
70. Dreher, M.R. et al., Evaluation of an elastin-like polypeptide-doxorubicin conjugate for cancer therapy, *J. Controlled Release*, 91, 31, 2003.
71. Urry, D.W. et al., Medical applications of bioelastic materials, in *Biotechnological Polymers: Medical, Pharmaceutical and Industrial Applications*, Gebelein, C.G., Ed., Technomic, Atlanta, 1993, p. 82.
72. Urry, D.W. et al., Elastic protein-based polymers in soft tissue augmentation and generation, *J. Biomater., Sci.-Polym. Ed.*, 9, 1015, 1998.
73. Panitch, A. et al., Design and biosynthesis of elastin-like artificial extracellular matrix proteins containing periodically spaced fibronectin CS5 domains, *Macromolecules*, 32, 1701, 1999.
74. Girotti, A. et al., Design and bioproduction of a recombinant multi(bio)functional elastin-like protein polymer containing cell adhesion sequences for tissue engineering purposes, *J. Mater. Sci. Mater. Med.*, 15, 479, 2004.

Chapter 7

Imprinting Using Smart Polymers

Carmen Alvarez-Lorenzo, Angel Concheiro, Jeffrey Chuang, and Alexander Yu. Grosberg

CONTENTS

7.1 Introduction

The use of materials as so-called smart polymers is motivated by the simple observation that biological molecules perform incredibly complex functions. While the goal of engineering polymers as effectively as nature remains somewhat in the realm of science fiction, in recent years many researchers have sought to find or design synthetic polymeric materials capable of mimicking one or another "smart" property of biopolymers. Polymer gels are promising systems for such smart functions because of their volume phase transition, predicted theoretically by Dusek and Patterson[1] in 1968 and experimentally demonstrated by Tanaka[2] in 1978. Gel collapse can be driven by any one of the four basic types of intermolecular interactions operational in water solutions and in molecular biological systems,[3] namely, by hydrogen bonds and by van der Waals, hydrophobic, and Coulomb interactions between ionized (dissociated) groups. According to Flory's theory,[4] the degree of swelling of a hydrogel is the result of a competition between the entropy due to polymer conformations, which causes rubber elasticity, and the energy associated with internal attractions and repulsions between the monomers in the gel and the solvent. A change in the environmental conditions, such as temperature, pH, or composition, modifies the balance between the free energy of the internal interactions and the elasticity component, inducing a volume phase transition.[5] The variety of external stimuli that can trigger the phase transition as well as the present possibilities of modulating the rate and the intensity of the response to the stimulus guarantee smart gels a wealth of applications.

The interest in stimuli-sensitive hydrogels, especially in the biomedical field, could be remarkably increased if they could mimic the recognition capacity of certain biomacromolecules (e.g., receptors, enzymes, antibodies). The unique details of the protein's native state, such as its shape and charge distribution, enable it to recognize and interact with specific molecules. Proteins find their desired conformation out of a nearly infinite number of possibilities. In contrast, as known from recent theoretical developments, a polymer with a randomly made sequence will not fold in just one way (see, for example, the review by Pande et al.[6] and references therein and Tanaka and Annaka[7]). Therefore, the ability of a polymer (or polymer hydrogel) to always fold back into the same conformation after being stretched and unfolded, i.e., to thermodynamically memorize a conformation, should be related to properly selected or designed nonrandom sequences. To obtain, under proper conditions, synthetic systems with sequences able to adopt conformations with useful functions, the molecular imprinting technology can be applied.[8–10] The hydrogels can recognize a substance if they are synthesized in the presence of such a substance (which acts as a template) in a conformation that corresponds to the global minimum energy. The "memorization" of this conformation, after the swelling of the network and the washing of the template, will only be possible if the network is always able to fold into the conformation, upon synthesis, that can carry out its designated function[11] (Figure 7.1). This revolutionary idea is the basis of new approaches to the design of imprinted hydrogels and has been developed at different levels, as explained below.

7.1.1 Approaches to Molecular Imprinting

The concept of molecular imprinting was first applied to organic polymers in the 1970s, when covalent imprinting in vinyl polymers was first reported.[10,12] The noncovalent imprinting was introduced a decade later.[13,14] Both approaches are aimed at creating tailor-made cavities shaped with a high specificity and affinity for a target molecule inside or at the surface of highly cross-linked polymer networks. To carry out the process, the template is added to the monomers and cross-linker solution before polymerization,

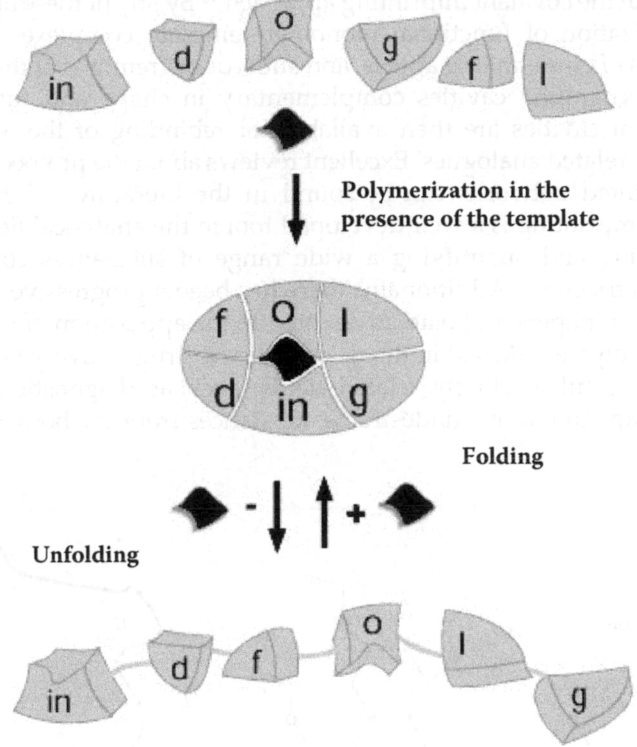

FIGURE 7.1
Intuitive view of the recognition process of a template by an imprinted smart hydrogel based on the educative model drawn by Grosberg and Khokhlov. (From Grosberg, A.Yu. and Khokhlov, A.R., *Giant Molecules*, Academic Press, San Diego, 1997. With permission.)

which allows some of the monomers (called "functional" monomers) to be arranged in a configuration complementary to the template. The functional monomers are arranged in position through either covalent bonds or noncovalent interactions, such as hydrogen bond or ionic, hydrophobic, or charge-transfer interactions (Figure 7.2). In the first case, the template is covalently bound to the monomers prior to polymerization; after synthesis of the network, the bonds are reversibly broken for removal of the template molecules and formation of the imprinted cavities. In the noncovalent or *self-assembly* approach, the template molecules and functional monomers are arranged prior to polymerization to form stable and soluble complexes of appropriate stoichiometry by noncovalent or metal coordination interactions. In this case, multiple-point interactions between the template molecule and various functional monomers are required to form strong complexes in which both species are bound as strongly as in the case of a covalent bond. The noncovalent imprinting protocol allows more versatile combinations of templates and monomers, and provides faster bond association and dissociation kinetics than the covalent imprinting approach.[15] By any of these approaches, copolymerization of functional monomer–template complexes with high proportions of cross-linking agents, and subsequent removal of the template, provides recognition cavities complementary in shape and functionality. These vacant cavities are then available for rebinding of the template or structurally related analogues. Excellent reviews about the protocols to create rigid imprinted networks can be found in the literature.[16, 17] Nowadays, molecular imprinting is a well-developed tool in the analytical field, mainly for separating and quantifying a wide range of substances contained in complex matrices.[18–20] Additionally, there has been a progressive increase in the number of papers and patents devoted to the application of molecularly imprinted polymers (MIPs) in the design of new drug-delivery systems and of devices useful in closely related fields, such as diagnostic sensors or chemical traps to remove undesirable substances from the body.[21,22]

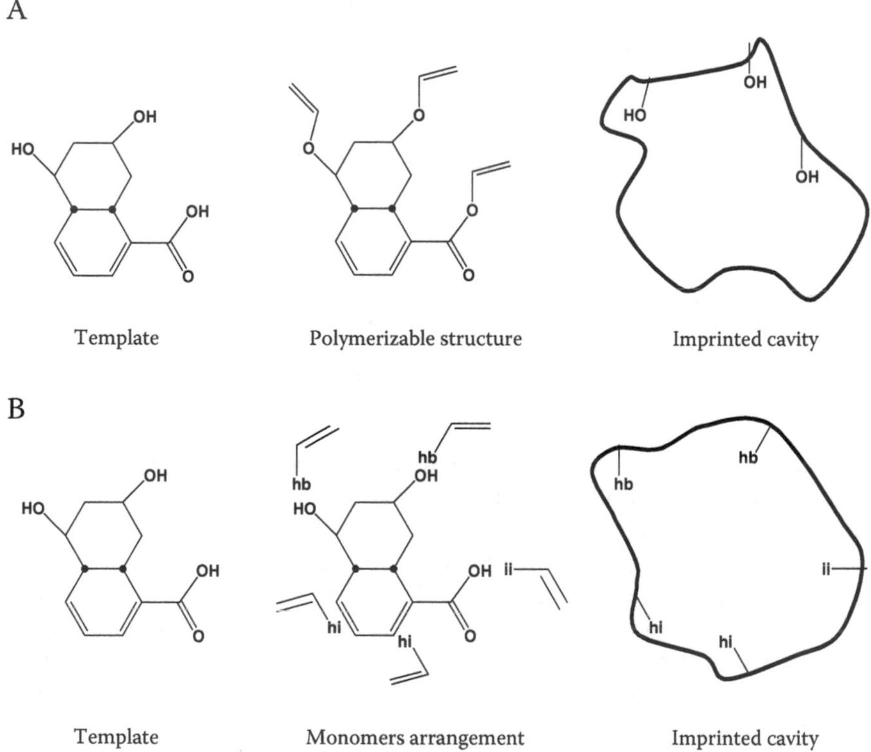

FIGURE 7.2
Schematic view of the imprinting process: (a) covalent approach, in which the template is covalently bound to polymerizable binding groups that are reversibly broken after polymerization; and (b) noncovalent approach, in which the template interacts with functional monomers through noncovalent interactions (e.g., ionic interaction, ii; hydrophobic interaction, hi; or hydrogen bond, hb) before and during polymerization.

To obtain functional imprinted cavities, several factors must be taken into account. It is essential to ensure that the template

- Does not bear any polymerizable group that could attach itself irreversibly to the polymer network
- Does not interfere in the polymerization process
- Remains stable at moderately elevated temperatures or upon exposure to UV irradiation

Other key issues are related to the nature and proportion of the monomers and to the synthesis conditions (solvent, temperature, etc.). The cavities should have a structure stable enough to maintain their conformation in the absence of the template and, at the same time, be sufficiently flexible to facilitate the attainment of a fast equilibrium between the release and re-uptake of the template in the cavity. The conformation and the stability of the imprinted cavities are related to the mechanical properties of the network and depend to a great extent on the cross-linker proportion. Most imprinted systems require 50 to 90% of cross-linker to prevent the polymer network from changing the conformation adopted during synthesis.[23] Consequently, the chances to modulate the affinity for the template are very limited, and it is not foreseeable that the network will have regulatory or switching capabilities. The lack of response to changes in the physicochemical properties of the medium within the biological range or to the presence of a specific substance notably limits their utility in the biomedical field. A high cross-linker proportion also considerably increases the stiffness of the network, making it difficult to adapt its shape to a specific device or to living tissues.

7.1.2 Imprinting in Gels

A further step in the imprinting technology is the development of stimuli-sensitive imprinted hydrogels. The low-cross-linked proportion required to achieve adequate viscoelastic properties can compromise the stability of the imprinted cavities in the hydrogel structure, resulting in some sacrifice of both affinity and selectivity. Strong efforts are being made to adapt the molecular imprinting technology to materials that are more flexible and thus more biocompatible, such as hydrogels for use in the biomedical field.[24–30] To produce low-cross-linked hydrogel networks capable of undergoing stimuli-sensitive phase transitions, the synthesis is carried out in the presence of template molecules, each able to establish multiple contact points with functional monomers.[31–35] Multiple contacts are the key for strong adsorption of the template molecules because of the larger energy decrease upon adsorption as well as the higher sensitivity due to the greater information provided for recognition. As in the classical noncovalent approach, the monomers and the template molecules are allowed to move freely and settle themselves into a configuration of thermodynamic equilibrium. The monomers are then

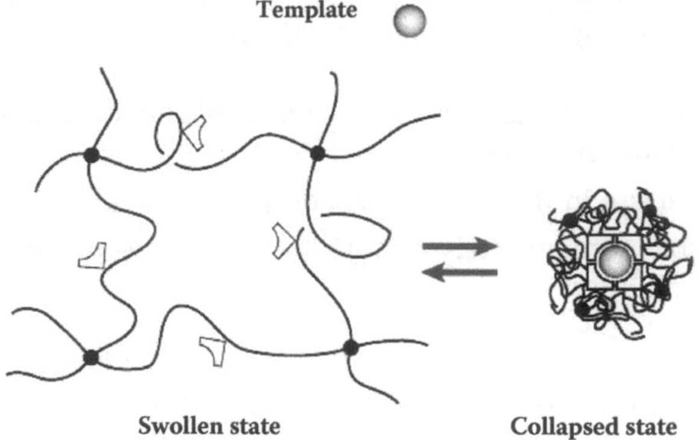

FIGURE 7.3
The volume phase transition of the hydrogel — induced by external stimuli such as a change in pH, temperature, or electrical field — modifies the relative distance of the functional groups inside the imprinted cavities. This alters their affinity for the template. The affinity is recalled when the stimulus reverses and the gel returns to its original conformation.

polymerized in this equilibrium conformation at the collapsed state. As the hydrogel is made from the equilibrium system by freezing the chemical bonds forming the sequence of monomers, we might expect such a hydrogel to be able to return to its original conformation, at least to some degree of accuracy, upon swelling–collapse cycles in which the polymerized sequence remains unchanged (Figure 7.3). If the memory of the monomer assembly at the template adsorption sites is maintained, truly imprinted hydrogels will result. The combination of stimuli sensitivity and imprinting can have considerable practical advantages: the imprinting provides a high loading capacity of specific molecules, while the ability to respond to external stimuli modulates the affinity of the network for the target molecules, providing regulatory or switching capability of the loading/release processes.

This chapter focuses on the sorption/release properties of imprinted smart gels compared with those of random (nonimprinted) heteropolymer gels. We first discuss the theoretical ideas behind the adsorption properties of random heteropolymer gels and review the experiments that have been done on adsorption of target molecules by random gels, detailing the dependence of the affinity on structural and environmental factors.[36–39] We then show the early successes of the imprinting method and highlight the advantages of the imprinted gels compared with random gels.

7.2 The Tanaka Equation

Tanaka and coworkers[31–35] were pioneers in proposing the creation of stimuli-sensitive gels with the ability to recognize and capture target molecules using polymer networks consisting of at least two species of monomers, each having a different role. One forms a complex with the template (i.e., the functional or absorbing monomers capable of interacting ionically with a target molecule), and the other allows the polymers to swell and shrink reversibly in response to environmental changes (i.e., a smart component such as *N*-isopropylacrylamide, NIPA) (Figure 7.4). The gel is synthesized in the collapsed state and, after polymerization, is washed in a swelling medium. The imprinted cavities develop affinity for the template molecules when the functional monomers come into proximity, but when they are separated, the affinity diminishes. The proximity is controlled by the reversible phase transition that consequently controls the adsorption/release of the template (Figure 7.3). A systematic study of the effects of the functional monomer concentration and the cross-linker proportion of the hydrogels (for both imprinted and nonimprinted gels) and of the ionic strength of the medium on the affinity of the hydrogels for different templates led to the development of an equation, called the Tanaka equation in memory of the late Professor Toyoichi Tanaka (1946–2000).[39]

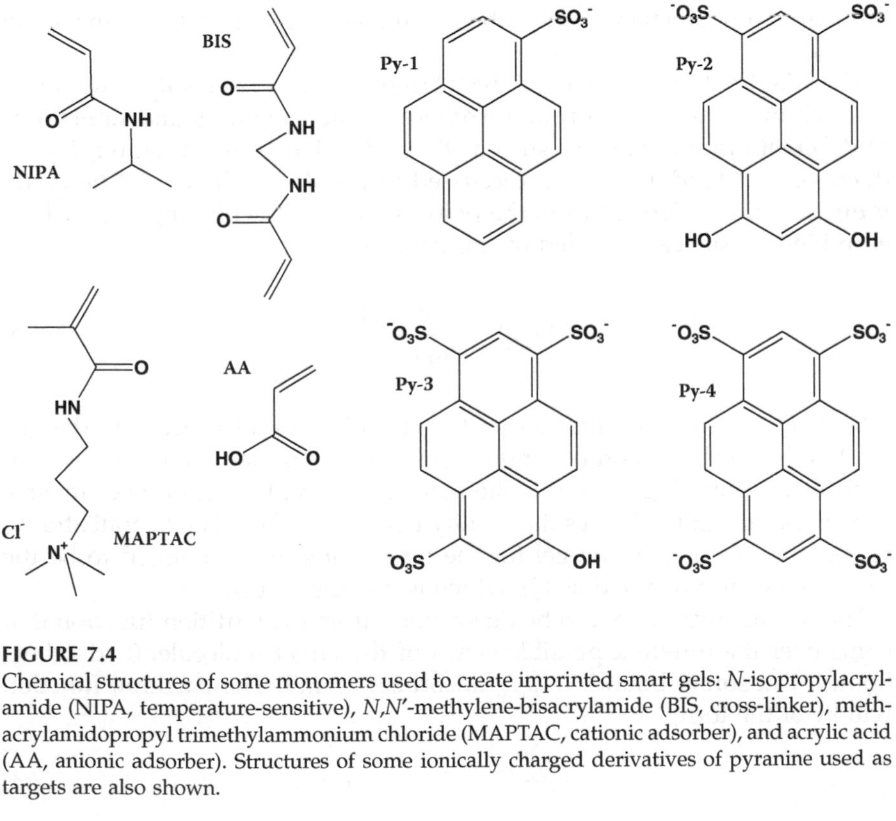

FIGURE 7.4
Chemical structures of some monomers used to create imprinted smart gels: *N*-isopropylacrylamide (NIPA, temperature-sensitive), *N,N′*-methylene-bisacrylamide (BIS, cross-linker), methacrylamidopropyl trimethylammonium chloride (MAPTAC, cationic adsorber), and acrylic acid (AA, anionic adsorber). Structures of some ionically charged derivatives of pyranine used as targets are also shown.

7.2.1 Theoretical Considerations

The Tanaka equation[39] relates the gel affinity for the target molecules with the aforementioned variables as follows:

$$\text{Affinity} \approx \frac{[\text{Ad}]^p}{p[\text{Re}]^p}\exp(-p\beta\varepsilon)\exp\left(-(p-1)c\frac{[\text{Xl}]}{[\text{Ad}]^{2/3}}\right) \tag{7.1}$$

where [Ad] represents the concentration of functional monomers in the gel; [Re] is the concentration of replacement molecules, i.e., ions that are bound to the target molecule when it is not bound to the functional monomers (in cases where they have ionic or protonized groups); [Xl] is the concentration of cross-linker; p is the number of bonds that each template can establish with the functional monomers; β is the Boltzmann factor ($1/k_BT$); ε is the difference between the binding energy of an adsorbing monomer to the target molecule and that of a replacement molecule to the target molecule; and c is a constant that can be estimated from the persistence length and concentration of the main component of the gel chains (e.g., NIPA). The main assumption in Equation 7.1 is that the adsorption of target molecules is dominated by one value of p at each state of the gel. The value of p changes from 1 in the swollen state to p_{max} in the collapsed state, where p_{max} is the number of functional monomers that simultaneously interact with the target molecule.

The affinity of a network for a given target molecule is usually determined through the analysis of sorption isotherms. The suitability and limitations of different binding models, such as those developed by Langmuir, Freundlich, or Scatchard, have been discussed in detail elsewhere.[40–42] The Langmuir isotherm is derived from the point of view of the binding sites, where each binding site can be filled or unfilled:

$$[T_{ads}] = S\frac{K[T_{sol}]}{K[T_{sol}]+1} \tag{7.2}$$

where $[T_{ads}]$ is the concentration of target molecules adsorbed into the gel, $[T_{sol}]$ is the concentration of target molecules in solution, and S is the concentration of binding sites. K is the binding constant, with units of inverse concentration, and indicates the affinity per binding site. The overall affinity of the binding sites in the gel for the target molecule is defined to be the product SK (also denoted as Q), which is dimensionless.

The overall affinity SK can be determined from the partition function that sums over the different possible states of the target molecule: 0 adsorbers bound, 1 adsorber bound,..., p_{max} adsorbers bound. The partition function will be of the form:

$$Z = Z_0 + Z_1 + Z_2 + \cdots + Z_{p_{max}} \tag{7.3}$$

where Z_p indicates the term of the partition function in which a target molecule is bound by p adsorbing monomers, and Z_0 corresponds to the case of the target molecule being completely unbound. In the Langmuir equation, the term $K[T_{sol}]$ is proportional to the fraction of filled binding sites, and $SK[T_{sol}]$ is proportional to the concentration of bound target molecules. Compared with Equation 7.3, we see that

$$SK[T_{sol}] \propto Z_0 + Z_1 + Z_2 + \cdots + Z_{P\max} \tag{7.4}$$

The partition function component Z_0 must be proportional to the number of target molecules in solution (i.e., $[T_{sol}]$). Therefore

$$SK \propto (Z_1 + Z_2 + \cdots + Z_{P\max})/Z_0 \tag{7.5}$$

To calculate each of the terms Z_p, a "fixed-point model" was developed. We can consider the case of a cross-linked gel made of a major component (i.e., a smart one, such as NIPA) and some functional monomers, prepared in the absence of templates (i.e., nonimprinted). The polymerized monomers can move in the gel to some extent, but are also constrained by the connectivity of the chains. When the cross-linking density is high, there are many constraints on the motion of the monomers. Conversely, at low cross-linker densities, monomers can diffuse more freely. The length scale of monomer localization is determined by the concentration of cross-linker in the gel.

If the gel is immersed in the solution of target molecules and these can diffuse into and out of the gel to form an adsorption binding site of p adsorbing monomers, the monomers have to move in space to properly group together. However, their motions are severely restricted because almost every adsorbing monomer in the gel belongs to a subchain, which means it is connected by the polymer to two cross-links (there might also be a few adsorbing groups on the dangling ends, connected to only one cross-link). Apart from real cross-links, the freedom of subchains is also restricted by the topological constraints, such as entanglements between polymers.

From a qualitative and very simplified approach, one particular adsorbing monomer in the gel can only access some relatively well-defined volume. It is reasonable to assume that in the center of this volume the adsorbing group is free to move, but approaching the periphery of its spherical cage the group feels increasing entropic restrictions. This can be modeled by saying that for each adsorbing monomer in the gel there is a point fixed in space, which is the center of the cage, and then there is a free-energy-potential well (of entropic origin) around this center. Every adsorbing monomer is attached to its corresponding self-consistent center by an effective polymer chain, like a dog on a leash (Figure 7.5). The length of the effective chain must be of the same order as the distance between cross-link points. Thus, it will be a decreasing function of the cross-link density. In a simplified way, we can imagine that each one of the adsorbing monomers is at the end of one of these effective chains. At the other end of the chain is one of these points

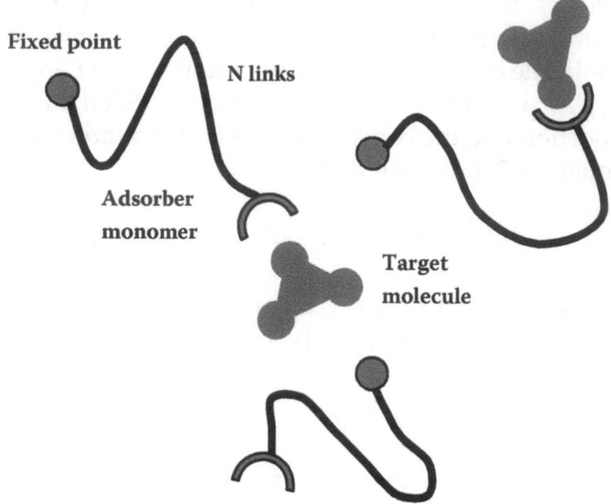

FIGURE 7.5
Schematic drawing of the fixed-point model that places each adsorber monomer at the end of a finite chain of length n, the value of which is inversely proportional to the cross-linker density. In this example, the target molecule can interact with 1 to 3 adsorber monomers ($p_{max} = 3$).

fixed in space, the positions of which are distributed randomly in the gel. Each chain is assumed to be made of n links, where n is inversely proportional to the cross-linking density of the gel, based on the concept that additional cross-links increase the frustration in the gel. The parameter n should be proportional to the ratio of main-component monomers to cross-linker monomers.

An advantage of the fixed-point model is that it allows one to determine the entropic properties of the network using the well-known statistics of polymer chains. Adsorption of target molecules in the gel will deform the chain network, and the accompanying entropy loss can be analyzed via the entropy of Gaussian chains. This entropic effect will be affected not only by the cross-linker density, but also by the density of adsorber monomers, which are implicit in the definition of a fixed point. Note that this density can be adjusted via the gel-swelling phase transition. In the swollen state there will be a low density of fixed points, and in the collapsed state there will be a high density.

The dependences shown in Equation 7.1 can be qualitatively explained using this model as follows (Figure 7.5 and Figure 7.6).

1. Power-law dependence of the affinity on [Ad]. For a target molecule to be adsorbed, the p adsorbing monomers must be clustered together to simultaneously bind it. The probability of such a cluster existing at a given point in a random (nonimprinted) gel is a product of the probabilities for each of the adsorbing monomers. Therefore, the dependence goes as $[Ad]^p$. Each of these clusters requires p adsorbing monomers; hence SK is proportional to $1/p$.

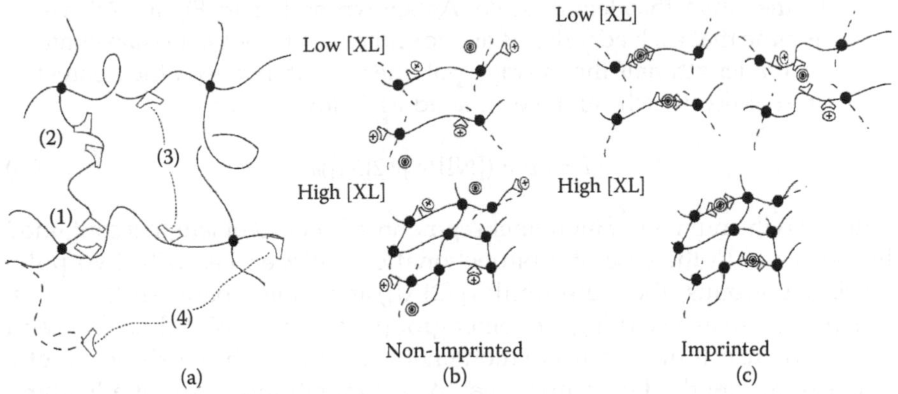

FIGURE 7.6
Binding site formation and gel microstructure. (a) Candidate adsorber pairs can be located near a cross-link (1), on a single chain (2), and on distant chains without (3) and with (4) intervening entanglements. (b) With increasing cross-linking density [XL], adsorbers are frustrated in non-imprinted gels. (c) In imprinted gels, flexibility at low [XL] leads to competition for targets, with possible topological consequences (mispairings) that diminish at high [XL]. (Reproduced from Stancil, K.A. et al., *J. Phys. Chem. B*, 109, 6636, 2005. With permission from the American Chemical Society.)

2. Power-law dependence of the affinity on [Re]. The replacement molecules (typically salt ions) act as competitors to the adsorbing monomers. In solution, a target molecule can either be adsorbed into the gel or bound by p replacement molecules. Binding to the replacement molecules prevents adsorption by the gel. Similar to a mass action law, the p replacement molecules must cluster around the target. This creates a power-law dependence similar to that for target molecule adsorption by adsorbing monomers, but with an opposite sign exponent ($[Re]^{-p}$). Note that it was assumed that each replacement molecule binds to one site on the target molecules, as the adsorbing monomers do. If the adsorber monomers and the replacement molecules have different valences, these exponents should be modified.
3. The term $\exp(-\beta p\varepsilon)$ represents the enthalpy contribution to the sorption, i.e., the attraction energy of a target molecule to p adsorbing monomers. This is a Boltzmann probability based on a binding energy ε per adsorbing monomer.
4. The term $\exp[-c(p-1)[Xl]/[Ad]^{2/3}]$ summarizes the entropy restrictions to the sorption, which are mainly related to the cross-linking density, as mentioned above. The adsorber units in the gel can move rather freely within a certain volume determined by the cross-linking density. Below a certain length scale associated with the cross-linking density, the gel behaves like a liquid, allowing the adsorber groups to diffuse almost freely. Beyond that length scale, however, the gel behaves as an elastic solid body. The adsorber units cannot diffuse

further than that length scale. As shown earlier in Figure 7.5, we assume that each adsorber is at one end of a fictitious Gaussian chain with a length half the average polymer length between the nearest cross-links, which can be estimated as follows:

$$l = nb = ([\text{NIPA}]/2[\text{Xl}])a \tag{7.6}$$

where n is the number of monomer segments of persistent length b contained in the chain. In the case of a bifunctional cross-linker, i.e., with two polymerizable groups, there are ([NIPA]/2[Xl]) monomers between the cross-link point and an adsorbing monomer group. Then, $n = ([\text{NIPA}]/2m[\text{Xl}])$ and $b = ma$, where m is the number of monomers involved in the persistent length, and a is the length of each monomer. At a concentration [Ad] of adsorbing monomers, the average spatial distance between adsorbing monomers is $R = [\text{Ad}]^{-1/3}$. For a molar concentration C_{ad}, this corresponds to $R = 1\ \text{cm}/(C_{ad}N_A)^{1/3}$, where N_A is the Avogadro number. This fictitious Gaussian chain represents the restricted ability of the adsorber groups to diffuse within a certain volume in the gel. We expect that the probability for two adsorber monomers to meet should be proportional to the Boltzmann factor of the entropy loss associated with the formation of one pair of adsorbers

$$P = P_0 \exp(-R^2/nb^2) = P_0 \exp(-c[\text{Xl}]/[\text{Ad}]^{2/3}) \tag{7.7}$$

where the quantity c is determined by the persistence length, the number of monomers in a persistence length, and the concentration of the main component of the chains through the relation

$$c = 2m/([\text{NIPA}]b^2) \tag{7.8}$$

Since the adsorption of a divalent target by two adsorbers brings together each end from two fictitious Gaussian polymers, the affinity should be proportional to this probability. If more than two contact points are expected, the equation can be generalized as

$$SK \propto \exp[-c(p-1)[\text{Xl}]/[\text{Ad}]^{2/3}] \tag{7.9}$$

where p is the number of contact points. If the target molecule is adsorbed only by a single contact ($p = 1$), the affinity should be independent of the cross-linker concentration. In contrast, if the target binding site requires several adsorber monomers ($p > 1$), the cross-links may frustrate the formation of the binding site, and the frustration will increase with p. This will be particularly significant at low concentrations of [Ad] (since only a small fraction of the adsorber monomers will be close enough together to participate in multiple binding sites) and high cross-linking concentration. It should also be taken into account that the adsorption process itself can increase the cross-linking density, since the complexes will act as tie junctions of polymeric chains. This could modify the value of n as the sorption is going on. Nevertheless, usually this last contribution to the cross-linking is low enough to be negligible.

It can be shown by consideration of the relative weights of the terms Z_p that for a given set of experimental parameters, the affinity SK will almost always be dominated by a single value of p, either $p = 1$ or $p = p_{max}$. This can be understood by considering whether the attraction of the target molecule to adsorber monomers (due to energetic and concentration effects) is stronger than the repulsion due to the entropy loss required to deform the gel. If attraction is sufficiently favored, then the target will be bound by as many adsorber monomers as possible and, therefore, $p = p_{max}$. However, if the entropy loss to deform the gel is stronger, then the adsorption will only be possible by single adsorber monomers, i.e., $p = 1$, which would not require deformation of the network (no chain entropy penalty).

The basic concept of gels as smart materials is that they will have high affinity for the target in the collapsed state, but low affinity in the swollen state. By controlling the phase transition of the gel, one will be able to create a switchlike behavior in the affinity. The Tanaka equation allows one to predict the composition of gels that will drastically change affinity during the gel phase transition. If the gel is to have a low affinity in the swollen state, the adsorber monomers should have only a weak attraction to the target molecules, i.e., any adsorption should be single-handed ($p = 1$). To have a high gel affinity, adsorption in the collapsed phase should involve as many adsorber monomers as possible ($p = p_{max}$). The p value transition should occur where the entropic and energetic contributions to the affinity are equal, i.e., the crossover should occur when

$$\ln([\mathrm{Ad}]/[\mathrm{Re}]) \approx ([\mathrm{Ad}]^{2/3}\, nb^2)^{-1} + \beta\varepsilon \tag{7.10}$$

A more detailed description of the Tanaka theory has been published by Ito et al.[39,40]

The experiments discussed below use gels in which p changes across the phase transition. However, the design of such gels still requires a significant amount of research, since it is difficult to know *a priori* the exact value for the binding energy ε.

Since the Tanaka equation is based on the idea that there are many sets of adsorber monomers that work together locally to form binding sites for the target molecules and that the probability for chain stretching drops off exponentially with distance, it can also explain the greater affinity of imprinted gels compared with the nonimprinted ones. The synthesis in the presence of the target molecules leads to the distribution of the adsorbing monomers in groups of the p members required for the binding of each target molecule. In the imprinted gels, the p members are closely fixed during polymerization in the collapsed state due to the template. The template is removed from the gel in the swollen state because of the deformation of the binding sites. Once again, in the collapsed state, the binding sites can be reconstructed, and since each binding site possesses the needed *p* adsorbing monomers close together, the entropic restrictions to the sorption should be minimized (Figure 7.6). Representative examples of the enhancement in affinity observed for imprinted smart gels are given in the following subsections.

7.2.2 Experimental Assessment of the Tanaka Equation

Among the studies carried out to assess the Tanaka equation in nonimprinted heteropolymer gels,[36–38,43] those carried out with pyrene derivatives as target molecules are particularly representative.[36–39] The gels were prepared by free-radical polymerization using 6*M* *N*-isopropylacrylamide (NIPA) as the stimuli-sensitive component, 0 to 120m*M* methacrylamidopropyl ammonium chloride (MAPTAC) as functional monomer, and 5 to 200m*M* *N,N'*-methylenebis(acrylamide) (BIS) as cross-linker. The monomers were dissolved in dimethylsulfoxide and polymerized inside micropipettes (i.d. 0.5 mm). Once washed with water, all gels were collapsed at 60°C and showed the same degree of swelling as during polymerization. As adsorbates or target molecules, several different types of pyrene sulfonate derivatives were used: 1-pyrene sulfonic acid sodium salt (Py-1·Na), 6,8-dihydroxy-pyrene-1,3-disulfonic acid disodium salt (Py-2·2Na), 8-methoxy pyrene-1,3,6-trisulfonic acid trisodium salt (Py-3·3Na), and 1,3,6,8-pyrene tetrasulfonic acid tetrasodium salt (Py-4·4Na), portrayed in Figure 7.4. These chemicals present 1 (Py-1), 2 (Py-2), 3 (Py-3), or 4 (Py-4) anionic charges, which can interact electrostatically with a cationic charged site such as on MAPTAC. Pieces of cylindrical gel (5 to 20 mg dry weight) were placed in 2- or 4-ml target aqueous solution, the concentration of which ranged from 2 to 0.5m*M*. The solutions also contained NaCl of a prescribed concentration (27 to 200m*M*) to provide chloride ions to replace the target molecules. The samples were kept swollen (20°C) or shrunken (60°C) for 48 h, and the adsorption isotherms were analyzed in terms of the Langmuir equation (Equation 7.2) to calculate the affinity *SK*.

Figure 7.7 shows the dependence of affinity for Py-3 and Py-4 on the MAPTAC concentration. Above a certain MAPTAC concentration (20m*M*) and in the collapsed state, both the log–log plots show a straight line, with slope 3 for Py-3 and with slope 4 for Py-4. These power-law relationships are due to three and four adsorption points, respectively. Adsorption sites are formed when three (or four) equivalent adsorbing molecules (MAPTAC) gather to capture one Py-3 (or Py-4) molecule. The obtained power laws are in agreement with the Tanaka equation. At MAPTAC concentrations below 10m*M*, the major component of the gel (NIPA) contributes more to the adsorption of pyranine (due to a hydrophobic interaction) than do the MAPTAC groups, and the power-law exponent becomes zero. In the swollen state, the log–log slope becomes 1, indicating that MAPTAC adsorbs the target molecule with a single contact. Single-point adsorption is favored because the MAPTAC monomers are well separated one from another, and it becomes entropically unfavorable for the multipoint adsorption complex to assemble.

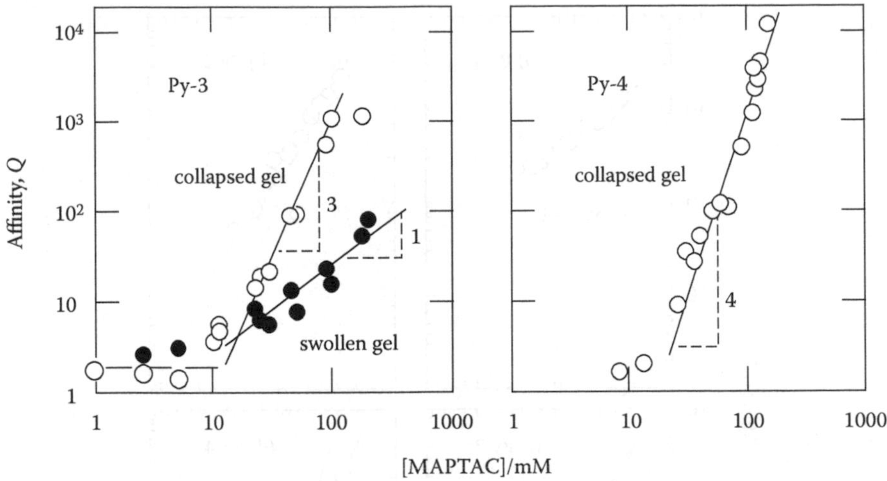

FIGURE 7.7
Dependence of the affinity for Py-3 and Py-4 on the concentration of adsorber monomer. In this experiment, an amount (100m*M*) of salt was added to the system to sweep out the Donnan potential. (Reproduced from Oya, T. et al., *Science*, 286, 1543, 1999. With permission from the American Association for the Advancement of Science.)

The slope returns to 3 or 4 upon shrinking, indicating recovery of the multipoint binding sites. The reversible adsorption ability is controlled by the volume phase transition.

The effect of external salt concentration on the binding of target molecules to the gel is shown in Figure 7.8. As mentioned above, when these low-cross-linked gels are in the collapsed state, a number of MAPTAC groups equal to the number of charges of the target (p_{max}) can gather to form a binding site. Since the attraction is electrostatic, coexistent ions in the solution can make the target molecule adsorption difficult. Ions with the same charge sign as the target molecules should compete with the target molecules for binding with the adsorber monomers. In the cases Py-1, -2, and -3, the log–log plot showed slopes of –1, –2, and –3, respectively, i.e., $-p_{max}$, where p_{max} is the number of charged groups on the target molecule (affinity $\propto$ $[Re]^{-p}$). Py-4 followed a similar behavior, though there was a small discrepancy below 40m*M* of salt concentration, probably because of the increasing effect of the Donnan potential at lower salt concentrations. In the swollen state ($p = 1$), the log–log plots showed slopes of –1. Therefore, the Tanaka equation fits the results quite well in a wide range of salt concentrations.

As explained above, the multipoint adsorption should lead to an entropy loss in the polymer chains that are bound together by the target molecules. The effect of this entropy loss, which is a function of the concentration of cross-linker and adsorber, is to reduce the affinity of the gel for the target. Several experimental studies have shown that the affinity of a heteropolymer gel involved in multiple contact adsorption decreases with its cross-linking degree.

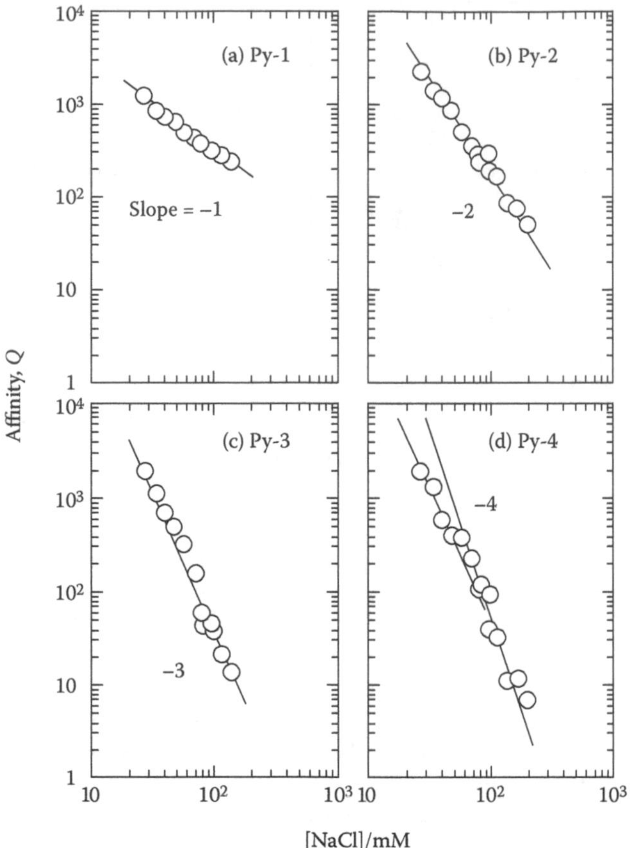

FIGURE 7.8
Dependence of the affinity for pyranines on the replacement ion concentration. The solid line displays a slope with the value shown in each plot. In this experiment, the concentrations of the adsorber (MAPTAC) and the cross-linker (BIS) were fixed at 40 and 10m*M*, respectively. (Reproduced from Watanabe, T. et al., *J. Chem. Phys.*, 115, 1596, 2001. With permission from the American Institute of Physics.)

Hsein and Rorrer[44] showed an exponential decrease in calcium adsorption by chitosan as the extent of cross-linking increases. Eichenbaum et al.[45] found that, for alkali earth metal binding in methacrylic acid-*co*-acrylic acid microgels, the cross-links prevent the carboxylic groups from achieving the same proximity as in a linear polymer, which affects the binding properties of the metals. In the case of a thermosensitive gel, the influence of the degree of cross-linking has also been shown.[24,31]

The volume phase transition of a NIPA gel induced by a stimulus is responsible for separating the adsorber monomers (e.g., MAPTAC) in the swollen state, which decreases their probability of coming close to each other to adsorb a multicharged target. Consequently, in the swollen state, the affinity for the target increases slightly with the degree of cross-linking, since the degree of swelling is reduced. In contrast, in the collapsed state an exponential decrease of the affinity with the cross-linking, which was especially significant for the cases of contact numbers above 2, was observed (Figure 7.9). The affinity values predicted by the Tanaka equation agree well with those experimentally obtained. This unfavorability for the affinity has been understood to be a "frustration" to the mobility of the adsorbing sites.[31,46,47] The frustration can also be viewed in terms of the "flexibility" of the polymer chains, which is critical for allowing the adsorber groups to come into proximity for a multipoint adsorption. A detailed analysis of the data shown in Figure 7.9 revealed that a plot of the exponential decay rate vs. the number of contact points gives a linear relationship in which the slope is 0.32. This slope associates with the parameter *c* in Equation 7.1, and

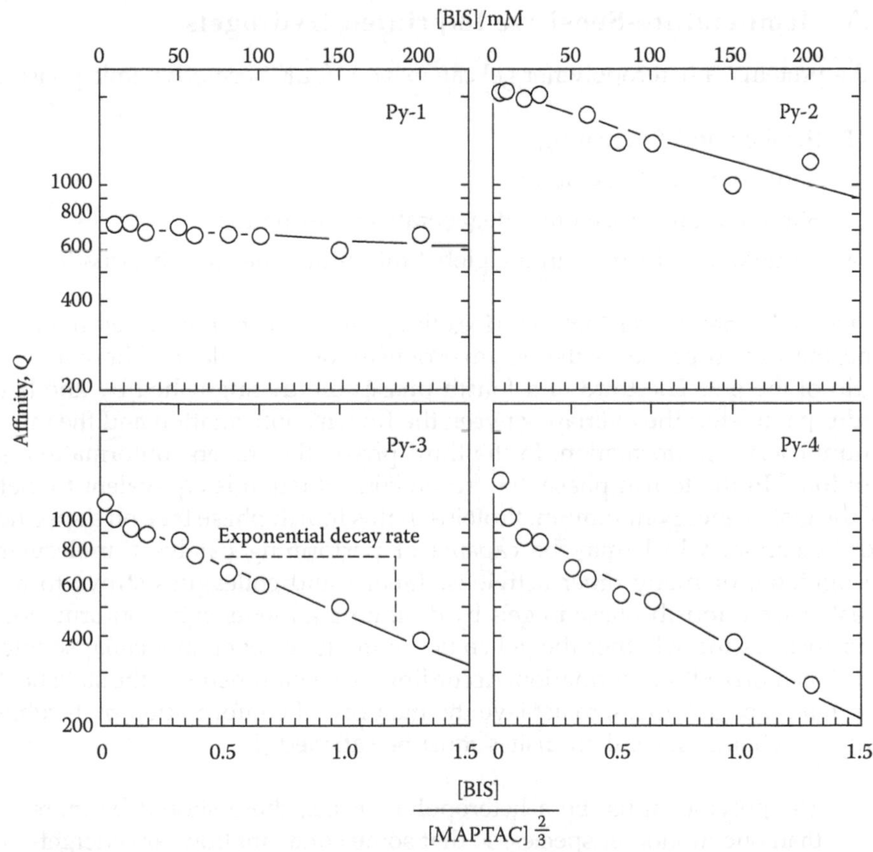

FIGURE 7.9
Dependence of the affinity of gels at 60°C for pyranines on the concentration of cross-linker BIS (upper *x*-scale) or on the ratio of cross-linker concentration to that of adsorber monomer to the two-thirds power (lower *x*-scale). (Reproduced from Ito, K. et al., *Prog. Polym. Sci.*, 28, 1489, 2003. With permission from Elsevier.)

can be used to calculate the persistence length $b = ma$ for the polymer chains, applying Equation 7.8. For c = 0.32, the value of b equals 2.9 nm. The persistent length is theoretically predicted to be around 2 nm, e.g., ≈10 monomers (m = 10) of 2 Å ($a = 2 \times 10^{-10}$ m).[48] Thus, the obtained result is somewhat reasonable. Therefore, the theory predicts and explains well the exponential decay with concentration of the cross-linker. The cross-links and polymer connections create frustrations so that the adsorber groups (MAPTAC) cannot lower the energy of the polymer by forming pairs, triplets, or groups of p members for capturing target molecules. As will be shown below, such frustrations can be overcome using molecular imprinting.

7.3 Temperature-Sensitive Imprinted Hydrogels

Like proteins, a heteropolymer gel can exist in four thermodynamic phases:

1. Swollen and fluctuating
2. Shrunken and fluctuating
3. Shrunken and frozen in a degenerate conformation
4. Shrunken and frozen in the global minimum energy conformation

The order parameter that describes the phase transition between the first and the second phases is the polymer density or, equivalently, the swelling ratio of the gel. The third and fourth phases are distinguished by another order parameter: the overlap between the frozen conformation and the minimum energy conformation. In the third phase, the frozen conformation is random.[7] In the fourth phase, the frozen conformation is equivalent to that of the global energy minimum. Proteins in this fourth phase take on a specific conformation, which may be capable of performing catalysis, molecular recognition, or many other activities. Tanaka and colleagues strove to recreate such a fourth phase in gels by designing a low-energy conformation and then testing whether the gel could be made to reversibly collapse into this "memorized" conformation. According to developments in the statistical mechanics of polymers, to achieve the memory of conformation by flexible polymer chains, several requisites must be satisfied.[49]

1. The polymer must be a heteropolymer, i.e., there should be more than one monomer species, so that some conformations are energetically more favorable than others.
2. There must be frustrations that hinder a typical polymer sequence from being able to freeze to its lowest energy conformation (as considered in absence of the frustration). Such frustrations may be due to the interplay of chain connectivity and excluded volume, or

they may be created by cross-links. For example, a cross-linked polymer chain will not freeze into the same conformation as the non-cross-linked polymer chain, at least for most polymer sequences.

3. The sequence of monomers must be selected so as to minimize these frustrations,[50] i.e., a particular polymer sequence should be designed such that the frustrating constraints do not hinder the polymer from reaching its lowest energy conformation.

These three conditions allow the polymer to have a global free-energy minimum at one designed conformation.

The nonimprinted gels described in previous sections satisfy the first two conditions, and can be engineered to satisfy the third. The adsorber and the main-component monomers provide heterogeneity in interaction energies, since adsorber interactions are favorably mediated by the target molecules.[51] Frustrations to the achievement of the global energy minimum exist due to the cross-links in the gel as well as chain connectivity. To meet the third condition, the minimization of the frustration, the molecular imprinting technique can be particularly useful. In this section, we discuss the early experimental successes of the imprinting method and the application of the Tanaka equation to imprinted gels.

Two approaches were investigated to achieve elements of conformational memory in gels:

1. Cross-linking of a polymer dispersion or preformed gel in the presence of the target molecule (two-step imprinting or post-imprinted)
2. Simultaneous polymerization and cross-linking of the monomer solution (one-step imprinting)

The first approach was studied in heteropolymer gels consisting of NIPA as the major component sensitive to stimuli and MAPTAC as the charged monomer able to capture pyranine target molecules.[24] The polymer networks were randomly copolymerized, in the absence of pyranines, with a small quantity of permanent cross-links and thiol groups (–SH). The gels were then further cross-linked by connecting thiol group pairs into disulfide bonds (–S–S–). The process was carried out directly (nonimprinted gels) or after the gels were immersed in a solution of pyranines and had all charged groups forming complexes with the target molecules (postimprinting). These post-cross-links were still in very low concentrations, in the range 0.1 to 3 mol%, and therefore the gels could freely swell and shrink to undergo the volume phase transition. The postimprinted gels showed higher affinity for the target than those that were randomly post-cross-linked (Figure 7.10). However, this post-cross-linking approach has a fundamental drawback. Before the post-cross-linking, the sequence of the components has already been determined and randomly quenched. The minimization of the frustration is allowed only in the freedom of finding best partners among –SH groups. For this reason, the imprinting using a post-cross-linking technique can give only a partial success.

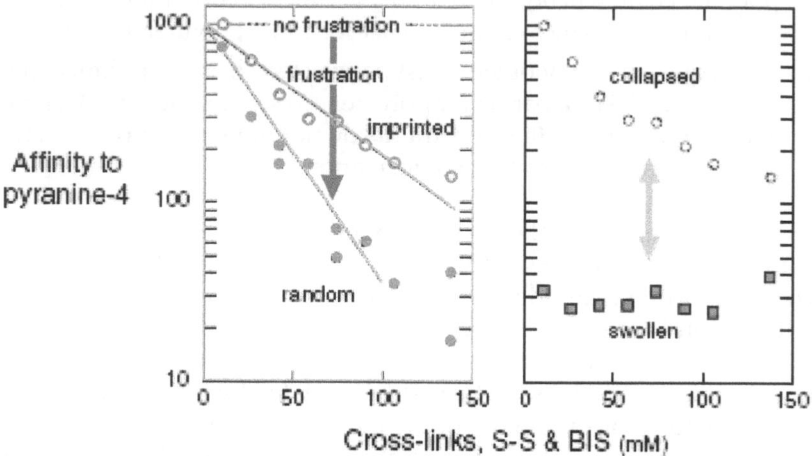

FIGURE 7.10
The affinity for Py-4 is shown as a function of S–S bonds plus BIS (10m*M*) concentration within the gels in the collapsed state at 60°C. For random gels, the affinity decays exponentially with S–S concentration (the solid line is a guide to the eye). The loss of ability to form complexes indicates the frustration due to cross-links. For the postimprinted gels, the affinity increases 3 to 5 times at higher S–S concentrations. This indicates the partial resolution of frustration. When the gels are allowed to swell at 25°C, the affinity becomes very low and independent of S–S concentration, indicating the destruction of the adsorption sites upon gel swelling (right panel). When the imprinted gels are allowed to collapse again, the pyranine adsorption resumes its original high values, indicating the restoration of the original complexes with the same MAPTAC monomers. (Reproduced from Enoki, T. et al., *Phys. Rev. Lett.*, 85, 5000, 2000. With permission from the American Institute of Physics.)

Ideally, the entire sequence of all monomers should be chosen so that the system will be in its global energy minimum. The complete minimization of the frustration can be achieved by polymerizing monomers while they self-organize in space at a low-energy spatial arrangement.[31,32,35] This second approach controls the sequence formation, allowing the monomers to equilibrate in the presence of a target molecule. These monomers are then polymerized with a cross-linker. It is hoped that, upon removal of the template species, binding sites with the spatial features and binding preferences for the template are formed in the polymer matrix. The choice of functional monomers and the achievement of an adequate spatial arrangement of functional groups are two of the main factors responsible for specificity and reversibility of molecular recognition.

The first experiments carried out by Alvarez-Lorenzo et al.[31,32] used gels prepared by polymerization of NIPA, small amounts of methacrylic acid (MAA), and BIS in the absence (nonimprinted) or the presence of divalent cations. MAA was used as the functional monomer able to form complexes in the ratio 2:1 with divalent ions. The effect of temperature on the adsorption capacity of the imprinted copolymers prepared with different template ions and in different organic solvents was compared with that of the nonimprinted ones. Successful imprinting was obtained using calcium or lead ions as template. After removing the template and swelling in water at room temperature, the affinity for divalent ions notably decreased. When the gels were shrunken by an increase in temperature, the affinity was recovered (Figure 7.11). The measurements of the affinity suggested that multipoint adsorption occurs for both imprinted and nonimprinted gels in the collapsed state (saturation $S \approx [Ad]/2$), but that in the imprinted gel, the multipoint

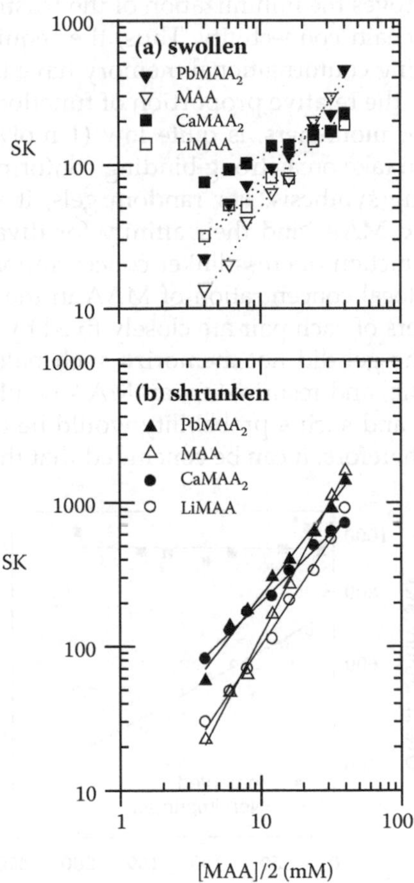

FIGURE 7.11
The overall affinity *SK* of the nonimprinted and imprinted gels for calcium ions is plotted as a function of methacrylic acid monomer concentration for (a) the swollen state and (b) the shrunken state. The values of the slope for nonimprinted gels duplicate the slope of the imprinted ones prepared with Ca^{2+} as template or with the alternative template Pb^{2+}. (Reproduced from Alvarez-Lorenzo, C. et al. , *J. Chem. Phys.*, 114, 2812, 2001. With permission from the American Institute of Physics.)

adsorption is due to memorized binding sites. After recollapsing, two-point adsorption is recovered in the random gels with a power law of $SK \propto [Ad]^2$, while the imprinted gels showed a stronger affinity with a power law of $SK \propto [Ad]^1$. The difference in the power dependence of the affinity is due to a different [Ad] dependence for K. The affinity per adsorption site is actually independent of [Ad] for imprinted gels. This is because the adsorbers are ordered such that they already form sites with their unique partners. Hence, the adsorbers are ordered in groups of two to give the highest possible K at all values of [Ad]. Furthermore, the affinity of the imprinted gels does not change at all with an increase of the cross-linking density, while the affinity decreases for random gels (Figure 7.12). The nondependence of the affinity on [Xl] proves the minimization of the frustration with regard to the cross-links and chain connectivity. Thus, the requirements 1, 2, and 3 necessary for obtaining conformational memory have been achieved.

In these hydrogels the relative proportion of functional monomers, compared with the other monomers, is quite low (1 mol%), and it should be difficult for the gel to take on a strong-binding conformation other than the one imprinted during synthesis. For random gels, it was difficult to pair randomly distributed MAA, and their affinity for divalent ions decreased exponentially as a function of cross-linker concentration. In contrast, in the imprinted gels, the local concentration of MAA in the binding site is very high, i.e., the members of each pair are closely fixed by the template during polymerization. If the gel did not memorize such pairs after washing the template out, swelling, and reshrinking, an MAA would have to look for a new partner nearby, and such a probability would be the same as that in a nonimprinted gel. Therefore, it can be concluded that the greater adsorption

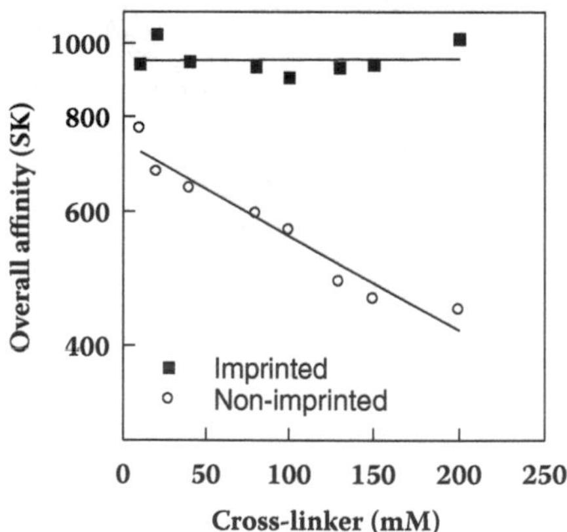

FIGURE 7.12
Influence of the cross-linker (BIS) concentration on the overall affinity for calcium ions of the imprinted (full symbols) and nonimprinted (open symbols) NIPA (6*M*) gels in the shrunken state in water. The concentration of functional monomers (MAA) was fixed at 32m*M*. (Reproduced from Alvarez-Lorenzo, C. et al., *Macromolecules*, 33, 8693, 2000. With permission from the American Chemical Society.)

capacity of the imprinted gels comes from the successfully memorized MAA pairs.[32,35,52]

Güney et al.[53,54] followed a similar procedure to prepare temperature-sensitive gels that specifically recognize and sorb heavy metal ions from aqueous media in an effort to develop chemosensors. Kanazawa et al.[55] used another functional monomer *N*-(4-vinyl)benzyl ethylenediamine) (Vb-EDA) that contains two nitrogen groups that can specifically form coordination bonds with one copper ion; each copper ion requires two functional monomers to complete its bonding capacity. This occurs when the gel is formed at the shrunken state. At the swollen state, the bonds are broken and the affinity for the ions disappears. These gels showed a high specificity for Cu^{2+} compared with Ni^{2+}, Zn^{2+}, or Mn^{2+}. The different coordination structure — square planar for Cu^{2+} and Ni^{2+}, tetrahedral for Zn^{2+}, and octahedral for Mn^{2+} — together with the differences in ion radii explains this specificity, which was not observed in the nonimprinted gels. The amphiphilic character of the functional monomer Vb-EDA made it possible to develop an emulsion polymerization procedure to prepare imprinted microgels. The main advantage of these microgels is that they can adsorb and release the copper ions faster because of their quick volume phase transitions.[56]

Detailed calorimetric studies of the thermal volume transition of polyNIPA (PNIPA) hydrogels and the influence of ligand binding on the relative stability of subchain conformations have been carried out by Grinberg and coworkers.[57–59] The dependence of the critical temperature of PNIPA hydrogels on the proportion of ionic co-monomers is an obstacle to obtaining devices with a high loading capability while still maintaining the PNIPA temperature-sensitive range.[2] This can be overcome by synthesizing interpenetrated polymer networks (IPN) of PNIPA with ionizable hydrophilic polymers. Although only a few papers have been devoted to this topic, the results obtained in those studies have shown the great potential of IPNs.[60,61] A two-step approach to imprint interpenetrated gels with metal ions was identified by Yamashita et al.[62] It basically consists in (a) polymerization of AA monomers to have a loosely cross-linked (1 mol%) polyAA network; (b) immersion of polyAA in copper solution to enable the ions to act as junction points between different chains; and (c) transfer of polyAA-copper ion complexes to a NIPA solution containing cross-linker (9.1 or 16.7 mol%) and synthesis of the NIPA network in the collapsed state (Figure 7.13). The nonimprinted IPNs (i.e., prepared in the absence of copper ions) showed a similar affinity for Cu^{2+} and for Zn^{2+}. In contrast, the imprinted IPNs in the collapsed state could discriminate between the square planar structure of Cu^{2+} and the tetrahedral structure of Zn^{2+}.

NIPA-based imprinted hydrogels have also been prepared using organic molecules as templates. Watanabe et al.[63] observed that NIPA (16 mmol)–acrylic acid (4 mmol) cross-linked (1 mmol) polymers synthesized in dioxane and in the presence of norephedrine (2 mmol) or adrenaline (2 mmol) showed, after template removal, an increase in the swelling ratio in the collapsed state as the target molecule concentration in water increases

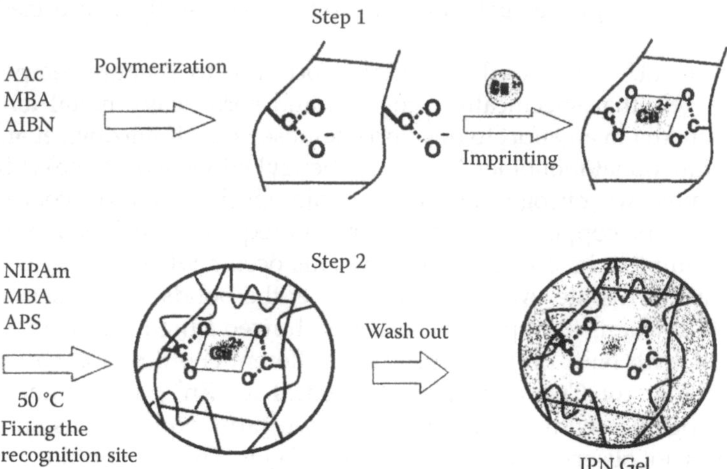

FIGURE 7.13
Two-step procedure to obtain an interpenetrated system comprising a Cu^{2+} imprinted poly(acrylic acid) hydrogel and a poly(*N*-isopropylacrylamide) temperature-sensitive hydrogel. (Reproduced from Yamashita, K. et al., *Polym. J.*, 35, 545, 2003. With permission from the Society of Polymer Science of Japan.)

(Figure 7.14). Since the molar concentration of the adsorbing monomers was much higher than the cross-linking density, the cross-links should not have created frustration. Thus the imprinting effect may only be due to the phase separation caused by the template during polymerization, which is macroscopically evidenced as a change in the conformation when the network collapses. Liu et al.[64,65] obtained temperature-sensitive imprinted gels for 4-aminopyridine and L-pyroglutamic, which showed the same ability to sorb and release the drug after several shrinking–swelling cycles. These gels had significantly higher saturation and affinity constants than the nonimprinted ones and were also highly selective. These results indicate that temperature-sensitive imprinted gels have a potential application in drug delivery.

Natural polymers, such as chitosan, have also been evaluated as a basis for temperature-sensitive hydrogels instead of synthetic monomers.[66] Chitosan is an aminopolysaccharide (obtained from chitin) that can be chemically cross-linked through the Schiff base reaction between its amine groups and the aldehyde ends of some molecules, such as glutaraldehyde.[67] A recent study has shown that if the reaction is carried out in the presence of target molecules, such as dibenzothiophenes (DBT), imprinted networks with a remarkably greater adsorption capability than nonimprinted ones can be obtained. This effect was particularly important when the gel was collapsed in the same solvent (acetonitrile) and at the same temperature (50°C) as were used during the cross-linking.[66] Additionally, the DBT-imprinted gels showed a high selectivity for the target molecules compared with other

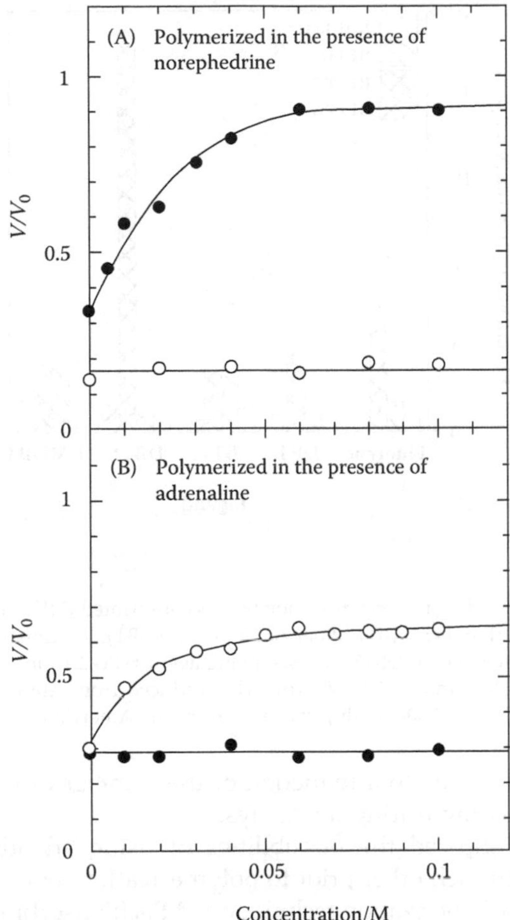

FIGURE 7.14
Equilibrium swelling ratios at 50°C as a function of concentration of either norephedrine (solid symbols) or adrenaline (open symbols) for imprinted gels prepared in the presence of norephedrine (a) or adrenaline (b). (Reproduced from Watanabe, M. et al., *J. Am. Chem. Soc.*, 120, 5577, 1998. With permission from the American Chemical Society.)

structurally related compounds (Figure 7.15). These gels have been proposed as traps for organosulfur pollutants.

In general, stimuli-sensitive imprinted gels are very weakly cross-linked (less than 2 mol%) and, therefore, the success of the imprinting strongly depends on the stability of the complexes of template/functional monomers during polymerization and after the swelling of the gels. If the molar ratio in the complex is not appropriate or if the complex dissociates to some extent during polymerization, the functional monomers will be far apart from both the template and each other, and the imprinting will be thwarted. However if the interaction is too strong, it may be difficult to remove the templates

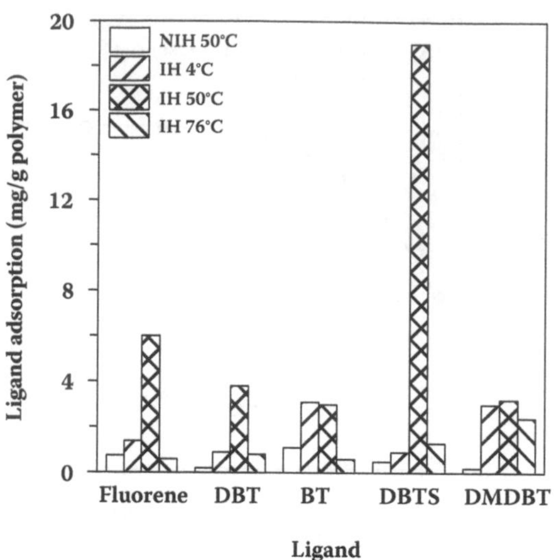

FIGURE 7.15
Effect of temperature and ligand on adsorption by a nonimprinted (NIH) and DBTS-imprinted hydrogel (IH): benzothiophene (BT), dibenzothiophene (DBT), 4,6-dimethyl-DBT (DMDBT), and dibenzothiophene sulfone (DBTS). Cross-linking agent ratio 2:1 mol, 50°C. Uptake conditions: [ligand]/CH_3CN = 4m*M*; 16 h; 300 rpm. (Reproduced from Aburto, J. and Le Borgne, S., *Macromolecules*, 37, 2938, 2004. With permission from the American Chemical Society.)

completely, which leads to a reduction of the number of free binding sites and template bleeding during the assays.

Recent efforts showed the possibilities of using adsorbing monomers directly bonded to each other prior to polymerization, which avoids the use of the template polymerization technique.[33,34] Each adsorbing monomer can be broken after polymerization to obtain pairs of ionic groups with the same charge. Since the members of each pair are close together, they can capture target molecules through multipoint ionic interactions (Figure 7.16). The adsorption process was found to be independent of the cross-linking density, and the entropic frustrations were completely resolved. Divalent ions or molecules with two ionic groups in their structures can be loaded in a greater amount and with higher affinity by these "imprinted" hydrogels.[34] Furthermore, the hydrogels prepared with PNIPA and imprinters were gifted with a new ability not observed in common stimuli-sensitive gels: they can re-adsorb, at the shrunken state, a significantly high amount of the templates previously released at the swollen state. Common (nonimprinted) temperature-sensitive PNIPA hydrogels have a pulsate release behavior that allows a substance entrapped in the polymer network to diffuse out of the hydrogel in the swollen state, but then stops the release when the temperature increases and the network collapses.[68,69] In contrast, a change from the swollen to the shrunken state of the imprinted PNIPA hydrogels not only stops the release, but also promotes a re-adsorption process. This process occurs

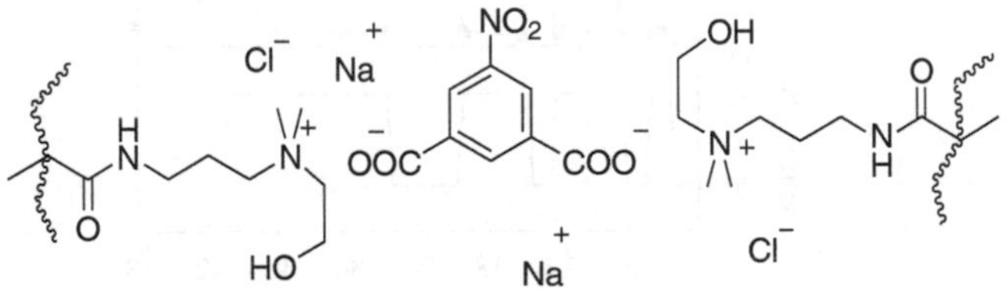

FIGURE 7.16
Structure of Imprinter-Q monomer (a) and of the binding sites of the gel made with Imprinter-Q after breakage of the 1,2-glycol bond (b), and schematic representation of the capture of a target molecule (c). (Reproduced from Moritani, T. and Alvarez-Lorenzo, C., *Macromolecules*, 34, 7796, 2001. With permission from the American Chemical Society.)

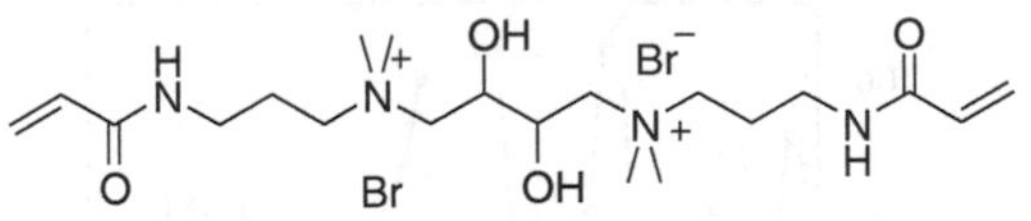

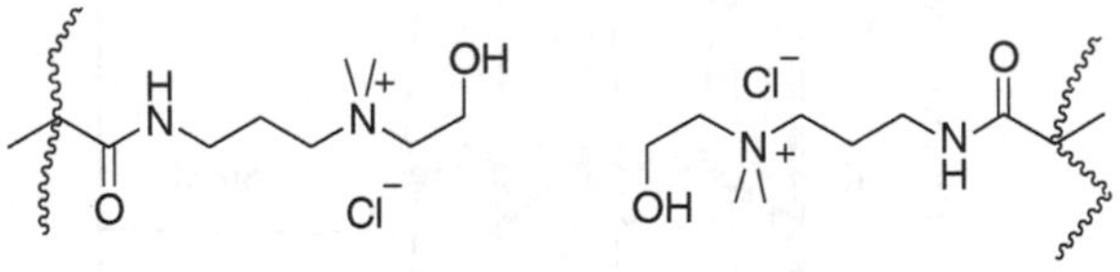

quickly and in a way that can be reproduced after several temperature cycles (Figure 7.17). At 37°C, the high affinity for the template provokes a stop in the release when an equilibrium concentration between the surrounding solution and the hydrogel is reached.

This type of hydrogel has a great potential for development of drug-delivery devices capable of maintaining stationary drug levels in their environment. The gel would stop the release while the released drug has not yet been absorbed or distributed but remains near the hydrogel.

7.4 pH-Sensitive Imprinted Gels

The term "pH-sensitive imprinted gel" can comprise two different behaviors:

1. Imprinted gels with adsorber monomers that show an affinity for the target molecule that can be tuned by changes in pH, but that do not undergo phase transitions
2. Imprinted gels that contain protonizable groups capable of causing pH-induced phase transitions and adsorber groups responsible for the interactions with the target molecule

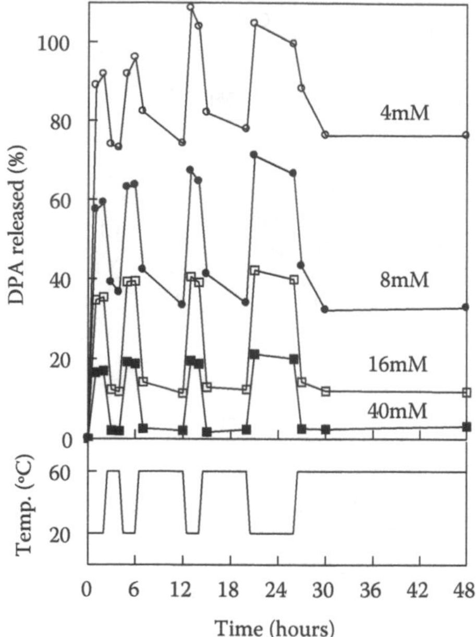

FIGURE 7.17
Influence of temperature on the release and readsorption of disodium 5-nitroisophthalate (DPA) in water by imprinted NIPA (6*M*) gels prepared with different concentrations of Imprinter-Q. Cross-linker concentration was 40m*M*. Degrees of swelling at 20°C and 60°C were 6.0–6.5 and 0.9, respectively. (Reproduced from Moritani, T. and Alvarez-Lorenzo, C., *Macromolecules*, 34, 7796, 2001. With permission from the American Chemical Society.)

In the first group we could insert almost all temperature-sensitive gels described in the previous section, as well as many other examples of non-smart imprinted networks,[70] since the recognition relies upon strong ionic interactions between the functional monomers and the template. If a change in pH modifies the degree of protonization of their chemical groups (although no change in degree of swelling takes place), a strong change in binding energy and, therefore, in affinity will occur.[71] In this section, we will mainly focus on gels that undergo pH-sensitive phase transitions, these changes in volume being mainly responsible for the control of the sorption process. Nevertheless, in some cases, the pH can alter both the swelling and the binding energy.

Kanekiyo et al.[72,73] have developed pH-sensitive imprinted gel particles using hydrophobic interactions to sorb the target molecules. The method consists of using a polymerizable derivative of amylose capable of wrapping around a hydrophobic template, such as bisphenol-A. The helical inclusion complex formed is then copolymerized with a cross-linker and a monomer having ionizable groups (e.g., acrylic acid, AA) (Figure 7.18). Similarly, other imprinted polymers were prepared with acrylamide instead of acrylic acid. The rebinding ability of MIPs prepared with AA showed a strong dependence

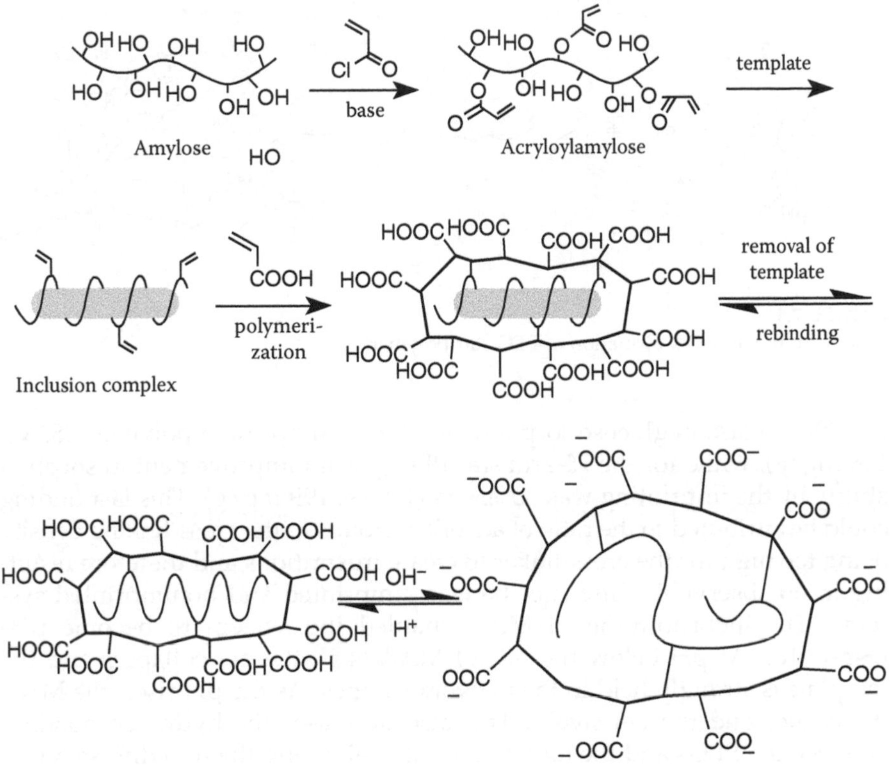

FIGURE 7.18
Synthesis of amylose-based imprinted polymer and its pH-responsive structural change. (Reproduced from Kanekiyo, Y. et al., *Angew. Chem.*, 42, 3014, 2003. With permission from Wiley-VCH Verlag GMbH & Co. KG.)

on pH: the greater the pH, the lower the binding of bisphenol-A. These results indicate that the binding cavity created through the imprinting process is disrupted by a conformational change in the amylose chain arising from the electrostatic repulsion between the anionic groups. A decrease in pH restores the cavities and the binding affinity.

The group of Peppas has developed an imprinting procedure using star-shaped polyethylene glycols (PEGs) copolymerized with methacrylic acid (MAA) for the recognition of sugars in aqueous solution.[74] Star polymers, also called hyperbranched polymers, have a large number of arms emanating from a central core and, therefore, they can contain a large number of functional groups in a small volume. The star-shaped PEGs provide the multiple hydrogen bonds required to interact with sufficient strength in water with sugars or proteins (Figure 7.19). Two different star-PEGs were evaluated. Grade 423 with 75 arms (MW = 6,970) and Grade 432 with 31 arms (MW = 20,000). Imprinted networks were obtained by carrying out the polymerization in the presence of glucose. The imprinted 31-arm star-PEG gel showed

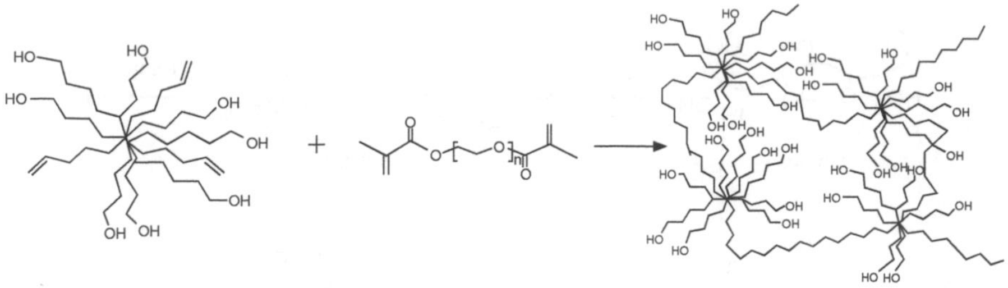

FIGURE 7.19
Scheme of the network formation of PEG star polymers.

a 213% increase in glucose sorption over the nonimprinted polymer (253 vs. 118 mg/g), while for the 75-arm star-PEG gels no improvement in sorption ability by the imprinting was observed (198 vs. 199 mg/g). This last finding could be attributed to the ratio of adsorber groups to the cross-linking density being too high for the cross-linker to create frustrations and therefore resulting in no observable difference between imprinted and nonimprinted systems. The incorporation of MAA enabled the system to become pH-responsive. At pH below the pK_a of MAA (4.5), the gel collapses and the template is strongly held in the network cavities. As the pH rises, the MAA dissociates and the gel swells. This also decreases the hydrogen bonding between the PEGs and the target molecules, allowing them to diffuse out of the gel.

The synthesis of MIPs selective to natural macromolecules, such as peptides and proteins, is not common. It is difficult because bulky protein cannot easily move in and out through the mesh of a polymer network. The attempts to overcome these limitations have been focused on synthesizing macroporous MIPs[75,76] or creating imprinted cavities at the surface of the network.[77,78] Stimuli-sensitive networks can be particularly useful for overcoming steric impediments, as recently shown by Demirel et al.[79] using pH- and temperature-sensitive gels imprinted for serum albumin (BSA). The ionic poly(*N*-tertbutylacrylamide-*co*-acrylamide/maleic acid) hydrogels synthesized in the presence of BSA showed a remarkably greater affinity for the protein compared with the nonimprinted ones, the adsorption being dependent on both pH and temperature. The hydrogels were synthesized at 22.8°C, at the swollen state. At this temperature, the adsorption is maximal. In contrast, when the gel collapses it is difficult for the protein to diffuse into the gel, the imprinted cavities are distorted, and the nature of the interactions can also be altered. At low temperature, the interactions between BSA and the hydrogel are based on hydrogen bonds. As the temperature rises, hydrogen bonds become weaker, while hydrophobic interactions get stronger. These results clearly highlight the relevance of the memorization of the conformation achieved during polymerization, which provides the gel with the ability to recognize a given template.

7.5 Conclusions

The synthesis of stimuli-sensitive hydrogels in the presence of target molecules enables one to design the sequence of adsorbing or functional monomers in the polymer network. Through such design, cavities with size and chemical properties complementary to the target can be imprinted in the hydrogel. The appropriate polymer sequence is fixed during polymerization, which allows the hydrogel to memorize the desired conformation. After undergoing cycles of swelling and collapse, the gel maintains the ability to recognize and host the target molecules. Adsorption occurs through multiple contacts in the collapsed state, but through single contacts in the swollen state. This multipoint adsorption makes the collapsed-state affinity significantly larger than that in the swollen state. The abrupt change in affinity during the gel volume phase transition allows one to turn gel adsorption on and off. The Tanaka equation can theoretically explain and predict the effects on a gel's affinity for the target molecules of the adsorbing monomer and cross-linker proportions in the network and of the salt or replacement molecules' concentration in the sorption medium. The Tanaka equation is also a useful tool for better understanding the advantages of imprinted stimuli-sensitive hydrogels, compared with nonimprinted ones, and for optimizing their properties for application in different fields.

Acknowledgments

This work was financed by the Ministerio de Ciencia y Tecnología, FEDER (RYC2001-8; SAF2005-01930), and Xunta de Galicia (PGIDIT03PXIC20303PN), Spain.

References

1. Dusek, K. and Patterson, D., Transition in swollen polymer networks induced by intramolecular condensation, *J. Polym. Sci. A2*, 6, 1209, 1968.
2. Tanaka, T., Collapse of gels and the critical endpoint, *Phys. Rev. Lett.*, 40, 820, 1978.
3. Ilmain, F., Tanaka, T., and Kokufuta, E., Volume transition in a gel driven by hydrogen bonding, *Nature*, 349, 400, 1991.
4. Flory, P.J., *Principles of Polymer Chemistry*, Cornell, New York, 1953.
5. Shibayama, M. and Tanaka, T., Volume phase transition and related phenomena of polymer gels, in *Advances in Polymer Science, Responsive Gels: Volume Transitions I*, Dusek, K., Ed., Springer, Berlin, 1993, vol. 109, pp. 1–62.
6. Pande, V.S., Grosberg, A.Yu., and Tanaka, T., Statistical mechanics of simple models of protein folding and design, *Biophys. J.*, 73, 3192, 1997.

7. Tanaka, T. and Annaka, M., Multiple phases of gels and biological implications, *J. Intel. Mat. Syst. Struct.*, 4, 548, 1993.
8. Pande, V.S., Grosberg, A.Yu., and Tanaka, T., Folding thermodynamics and kinetics of imprinted renaturable heteropolymers, *J. Chem. Phys.*, 101, 8246, 1994.
9. Pande, V.S., Grosberg, A.Yu., and Tanaka, T., Thermodynamic procedure to synthesize heteropolymers that can renature to recognize a given target molecule, *Proc. Natl. Acad. Sci.*, 91, 12976, 1994.
10. Wulff, G., Molecular imprinting in cross-linked materials with the aid of molecular templates: a way towards artificial antibodies, *Angew. Chem. Int. Ed. Engl.*, 34, 1812, 1995.
11. Grosberg, A.Yu. and Khokhlov, A.R., *Giant Molecules*, Academic Press, San Diego, 1997.
12. Wulff, G. and Biffis, A., Molecularly imprinting with covalent or stoichiometric non-covalent interactions, in *Molecularly Imprinted Polymers*, Sellergren, B., Ed., Elsevier, Amsterdam, 2001, pp. 71–111.
13. Arshady, R. and Mosbach, K., Synthesis of substrate-selective polymers by host-guest polymerization, *Makromol. Chem.*, 182, 687, 1981.
14. Sellergren, B., The non-covalent approach to molecular imprinting, in *Molecularly Imprinted Polymers*, Sellergren, B., Ed., Elsevier, Amsterdam, 2001, pp. 113–184.
15. Ansell, R.J., Molecularly imprinted polymers in pseudoimmunoassay, *J. Chromatogr. B*, 804, 151, 2004.
16. Yan, M. and Ramström, O., Molecular imprinting: an introduction, in *Molecularly Imprinted Materials*, Yan, M. and Ramström, O., Eds., Marcel Dekker, New York, 2005, pp. 1–12.
17. Mayes, A.G. and Whitcombe, M.J., Synthetic strategies for the generation of molecularly imprinted organic polymers, *Adv. Drug Del. Rev.*, 57, 1742, 2005.
18. Andersson, L.I., Molecular imprinting for drug bioanalysis: a review on the application of imprinted polymers to solid-phase extraction and binding assay, *J. Chromatogr. B*, 739, 163, 2000.
19. Kandimalla, V.B. and Ju, H.X., Molecular imprinting: a dynamic technique for diverse applications in analytical chemistry, *Anal. Bioanal. Chem.*, 380, 587, 2004.
20. Xu, X., Zhu, L., and Chen, L., Separation and screening of compounds of biological origin using molecularly imprinted polymers, *J. Chromatogr. B*, 804, 61, 2004.
21. Piletsky, S.A. and Turner, A.P.F., Electrochemical sensors based on molecularly imprinted polymers, *Electroanalysis*, 14, 317, 2002.
22. Hillberg, A.L., Brain, K.R., and Allender, C.J., Molecular imprinted polymer sensors: implications for therapeutics, *Adv. Drug Del. Rev.*, 57, 1875, 2005.
23. Sibrian-Vazquez, M. and Spivak, D.A., Improving the strategy and performance of molecularly imprinted polymers using cross-linking functional monomers, *J. Org. Chem.*, 68, 9604, 2003.
24. Enoki, T., Tanaka, K., Watanabe, T. et al., Frustrations in polymer conformation in gels and their minimization through molecular imprinting, *Phys. Rev. Lett.*, 85, 5000, 2000.
25. Byrne, M.E., Park, K., and Peppas, N.A., Molecular imprinting within hydrogels, *Adv. Drug Del. Rev.*, 54, 149, 2002.
26. Miyata, T., Uragami, T., and Nakamae, K., Biomolecule-sensitive hydrogels, *Adv. Drug Del. Rev.*, 54, 79, 2002.

27. Alvarez-Lorenzo, C. and Concheiro, A., Molecularly imprinted polymers for drug delivery, *J. Chromatogr. B*, 804, 231, 2004.
28. Cunliffe, D., Kirby, A., and Alexander, C., Molecularly imprinted drug delivery systems, *Adv. Drug Del. Rev.*, 57, 1836, 2005.
29. Alvarez-Lorenzo, C. and Concheiro, A., Molecularly imprinted materials as advanced excipients for drug delivery systems, in *Biotechnology Annual Review*, vol. 12, El-Gewely, M.R., Ed., Elsevier, Amsterdam, 225, 2006.
30. Alvarez-Lorenzo, C. and Concheiro, A., Molecularly imprinted gels and nano- and microparticles: manufacture and applications, in *Smart Assemblies and Particulates*, vol. 7, Arshady, R. and Kono, K., Eds., Kentus Books, London, 2006, pp. 275–336.
31. Alvarez-Lorenzo, C. et al., Polymer gels that memorize elements of molecular conformation, *Macromolecules*, 33, 8693, 2000.
32. Alvarez-Lorenzo, C. et al., Reversible adsorption of calcium ions by imprinted temperature sensitive gels, *J. Chem. Phys.*, 114, 2812, 2001.
33. D'Oleo, R., Alvarez-Lorenzo, C., and Sun, G., A new approach to design imprinted polymer gels without using a template, *Macromolecules*, 34, 4965, 2001.
34. Moritani, T. and Alvarez-Lorenzo, C., Conformational imprinting effect on stimuli-sensitive gels made with an imprinter monomer, *Macromolecules*, 34, 7796, 2001.
35. Stancil, K.A., Feld, M.S., and Kardar, M., Correlation and cross-linking effects in imprinting sites for divalent adsorption in gels, *J. Phys. Chem. B*, 109, 6636, 2005.
36. Oya, T. et al., Reversible molecular adsorption based on multiple point interaction by shrinkable gels, *Science*, 286, 1543, 1999.
37. Watanabe, T. et al., Affinity control through salt concentration on multiple contact adsorption into heteropolymer gel, *J. Chem. Phys.*, 115, 1596, 2001.
38. Ito, K. et al., Multiple contact adsorption of target molecules, *Macromol. Symp.*, 207, 1, 2004.
39. Ito, K. et al., Multiple point adsorption in a heteropolymer gel and the Tanaka approach to imprinting: experiment and theory, *Prog. Polym. Sci.*, 28, 1489, 2003.
40. Rampey, A.M. et al., Characterization of the imprinting effect and the influence of imprinting conditions on affinity, capacity, and heterogeneity in molecularly imprinted polymers using the Freundlich isotherm-affinity distribution analysis, *Anal. Chem.*, 76, 1123, 2004.
41. Pap, T. and Horvai, G., Binding assays with molecularly imprinted polymers: why do they work? *J. Chromatogr. B*, 804, 167, 2004.
42. Umpleby II, R.J. et al., Characterization of the heterogeneous binding site affinity distributions in molecularly imprinted polymers, *J. Chromatogr. B*, 804, 141, 2004.
43. Alvarez-Lorenzo, C. and Concheiro, A., Reversible adsorption by a pH- and temperature-sensitive acrylic hydrogel, *J. Control. Rel.*, 80, 247, 2002.
44. Hsein, T.Y. and Rorrer, G.L., Heterogeneous cross-linking of chitosan gel beads: kinetics, modeling, and influence on cadmium ion adsorption capacity, *Ind. Eng. Chem. Res.*, 36, 3631, 1997.
45. Eichenbaum, G.M. et al., Alkali earth metal binding properties of ionic microgels, *Macromolecules*, 33, 4087, 2000.
46. Takeoka, Y. et al., First order phase transition and evidence for frustration in polyampholytic gels, *Phys. Rev. Lett.*, 82, 4863, 1999.
47. Alvarez-Lorenzo, C. et al., Simultaneous multiple-point adsorption of aluminum ions and charged molecules in a polyampholyte thermosensitive gel: controlling frustrations in a heteropolymer gel, *Langmuir*, 17, 3616, 2001.

48. Grosberg, A.Yu. and Khokhlov, A.R., *Statistical Physics of Macromolecules*, AIP, New York, 1994.
49. Pande, V.S., Grosberg, A.Yu., and Tanaka, T., Heteropolymer freezing and design: towards physical models of protein folding, *Rev. Mod. Phys.*, 72, 259, 2000.
50. Bryngelson, J.D. and Wolynes, P.G., Spin glasses and the statistical mechanics of protein folding, *Proc. Natl. Acad. Sci. USA*, 2, 7524, 1987.
51. Tanaka, T. et al., Reversible molecular adsorption as a tool to observe freezing and to perform design of heteropolymer gels, *Ber. Bunsenges. Phys. Chem.*, 102, 1529, 1998.
52. Hiratani, H. et al. Effect of reversible cross-linker, *N,N'*-bis(acryloyl)cystamine, on calcium ion adsorption by imprinted gels, *Langmuir*, 17, 4431, 2001.
53. Güney, O., Yilmaz, Y., and Pekcan, O., Metal ion templated chemosensor for metal ions based on fluorescence quenching, *Sensors Actuators B*, 85, 86, 2002.
54. Güney, O., Multiple-point adsorption of terbium ions by lead ion templated thermosensitive gel: elucidating recognition of conformation in gel by terbium probe, *J. Molecular Recogn.*, 16, 67, 2003.
55. Kanazawa, R., Yoshida, T., Gotoh, T., and Sakohara, S., Preparation of molecular imprinted thermosensitive gel adsorbents and adsorption/desorption properties of heavy metal ions by temperature swing, *J. Chem. Eng. Japan*, 37, 59, 2004.
56. Kanazawa, R., Mori, K., Tokuyama, H., and Sakohara, S., Preparation of thermosensitive microgel adsorbent for quick adsorption of heavy metal ions by a temperature change, *J. Chem. Eng. Japan*, 37, 804, 2004.
57. Grinberg, N.V. et al., Studies of the thermal volume transition of poly(*N*-isopropylacrylamide) hydrogels by high-sensitivity differential scanning microcalorimetry, 1: dynamic effects, *Macromolecules*, 32, 1471, 1999.
58. Grinberg, V.Ya. et al., Studies of the thermal volume transition of poly(*N*-isopropylacrylamide) hydrogels by high-sensitivity microcalorimetry, 2: thermodynamic functions, *Macromolecules*, 33, 8685, 2000.
59. Burova, T. et al., Effects of ligand binding on relative stability of subchain conformations of weakly charged *N*-isopropylacrylamide gels in swollen and shrunken states, *Macromolecules*, 36, 9115, 2003.
60. Alvarez-Lorenzo, C. et al., Temperature-sensitive chitosan-poly(*N*-isopropylacrylamide) interpenetrated networks with enhanced loading capacity and controlled release properties, *J. Control. Rel.*, 102, 629, 2005.
61. Yamashita, K., Nishimura, T., and Nango, M., Preparation of IPN-type stimuli responsive heavy-metal-ion adsorbent gel, *Polym. Adv. Technol.*, 14, 189, 2003.
62. Yamashita, K. et al., Two-step imprinting procedure of inter-penetrating polymer network-type stimuli-responsive hydrogel adsorbents, *Polym. J.*, 35, 545, 2003.
63. Watanabe, M., Akahoshi, T., Tabata, Y., and Nakayama, D., Molecular specific swelling change of hydrogels in accordance with the concentration of guest molecules, *J. Am. Chem. Soc.*, 120, 5577, 1998.
64. Liu, X.Y. et al., Fabrication of temperature-sensitive imprinted polymer hydrogel, *Macromol. Biosci.*, 4, 412, 2004.
65. Liu, X.Y. et al., Design of temperature sensitive imprinted polymer hydrogels based on multiple-point hydrogen bonding, *Macromol. Biosci.*, 4, 680, 2004.
66. Aburto, J. and Le Borgne, S., Selective adsorption of dibenzothiophene sulfone by an imprinted and stimuli-sensitive chitosan hydrogel, *Macromolecules*, 37, 2938, 2004.

67. Berger, J. et al., Structure and interactions in covalently and ionically crosslinked chitosan hydrogels for biomedical applications, *Eur. J. Pharm. Biopharm.*, 57, 19, 2004.
68. Qiu, Y. and Park, K., Environment-sensitive hydrogels for drug delivery, *Adv. Drug Del. Rev.*, 53, 321, 2001.
69. Kikuchi, A. and Okano, T., Pulsatile drug release control using hydrogels, *Adv. Drug Del. Rev.*, 54, 53, 2002.
70. Puoci, F. et al., Spherical molecularly imprinted polymers (SMIPs) via a novel precipitation polymerization in the controlled delivery of salazine, *Macromol. Biosci.*, 4, 22, 2004.
71. Chen, Y.B. et al., Influence of the pH on the behavior of an imprinted polymeric stationary phase: supporting evidence for a binding site model, *J. Chromatog. A*, 927, 1, 2001.
72. Kanekiyo, Y., Naganawa, R., and Tao, H., Molecular imprinting of bisphenol-A and alkylphenols using amylose as a host matrix, *Chem. Commun.*, 22, 2698, 2002.
73. Kanekiyo, Y., Naganawa, R., and Tao, H., pH-responsive molecularly imprinted polymers, *Angew. Chem.*, 42, 3014, 2003.
74. Oral, E. and Peppas, N.A., Responsive and recognitive hydrogels using star polymers, *J. Biomed. Mater. Res.*, 68A, 439, 2004.
75. Guo, T.Y. et al., Chitosan beads as molecularly imprinted polymer matrix for selective separation of proteins, *Biomaterials*, 26, 5737, 2005.
76. Hawkins, D.M., Stevenson, D., and Reddy, S.M., Investigation of protein imprinting in hydrogel-based molecularly imprinted polymers (HydroMIPs), *Anal. Chim. Acta*, 542, 61, 2005.
77. Shnek, D.R. et al., Specific protein attachment to artificial membranes via coordination to lipid-bound copper(II), *Langmuir*, 10, 2382, 1994.
78. Rachkov, A. and Minoura, N., Towards molecularly imprinted polymers selective to peptides and proteins: the epitope approach, *Biochim. Biophys. Acta*, 1544, 255, 2001.
79. Demirel, G. et al., pH/temperature-sensitive imprinted ionic poly (*N*-tert-butylacrylamide-*co*-acrylamide/maleic acid) hydrogels for bovine serum albumin, *Macromol. Biosci.*, 5, 1032, 2005.

Chapter 8

Smart Hydrogels

Kong Jilie and Mu Li

CONTENTS

8.1 Introduction

Hydrogels are three-dimensional networks of hydrophilic polymers that hold a large amount of water while maintaining the solid state. So, hydrogels can swell dramatically in the presence of aqueous solution and can also deswell upon certain stimulation. Although the definition of "macroporous hydrogels" has been extensively debated, generally they are defined as hydrogels having pores larger than 50 nm.[1] Compared with conventional hydrogels, macroporous hydrogels have a much higher swelling/deswelling rate. For instance, their deswelling rates (measured as the time needed to achieve a certain level of maximum swelling) are within minutes, while for conventional hydrogels it is a matter of hours or days. Because swelling and deswelling of hydrogels are controlled by diffusion of solute into and out of the gel matrix, apparently it is much easier and faster for solute to pass through larger pores. It is also possible that solvent molecules are taken up into the gels by capillary forces, which is much faster than diffusion.[2]

Smart hydrogels are hydrogels that reversibly swell and shrink sharply upon a small change in environmental conditions, such as temperature, pH, electric field, light, pressure, ionic strength, solvent, etc. Temperature and pH are widely used stimuli because they are easy to control. Poly(*N*-isopropylacrylamide) (PNIPAAm) is the most widely studied temperature-responsive polymer. Hydrogels from PNIPAAm exhibit a transition temperature of approximately 32°C in aqueous solution. Below the transition temperature, the hydrogel network absorbs and retains a large amount of water, while above this temperature it shrinks to a dense state and loses much water. By incorporating some stimuli-responsive co-monomers either into the backbone of the network or as pendant groups, one can prepare smart hydrogels responsive to various stimuli. The stimuli-responsive properties offer the possibility of gel-based technologies, such as separation and purification, molecular recognition, flow control, controlled drug release, tissue engineering (e.g., artificial muscles), soft machines, etc.

Dynamic response is critical to smart-hydrogel applications, but conventional hydrogels have low response rates. Many efforts have been made to create smart hydrogels with fast response. For example, hydrophilic side chains grafted to the hydrogel backbone can form a hydrophilic channel in the network for quick release of water.[3–5] In addition, because of their intrinsic swift response property, macroporous smart hydrogels have attracted much attention.

8.2 Progress in Synthesis of Fast-Responsive Hydrogels from Smart Polymers

To obtain a fast-responsive hydrogel with macroporous structures, a phase separation must occur during the network formation process. Many methods — the use of additives, special solvents, and templates,[6–10] cross-linking polymerization, controlling the preparation temperature, freeze-drying[11] — have been employed to prepare macroporous hydrogels. The macroporosity of the resulting networks depends on the gel synthesis parameters, such as temperature, cross-linker, monomer concentrations, and the amount and type of the pore-forming agents presented during the network formation process. Great efforts have been taken to optimize the synthesis process to create macroporous smart hydrogels with better mechanical properties[12] and, more importantly, with enhanced responsive properties, e.g., swelling ratio, deswelling rate, etc.

This section discusses recent developments in polymerizing fast-responsive hydrogels by using additives and special solvents, via cross-linking polymerization, and by controlling preparation temperature. Some typical examples of these studies are summarized in Table 8.1.

8.2.1 Use of Additives

Additives such as pore-forming agents and foaming reagents are typically used to synthesize macroporous hydrogels. Many kinds of pore-forming agents, including sucrose-modified starches,[13] fractionated particles of sodium chloride,[14] ice crystals,[15] hydroxypropyl cellulose, acetone, 1,4-dioxane, silica particles,[16] and poly(ethylene glycol) (PEG), have been used during the polymerization of macroporous hydrogels. After polymerization, these porogens can be leached out, leaving an interconnected pore network.

Most of the pore-forming agents used to date have been hydrophilic. For example, PEG with different molecular weights has been investigated as a

TABLE 8.1

Polymerization Methods of Macroporous Smart Hydrogels and Their Properties

Polymerization Methods	Examples Reagents	Hydrogel	Pore Size (μm[a])	Shrinking Time (min)[b]	Swelling Ratio[c]	Refs.
Using additives	PEG	PNIPAAm	10–80	≈3–20	≈80	Zhang and Zhuo[17] Zhang et al.[18] Zhuo and Li[19] Zhang et al.[20] Dogu and Okay[21]
	PLA nanospheres	PNIPAAm	≈8.2	37	≈25	Zhang et al.[22]
	THF	PNIPAAm	—	1.5	≈51	Zhang and Chu[23]
Using special solvent	water/acetone	PNIPAAm	—	<10	50–60	Zhang et al.[26]
	water/1,4-dioxane	P(NIPAAm-co-ACD)	—	<20	≈55	Zhang et al.[27]
	NaCl solution (1.0*M*)	PNIPAAm	—	5	≈60	Cheng et al.[30]
Cross-linking	3-chloropentane-2,4-dione	PAL	21–100	—	130	Das et al.[32]

Note: PEG = polyethylene glycol, PNIPAAm = poly(*N*-isopropylacrylamide), PLA = polylactide, THF = tetrahydrofuran, P(NIPAAm-*co*-ACD) = poly(*N*-isopropylacrylamide-*co*-acryloyl-β-cyclodextrin, PAL = polyallylamine.

[a] Some literature reports just showed SEM photos of hydrogels without giving exact data. So in the table, some data of pore sizes are not given.

[b] Shrinking time defines the time for the hydrogels to lose 90% retained water.

[c] Normally, the swelling ratio (SR) is defined as SR = $(W_s - W_d)/W_d$, where W_s and W_d represent the weight of the wet and dry hydrogel, respectively.

pore-forming agent to obtain macroporous PNIPAAm hydrogels.[17–21] The hydration and exclusion volume of PEGs can provide spatial hindrance during the polymerization and cross-linking process, resulting in hydrogels with macroporous structure and fast-response kinetics. Using PEG with a molecular weight of 2000 as a pore-forming agent, the pore size of the prepared PNIPAAm hydrogel is 10 to 20 μm, and the swelling ratio at room temperature is 60 to 80. In contrast, the swelling ratio for control PNIPAAm hydrogels prepared without PEG is only about 15. The macroporous network structure of the hydrogels can be adjusted by applying PEGs with different molecular weights. An increase in the molecular weight of the selected PEG enlarges the average pore size of the prepared hydrogels. For example, when PEGs with molecular weights of 2000, 4000, and 6000 are used to prepare PNIPAAm hydrogels, the average pore sizes obtained are 10 to 20, 30 to 40, and 60 to 80 μm, respectively. When the molecular weight of the PEG additive is less than 2000, the response rate of the resulting PNIPAAm hydrogel is proportionally related to the molecular weight of the PEG. However, further increasing the molecular weight of PEG (molecular weight > 2000) does not improve the hydrogel's response rate. In addition, the PEG content, preparation temperature, and cross-linker content can also affect the resulting hydrogel's response properties.[21]

The application of a hydrophobic additive, poly(lactic acid) (PLA) nanospheres, was described by Zhang et al.[22] Micrographs taken by scanning electron microscopy (SEM) revealed that PLA induced a macroporous structure in PNIPAAm hydrogels that made the hydrogel pore size (≈8.2 μm) a few microns larger than that of a normal gel (≈3.4 μm). As a result, the equilibrium swelling ratio for the PLA-treated hydrogel was 25, compared with a ratio of 15 for a normal hydrogel. Temperature-responsive deswelling of the PLA-treated hydrogels from swollen (22°C) to shrunken (50°C) states was much faster, too. They lost around 90% water in 7 min and approximately 95% in 37 min at 50°C, while the control hydrogels prepared without PLA lost only around 30% water in 7 min and 50% in 37 min at 50°C.

Organic solvents such as dichloromethane, acetone, and methanol can be used as foaming reagents during the polymerization of macroporous hydrogels. A novel PNIPAAm hydrogel prepared with an organic solvent, tetrahydrofuran (THF), as both a foaming and precipitation reagent during polymerization/cross-linking in water at room temperature exhibited a superfast response to temperature change.[23] The macroporous network was generated due to THF evaporation and simultaneous evaporation of water at an increased reaction temperature due to the highly exothermic gelation process. The macroporous hydrogels lost 90% water in 90 sec and reached a stable shrunken state in about 10 min, whereas normal hydrogels needed weeks to reach this state (Figure 8.1).

Poly(*N*-isopropylacrylamide)/poly(ethylene glycol) diacrylate microgel particles were used as novel pore-forming additives. Although the pore size (2 to 7 μm) was less than that obtained using conventional pore-forming agents like PEG, they demonstrated much better thermosensitive characteristics in

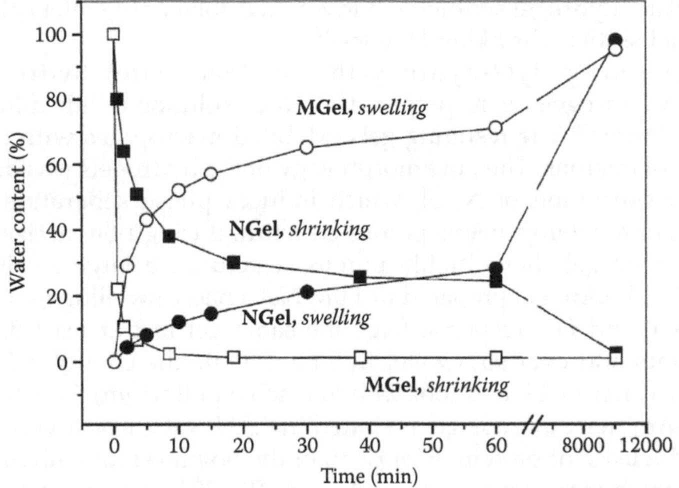

FIGURE 8.1
Shrinking (□, ■) and swelling (○, ●) kinetics of modified (MGel) and normal (NGel) PNIPAAm hydrogels in water at 48°C and 22°C. (Modified from Zhang, X.Z. and Chu, C.C., *Chemical Commun.*, 3, 350, 2004. With permission.)

terms of higher equilibrium swelling ratio (≈70) as well as higher response rates (lost over 90% water after 18 min at 48°C). They also significantly improved the mechanical properties of the hydrogels, with a compression modulus of 20 to 34 kPa, compared with 12.7 kPa for the normal PNIPAAm hydrogels.[24] A similar study reported the use of poly(NIPAAm-*co*-acrylic acid) microgel particles as additives to prepare fast-responsive bulk hydrogels, which exhibited improved swelling capability.[25]

8.2.2 Utilization of Special Solvents

Special solvents have wide applications in preparing fast-responsive hydrogels. A hydrophilic/hydrophobic balance exists in PNIPAAm hydrogel networks because of the presence of the –CONH– hydrophilic groups and the $-CH(CH_3)_2$ hydrophobic groups in the side chains. Consequently, an appropriate mixture of hydrophobic and hydrophilic solvents can help in the production of fast-responsive hydrogels. Zhang et al. reported using a water–acetone mixture to polymerize macroporous PNIPAAm hydrogels.[26] Alternatively, a mixture of water/1,4-dioxane was used as solvent to prepare temperature-sensitive poly(NIPAAm-*co*-acryloyl-β-cyclodextrin) hydrogels. Because the mixed solvent was a poor solvent for the copolymers, phase separation occurred during the polymerization, which resulted in a heterogeneous, porous structure of the hydrogels. The prepared hydrogels have fast shrinking characteristics, returning to the equilibrium state in less than 5 min and losing at least 80% water within 20 min.[27] A poly (NIPAAm-*co*-methyl methacrylate) hydrogel synthesized in a mixed solvent of water/glacial acetic acid had similar shrinking kinetics.[28]

Macroporous poly(2-hydroxyethyl methacrylate) hydrogels and PNIPAAm hydrogels were prepared in NaCl solutions with different salt concentrations.[29,30] The resulting gels exhibited macropores with diameters up to tens of microns. The pore morphology of the hydrogels was dependent on the concentration of NaCl, which induces phase separation and the formation of a heterogeneous porous structure during polymerization. The resulting hydrogels have highly porous structures, a large swelling ratio (e.g., the PNIPAAm gel prepared in 1.0*M* NaCl had a swelling ratio of about 70 at 15°C), and fast response (e.g., the same gel lost at least 85% water within 1 min and over 95% water in 5 min, while the control gel lost only about 50% water in 2 h and took 24 h to reach equilibrium). The application of PNIPAAm macroporous gel prepared in 0.3*M* NaCl was investigated for controlled release of protein. Almost all of the bovine serum albumin (BSA) loaded in the hydrogel matrix was released after 70 h at a temperature lower than the transition temperature, while the conventional gel released only 40% BSA after 70 h.

Macroporous PNIPAAm hydrogels with improved sensitivity were prepared in 2*M* aqueous glucose solution. The hydrogels exhibited a much faster response and underwent full deswelling in 2 min (lost around 95% water) when the temperature was raised above the transition temperature.[31]

8.2.3 Cross-Linking Polymerization

Cross-linking polymerization with appropriate cross-linkers is widely used to prepare macroporous hydrogels. Macroporous structures can be designed by controlling the content of cross-linker and the conditions of polymerization.

pH-responsive hydrogels of poly(acrylamide-*co*-acrylic acid) have been synthesized with *N,N'*-methylene-bisacrylamide as the cross-linker, resulting in a "superporous" structure with pore size up to 100 μm.[2] Another inherently luminescent pH-responsive macroporous hydrogel was produced from water-soluble poly(allylamine) using 3-chloropentane-2,4-dione as a cross-linker.[32] This hydrogel has oval pores with their long axis varying in the range of 21 to 100 μm. The hydrogels produced had a swelling ratio up to 130 at 20°C. The hydrogels swelled in the pH range from 5 to 10, and deswelled sharply at pH values below 5 or above 10.

Macroporous PNIPAAm hydrogels were generated using ethylene triethoxy silane as the cross-linking agent at two temperatures (18 and 60°C) under acidic conditions. The resulting hydrogels reached stable water retention in less than 5 min and reswelled quickly. The gels absorbed over 90% of the total reswelling water within 1 min and reached stable water uptake in less than 5 min, while conventional hydrogels absorbed only about 35% after 21 min (Figure 8.2). The fast kinetics was attributed to an interconnected macroporous structure, as evidenced by SEM. It was suggested that macropores would aid in transporting water into or out of the network of

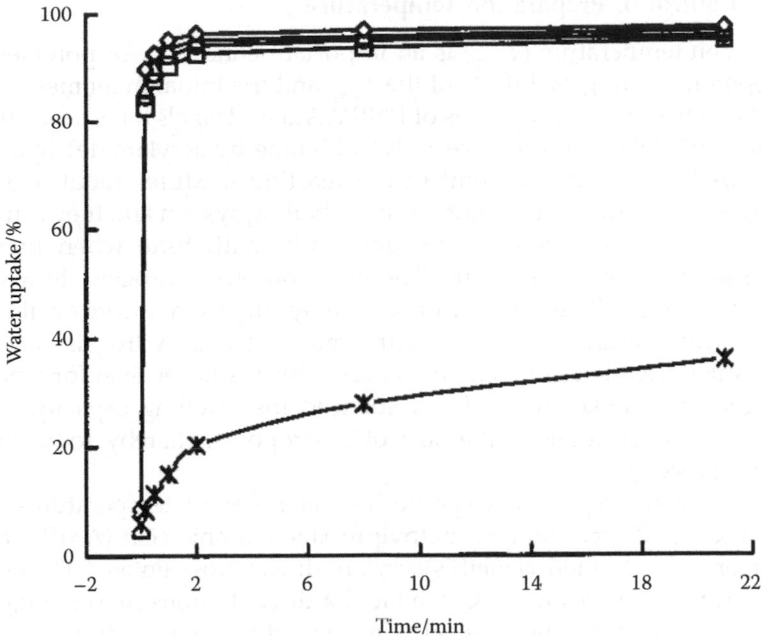

FIGURE 8.2
Reswelling kinetics of the PNIPAAm gels at 18°C: normal PNIPAAm gels (*) and macroporous PNIPAAm gels produced from the polymerization feed containing 10 (□), 20 (○), 30 (△), and 40 (◇) mg of ethylene triethoxy silane per 100 mg of NIPAAm. (Modified from Zhang, X.Z. and Zhuo, R.X., *Langmuir*, 17, 12, 2001. With permission.)

the macroporous PNIPAAm hydrogels during the deswelling or reswelling process and prevent the development of a dense, thick skin layer during the hydrogel shrinking. The high reversibility of this macroporous hydrogel could be very useful in the field of biomaterials and biotechnology.[33]

Smart macroporous resins were produced by inverse suspension polymerization.[34,35] For example, thermosensitive poly(*N*,*N*-diethylacrylamide-*co*-2-hydroxyethyl methacrylate) resins were prepared by inverse suspension polymerization with *N*,*N*′-methylenebis(acrylamide) as a cross-linker.[35] The hydrogels had a macroporous network structure with good thermo-induced swelling properties in a variety of solvents. The hydrogels had a high concentration of functional hydroxyl groups and were employed as supports in solid-phase peptide synthesis.

Radiation cross-linking polymerization without cross-linker is an alternative way to prepare macroporous hydrogels. For example, gamma radiation and electron-beam irradiation have been employed to initiate synthesis of macroporous PNIPAAm hydrogels.[36,37] The resulting hydrogels have very fast deswelling kinetics, about 100 times faster than conventionally cross-linked gels. The microstructures of the gels were observed to be greatly influenced by the irradiating dose, dose rate, and temperature.

8.2.4 Control of Preparation Temperature

Preparation temperature (T_{prep}) is an important condition for polymerizing macroporous hydrogels. Effects of the T_{prep} and the initial monomer concentration on the porosity properties of PNIPAAm hydrogels have been studied intensively.[38] Solvent-responsive poly(*N*,*N*-dimethylacrylamide) hydrogels synthesized at the boiling point of the reacting mixture, about 80°C, are macroporous because of the bubbles of the boiling system and they can swell in water, while they deswell exponentially with time when they are immersed in acetone or dioxane. The macroporous hydrogels show about three times the swelling capacity of normal hydrogels that were synthesized at room temperature. Furthermore, the macroporous hydrogels can reach stable water retention in less than 20 min, which is faster than for a normal hydrogel. Cross-linker ratio also influenced the swelling capacity. Larger cross-linker ratios reduced the size of macropores, thereby reducing the swelling capacity.[39]

A series of hydrogels was synthesized at different temperatures using sodium salt of 2-acrylamido-2-methylpropane sulfonic acid (AMPS) as the monomer and *N*,*N*′-methylene(bis)acrylamide as a cross-linker.[40] The PAMPS hydrogels prepared below –8°C exhibited a discontinuous morphology consisting of 30- to 50-μm polyhedral pores and showed a superfast response to different solvents. As shown in Figure 8.3, the swelling in water for the PAMPS hydrogels reached equilibrium in less than 30 sec and deswelling in

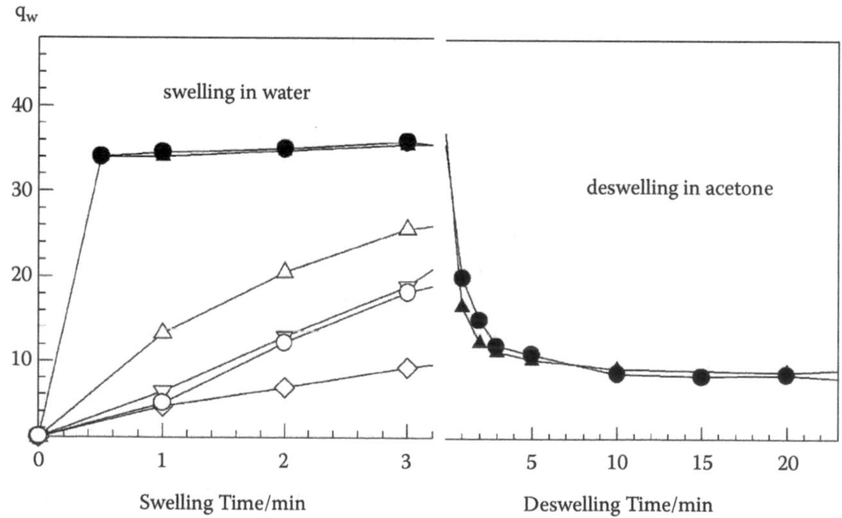

FIGURE 8.3
The initial period of swelling and deswelling of hydrogels prepared at various temperatures, depicted as the variation of the weight swelling ratio q_w with the time of swelling or deswelling. T_{prep} = –22 (●), –18 (▲), –8 (∞), 0 (△), 15 (▽), and 25°C (○). (Modified from Ozmen, M.M. and Okay, O., *Polymer*, 46, 8119, 2005. With permission.)

acetone in 5 to 10 min. Nonporous structures were observed for the gels formed at higher temperatures.

8.3 Applications of Fast-Responsive Hydrogels

Fast-responsive hydrogels have displayed various spectacular applications. The large pores in the macroporous smart hydrogels allow molecules to enter the matrix, and their stimulus response enables reversible immobilization of molecules. Because of their high biocompatibility, fast-responsive hydrogels can also be used for tissue engineering and cell culture. The special three-dimensional structure of the macroporous smart hydrogels and the possibility of attaching functional side chains onto the pore surface make it possible to recognize target molecules and selectively separate molecules from mixture solutions. Fast-responsive hydrogels are promising candidates for controlled drug delivery because of their controlled dimensional transition. The sharp shape change of smart hydrogels in response to external stimuli raises the possibility of translating environmental signals to mechanical energy, leading to the development of unique actuators, e.g., artificial muscles.

8.3.1 Reversible Immobilization and Separation of (Bio)macromolecules

Because of their large pore size, reversible volume transition, and swift response, the fast-responsive hydrogels are suitable for reversible immobilization of (bio)macromolecules and for protein separation (Figure 8.4). Most thermoresponsive hydrogels have transition temperatures under 50°C, so they are reasonably acceptable for protein immobilization and separation. For example, macroporous PNIPAAm hydrogels have been applied for reversible adsorption of bovine serum albumin (BSA, $M = 69{,}000$ g/mol).[30,41] During temperature-induced swelling, the hydrogels showed a quantitative uptake of BSA molecules from aqueous solutions, and a complete unloading was achieved by temperature induced deswelling.[41]

Three ionic and nonionic temperature-sensitive hydrogels — PNIPAAm, poly(*N*-isopropylacrylamide-*co*-acrylic acid), and poly(*N*-isopropylacrylamide-*co*-hydroxypropylmethacrylate) — were synthesized and used to separate and concentrate lignin from aqueous solutions.[42] After being used for three cycles, the hydrogels still retained most of their water uptake capacities, confirming that the hydrogels exhibit temperature reversibility. Hydrogels prepared from the copolymer of NIPAAm and 2-acrylamide-2-methylpropane sulfonate (P[NIPAAm-*co*-AMPS]) were employed to concentrate aqueous BSA solution with low cost and low energy consumption. The separation efficiency was up to more than 80%, even when aqueous BSA concentration was below 5%.[43,44]

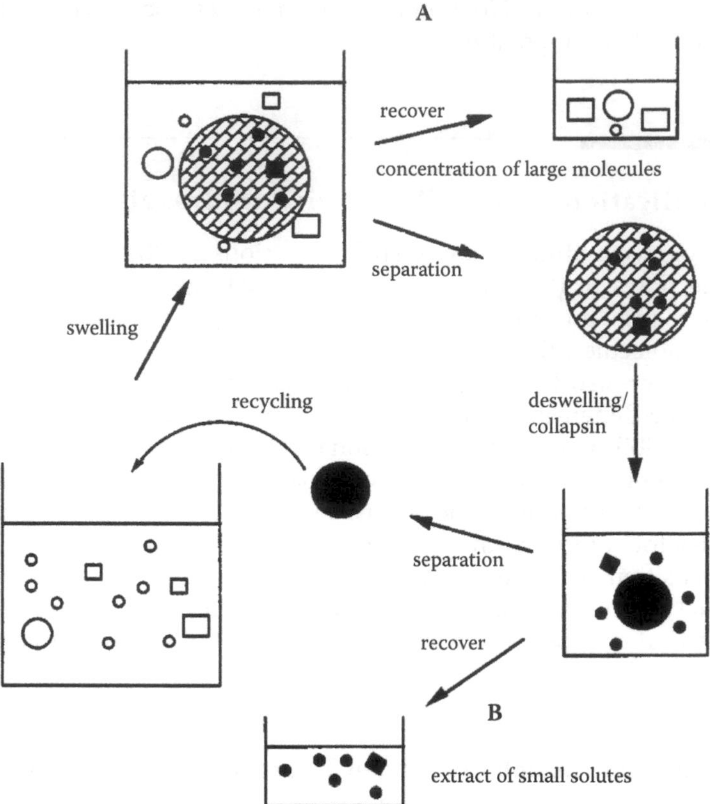

FIGURE 8.4
An example of hydrogel-based bioseparation. In process A, large molecules can be selectively excluded from the stimuli-sensitive hydrogels. In an extraction process B, small solutes entrapped inside the hydrogels can be extracted by deswelling of the hydrogels. (From Kim, J.J. and Park, K., *Bioseparation*, 7, 177, 1998. With permission.)

Bioseparation using hydrogels is based on the size selectivity of the hydrogels to the target molecules. Modulation of the mesh size of the hydrogel is thus crucial. This can generally be achieved by adjusting cross-linking density and monomer concentration in the feed solution. Comb-type PNIPAAm gel beads were synthesized via this approach.[45] Their swelling capacity, swelling kinetics, and separation efficiencies in concentrating albumin, γ-globulin, and vitamin B_{12} solutions proved to be much better than that of normal-type PNIPAAm beads. The incorporation of grafted chains into the gel makes the effective mesh size smaller, thereby inducing an additional obstruction for the solutes within the network and excluding the high-molecular-weight solutes.[45]

For temperature-sensitive hydrogels containing ionizable groups (e.g., carboxyl or amine groups), pH can also be used to control the swelling property. For these types of smart hydrogels, the separation efficiency can be influenced by temperature, pH ionic strength, operation time, and hydrogel composition.

The use of fast-responsive hydrogels in electrophoresis has also been demonstrated. Electrophoresis through agarose or polyacrylamide gels is the standard method applied in the separation, purification, and identification of DNA or RNA fragments. The use of temperature-sensitive hydrogels in electrophoresis made it easier to recover separated DNA fragments. For example, a thermoreversible hydrogel of multiplied block copolymers, composed of poly(ethylene oxide) and poly(propylene oxide), has been applied for DNA electrophoresis.[46] The gel slices containing separated DNA fragments were liquefied by cooling on ice, and the DNA could be precipitated with ethanol without any of the contaminants that inhibit the subsequent enzymatic reactions. The resolving range of this smart electrophoresis lies in between the effective ranges of polyacrylamide and agarose gel electrophoreses.

8.3.2 Molecular Recognition

In addition to the previously cited environmental factors, some specific molecules can also initiate a swelling or shrinking response of smart hydrogels. For example, a smart hydrogel responds to glucose because of its response to gluconic acid produced from glucose by the enzymatic reaction of glucose oxidase. There are also smart hydrogels that can recognize and capture target molecules by multiple-point interaction, and their affinity to the targets can be reversibly changed by more than one order of magnitude.[47] Among the molecule-responsive hydrogels, hybrid hydrogels that contain biomacromolecules and exhibit swelling or shrinking changes in response to macromolecules such as proteins, DNA, antigens, and other biomolecules have attracted much attention.[48–50]

Antigen-sensitive hydrogels have been created for the further development of immunoassays. Antigen-responsive volume changes were observed and employed as the quantitative criterion when an antibody and its complementary antigen were grafted to a hydrogel network. Several approaches can be used to prepare the antigen-responsive hydrogels, including physical entrapment or chemical conjugation of antibodies or antigens to the network.[48] A hydrogel with rabbit IgG-sensitive cross-linking structure has been demonstrated.[49] The hydrogel consists of a copolymer of acrylamide and *N,N′*-methylenebis (acrylamide) with antigen–antibody (rabbit IgG and goat anti-rabbit IgG) binding at cross-linking points in the hydrogels. The addition of free rabbit IgG induces swelling of the hydrogels. Because of the exchange between the polymerized antigen and free antigen, the complex between the goat anti-rabbit IgG and rabbit IgG immobilized in the hydrogel can be dissociated by the added free native rabbit IgG. Consequently, the cross-linking density decreased and the gel swelled (Figure 8.5).

Based on the principle of the antigen-sensitive hydrogels, DNA-sensitive hydrogels displaying DNA-induced swelling or shrinking have been reported. Smart hydrogels that contain rationally designed ssDNA as the cross-linker can shrink or swell in response to ssDNA samples.[50] The ssDNA probe was designed to be complementary to the ssDNA sample. Their

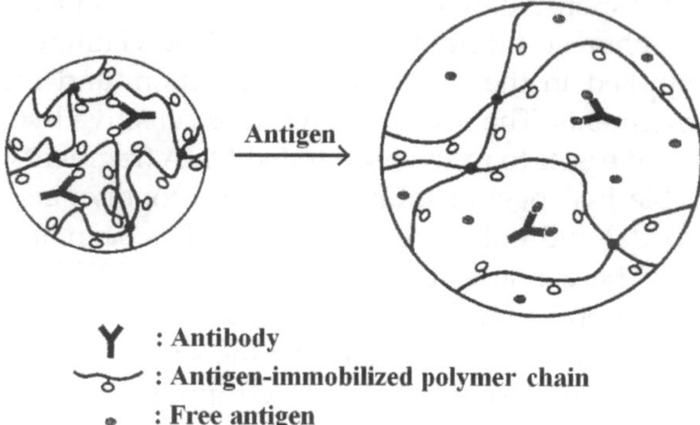

FIGURE 8.5
Proposed mechanism for swelling changes to an antigen–antibody hydrogel in response to a free antigen. (From Miyata, T., Asami, N., and Uragami, T., *Macromolecules*, 32, 2082, 1999. With permission.)

duplex formation led to structural changes in the hydrogels, as shown in Figure 8.6. The hydrogels could recognize even single base differences in the ssDNA sample. The DNA-sensitive hydrogels have potential applications in fabricating DNA-sensing devices or DNA-triggered actuators.

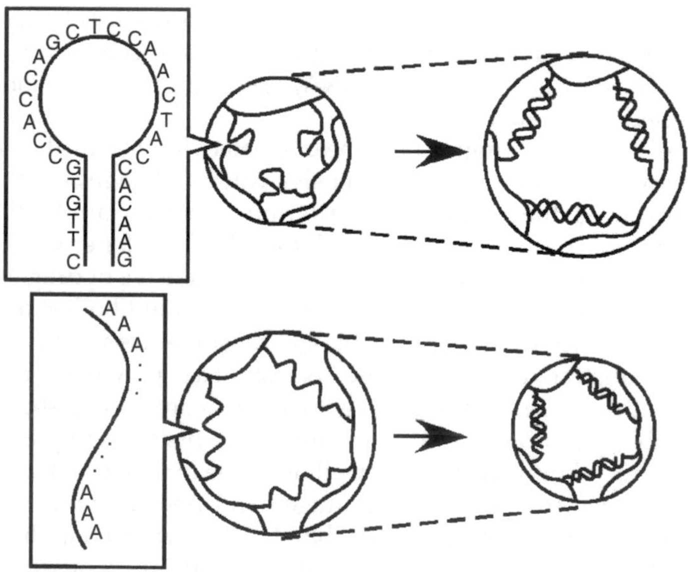

FIGURE 8.6
The response of novel hybrid hydrogels containing ssDNA as a cross-linker to ssDNA. Hydrogel containing the ssDNA with a stem-loop structure like a "molecular beacon" molecule is expected to swell (top), whereas that containing the ssDNA without an intramolecular base pair is expected to shrink (bottom) on the basis of the binding with its complementary ssDNA samples. (From Murakami, Y. and Maeda, M., *Biomacromolecules*, 6, 2927, 2005. With permission.)

Hybrid hydrogels containing peptides or proteins have also been described. In a hybrid hydrogel system assembled from water-soluble synthetic polymers with a well-defined protein-folding motif, the coiled coil undergoes temperature-induced collapse (at ≈40°C) owing to the cooperative conformational transition of the coiled-coil protein domain, as illustrated in Figure 8.7.[51]

One obvious problem for molecule-sensitive hydrogels is that their response rates are generally lower than that of normal smart macroporous hydrogels, so more efforts should be made to enhance the responsive rates of the molecule-sensitive hydrogels.

8.3.3 Smart Hydrogels as Actuators

As wet and soft materials, hydrogels have attracted much attention as actuators for flow control, drug delivery, artificial muscles, etc., since they are driven by an extremely low energy compared with conventional actuator systems. Research interests have focused on electroconductive[52] and ion-conductive hydrogels[53] because their bending motion could easily be driven electronically. When a water-swollen polyelectrolyte gel is interposed between a pair of plate electrodes and a dc current is applied, the gel undergoes electrically induced chemomechanical contraction and concomitant water exudation in the air. This volumetric change is induced by the oxidation of electroactive polymers upon external electric potential (Figure 8.8). Magnetic field[54] and thermal energy[55] were also used as driving forces of the actuators. Thermosensitive actuator systems with transitions in the body-temperature range (≈34 to 42°C) have a special potential for medical devices.

8.3.3.1 Flow Control

Smart hydrogels can change their volume and elasticity reversibly in response to a change in the properties of the surrounding liquid, such as temperature, pH, and ionic strength. Consequently, they can be used to regulate solution flow. Furthermore, they have promising potential to act as smart chemomechanical valves, which have important applications in micro total analysis systems (μ-TAS) or drug-delivery systems without any other external source of energy.[56]

For example, hydrogels of copolymers of PNIPAAm and acrylamide have been used to control the flow in response to both temperature and pH. Another pH-sensitive hydrogel, cross-linked poly(vinyl alcohol)/poly(acrylic acid), has been demonstrated to perform as a valve responsive to alcohol. The cycling times of the hydrogel valve were 25 sec (on) and 35 sec (off).[57] Electronically controllable microvalves based on temperature-sensitive PNIPAAm hydrogels have also been addressed. Upon electronic control achieved by a heating element, the valve built in the channel operates via reversible swelling and deswelling, consequently preventing or allowing the fluid to flow through the channel. Two different designs of valves, a

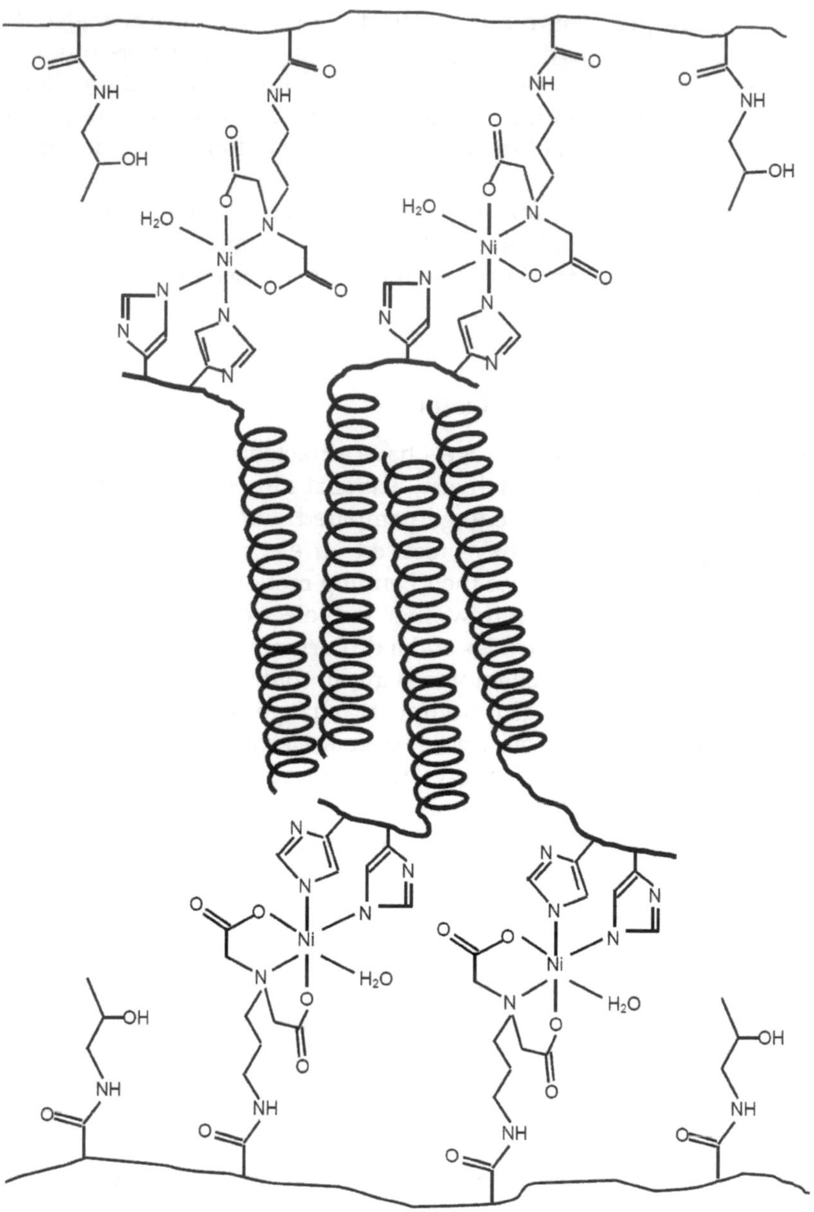

FIGURE 8.7
Structural representation of the hybrid hydrogel primary chains and the attachment of His-tagged coiled-coil proteins. Poly[N-(2-hydroxypropyl) methacrylamide-*co*-N-(N,N′-dicarboxymethylaminopropyl) methacrylamide is shown here as the primary chains. The pendant iminodiacetate groups form complexes with transition-metal ions, such as Ni^{2+}, to which the terminal histidine residues of the coiled coils are attached. A tetrameric coiled coil (not drawn to scale), consisting of two parallel dimers associating in an antiparallel fashion, is shown here as an example of many of the possible conformations. (From Wang, C., Stewart, R.J., and Kopeček, J., *Nature*, 397, 417, 1999. With permission.)

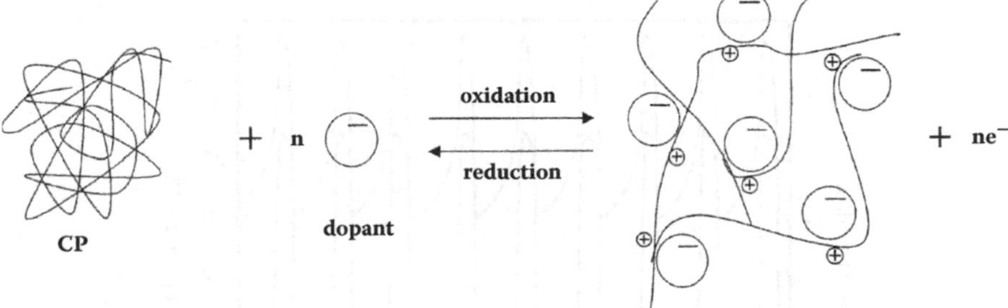

FIGURE 8.8
Conformation changes and subsequent volume changes occurring during redox reaction in conducting polymer. (From Onoda, M. et al., *Electrical Eng. Jpn.*, 149, 7, 2004. With permission.)

microvalve with an actuator chamber containing PNIPAAm hydrogel particles and a photopatterned microvalve, were introduced. The switching time was adjustable from 0.3 sec to 10 sec (Figure 8.9).[58] In addition, glucose-sensitive hydrogel membranes have been demonstrated to change the delivery rate of macromolecules in response to glucose. Typically, 0.1*M* D-glucose caused a substantial increase in diffusion rate of both insulin and lysozyme, while for 0.1*M* glycerol or L-glucose, no such effect was observed. Sequential addition and removal of external glucose in a stepwise manner showed that the rate changes were reversible and tunable.[59]

8.3.3.2 Controlled Drug Release

Based on the developments mentioned above, smart hydrogels may have applications in intra- or extracorporeal drug delivery systems. The swelling or shrinking of smart hydrogel networks containing a drug in response to small changes in environmental factors can successfully control drug release because the diffusion of the drug out of the beads depends on the gel state.

The release behavior of 5-fluorouracil from the PNIPAAm/PNIPAAm interpenetrating polymer networks (IPNs) was controlled by the diffusion of water molecules in the gel network at low temperature, while 5-fluorouracil inside the gel would not diffuse at a higher temperature (Figure 8.10).[60] As another interesting example, hydrogels based on *n*-alkyl methacrylate esters, acrylic acid, and acrylamide cross-linked with 4,4-di(methacryloylamino)azobenzene exhibited pH-controlled drug release swelling in a basic environment. A drug carried in such a hydrogel could be protected from being released in the stomach because of the low pH there (1.0 to 3.5), but be completely released in the small intestine and the colon, where the pH is between 5.0 and 7.0.[61] Biodegradable hydrogels show promise for use as implantable carriers for drug-delivery systems because of their ability to respond to environmental stimuli and their good biocompatibility. For example, biodegradable hydrogels consisting of gelatin and dextran with an IPN structure were prepared and then shown to degrade only in the presence of

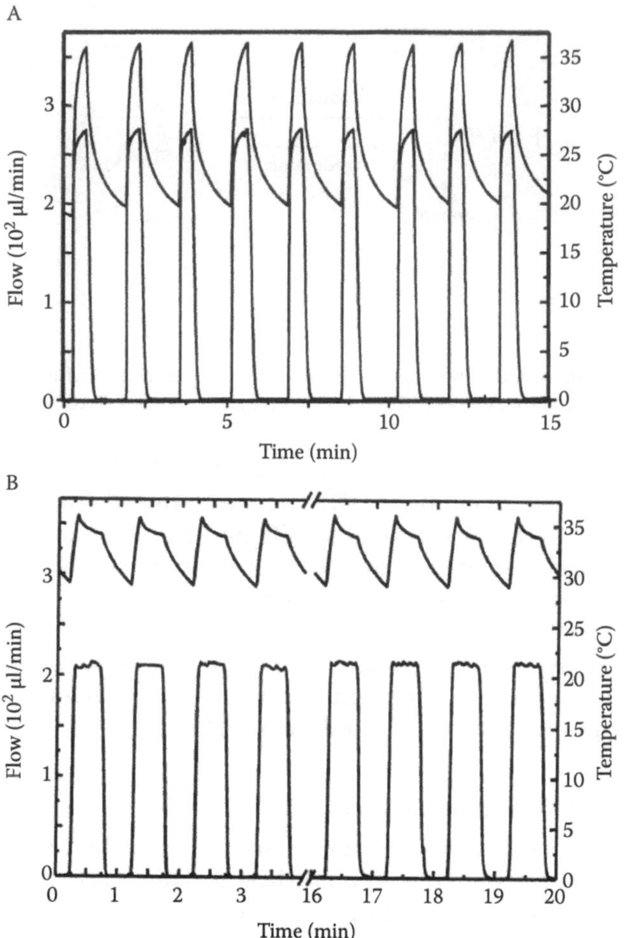

FIGURE 8.9
(a) Behavior of a microvalve with photopatternable hydrogel actuator (size 250 × 250 × 50 µm³) at nonpower-modulated work (heating power 300 mW). Upper curve: temperature vs. time; lower curve: flow rate vs. time. The fastest switching times for photopatterned actuator dots are presently 4 sec for opening and 10 sec for closing. (b) Behavior of a microvalve with PNIPAAm particle actuator (size 500 × 500 × 50 µm³) at power-modulated work (power peak at 380 mW for 500 msec, retaining power 150 mW). Upper curve: temperature vs. time; lower curve: flow rate vs. time. The fastest switching times for actuators based on PNIPAAm particles are presently 300 msec for opening and about 2 sec for closing. The operating pressure was about 1.5 bar. (Modified from Richter, A. et al., *J. Microelectromechanical Syst.*, 12, 748, 2003. With permission.)

both α-chymotrypsin and dextranase. This dual enzymatic degradation of hydrogels can be used to achieve dual-stimuli-responsive drug release. Hydrogels were prepared with lipid microspheres as drug microreservoirs. In the presence of α-chymotrypsin and dextranase, the hydrogel structure could be completely destroyed, thereby releasing the drug.[62]

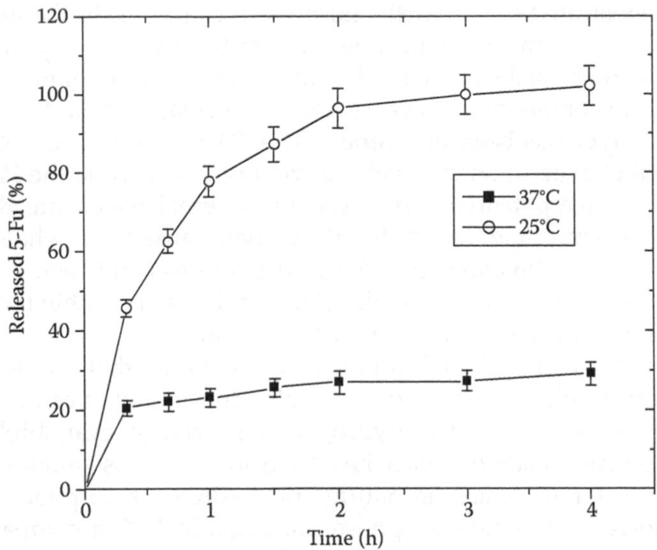

FIGURE 8.10
Cumulative release of 5-Fu from PNIPAAm hydrogels at different temperatures as a function of time. (Modified from Zhang, J.T. et al., *J. Polym. Sci. A: Polym. Chem.*, 42, 1249, 2004. With permission.)

8.3.3.3 *Artificial Muscles*

Artificial muscles are materials that can directly convert chemical energy into mechanical work.[63] Progress toward artificial muscles[64,65] will lead to great benefits, particularly in the medical area, such as prosthetics that use electroactive polymer hydrogels to replace damaged human muscles.[66]

Smart hydrogels of electroactive polymers have the potential to perform as artificial muscles that could match the force and energy density of biological muscles because such hydrogels show a remarkable volumetric change in response to external stimuli, and can consequently convert chemical energy into mechanical work. The idea of preparing artificial muscles with smart electroactive polymer hydrogels is attractive because an electric field is easily controlled and manipulated. The first generation of artificial muscles was developed by Katchalsky and coworkers in the 1950s using polymeric gels. Since then, many interesting works have been presented. A cross-linked gel of poly(vinyl alcohol) chains entangled with polyacrylic-acid chains shows rapid electric-field-associated bending deformation: a gel rod of 1-mm diameter bends semicircularly within 1 sec upon the application of an electric field. An artificial fish with a gel tail swam forward at a velocity of 2 cm·sec^{-1} as the gel oscillated under sinusoidally varied electric fields. A mechanical hand composed of four gel fingers could pick up a fragile quail egg (9 g) from a sodium-carbonate solution and hold it without breaking it, controlled by an electrical signal.[67] This device performs in solution. Recently, it was demonstrated that an artificial muscle can work in the air. The gel expanded and contracted along one axis, rather than simply bending, in response to an applied electric field without any overall volume change. A bilayer system containing a polyacrylamide (PAAM) layer and a polyacrylic acid (PAA) layer has been developed. The PAAM layers do not undergo dimensional change upon external electric potential, while the PAA layers shrank at positive potential and forced the internal water into the PAAM layer. At the same time, the stiffer PAA layer caused the whole stack to contract along the directions parallel to the interface between layers. This system provides linear motion while retaining the water within the actuator; hence the system works in air as an artificial muscle.[68]

Polymeric networks with a liquid crystalline nature that can be triggered by the thermotropic phase transition are also considered as candidates for a useful muscle system.[69,70] The hydrogel of poly(11-(4′-cyanobiphenyloxy) undecyl acrylate-*co*-acrylic acid) has hydrophobic cross-linking junction zones with a liquid crystalline nature and hydrated amorphous domains. The gel showed a smectic mesophase transition at 40°C accompanied by a drastic shrinkage of the oriented hydrogel fibers. Thus this hydrogel can be considered to be a new thermally controllable actuator in the aqueous milieu at ambient conditions.[70]

8.4 Conclusion

Fast-responsive hydrogels undergo tremendous and rapid dimensional changes in response to environmental stimuli such as temperature, pH, solvent, electric field, special reagents, etc. This property makes them promising materials for bioseparation, molecular recognition, actuators, etc. Because of their biocompatibility and special three-dimensional structure with large pores, macroporous smart hydrogels can be employed as tissue-engineering materials. Given these attractive prospects, efforts are being made to explore novel and better methods to prepare fast-responsive hydrogels and expand their applications.

References

1. Okay, O., Macroporous copolymer networks, *Prog. Polym. Sci.*, 25, 711, 2000.
2. Gemeinhart, R.A. et al., pH-sensitivity of fast responsive superporous hydrogels, *J. Biomater. Sci.: Polym. Ed.*, 11, 1371, 2000.
3. Zhang, J.T., Huang, S.W., and Zhuo, R.X., Temperature-sensitive polyamidoamine dendrimer/poly(*N*-isopropylacrylamide) hydrogels with improved responsive properties, *Macromol. Biosci.*, 4, 575, 2004.

4. Zhang, J.T., Cheng, S.X., and Zhuo, R.X., Poly(vinyl alcohol)/poly(*N*-isopropylacrylamide) semi-interpenetrating polymer network hydrogels with rapid response to temperature changes, *Colloid Polym. Sci.*, 281, 580, 2003.
5. Khorram, M., Vasheghani-Farahani, E., and Ebrahimi, N.G., Fast responsive thermosensitive hydrogels as drug delivery systems, *Iranian Polym. J.*, 12, 316, 2003.
6. Ford, M.C. et al., A macroporous hydrogel for the coculture of neural progenitor and endothelial cells to form functional vascular networks *in vivo*, *Proc. Natl. Acad. Sci. U.S.A.*, 103, 2512, 2006.
7. Bu, H., Rong, J.H., and Yang, Z.Z., Template synthesis of polyacrylonitrile-based ordered macroporous materials and their derivatives, *Macromol. Rapid Commun.*, 23, 460, 2002.
8. Takeoka, Y. and Watanabe, M., Template synthesis and optical properties of chameleonic poly(*N*-isopropylacrylamide) gels using closest-packed self-assembled colloidal silica crystals, *Adv. Mater.*, 15, 199, 2003.
9. Takeoka, Y. and Watanabe, M., Polymer gels that memorize structures of mesoscopically sized templates: dynamic and optical nature of periodic ordered mesoporous chemical gels, *Langmuir*, 18, 5977, 2002.
10. Rong, J.H. et al., Template-synthesis and characterization of responsive ordered macroporous hydrogels, *Chemical J. Chin. Univs.: Chin.*, 25, 1771, 2004.
11. Zhang, X.Z. and Zhuo, R.X., Preparation of fast responsive, temperature-sensitive poly(*N*-isopropylacrylamide) hydrogel, *Macromol. Chem. Phys.*, 200, 2602, 1999.
12. Liu, L.S. et al., Pectin and polyacrylamide composite hydrogels: effect of pectin on structural and dynamic mechanical properties, *J. Appl. Polym. Sci.*, 92, 1893, 2003.
13. Zhang, J.T. et al., Temperature-sensitive poly(*N*-isopropylacrylamide) hydrogels with macroporous structure and fast response rate, *Macromol. Rapid Commun.*, 24, 447, 2003.
14. Pradny, M. et al., Macroporous hydrogels based on 2-hydroxyethyl methacrylate, part 1: copolymers of 2-hydroxyethyl methacrylate with methacrylic acid, *Coll. Czech. Chem. Commun.*, 68, 812, 2003.
15. Zeng, Y., Wang, R.C., and Zhou, B.L., Mechanical sensitivity of macroporous poly(vinyl formal) hydrogels to water content, *Acta Polymerica Sinica*, 3, 363, 2005.
16. Serizawa, T., Wakita, K., and Akashi, M., Rapid deswelling of porous poly(*N*-isopropylacrylamide) hydrogels prepared by incorporation of silica particles, *Macromolecules*, 35, 10, 2002.
17. Zhang, X.Z. and Zhuo, R.X., Preparation of fast responsive, thermally sensitive poly(*N*-isopropylacrylamide) gel, *Eur. Polym. J.*, 36, 2301, 2000.
18. Zhang, X.Z. et al., Preparation and characterization of fast response macroporous poly(*N*-isopropylacrylamide) hydrogels and (*N*-isopropylacrylamide) hydrogels, *Langmuir*, 17, 6094, 2001.
19. Zhuo, R.X. and Li, W., Preparation and characterization of macroporous poly (*N*-isopropylacrylamide) hydrogels for the controlled release of proteins, *J. Polym. Sci. A: Polym. Chem.*, 41, 152, 2003.
20. Zhang, J.T. et al., Preparation and characterization of fast-responsive porous poly(*N*-isopropylacrylamide-*co*-acrylic acid) hydrogels, *Chem. J. Chin. Univs.: Chin.*, 25, 2370, 2004.
21. Dogu, Y. and Okay, O., Swelling–deswelling kinetics of poly(*N*-isopropylacrylamide) hydrogels formed in PEG solutions, *J. Appl. Polym. Sci.*, 99, 37, 2006.

22. Zhang, X.Z., Chu, C.C., and Zhuo, R.X., Using hydrophobic additive as pore-forming agent to prepare macroporous PNIPAAm hydrogels, *J. Polym. Sci. A: Polym. Chem.*, 43, 5490, 2005.
23. Zhang, X.Z. and Chu, C.C., Preparation of thermosensitive PNIPAAm hydrogels with superfast response, *Chem. Commun.*, 3, 350, 2004.
24. Zhang, X.Z. and Chu, C.C., Fabrication and characterization of microgel-impregnated, thermosensitive PNIPAAm hydrogels, *Polymer*, 46, 9664, 2005.
25. Cai, W. and Gupta, R.B., Fast-responding bulk hydrogels with microstructure, *J. Appl. Polym. Sci.*, 83, 169, 2002.
26. Zhang, X.Z., Zhuo, R.X., and Yang, Y.Y., Using mixed solvent to synthesize temperature sensitive poly(*N*-isopropylacrylamide) gel with rapid dynamics properties, *Biomaterials*, 23, 1313, 2002.
27. Zhang, J.T., Huang, S.W., and Zhuo, R.X., Preparation and characterization of novel temperature sensitive poly(*N*-isopropylacrylamide-*co*-acryloyl beta-cyclodextrin) hydrogels with fast shrinking kinetics, *Macromol. Chem. Phys.*, 205, 107, 2004.
28. Zhang, X.Z. and Zhuo, R.X., Synthesis and properties of thermosensitive poly *N*-isopropylacrylamide-*co*-methyl methacrylate hydrogel with rapid response, *Mater. Lett.*, 52, 5, 2002.
29. Liu, Q. et al., Preparation of macroporous poly(2-hydroxyethyl methacrylate) hydrogels by enhanced phase separation, *Biomaterials*, 21, 2136, 2000.
30. Cheng, S.X. et al., Macroporous poly(*N*-isopropylacrylamide) hydrogels with fast response rates and improved protein release properties, *J. Biomedical Mater. Res.*, 67A, 96, 2003.
31. Zhang, J.T., Cheng, S.X. and Zhuo, R.X., Preparation of macroporous poly(*N*-isopropylacrylamide) hydrogel with improved temperature sensitivity, *J. Polym. Sci. A: Polym. Chem.*, 41, 2390, 2003.
32. Das, R.R. et al., A new macroporous luminescent hydrogel from poly(allylamine), *Macromol. Rapid Commun.*, 22, 850, 2001.
33. Zhang, X.Z. and Zhuo, R.X., Dynamic properties of temperature-sensitive poly(*N*-isopropylacrylamide) gel cross-linked through siloxane linkage, *Langmuir*, 17, 12, 2001.
34. Ni, C.H., Wang, Z., and Zhu, X.X., Preparation and characterization of thermosensitive beads with macroporous structures, *J. Appl. Polym. Sci.*, 91, 1792, 2004.
35. Wang, Z., Song, Z.J., and Zhu, X.X., Preparation and characterization of crosslinked poly(*N,N*-diethylacrylamide-*co*-2-hydroxyethyl methacrylate) resins and their application as support in solid-phase peptide synthesis, *J. Polym. Sci. A: Polym. Chem.*, 41, 1681, 2003.
36. Safrany, A., Macroporous gels with fast response prepared by e-beam crosslinking of poly(*N*-isopropylacrylamide) solution, *Nucl. Instruments Methods Phys. Res. B*, 236, 587, 2005.
37. Panda, A. et al., Synthesis and swelling characteristics of poly(*N*-isopropylacrylamide) temperature sensitive hydrogels crosslinked by electron beam irradiation, *Radiat. Phys. Chem.*, 58, 101, 2000.
38. Sayil, C. and Okay, O., Macroporous poly(*N*-isopropylacrylamide) networks, *Polym. Bull.*, 48, 499, 2002.
39. Pastoriza, A., Pacios, I.E., and Pierola, I.F., Kinetics of solvent responsiveness in poly(*N,N*-dimethylacrylamide) hydrogels of different morphology, *Polym. Int.*, 54, 1205, 2005.

40. Ozmen, M.M. and Okay, O., Superfast responsive ionic hydrogels with controllable pore size, *Polymer*, 46, 8119, 2005.
41. Fanger, C., Wack, H., and Ulbricht, M., Macroporous poly(*N*-isopropylacrylamide) hydrogels with adjustable size "cut-off" for the efficient and reversible immobilization of biomacromolecules, *Macromol. Biosci.*, 6, 393, 2006.
42. Cai, W.S., Anderson, E.C., and Gupta, R.B., Separation of lignin from aqueous mixtures by ionic and nonionic temperature-sensitive hydrogels, *Industrial Eng. Chem. Res.*, 40, 2283, 2001.
43. Zhuang, Y.F. et al., Radiation polymerization and concentration separation of P(NIPA-*co*-AMPS) hydrogels, *Polym. Int.*, 54, 617, 2005.
44. Yang, H. et al., Synthesis of strong electrolyte temperature-sensitive hydrogels by radiation polymerization and application in protein separation, *Macromol. Biosci.*, 3, 400, 2003.
45. Annaka, M. et al., Preparation of comb-type *N*-isopropylacrylamide hydrogel beads and their application for size-selective separation media, *Biomacromolecules*, 4, 395, 2003.
46. Yoshioka, H., Mori, Y., and Shimizu, M., Separation and recovery of DNA fragments by electrophoresis through a thermoreversible hydrogel composed of poly(ethylene oxide) and poly(propylene oxide), *Anal. Biochem.*, 323, 218, 2003.
47. Oya, T. et al., Reversible molecular adsorption based on multiple-point interaction by shrinkable gels, *Science*, 286, 1543, 1999.
48. Lu, Z.R., Kopeckova, P., and Kopeček, J., Antigen responsive hydrogels based on polymerizable antibody Fab′ fragment, *Macromol. Biosci.*, 3, 296, 2003.
49. Miyata, T., Asami, N., and Uragami, T., Preparation of an antigen-sensitive hydrogel using antigen-antibody bindings, *Macromolecules*, 32, 2082, 1999.
50. Murakami, Y. and Maeda, M., DNA-responsive hydrogels that can shrink or swell, *Biomacromolecules*, 6, 2927, 2005.
51. Wang, C., Stewart, R.J., and Kopecek, J., Hybrid hydrogels assembled from synthetic polymers and coiled-coil protein domains, *Nature*, 397, 417, 1999.
52. Lu, W. et al., Use of ionic liquids for pi-conjugated polymer electrochemical devices, *Science*, 297, 983, 2002.
53. Shahinpoor, M., Ionic polymer-conductor composites as biomimetic sensors, robotic actuators and artificial muscles: a review, *Electrochimica Acta*, 48, 2343, 2003.
54. Juliac, E. et al., Ultrasonic investigations of hydrogels containing barium ferrite particles, *J. Phys. Chem. B*, 107, 5426, 2003.
55. Tajbakhsh, A.R. and Terentjev, E.M., Spontaneous thermal expansion of nematic elastomers, *Eur. Phys. J. E*, 6, 181, 2001.
56. Kuckling, D., Richter, A., and Arndt, K.F., Temperature and pH-dependent swelling behavior of poly(*N*-isopropylacrylamide) copolymer hydrogels and their use in flow control, *Macromol. Mater. Eng.*, 288, 144, 2003.
57. Arndt, K.F., Kuckling, D., and Richter, A., Application of sensitive hydrogels in flow control, *Polym. Adv. Technol.*, 11, 496, 2000.
58. Richter, A. et al., Electronically controllable microvalves based on smart hydrogels: magnitudes and potential applications, *J. Microelectromechanical Syst.*, 12, 748, 2003.
59. Tang, M. et al., A reversible hydrogel membrane for controlling the delivery of macromolecules, *Biotechnol. Bioeng.*, 82, 47, 2003.
60. Zhang, J.T. et al., Preparation and properties of poly(*N*-isopropylacrylamide)/poly(*N*-isopropylacrylamide) interpenetrating polymer networks for drug delivery, *J. Polym. Sci. A: Polym. Chem.*, 42, 1249, 2004.

61. Yin, Y.H., Yang, Y.J., and Xu, H.B., Swelling behavior of hydrogels for colon-site drug delivery, *J. Appl. Polym. Sci.*, 83, 2835, 2002.
62. Kurisawa, M. and Yui, Nobuhiko, Gelatin/dextran intelligent hydrogels for drug delivery: dual-stimuli-responsive degradation in relation to miscibility in interpenetrating polymer networks, *Macromol. Chem. Phys.*, 199, 1547, 1998.
63. Schreyer, H.B. et al., Electrical activation of artificial muscles containing polyacrylonitrile gel fibers, *Biomacromolecules*, 1, 642, 2000.
64. Onoda, M. et al., Artificial muscle using conducting polymers, *Electrical Eng. Jpn.*, 149, 7, 2004.
65. Careem, M.A. et al., Dependence of force produced by polypyrrole-based artificial muscles on ionic species involved, *Solid State Ionics*, 175, 725, 2004.
66. Yoseph, B.H., Proceedings of the 42nd AIAA Structures, Structural Dynamics, and Materials Conference (SDM), Gossamer Spacecraft Forum (GSF), Seattle, WA, April 16–19, 2001.
67. Shiga, T., *Neutron Spin Echo Spectroscopy: Viscoelasticity, Rheology,* Abe, A., Ed., Springer-Verlag, Heidelberg, 1997, pp. 133–163.
68. Liu, Z.S. and Calvert, P., Multilayer hydrogels as muscle-like actuators, *Adv. Mater.*, 12, 288, 2000.
69. Camacho-Lopez, M. et al., Fast liquid-crystal elastomer swims into the dark, *Nat. Mater.*, 3, 307, 2004.
70. Kaneko, T. et al., Thermoresponsive shrinkage triggered by mesophase transition in liquid crystalline physical hydrogels, *Macromolecules*, 37, 5385, 2004.

Chapter 9

Macroporous Hydrogels from Smart Polymers

Oguz Okay

CONTENTS

9.1 Introduction

Hydrophilic gels, also called "hydrogels," are hydrophilic polymer networks swollen in water. Water inside the hydrogel allows free diffusion of some solute molecules, while the polymer serves as a matrix to hold water together. Hydrogels are mainly prepared by free-radical cross-linking copolymerization of acrylamide-based monomers with a divinyl monomer (cross-linker) in aqueous solutions. To increase their swelling capacity, an ionic co-monomer is also included in the monomer mixture. The desired properties of hydrogels—increased swelling capacity, modulus of elasticity, and degree of heterogeneity—are obtained by adjusting both the concentration and the composition of the initial monomer mixture.

Hydrogels are able to absorb 10 to 1000 times their dry volume of water without dissolving. They also exhibit drastic volume changes in response to specific external stimuli, such as the temperature, solvent quality, pH, electric field, etc.[1] Depending on the design of the hydrogel matrices, this volume change can occur continuously over a range of stimulus level or discontinuously at a critical stimulus level. These properties of hydrogels have received considerable interest over the last three decades. Today, these soft and smart materials belong to the most important class of functional polymers in modern biotechnology. They are useful materials for drug-delivery systems, artificial organs, separation operations in biotechnology, processing of agricultural products, on–off switches, sensors, and actuators. Despite this fact and considerable research in this field, the design and control of hydrogel-based devices still present some problems, since a number of network properties are inversely coupled. For example, decreasing the degree of cross-linking of hydrogels to increase their mesh size ("molecular porosity") results in their accelerated chemical degradation. Further, loosely cross-linked hydrogels are soft materials when handled in the swollen state; typically, they exhibit moduli of elasticity on the order of kilopascals. The poor mechanical performance of highly swollen hydrogels limits their technological applications.

A fast response of hydrogels to the external stimuli is also a requirement in many application areas of these materials. However, the kinetics of hydrogel volume change involves absorbing or desorbing solvent by the polymer network, which is a diffusive process. This process is slow and even slower near the critical point. Increasing the response rates of hydrogels has been one of the challenging problems in the last 25 years. Several techniques have been proposed to increase their response rates. The simplest technique is to reduce the size of the hydrogel particles.[2] Since the rate of response is inversely proportional to the square of the size of the gel,[3] small hydrogel particles respond to the external stimuli more quickly than bulk gels. However, this approach has the drawback that, with a reduction of the gel size to submicrometer range, separation of gel particles from the surrounding liquid medium requires additional efforts. Attachment of linear polymer chains on the gel particles is another approach to increase the response rates of hydrogels.[4–6] The increased response rate of these porcupine polymers is due to the fact that the dangling chains in a gel easily collapse or expand in response to an external stimulus because one side of the dangling chain is free.

Another technique to obtain fast-responsive hydrogels is to create voids (pores) inside the hydrogel matrix, so that the response rate becomes a function of the microstructure rather than the size or the shape of the gel samples. For a polymer network having an interconnected pore structure, absorption or desorption of water occurs through the pores by convection, which is much faster than the diffusion process that dominates the nonporous hydrogels. Macroporous hydrogels contain nanometer- to micron-size liquid channels separated by cross-linked polymer regions, which provide sufficient mechanical stability. It should be noted that the term "macroporous" does not mention the size of the pores in these materials; it is used to distinguish these gels from those exhibiting a swollen-state porosity, i.e., a molecular porosity due to the space between the network chains in their swollen states. Thus the term "macroporous hydrogel" refers to materials exhibiting a porosity in their dry state, which is characterized by a lower density of the network due to the voids than that of the matrix polymer. In contrast to inorganic porous gels such as silica gel carriers, macroporous gels offer flexibility, ductility, and the ability to react with a large number of organic molecules.

The technique to obtain cross-linked polymers with a macroporous structure was discovered at the end of 1950s.[7–15] Since then, the procedures to make such gels have greatly improved. The basic technique to produce macroporous hydrogels involves the free-radical cross-linking copolymerization of the monomer-cross-linker mixture in the presence of an inert substance (the diluents), which is soluble in the monomer mixture. To obtain macroporous structures, a phase separation must occur during the course of the network formation process so that the two-phase structure formed is fixed by the formation of additional cross-links. After the polymerization, the diluent is removed from the network, leaving a porous structure within the highly cross-linked polymer network. Thus, the inert diluent acts as a pore-forming agent and plays an important role in the design of the pore structure of cross-linked materials. Today, the formation of macroporous polymers by the phase-separation technique is qualitatively understood, and several reviews have been published covering the developments in this area.[16–19] Another technique to create a macroporous network structure is the use of inert templates in the preparation of hydrogels.[20,21] In this technique, the polymer-formation reactions are carried out in the presence of templates; a macroporous structure in the final hydrogel matrix appears after extraction of the template materials. For example, by the cryogelation technique, the polymer-formation reactions are carried out below the bulk freezing temperature of the reaction system.[22–26] Thus, the essential feature of such reaction systems is that the monomers and the initiator are concentrated in the unfrozen microzones of the apparently frozen system. The polymerization and cross-linking reactions proceed in the unfrozen microzones of the reaction system. A macroporous structure in the final material appears due to the existence of ice crystals acting as a template for the formation of the pores. Macroporous hydrogels can also be prepared by foaming techniques, making use of gaseous pore-forming agents. These can be generated by evaporation of solvents or produced by chemical reactions.[27]

This chapter discusses the general principles for the formation of macroporous hydrogels by phase-separation and cryogelation techniques exhibiting fast-responsive properties against the external stimuli. The formation conditions of macroporous hydrogels starting from vinyl–divinyl monomers as well as their properties are discussed using selected examples from the literature.

9.2 Synthetic Strategies to Macroporous Hydrogels

9.2.1 Geometry of Hydrogels

Macroporous hydrogels made in the form of beads have attracted attention as carrier matrices in a wide variety of applications. Beads of hydrogels with diameters of 0.1 to 1.5 mm can be prepared by the classical suspension-polymerization technique, in which water-insoluble derivatives of the monomers are used for the polymerization and the beads formed are subsequently hydrolyzed or aminolyzed.[28–32] The direct synthesis of such hydrophilic particles is also possible starting from water-soluble co-monomer systems such as acrylamide (AAm)–*N*,*N*-methylene(bis)acrylamide (BAAm) or 2-hydroxyethylmethacrylate (HEMA)–ethylene glycol dimethacrylate (EGDM) systems. For this purpose, various salts were added into the water phase to diminish the water solubility of the monomers (salting-out effect). For example, sodium chloride, calcium chloride, or alcohols were used to reduce the monomer solubility and thus to allow formation of spherical, hydrophilic beads.[33–41] Another approach to prepare hydrogel beads is the inverse suspension-polymerization technique by which the water-soluble monomers or their aqueous solutions are suspended in an organic phase and polymerized therein to give copolymer beads having a controlled size.[42–47] For example, the cross-linking polymerization of an aqueous solution of *N*-isopropylacrylamide (NIPA) and BAAm in paraffin oil as the continuous phase results in cross-linked poly(NIPA) (PNIPA) beads of sizes 0.25 to 2.8 mm in diameter.[44–45] Macroporous hydrogels can also be prepared as monolithic objects directly in the final container, e.g., in glass capillaries or in thin plates.[48] Almost any monomer, including water-soluble monomers that are not suitable for classical suspension polymerization, can be used to form a monolith.

9.2.2 Formation Mechanism of Macroporous Structures by Reaction-Induced Phase Separation

The free-radical cross-linking copolymerization system for the production of macroporous hydrogels usually includes a monovinyl monomer, a divinyl monomer (cross-linker), an initiator, and the inert diluent. The inert diluent must be soluble in the monomer mixture but insoluble in the continuous phase of the suspension polymerization. The decomposition of the initiator produces free radicals that initiate the polymerization and cross-linking reactions. After a certain reaction time, called the gelation time, a three-dimensional network can start to form, so that the reaction system changes from liquid to solidlike state. Continuing polymerization and cross-linking reactions beyond the gel point decrease the amount of soluble reaction components by increasing both the amount and the cross-linking density of the network. After complete conversion of monomers to polymer, only the network and the diluent remain in the reaction system. The polymer network thus obtained exhibits different structures, depending on the gel synthesis parameters, as detailed in the following sections.

9.2.2.1 Expanded Gels

In the presence of a good solvent as the inert diluent, if a small amount of a cross-linker is used in the network synthesis, the hydrogels formed exhibit an expanded structure. Thus, during the formation of expanded gels, all the diluent molecules remain in the gel phase, i.e., no macroscopic phase separation takes place throughout the reactions. The expanded hydrogel collapses during the removal of the diluent after its synthesis and, therefore, it is nonporous in the glassy state. Expanded gels swell in good solvents much more than those synthesized in bulk, and they therefore exhibit a higher degree of molecular porosity than the corresponding bulk gels. It must be pointed out that a micro phase separation occurs during the formation of both bulk and expanded hydrogels due to the inhomogeneous distribution of cross-links throughout the gel sample, known as the spatial gel inhomogeneity. This is due to the fact that the cross-linker has at least two vinyl groups and, therefore, if one assumes equal vinyl group reactivity, the reactivity of the cross-linker is twice that of the monovinyl monomer. As a consequence, the cross-linker molecules are incorporated into the growing copolymer chains much more rapidly than the monomer molecules, so that the final network exhibits a cross-link density distribution. The network regions formed earlier are more highly cross-linked than those formed later. Moreover, cyclization and multiple cross-linking reactions also contribute to the spatial inhomogeneity in the hydrogels.[19] Thus, the unequal distribution of cross-links creates frozen concentration fluctuations in gels.

9.2.2.2 Heterogeneous Gels Formed by ν-Induced Syneresis

If the amount of the cross-linker in the reaction mixture is increased while the amount of the good solvent (diluent) remains constant, the highly cross-linked network formed cannot absorb all the diluent molecules present in the reaction mixture. As a result, a phase separation occurs during the gel formation process. The process of phase separation can proceed as follows: the growing gel deswells with the onset of phase separation and becomes a microgel (nucleus), whereas the separated liquid remains as a continuous phase in the reaction mixture. As the polymerization and cross-linking proceed, new nuclei are continuously generated due to the successive separation of the growing polymers, which react with each other through their pendant vinyl groups and radical centers located at their surfaces. These agglomeration processes result in the formation of a heterogeneous gel consisting of two continuous phases: a gel and a diluent phase. Removing of the diluent from the gel after synthesis creates voids (pores) of various sizes. This material is a macroporous copolymer network. It is seen that, in the presence of a good solvent as a diluent, the porous structures form due to the effect of monomer dilution, which is higher than the swelling capacity of the network This type of porosity formation in polymeric materials is called by Důsek as ν-induced syneresis, where ν refers to the cross-link density of the network.[16,49] It should be noted that, if the amount of the diluent is further increased, a critical point is passed at which the system becomes discontinuous, because the amount of the monomer is not sufficient and the growing chains cannot occupy the whole available volume. Consequently, a dispersion of macrogel particles in the solvent is produced. Increasing the amount of solvent decreases the size of the gel particles, and finally they are as small as ordinary macromolecules. These gel particles are microgels that are dissolved as a colloidal solution.[50]

9.2.2.3 Heterogeneous Gels Formed by χ-Induced Syneresis

If a nonsolvent or a linear polymer is used as the diluent of the monomer mixture, a phase separation can occur in the reaction system before the gel point. This results in the formation of a dispersion of separated (discontinuous) polymer phase in the continuous monomer + diluent phase. Continuing the polymerization increases the amount of the polymer. As a result, the first-phase separated and intramolecularly cross-linked particles (nuclei) agglomerate into larger clusters called microspheres. Continuing the reactions increases the number of clusters in the reaction system so that the polymer phase becomes continuous. Thus, a system consisting of two continuous phases results. Again, removing the diluent from the gel produces a macroporous copolymer. It is seen that, in the presence of nonsolvents or linear polymers as a diluent, the incompatibility between the network segments and the diluent molecules is responsible for the porosity formation. This mechanism is called χ-induced syneresis,[16] where χ is the polymer–solvent interaction parameter. Comparison of the mechanisms of ν- and χ-induced syneresis suggests that the pore structure formed by the first mechanism should result in more-ordered and smaller agglomerates than those formed in the latter mechanism. This is in accord with the experimental observations.[19]

According to the mechanistic picture of the gel formation presented above, various network structures can be obtained by free-radical cross-linking copolymerization of vinyl–divinyl monomers in the presence of inert diluents. Therefore, the preparation of macroporous hydrogels requires a careful choice of the experimental parameters.

9.2.3 Macroporous Hydrogels Prepared by Reaction-Induced Phase Separation

As seen in the previous section, a phase separation during the hydrogel preparation is mainly responsible for the formation of macroporous structures in the final hydrogels. The extent of phase separation during cross-linking, and thus the degree of heterogeneity of the resulting networks, mainly depends on the following gel synthesis parameters:

1. Cross-linker concentration
2. Amount and type of the inert diluent (pore-forming agent) present during the network formation process
3. Temperature

From these independent parameters, the amount of the cross-linker used in the hydrogel preparation plays a decisive role in obtaining a stable macroporous structure. A macroporous polymer network with a stable pore structure starts to form if a large amount of cross-linker is used during the hydrogel formation. The scanning electron micrographs (SEM) shown in Figure 9.1 illustrate the development of macroporosity in PNIPA networks, depending on the cross-linker (BAAm) content. Here, the hydrogels were prepared starting from NIPA and BAAm monomers at a total monomer concentration of 20 w/v% in water.[51] The onset reaction temperature T_0 was set to 22.5°C. It should be noted that, due to the exothermic reaction profiles, T_0 first increases and then decreases during the polymerization, so that the reactions

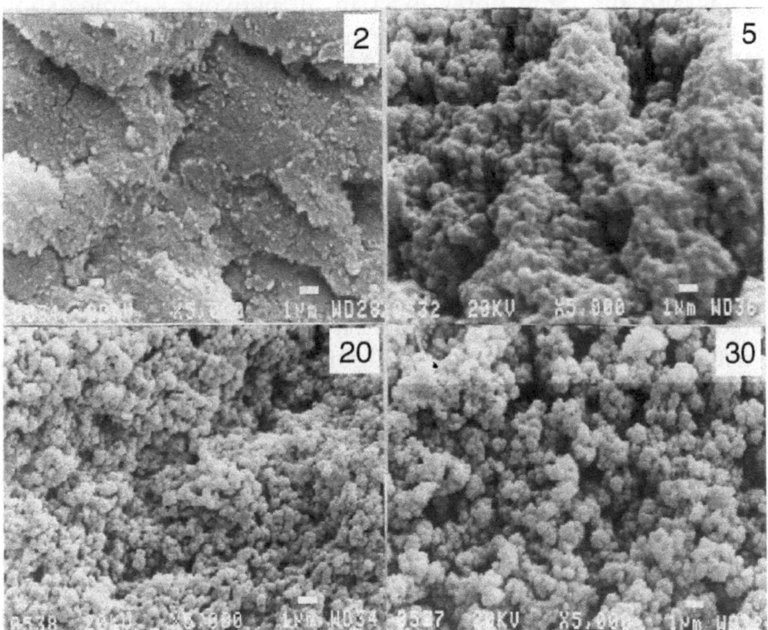

FIGURE 9.1
SEM images of PNIPA networks at 5000× (1-μm bar in the lower right corner of each image). BAAm contents of 2, 5, 20, and 30 wt% are indicated. Monomer concentration = 20%, T_0 = 22.5°C. (Reprinted from Sayil, C. and Okay, O., *Polymer*, 42, 7639, 2001. With permission from Elsevier.)

proceed nonisothermally. At 2% BAAm, the network consists of large polymer domains, and the discontinuities between the domains are large. When the cross-linker content increases from 2 to 5%, the morphology changes drastically, and a structure consisting of aggregates of spherical domains (microspheres) appears. At low cross-linker contents, the microspheres are more or less fused together to form large aggregates. As the cross-linker content increases from 5 to 30% BAAm, the morphology changes from a structure of large aggregates of poorly defined microspheres to one consisting of aggregates of 1- to 2-μm dimensions of well-defined microspheres. The microspheres are about 0.1 to 0.5 μm in diameter. At 30% BAAm, the structure looks like cauliflower, typical for a macroporous copolymer network. From the SEM images together with the pore-size distribution curves recorded using mercury porosimetry, three pore-size distributions in PNIPA networks were identified:[51]

1. Large pores with radius of 10 to 100 μm
2. Macropores with radius of 0.1 to 0.2 μm
3. Micropores with radius less than 60 Å

A conventional expanded PNIPA gel has a network mesh size in solvents of around 50 Å,[52] which corresponds to the micropores of the samples. The micropores and the large pores exist in all network samples shown in Figure 9.1. However, macropores exist in only the 30% cross-linker sample, and these pores contribute about 96% of the total porosity of this sample. The pore-size distribution of macropores is very narrow. Considering the SEM image of the 30% cross-linker sample, the macropores correspond to the interstices between the microspheres. This indicates that the interstices between the microspheres are accessible to mercury only for the 30% cross-linker sample. For this sample, the total porosity calculated from the intruded mercury volume was found to be 65%.

The appearance of macroporosity in PNIPA hydrogels with increasing cross-linker content can also be monitored from the dynamic swelling behavior of the networks immersed in water. Figure 9.2 shows the weight swelling ratio q_w of PNIPA hydrogels (mass of gel in water at time t/mass of dry gel) of various cross-linker (BAAm) contents plotted against the swelling time. The dependence of the swelling rate of the hydrogels on the cross-linker content shows two different regimes:

1. Below 5% BAAm, increasing cross-linker concentration decreases the rate of swelling of the hydrogels. This is expected due to the decreasing mesh size of the network with increasing cross-link density, which limits the diffusion of water molecules into the gel network.
2. Above 5 % BAAm, although the swelling capacities of the hydrogels continue to decrease, the rate of swelling rapidly increases with increasing cross-linker content, which is opposite to what is observed below 5% cross-linker.

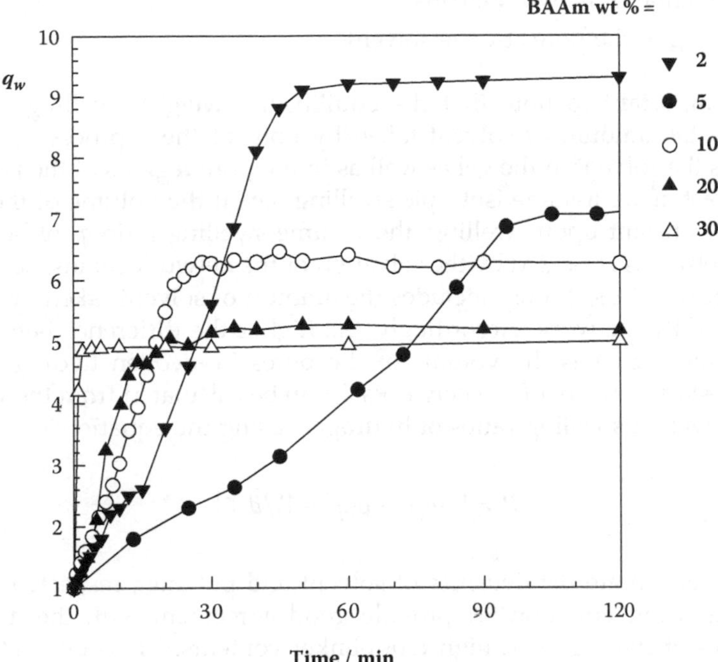

FIGURE 9.2
Weight swelling ratio of PNIPA networks q_w shown as a function of the swelling time at various levels of BAAm content. Monomer concentration = 20%, T_o = 22.5°C. (Reprinted from Sayil, C. and Okay, O., *Polymer*, 42, 7639, 2001. With permission from Elsevier.)

The increasing rate of swelling of PNIPA hydrogels above 5% BAAm is an indication of the appearance of a macroporous structure in these networks, as also observed from the SEM images in Figure 9.1. The existence and the connectivity of the pores in PNIPA hydrogels with more than 5% cross-linker play a crucial role in their fast swelling kinetics. Water can enter or leave the hydrogel through the interconnected pores by convection. Moreover, increasing the internal surface area of the networks also increases the contact area between water and the polymer, which contributes to their accelerating swelling rates. The PNIPA network with 30% BAAm swells about 60 times faster than the network with 5% cross-linker. As a result, the gel has potential applications in gel sensors and devices.

The relative values of the weight and the volume-swelling capacities of the hydrogels provide information about their internal structure in the swollen state. During the swelling process of macroporous hydrogels, the pores inside the gel network are rapidly filled with the solvent; at the same time, the network region takes up solvent from the environment by the diffusion process. Therefore, the swelling of such hydrogels is governed by two separate processes:

1. Solvation of network chains
2. Filling of the pores by the solvent

It is important to note that the equilibrium weight swelling ratio q_w includes the amount of solvent taken by both of these processes, i.e., q_w includes the solvent in the gel as well as in the pore regions of the network. In contrast, if we assume isotropic swelling, i.e., if the volume of the pores remains constant upon swelling, the volume-swelling ratio q_v of heterogeneous networks is caused by the solvation of the network chains, i.e., by the first process. Thus, q_v only includes the amount of solvent taken by the gel portion of the network. Accordingly, the higher the difference between q_w and q_v, the higher is the volume of the pores in swollen hydrogels. The swollen-state porosity of the networks P_s can be calculated from the volume and the weight swelling ratios of hydrogels using the equation[19]

$$P_s = 1 - q_v[1 + \rho(q_w - 1)/d_1]^{-1} \tag{9.1}$$

where d_1 and ρ are the densities of solvent and polymer, respectively. Calculations using Equation 9.1 provide good agreement with the dry-state porosities of hydrogels at high cross-linker contents.[19] However, at a low cross-linker or monomer concentration, P_s values are considerably larger than the dry-state porosities due to the collapse of weak pore structure during drying.

The mechanical properties of PNIPA hydrogels formed at various cross-linker contents also reflect their internal structure.[51] At low cross-linker contents, the modulus of elasticity of the hydrogels increases with increasing cross-linker content. This is expected and is in accord with the statistical theory of rubber elasticity. However, as the network structure changes from a network consisting of large polymer domains to a heterogeneous network consisting of agglomerates of microspheres, the elastic modulus starts to decrease with the cross-linker content. This is due to the fact that the bending-type deformations also become operative with increasing cross-linker content due to the buckling of the pore walls or of chains connecting the microspheres.

According to the above results, in water as a diluent and at $T_o = 22.5°C$, a maximum degree of macroporosity in PNIPA hydrogels can be obtained at a cross-linker content of 30% BAAm and at a total monomer concentration of 20%. Theoretically, the macroporosity of PNIPA hydrogels can be further increased by decreasing the initial monomer concentration below 20% or by increasing the cross-linker content above 30%. However, experiments carried out at lower initial monomer concentration showed a decrease in the porosity of the networks due to the collapse of the weak pore structure.[53] On the other hand, increasing the cross-linker content above 30% is not possible due to the insolubility of BAAm in the aqueous solution.

The gel preparation temperature is another parameter determining the porosity of temperature-sensitive hydrogels. As is well known, PNIPA gel is a typical temperature-sensitive gel exhibiting a volume phase transition at its lower critical solution temperature (LCST), which is approximately 34°C.[54,55] Below this temperature, the gel is swollen, and it shrinks as the temperature is raised. This feature of PNIPA hydrogels has been used to induce phase separation and, thus, to construct porosity inside the PNIPA hydrogel matrix. Phase separation during the formation of PNIPA hydrogels induced by a temperature increase was recently investigated by using real-time photon transmission and temperature measurements.[56]

Figure 9.3 shows the transmitted light intensity I_{tr} (solid curves) and the temperature of the reaction system T (dotted curves) plotted as functions of the reaction time t during the cross-linking copolymerization of NIPA monomer in the presence of BAAm cross-linker in aqueous solution. Here, both the initial monomer concentration and the BAAm content of the monomer mixture were fixed at 6.9 w/v% and 2 mol%, respectively, while the onset reaction temperature T_o was varied between 20 and 28°C, as indicated in the figure. The LCST of PNIPA is also shown in the figures by the horizontal dotted lines. Following the induction period, during which the temperature of the solution remains constant, all of the gelation experiments result in exothermic reaction profiles. As a consequence, the temperature of the reaction system increases sharply with time, attains a maximum value, and then decreases continuously at longer reaction times due to dissipation of reaction heat to the surroundings. The maximum temperatures attained during the polymerization (T_{max}) increase with the onset temperature T_o. Figure 9.3 also shows that the temperature rise during the gel formation and growth process is accompanied by a simultaneous decrease in the transmitted light intensities I_{tr}, indicating the occurrence of phase separation during the reactions. The time at which the temperature reaches a maximum value corresponds to the time at which I_{tr} becomes a minimum. The higher the temperature of the reaction system, the lower are the I_{tr} intensities. Even at temperatures below the LCST, I_{tr} decreases rapidly during the reactions, indicating the occurrence of phase separation below the LCST. Another interesting point shown in Figure 9.3 is that, at onset reaction temperatures T_o above 25°C, I_{tr} attains a limiting value after a given reaction time, indicating formation of permanent heterogeneity in the final PNIPA hydrogels. However, if the onset temperature is below 25°C, I_{tr} starts to increase again at longer times, indicating that the heterogeneity that appeared due to the temperature rise disappears again as the system cools down later.

The results can be explained by the spatial inhomogeneity in PNIPA hydrogels. PNIPA gels consist of lightly cross-linked and highly cross-linked regions.[57,58] If the temperature of such a partially formed, inhomogeneous PNIPA gel is increased rapidly from T_o to various T_{max} below the LCST, the loosely cross-linked regions can adjust their orientation to fit a new equilibrium. However, the highly cross-linked regions require much more time to reorganize. The differences in the relaxation times of various gel regions result in density fluctuations along the gelling system that scatter light and decrease the I_{tr} intensities. Further polymerization and cross-linking reactions

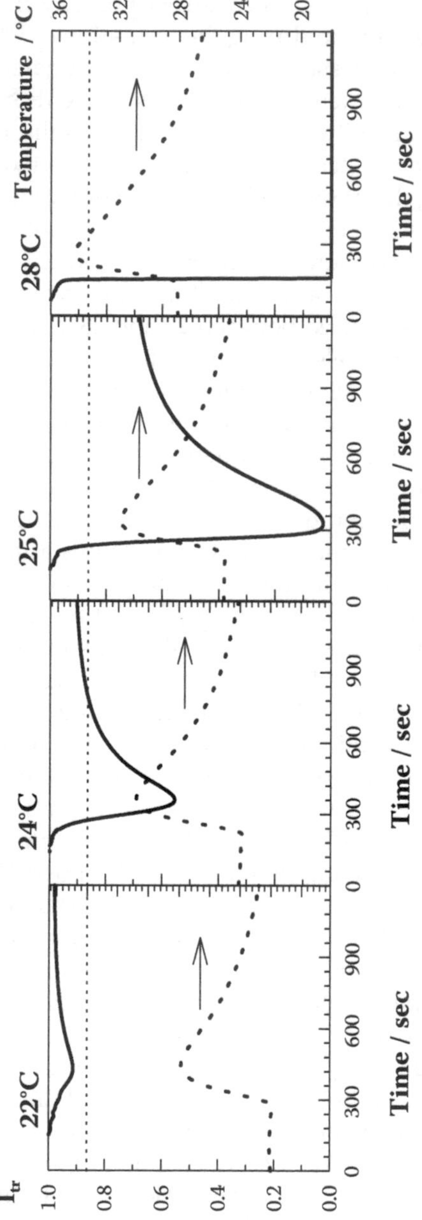

FIGURE 9.3
Variations of the transmitted light intensity I_{tr} and temperature T vs. reaction time during the cross-linking copolymerization of NIPA and BAAm at various onset reaction temperatures T_o. The dotted horizontal lines show the LCST of PNIPA. BAAm = 2 mol%.

fix these fluctuations in the final hydrogel. The observed decrease in I_{tr} at temperatures below the LCST can thus be explained by the concentration fluctuations due to the inhomogeneity within the gel network. Moreover, increasing cross-linker content is known to increase the degree of inhomogeneity in PNIPA gels. As a consequence, the I_{tr} intensities decrease further with increasing cross-linker content of the feed. The rate of change in I_{tr} with time corresponds to the growth rate of phase-separated domains in the reaction system. The growth rate of domains is slow if T_{max} < LCST, whereas it becomes rapid if T_{max} exceeds the LCST. This indicates that, if the temperature of the gel-forming system exceeds the LCST, as in the case of the experiments carried out at T_o = 28°C, a different mechanism of phase separation is operative in PNIPA gels. In these experiments, the sample temperature increases from T_o to various temperatures above the LCST within 1 min due to the exothermicity of the polymerization reactions (Figure 9.3). During this heating process, the interior of the gel system remains under constant-volume condition due to the slowness of the volume change of macroscopic gel samples near the critical point. As a result, the gel system undergoes a transition from the one-phase to the two-phase coexistence state along the critical isochore path via a spinodal decomposition process.[59]

Takata et al. showed that a rapid shrinking of PNIPA gels can be attained by increasing the onset reaction temperature T_o above the LCST due to the increasing degree of spatial gel inhomogeneity.[60] Safrany prepared fast-responsive PNIPA gels starting from 20 wt% aqueous PNIPA solutions by electron beam irradiation above the LCST.[61] Wu et al. prepared macroporous PNIPA gels above their LCSTs in the absence and presence of hydroxypropyl cellulose as a diluent.[62] Such PNIPA hydrogels were also prepared by starting the polymerization reactions of NIPA below the LCST of PNIPA and then elevating the temperature above it.[63] Sayil and Okay observed that PNIPA hydrogel prepared at T_o = 12°C swells in water very slowly (Figure 9.4).[53] As seen from the inset in Figure 9.4, this hydrogel, prepared with 30 wt% BAAm cross-linker, attains its equilibrium state in water after about 400 min. However, for the hydrogels prepared at higher temperatures (T_o = 22.5 to 50°C), the swelling process is very rapid; they reach the equilibrium state in water after about 2 min, i.e., they swell about 200 times faster than the hydrogel formed at 12°C. Thus, the results indicate that there is a critical gel preparation temperature, above which the rate of swelling of PNIPA hydrogels dramatically increases. Increasing porosity in PNIPA networks with an increase in the onset reaction temperature T_o is expected due to the simultaneous increase of the degree of phase separation during the gel formation process.

Figure 9.5 shows the integral size distributions of the pores in PNIPA networks formed at various T_o.[53] The porosity of the PNIPA network formed at T_o = 12°C is low, and it increases as the onset reaction temperature is increased to 22.5°C. However, further increase of the onset temperature decreases the porosity of the networks. This is rather contradictory to the expectation, since (a) the PNIPA gels are believed to be more heterogeneous

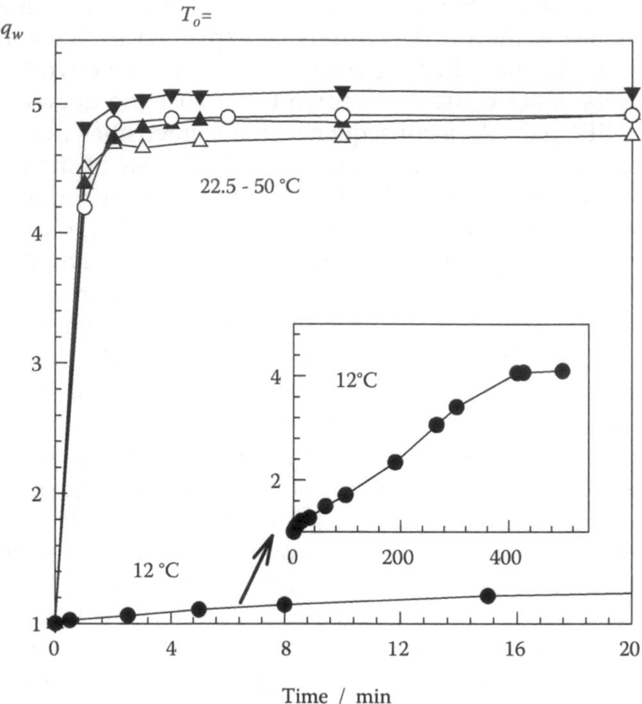

FIGURE 9.4
Weight swelling ratio q_w of PNIPA networks in water at 20°C shown as a function of the swelling time. T_o = 12 (●), 22.5 (○), 30 (▲), 40 (△), and 50°C (▼). BAAm = 30 wt%. (From Sayil, C. and Okay, O., *Polym. Bull.*, 48, 499, 2002. With kind permission of Springer Science and Business Media.)

as T_o increases due to the action of water as a poor solvent (pore-forming agent) at elevated temperatures, and (b) the opacity of gels formed above 22.5°C indicated the existence of phase-separated domains of sizes 10^{-1} μm. This result can be explained by the loose structure of the networks formed at higher temperatures. Indeed, PNIPA hydrogels formed at $T_o > 22.5$°C are weak, while those formed below this temperature are relatively hard. The pores in such weak matrices can collapse upon drying of the hydrogels from their swollen states. The collapse of the pores during drying has also been reported before in loosely cross-linked styrene–divinylbenzene copolymer networks.[19]

Another important parameter determining the degree of macroporosity in hydrogel networks is the type of the inert diluent used in the gel preparation. Addition of a poor solvent into the reaction system as an inert diluent further promotes the occurrence of a phase separation and, thus, leads to the formation of macroporous structures in hydrogels. Several diluents were used in the macroporous hydrogel preparation, such as solvents or nonsolvents for the growing polymer chains, or inert linear polymers.[64,65] Also, various

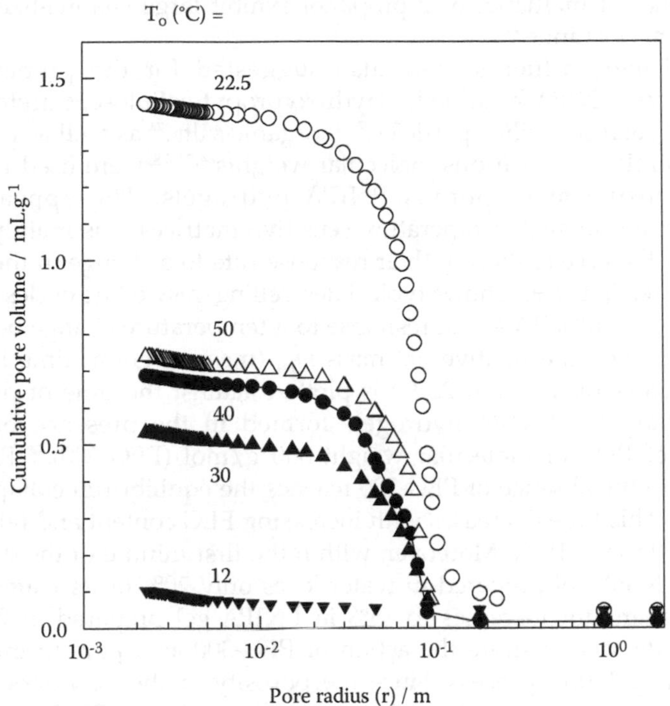

FIGURE 9.5
Integral size distributions of the pores in PNIPA networks at various onset reaction temperatures T_o. BAAm = 30 wt%. (From Sayil, C. and Okay, O., *Polym. Bull.*, 48, 499, 2002. With kind permission of Springer Science and Business Media.)

surfactants forming ordered structures in the aqueous monomer solution were reported to be efficient diluents for the synthesis of macroporous acrylamide-based hydrogels.[66,67] The cross-linking copolymerization of *N*-methylacrylamide with alkylene(bis)acrylamides in the presence of methanol, acetic acid, water, dimethylsulfoxide (DMSO), or dimethylformamide as diluents has been used for the preparation of macroporous polyamides.[68] It was shown that the more polar solvents (water, methanol) produce materials with higher specific surface area and internal pore volume than those prepared with less polar solvents such as DMSO. This finding is in accord with the formation mechanism of pores as detailed in the previous section. Xie et al. described the preparation conditions of macroporous networks from AAm and BAAm monomers with pore sizes up to 1000 nm using DMSO/alcohol mixtures as a diluent.[48] It was shown that, as the chain length of the alcohols used in combination with DMSO increases from C_1 to C_8, larger microspheres and thus larger pores form in the final material. Further increase in the chain length of alcohol decreased the size of the pores. Macroporous networks prepared from diethyleneglycol dimethacrylate in the presence of methanol or 2-propanol exhibit large channellike pores in the micrometer range.[69]

Several inert diluents were also suggested for the preparation of macroporous PNIPA hydrogels. Hydroxypropyl cellulose,[62] acetone,[70] 1,4-dioxane,[71] sucrose,[72] silica particles,[73] inorganic salts,[74] as well as poly(ethylene glycol) (PEG) of various molecular weights[64,75–78] were used to prepare fast-responsive macroporous PNIPA hydrogels. The appearance of macroporosity in such temperature-sensitive matrices was mainly demonstrated by the acceleration of their response rate to a change in the external temperature. Figure 9.6 shows typical deswelling–reswelling cycles of PNIPA gels with 1.2 mol% BAAm in response to a temperature change between 25 and 48°C. Here, the relative gel mass m_{rel} (mass of gel at time t/mass of equilibrium-swollen gel at 25°C) is plotted against the time of deswelling and swelling for PNIPA hydrogels formed in the presence of various amounts of PEG of molecular weight 300 g/mol (PEG-300).[64] PNIPA gel prepared in the absence of PEG-300 reaches the equilibrium collapsed state in 20 min. This time decreases with increasing PEG content and falls to only 4 min at 70 v/v% PEG. Moreover, within the first minute of the deswelling process, PNIPA gel prepared in water loses only 20% of its water content, while this amount increases to 85% in PNIPA gel prepared in 70% PEG. These results demonstrate the action of PEG-300 as a pore-forming agent during the gelation process. Since the porosity of the networks increases with the PEG concentration, the deswelling rate of PNIPA hydrogels increases with an increase in the amount of PEG used in the hydrogel preparation. Figure 9.6 also shows that the reswelling process of PNIPA gels occurs in two steps: a rapid reswelling step continuing for 10 min followed by a slow reswelling step until the equilibrium-swollen state, which needs a few days. Moreover, increasing concentration of PEG in the gel formation system slows down the response rate of the collapsed gels in cold water.

The two-step reswelling profile can be explained by the formation of a two-phase structure during swelling of PNIPA gels. When a collapsed PNIPA gel sample is immersed in cold water, the outer surface of the sample will first be in contact with water molecules, so the collapsed network chains at the gel surface will start to relax. Since the surface area of the collapsed gel is initially large, water absorption by the gel network occurs easily. However, as the gel swells, a two-phase structure forms: one containing solvated network chains at the gel surface and the other containing relatively unswollen network chains in the inner part of the gel sample. Since the surface area of the inner part of the gel becomes smaller with increasing degree of swelling, the reswelling process slows down after the initial rapid swelling period. Moreover, the decrease of the reswelling rate of gels with increasing PEG content can be explained by the collapse of the porous structure in PNIPA hydrogels. The collapse of the pores in the PNIPA network can be ascribed to the cohesional forces in play when the solvated PNIPA chains approach each other due to the loss of water. Since the pores in PNIPA networks with a low cross-link density are unstable, they disappear in the shrunken state of the gel.

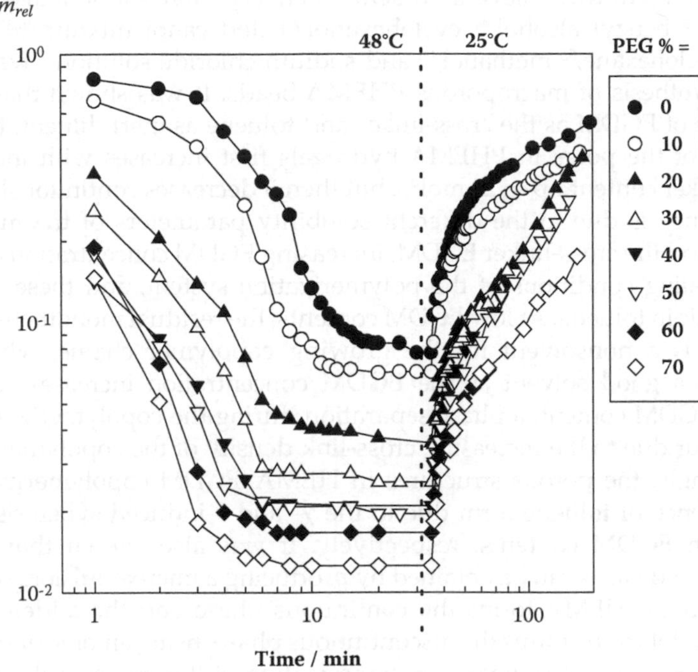

FIGURE 9.6
Deswelling–reswelling kinetics of PNIPA hydrogels shown as the variation of the relative gel mass m_{rel} (mass of gel at time t/mass of equilibrium-swollen gel at 25°C) with time at various levels of PEG-300 content used in hydrogel preparation. BAAm = 1.2 mol%, T_o = 9°C. (From Dogu, Y. and Okay, O., *J. Appl. Polym. Sci.*, 99, 37, 2006. Reprinted with permission of John Wiley & Sons, Inc.)

Cross-linked poly(2-hydroxyethylmethacrylate) (PHEMA) is an important class of hydrogels, widely used as a hydrogel for contact lenses.[79,80] The maximum equilibrium swelling of PHEMA gels in water is thermodynamically limited to 39 wt%. Thus, if HEMA is polymerized in the presence of more than 39 wt% water, a phase separation occurs due to the χ-induced syneresis, and an opaque heterogeneous gel forms. Huglin and Yip calculated the size and the concentration of phase-separated domains in PHEMA gels as 10^{-1} μm and 0.25 to 2.2% of the swollen gel, respectively.[81] By varying the amount of water as the diluent in the cross-linking polymerization of HEMA, a large range of porosities can be obtained in the final networks. Chirila et al. prepared macroporous PHEMA hydrogels from the aqueous solutions of HEMA together with EGDM and hexamethylene dimethacrylate (HDMA) as the cross-linker.[82] It was shown that at water concentrations above 60 wt%, macroporous PHEMA networks start to form with pore sizes of about 10 μm.

In addition to water, several diluents such as cyclohexanol/lauryl alcohol mixture,[83] benzyl alcohol,[41] cyclohexanol/1-dodecanol mixture,[36,37,84] toluene,[35] cyclohexane,[85] methanol,[86] and sodium chloride solution[87] were used in the synthesis of macroporous PHEMA beads. It was shown that, in the presence of EGDM as the cross-linker and toluene as inert diluent, the total volume of the pores in PHEMA hydrogels first increases with increasing cross-linker content up to 20 mol%, but then it decreases continuously.[35] This discrepancy is due to the different solubility parameters of the monomer HEMA and the cross-linker EGDM; increasing EGDM concentration changes the solvating conditions of the polymerization system, and these changes are drastic in toluene. At low EGDM contents, the residual monomer–toluene mixture is a nonsolvent for the growing copolymer chains, whereas it becomes a good solvent as the EGDM concentration increases. Thus, at higher EGDM content, a phase separation during the copolymerization can only occur due to the increasing cross-link density of the copolymer chains. Accordingly, the porous structures in HEMA–EGDM copolymerization in the presence of toluene form due to the χ- and ν-induced syneresis at low and high EGDM contents, respectively. It was also shown that porous PHEMA hydrogels can be obtained by producing a microemulsion, of which the monomer HEMA forms the continuous phase and the added diluent methylcyclohexane forms the discontinuous phase, using anionic or nonionic surfactants.[88] The cross-linking polymerization of the microemulsion using UV radiation results in a porous material that absorbs larger amount of water than the nonporous PHEMA. The size of the pores varies between 100 and 2000 nm and increases as the amount of the diluent increases.[88]

9.2.4 Formation of Macroporous Structures by Cryogelation

Another method for obtaining macroporous structures in hydrogels consists of freezing the initial polymerization mixture to an apparently solid monomer matrix during the gelation reactions.[22] Thus, the polymerization and cross-linking reactions are carried out below the bulk freezing temperature of the reaction solution containing the monomers and the initiator. During freezing of the reaction system, the unreacted monomers, initiator, as well as the polymer formed are expelled from the forming ice and entrapped within channels between the ice crystals. As a result, the polymerization reactions can only take place in spatially restricted reaction fields, which are the unfrozen microchannels of the apparently frozen system.[22–26] Since the actual concentrations of the monomer and the initiator in the microchannels are much larger than their nominal concentrations, the decrease of the rate constants of polymerization and cross-linking reactions at low temperatures is compensated by the increased monomer concentration in the reaction zones. Indeed, the critical monomer concentration for the onset of gelation is much lower in cryogelation systems compared with the conventional reaction systems, indicating the higher efficiency of cross-linking below the freezing temperature.

In these polymerization systems, although there is no phase separation during the course of the network formation process, the frozen zones of the reaction system act as templates during gelation, which can easily be removed from the gel by thawing, leading to a macroporous structure. The reason why water in aqueous solutions does not freeze at below the bulk freezing temperature is attributed, in the main, to the freezing-point depression of water due to the solutes. Although the usual solute concentrations can lower the freezing point by only a few degrees, once ice is present, the effect is enhanced. This is because solutes are excluded from the ice structure and become more concentrated in the remaining unfrozen regions. Thus, as water freezes (crystallizes), the solute concentration in the liquid phase rises continuously, so that successively greater osmotic pressure is required to keep the liquid phase in equilibrium with the pure ice phase.

Lozinsky and coworkers investigated in detail the polymerization reactions conducted below the freezing point of water.[22] They called such materials "cryogels." Cryogels formed from dilute aqueous solutions of gelatin, poly(vinyl alcohol), chitosan, as well as from cross-linking copolymerization of AAm and BAAm exhibit interconnected systems of macropores and sponge-like morphology.[22,23] Xue et al. prepared fast-responsive PNIPA gels by using a two-step polymerization method, the initial polymerization being carried out at 20°C, followed by cryopolymerization at –28°C for 24 h.[89] Zhang and Chu demonstrated that the polymerization reactions in DMSO at temperatures below the melting point of DMSO produce macroporous PNIPA hydrogels with very regular pores of sizes 1 μm.[90] It was shown that the hydrogels exhibit superfast and stable oscillatory swelling–deswelling behavior in solvents. Other temperature-sensitive hydrogels such as poly(*N*,*N*′-diethylacrylamide) hydrogels prepared by the cryogelation technique also exhibit superfast response rates against temperature changes.[91] Plieva et al. showed that the pore size and the thickness of pore walls in PAAm cryogels can be controlled by varying the initial monomer concentration of the reaction system.[92] Increasing monomer concentration increases the thickness of the pore walls but decreases the total porosity of the cryogels. Moreover, increasing the freezing rate of the reaction system during gelation also affects the size and the size distribution of pores in the cryogels.[93] As an alternative approach to the cryogelation technique, macroporous structures can also be obtained by freeze-drying and subsequent rehydration of expanded hydrogels.[94] In this case, a honeycomblike pore structure forms due to the freezing of free water in the gel, causing the polymer chains to condense.

The appearance of a porous structure by the cryogelation technique leads to a drastic change in both the swelling and mechanical properties of the hydrogels.[95] Figure 9.7 shows the equilibrium volume-swelling ratio V_{eq} (swollen gel volume/gel volume after preparation) and the elastic modulus G of the equilibrium-swollen hydrogels plotted against the gel preparation temperature T_{prep}. The hydrogels were prepared by free-radical cross-linking copolymerization of the sodium salt of 2-acrylamido-2-methylpropane sulfonic acid (AMPS) monomer with BAAm cross-linker at an initial monomer

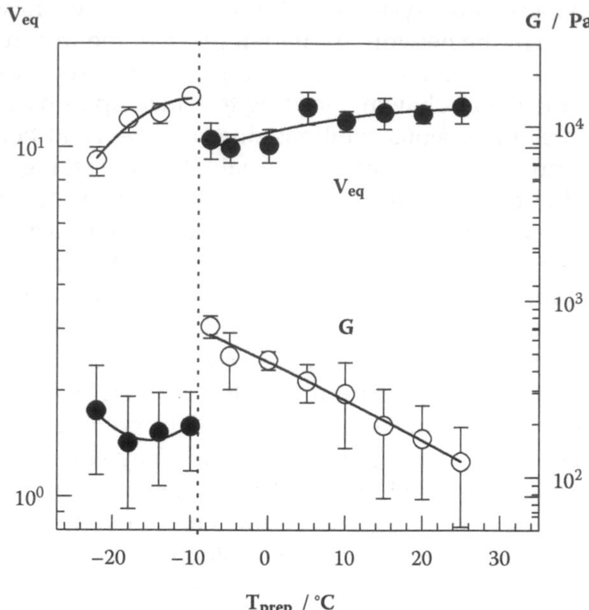

FIGURE 9.7
Volume-swelling ratio V_{eq} (filled symbols) and the elastic modulus G (open symbols) of equilibrium-swollen PAMPS hydrogels shown as a function of the gel preparation temperature T_{prep}. (Reprinted from Ozmen, M.M. and Okay, O., *Polymer*, 46, 8119, 2005. With permission from Elsevier.)

concentration of 5 w/v% in water. The cross-linker content in the monomer mixture was 17 mol%. Note that the gel preparation temperature T_{prep} is the temperature of the thermostated bath in which the reactions were carried out. Since the addition of the initiator into the monomer solution occurred at 0°C, the reactions proceed nonisothermally from the moment of the initiator addition to the moment when the temperature of the reaction system reaches T_{prep}. At T_{prep} = −8°C or above, poly(AMPS) (PAMPS) hydrogels exhibit relatively high swelling ratios V_{eq}, on the order of 10^1, and low moduli of elasticity G in the range of 10^2 to 10^3 Pa. However, decreasing T_{prep} below −8°C results in a tenfold decrease in the swelling ratio and about a tenfold increase in the elastic modulus of the gels. Thus, both the swelling and elastic properties of PAMPS hydrogels drastically change as T_{prep} is decreased below −8°C. The hydrogels formed at or above −8°C were transparent, while those formed at lower temperatures were opaque, indicating that these gels have separate domains in a spatial scale of submicrometers to micrometers.

Figure 9.8 shows the SEM image of PAMPS network formed at T_{prep} = −22°C. Although the geometry and size of the pores are quite irregular, one can identify regular assembly of polyhedral pores forming junctions with angles of about 120°. This type of microstructure is distinctly different from the macroporous networks formed by the reaction-induced phase-separation

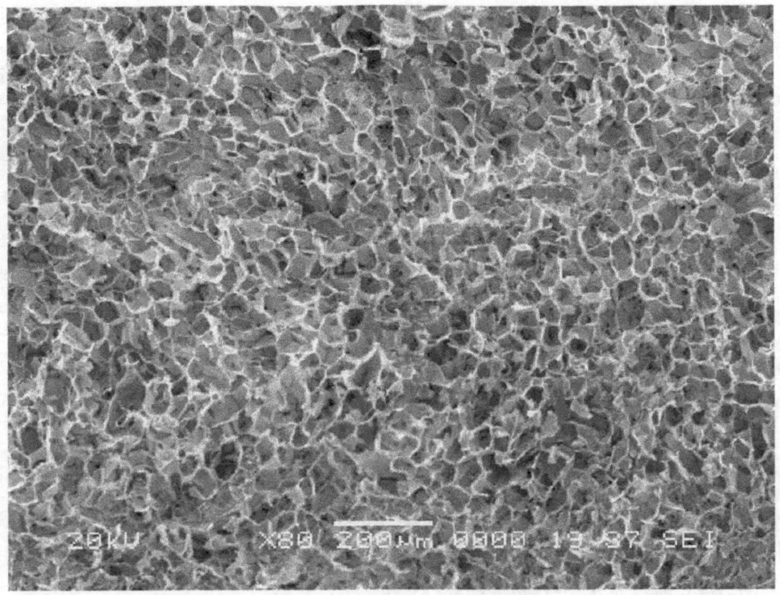

FIGURE 9.8
SEM of a PAMPS network formed at $T_{prep} = -22°C$. The scaling bar is 200 μm; magnification = ×55.

mechanism, where the structure looks like cauliflowers and consists of aggregates of various sizes (Figure 9.1). Figure 9.9, which shows SEM images of PAMPS networks formed at various T_{prep}, indicates that all of the polymer samples formed below –8°C have a porous structure with pore sizes of 30 to 50 μm, while those formed at or above –8°C exhibit a continuous morphology. At –10°C, the pore walls are too weak, such that they are more or less fused together to form large aggregates.

It was shown that completely reversible swelling–deswelling cycles can be obtained using PAMPS hydrogels prepared below –8°C.[65,95] The swelling–deswelling cycles of PAMPS hydrogels formed at various T_{prep} are shown in Figure 9.10, where the weight swelling ratio q_w of the gels is plotted against the time of swelling and deswelling in water and in acetone, respectively. Gels formed below –8°C attain their equilibrium-swollen volumes in less than 30 sec, while those formed at higher temperatures require about 1 h to reach their equilibrium state in water. Moreover, if swollen PAMPS hydrogels are immersed in acetone, those prepared below –8°C attain their equilibrium collapsed state in 5 to 10 min, while those formed at higher temperatures were too weak to withstand the volume changes. Thus, decreasing T_{prep} below –8°C results in the formation of superfast-responsive PAMPS hydrogels, which are also stable against volume changes.

The occurrence of the polymerization and cross-linking reactions below the freezing point of the reaction system is due to the presence of unfrozen regions in which the reactions proceed within the apparently frozen reaction system. Thus, even when cooled below the bulk freezing temperature, some

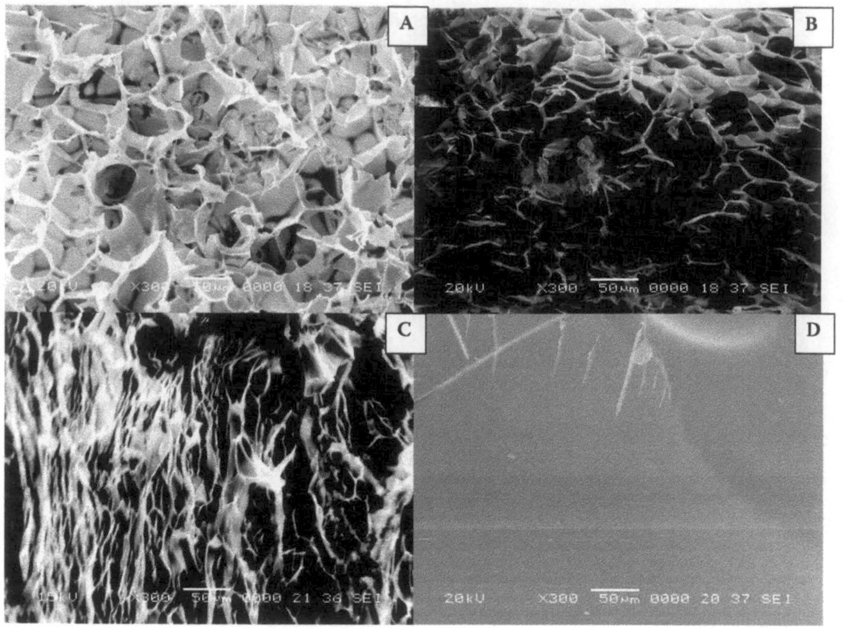

FIGURE 9.9
SEM of PAMPS networks formed at $T_{prep} = -22$ (A), -18 (B), -10 (C), and -8°C (D). The scaling bar is 50 μm. (Reprinted from Ozmen, M.M. and Okay, O., *Polymer*, 46, 8119, 2005. With permission from Elsevier.)

water in aqueous solutions remains unfrozen. The amount of unfrozen water depends on the temperature and on the amount and type of solute in the solution. Assuming a thermodynamic equilibrium condition between the ice and gel phases at the gel preparation temperature T_{prep}, the relationship between the polymer concentration in the unfrozen gel phase and the temperature was given by[95]

$$\frac{1}{T_{\text{prep}}} = \frac{1}{T_f^0} - \frac{R}{\Delta H_{\text{m}}}\left[\ln(1-\nu_2) + \nu_2 + \chi\nu_2^2 + 0.5\nu_{\text{e}} V_1 \nu_2 - f\nu_2\right] \tag{9.2}$$

where ΔH_m is the molar enthalpy of fusion of pure ice, T_f^0 is the normal freezing point of pure water, ν_2 is the volume fraction of polymer in the unfrozen zones of the reaction system, χ is the polymer–solvent interaction parameter, ν_e is the effective cross-link density of the network, V_1 is the molar volume of solvent, and f is the effective charge density, i.e., the fraction of charged units in the network chains that are effective in gel swelling. Note that the polymer volume fraction ν_2 relates to the polymer concentration c in the unfrozen reaction zone (in w/v%) by

$$c = \nu_2\,\rho\,10^2 \tag{9.3}$$

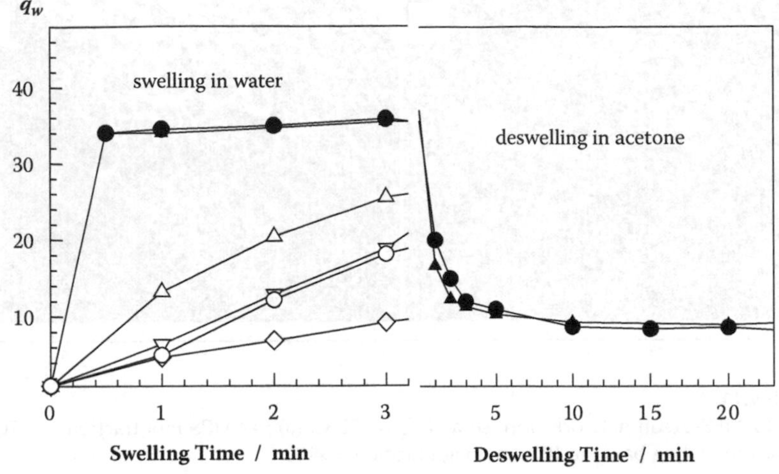

FIGURE 9.10
Swelling and deswelling kinetics of PAMPS hydrogels in water and acetone, respectively, shown as the variation of the weight swelling ratio q_w with the time of swelling or deswelling at various levels of T_{prep} = −22 (●), −18 (▲), −8 (◇), 0 (△), 15 (▽), and 25°C (○). (Reprinted from Ozmen, M.M. and Okay, O., *Polymer*, 46, 8119, 2005. With permission from Elsevier.)

where ρ is the polymer density. Further, assuming complete monomer conversion, the volume fraction of ice in the reaction system f_{ice} can be calculated as

$$f_{ice} = 1 - \frac{c_0}{c} \tag{9.4}$$

where c_0 is the initial monomer concentration.

Calculations using the above equations for various cross-link densities ν_e show that the effect of gel elasticity on the freezing-point depression is negligible.[95] However, the gel preparation temperature T_{prep} and the degree of ionization of the polymer chains f significantly affect the freezing point of the reaction system. For example, during the polymerization at an initial monomer concentration of c_0 = 5% and T_{prep} = −22, −10, and −5°C, the respective polymer concentrations in the reaction zones are 30.6, 13.5, and 6.7%, while the ice fractions, i.e., the porosities of the resulting networks after thawing, are 0.84, 0.63, and 0.25, respectively. Thus, the lower the T_{prep}, the higher is the polymer concentration in the reaction zones and the larger the ice fraction. Further, calculation results also predict that a decrease in the charge density f of the network chains would increase the volume fraction of ice in the reaction system, so that the pores in the final hydrogels would be larger.

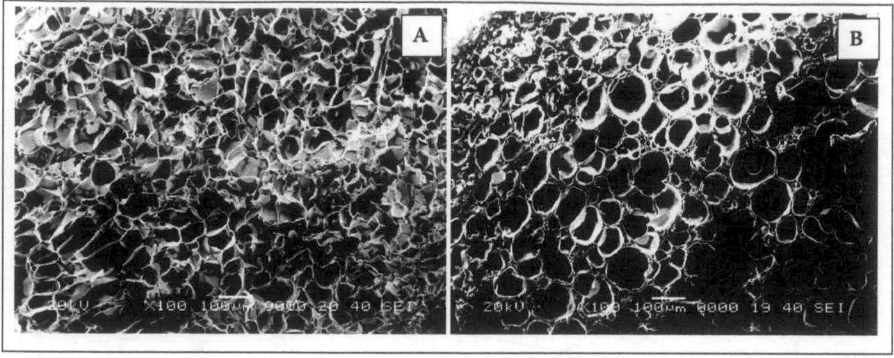

FIGURE 9.11
SEMs of AMPS–AAm networks formed at $T_{prep} = -22°C$. (a) f (AMPS mol fraction) = 0.70 and (b) $f = 0$. The scaling bars are 100 μm; magnification = ×100.

This prediction was also confirmed by experiments.[95] Figure 9.11 shows SEM images of two network samples prepared under identical conditions at $T_{prep} = -22°C$, except that their f values are different. At $f = 0$, the network consists of spherical pores of sizes 100 to 150 μm, which are much larger than those in the network prepared at $f = 0.7$.

9.3 Concluding Remarks

In recent years, synthesis strategies for macroporous hydrogels have been continually optimized, and the concept of cryogelation has become increasingly important for the synthesis of such smart materials. Current research in the field of macroporous hydrogels is focused on tailor-made designs of pore structure. The copolymerization system leading to macroporous hydrogels is a quasi-ternary system composed of a polymer network, soluble polymers, and low-molecular-weight compounds (monomers and diluent). All concentrations and properties of the system components change continuously during the cross-linking process. Moreover, various nonidealities also exist during the gelation reactions, such as cyclization, multiple cross-linking reactions, size-dependent reactivities of pendant vinyl groups, and diffusion-controlled reactions. A more thorough understanding of this complicated gel formation system is needed to improve structuring hydrogels over all length scales, starting from nanometer to micrometer. Moreover, given the fact that the volume-swelling ratio of the hydrogels and their macroporosity are inversely coupled, new synthetic approaches are needed for the preparation of macroporous, highly swollen hydrogels exhibiting drastic volume changes in response to external stimuli.

References

1. Shibayama, M. and Tanaka, T., Volume phase transition and related phenomena of polymer gels, *Adv. Polym. Sci.*, 109, 1, 1993.
2. Oh, K.S. et al., Effect of cross-linking density on swelling behavior of NIPA gel particles, *Macromolecules*, 31, 7328, 1998.
3. Tanaka, T. and Fillmore, D.J., Kinetics of swelling of gels, *J. Chem. Phys.*, 70, 1214, 1979.
4. Yoshida, R. et al., Comb-type grafted hydrogels with rapid de-swelling response to temperature changes, *Nature*, 374, 240, 1995.
5. Kaneko, Y. et al., Influence of freely mobile grafted chain length on dynamic properties of comb-type grafted poly(*N*-isopropylacrylamide) hydrogels, *Macromolecules*, 28, 7717, 1995.
6. Kaneko, Y. et al., Fast swelling/deswelling kinetics of comb-type arranged poly(*N*-isopropyl acrylamide) hydrogels, *Macromol. Symp.*, 109, 41, 1996.
7. Dŭsek, K., Macroporous resins, *Chem. Prumysl.*, 11, 439, 1961.
8. Dŭsek, K., Swelling of homogeneous and macroporous styrene-divinylbenzene copolymers, *Collection Czech. Chem. Commun.*, 27, 2841, 1963.
9. Dusek, K., Phase separation during the formation of three-dimensional polymers, *J. Polym. Sci.*, B3, 209, 1965.
10. Abrams, I.M., High porosity polystyrene cation exchange resins, *Ind. Eng. Chem.*, 48, 1469, 1956.
11. Meitzner, E.F. and Oline, J.A., Process for preparing macroreticular resins, copolymers and products of said process, U.S. patent 749,526, 1958.
12. Lloyd, W.G. and Alfrey, T., Network polymers, II: experimental study of swelling, *J. Polym. Sci.*, 62, 301, 1962.
13. Kunin, R., Meitzner, E.F., and Bortnick, N., Macroreticular ion exchange resins, *J. Am. Chem. Soc.*, 84, 305, 1962.
14. Kun, K.A. and Kunin, R., Pore structure of some macroreticular ion exchange resins, *J. Polym. Sci.*, B2, 587, 1964.
15. Millar, J.R. et al., Solvent modified polymer networks, part 1, *J. Chem. Soc.*, 218, 218, 1963.
16. Seidl, J. et al., Macroporose styrol-divinylbenzol copolymere und ihre anwendung in der chromatographie und zur darstellung von ionenaustauschern, *Adv. Polym. Sci.*, 5, 113, 1967.
17. Dŭsek, K., Network formation in chain crosslinking (co)polymerization, in *Developments in Polymerization 3*, Haward, R.N., Ed., Applied Science, London, 1982, p. 143.
18. Guyot, A. and Bartholin, M., Design and properties of polymers as materials for fine chemistry, *Prog. Polym. Sci.*, 8, 277, 1982.
19. Okay, O., Macroporous copolymer networks, *Prog. Polym. Sci.*, 25, 711, 2000.
20. Hentze, H.P. and Antonietti, M., Porous polymers and resins for biotechnological and biomedical applications, *Rev. Mol. Biol.*, 90, 27, 2002.
21. Alexander, C. et al., Molecular imprinting science and technology: a survey of the literature for the years up to and including 2003, *J. Mol. Recogn.*, 19, 106, 2006.
22. Lozinsky, V.I., Cryogels on the basis of natural and synthetic polymers: preparation, properties and application, *Russ. Chem. Rev.*, 71, 489, 2002.

23. Lozinsky, V.I. et al., The potential of polymeric cryogels in bioseparation, *Bioseparation*, 10, 163, 2002.
24. Arvidsson, P. et al., Direct chromatographic capture of enzyme from crude homogenate using immobilized metal affinity chromatography on a continuous supermacroporous adsorbent, *J. Chromatog.*, A986, 275, 2003.
25. Lozinsky, V.I. et al., Redox-initiated radical polymerisation of acrylamide in moderately frozen water solutions, *Macromol. Rapid Commun.*, 22, 1441, 2001.
26. Lozinsky, V.I. et al., Polymeric cryogels as promising materials of biotechnological interest, *Trends in Biotechnol.*, 21, 445, 2003.
27. Omidian, H., Rocca, J.G., and Park, K., Advances in superporous hydrogels, *J. Contr. Rel.*, 102, 3, 2005.
28. Kolarz, B.N., Trochimczuk, A., and Wojczynska, M., Surface sulphonated highly crosslinked styrene-divinylbenzene copolymers, *Angew. Makromol. Chem.*, 162, 193, 1988.
29. Kolarz, B.N. et al., Acrylic carriers for the immobilization of enzymes, *Angew. Makromol. Chem.*, 171, 201, 1989.
30. Kolarz, B.N., Wojczynska, M., and Herman, B., Polyacrylamide sorbents. Synthesis and sorption properties, *React. Polym. Ion Exch. Sorbents*, 11, 29, 1989.
31. Trochimczuk, A. and Kolarz, B.N., Acrylic polymers with improved hydrophilicity, *React. Polym. Ion Exch. Sorbents*, 11, 135, 1989.
32. Jelinkova, M. et al., Reactive polymers, 58: polyampholytes based on macroporous copolymers of glycidyl methacrylate with ethylene glycol dimethacrylate or divinylbenzene, *React. Polym. Ion Exch. Sorbents*, 11, 253, 1989.
33. Galina, H. and Kolarz, B.N., Studies of porous polymer gels, I: one-step suspension technique for preparation of moderately crosslinked methacrylic acid gels, *J. Appl. Polym. Sci.*, 23, 3017, 1979.
34. Galina, H. and Kolarz, B.N., Studies of porous polymer gels, II: structure and porosity of moderately crosslinked poly(methacrylic acid) gels, *J. Appl. Polym. Sci.*, 24, 891, 1979.
35. Okay, O. and Gurun, C., Synthesis and formation mechanism of porous 2-hydroxyethyl methacrylate-ethylene glycol dimethacrylate copolymer beads, *J. Appl. Polym. Sci.*, 46, 401, 1992.
36. Horak, D., Lednicky, F., and Bleha, M., Effect of inert components on the porous structure of 2-hydroxyethyl methacrylate-ethylene dimethacrylate copolymers, *Polymer*, 37, 4243, 1996.
37. Horak, D. et al., Hydrogels in endovascular embolization, I: spherical particles of poly(2-hydroxyethyl methacrylate) and their medico-biological properties, *Biomaterials*, 7, 188, 1986.
38. Robert, C.C.R., Buri, P.A., and Peppas, N.A., Effect of degree of crosslinking on water transport in polymer microparticles, *J. Appl. Polym. Sci.*, 30, 301, 1985.
39. Barr-Howell, B.D. and Peppas, N.A., Structural analysis of poly(2-hydroxyethyl methacrylate) microparticles, *Eur. Polym. J.*, 23, 591, 1987.
40. Scranton, A.B. et al., The physical mechanism for the production of hydrophilic polymer microparticles from aqueous suspensions, *J. Appl. Polym. Sci.*, 40, 997, 1990.
41. Jayakrishnan, A., Sunny, M.C., and Thanoo, B.C., Polymerization of 2-hydroxyethyl methacrylate as large size spherical beads, *Polymer*, 31, 1339, 1990.
42. Denizli, A., Kiremitci, M., and Piskin, E., Subcutaneous polymeric matrix system p(HEMA-BGA) for controlled release of an anticancer drug (5-fluorouracil), II: release kinetics, *Biomaterials*, 9, 363, 1986.

43. Patel, S.K., Rodriguez, F., and Cohen, C., Mechanical and swelling properties of polyacrylamide gel spheres, *Polymer*, 30, 2198, 1989.
44. Park, T.G. and Hoffman, A.S., Estimation of temperature-dependent pore size in poly(*N*-isopropylacrylamide) hydrogel beads, *Biotechnol. Prog.*, 10, 82, 1994.
45. Kayaman, N. et al., Structure and protein separation efficiency of poly(*N*-isopropylacrylamide) gels: effect of synthesis conditions, *J. Appl. Polym. Sci.*, 67, 805, 1998.
46. Wang, G., Li, M., and Chen, X., Inverse suspension polymerization of sodium acrylate, *J. Appl. Polym. Sci.*, 65, 789, 1997.
47. Melekaslan, D., Gundogan, N., and Okay, O. Elasticity of poly(acrylamide) gel beads, *Polym. Bull.*, 50, 287, 2003.
48. Xie, S., Svec, F., and Frechet, J.M.J., Preparation of porous hydrophilic monoliths: effect of the polymerization conditions on the porous properties of poly(acrylamide-*co*-*N*,*N*-methylenebisacrylamide) monolithic rods, *J. Polym. Sci. A. Polym. Chem.*, 35, 1013, 1997.
49. Dušek, K., in *Polymer Networks: Structure and Mechanical Properties*, Chompff, A.J. and Newman, S., Eds., Plenum Press, New York, 1971, pp. 245–260.
50. Funke, W., Okay, O., and Joos-Muller, B., Microgels: intramolecularly crosslinked macromolecules with a globular structure, *Adv. Polym. Sci.*, 136, 139, 1998.
51. Sayil, C. and Okay, O., Macroporous poly(*N*-isopropyl) acrylamide networks: formation conditions, *Polymer*, 42, 7639, 2001.
52. Appel, R. et al., Direct observation of polymer network structure in macroporous *N*-isopropylacrylamide gel by Raman microscopy, *Macromolecules*, 31, 5071, 1998.
53. Sayil, C. and Okay, O., Macroporous poly(*N*-isopropylacrylamide) networks, *Polym. Bull.*, 48, 499, 2002.
54. Hirokawa, T. and Tanaka, T., Volume phase transition in a nonionic gel, *J. Chem. Phys.*, 81, 6379, 1984.
55. Hirotsu, S., Coexistence of phases and the nature of first-order phase transition in poly-*N*-isopropylacrylamide gels, *Adv. Polym. Sci.*, 110, 1, 1993.
56. Kara, S., Okay, O., and Pekcan, O., Real time temperature and photon transmission measurements for monitoring phase separation during the formation of poly(*N*-isopropylacrylamide) gels, *J. Appl. Polym. Sci.*, 86, 3589, 2002.
57. Shibayama, M., Spatial inhomogeneity and dynamic fluctuations of polymer gels, *Macromol. Chem. Phys.*, 199, 1, 1998.
58. Kizilay, M.Y. and Okay, O., Effect of initial monomer concentration on spatial inhomogeneity in poly(acrylamide) gels, *Macromolecules*, 36, 6856, 2003.
59. Li, Y., Wang, G., and Hu, Z., Turbidity study of spinodal decomposition of an *N*-isopropylacrylamide gel, *Macromolecules*, 28, 4194, 1995.
60. Takata, S. et al., Dependence of shrinking kinetics of poly(*N*-isopropylacrylamide) gels on preparation temperature, *Polymer*, 43, 3101, 2002.
61. Safrany, A., Macroporous gels with fast response prepared by e-beam crosslinking of poly(*N*-isopropylacrylamide) solution, *Nucl. Inst. Meth. Phys. Res. B*, 236, 587, 2005.
62. Wu, X.S., Hoffman, A.S., and Yager, P., Synthesis and characterization of thermally reversible macroporous poly(*N*-isopropylacrylamide) hydrogels, *J. Polym. Sci. A, Polym. Chem.*, 30, 2121, 1992.
63. Kabra, B.G. and Gehrke, S.H., Synthesis of fast response, temperature-sensitive poly(*N*-isopropylacrylamide) gel, *Polym. Commun.*, 32, 322, 1991.

64. Dogu, Y. and Okay, O., Swelling-deswelling kinetics of poly(*N*-isopropylacrylamide) hydrogels formed in PEG solutions, *J. Appl. Polym. Sci.*, 99, 37, 2006.
65. Ceylan, D., Ozmen, M.M., and Okay, O., Swelling-deswelling kinetics of poly(acrylamide) hydrogels and cryogels, *J. Appl. Polym. Sci.*, 99, 319, 2006.
66. Antonietti, M. et al., Morphology variation of porous polymer gels by polymerization in lyotropic surfactant phases, *Macromolecules*, 32, 1383, 1999.
67. Pacios, I.E., Horta, A., and Renamayor, C.S., Macroporous gels of poly(*N,N*-dimethylacrylamide) obtained in the lamellar system AOT/water, *Macromolecules*, 37, 4643, 2004.
68. Shea, K.J. et al., Synthesis and characterization of highly crosslinked poly(acrylamides) and poly(methacrylamides): a new class of macroporous polyamides, *Macromolecules*, 23, 4497, 1990.
69. Safrany, A. et al., Control of pore formation in macroporous polymers synthesized by single-step γ-radiation-initiated polymerization and cross-linking, *Polymer*, 46, 2862, 2005.
70. Zhang, X., Zhuo, R., and Yang, Y., Using mixed solvent to synthesize temperature sensitive poly(*N*-isopropylacrylamide) gel with rapid dynamics properties, *Biomaterials*, 23, 1313, 2002.
71. Zhang, J., Huang, S., and Zhuo, R., Preparation and characterization of novel temperature sensitive poly(*N*-isopropylacrylamide-*co*-acryloyl beta-cyclodextrin) hydrogels with fast shrinking kinetics, *Macromol. Chem. Phys.*, 205, 107, 2004.
72. Zhang, J. et al., Temperature-sensitive poly(*N*-isopropylacrylamide) hydrogels with macroporous structure and fast response rate, *Macromol. Rapid Commun.*, 24, 447, 2003.
73. Serizawa, T., Wakita, K., and Akashi, M., Rapid deswelling of porous poly(*N*-isopropylacrylamide) hydrogels prepared by incorporation of silica particles, *Macromolecules*, 35, 10, 2002.
74. Cheng, S., Zhang, J., and Zhuo, R., Macroporous poly(*N*-isopropylacrylamide) hydrogels with fast response rates and improved protein release properties, *J. Biomed. Mater. Res.*, 67A, 96, 2003.
75. Cicek, H. and Tuncel, A., Preparation and characterization of thermoresponsive isopropylacrylamide-hydroxyethylmethacrylate copolymer gels, *J. Polym. Sci. A., Polym. Chem.*, 36, 527, 1998.
76. Zhang, X. and Zhuo, R., Preparation of fast responsive, thermally sensitive poly(*N*-isopropylacrylamide) gel, *Eur. Polym. J.*, 36, 2301, 2000.
77. Zhuo, R. and Li, W., Preparation and characterization of macroporous poly(*N*-isopropylacrylamide) hydrogels for the controlled release of proteins, *J. Polym. Sci. A., Polym. Chem.*, 41, 152, 2003.
78. Zhang, X. et al. Preparation and characterization of fast response macroporous poly(*N*-isopropylacrylamide) hydrogels, *Langmuir*, 17, 6094, 2001.
79. Wichterle, O. and Lim, D., Hydrophilic gels for biological use, *Nature (London)*, 185, 117, 1960.
80. Mathur, A.M., Moorjani, S.K., and Scranton, A.B., Methods for synthesis of hydrogel networks: a review, *J. Macromol. Sci.: Rev. Macromol. Chem. Phys.*, C36, 405, 1996.
81. Huglin, M.B. and Yip, D.C.F., Microsyneresis region in poly(2-hydroxyethyl methacrylate) hydrogels, *Macromolecules*, 25, 1333, 1992.
82. Chirila, T.V. et al., Hydrophilic sponges based on 2-hydroxyethyl methacrylate, 1, *Polym. Int.*, 32, 221, 1993.

83. Coupek, J., Krivakova, M., and Pokorny, S., Poly[(ethylene glycol dimethacrylate)-*co*-acrylamide] based hydrogel beads by suspension copolymerization, *J. Polym. Sci.*, C 42, 185, 1973.
84. Hradil, J. and Horak, D., Characterization of pore structure of PHEMA-based slabs, *React. Func. Polym.*, 62, 1, 2005.
85. Gomez, C.G., Igarzabel, C.I.A., and Strumia, M.C., Effect of the crosslinking agent on porous networks formation of hema-based copolymers, *Polymer*, 45, 6189, 2004.
86. Aroca, A.S. et al., Porous poly(2-hydroxyethyl acrylate) hydrogels prepared by radical polymerisation with methanol as diluent, *Polymer*, 45, 8949, 2004.
87. Liu, Q. et al., Preparation of macroporous poly(2-hydroxyethyl methacrylate) hydrogels by enhanced phase separation, *Biomaterials*, 21, 2163, 2000.
88. Bennett, D.J. et al., Synthesis of porous hydrogel structures by polymerizing the continuous phase of a microemulsion, *Polymer Int.*, 36, 219, 1995.
89. Xue, W., Hamley, I., and Huglin, M.B., Rapid swelling and deswelling of thermoreversible hydrophobically modified poly(*N*-isopropylacrylamide) hydrogels prepared by freezing polymerization, *Polymer*, 43, 5181, 2002.
90. Zhang, X.Z. and Chu, C.C., Thermosensitive PNIPAAm cryogel with superfast and stable oscillatory properties, *Chem. Commun.*, 12, 1446, 2003.
91. Lozinskii, V.I. et al., Thermoresponsible cryogels based on cross-linked poly(*N,N*-diethylacrylamide), *Polymer Science, Ser. A.*, 39, 1300, 1997.
92. Plieva, F.M. et al., Pore structure in supermacroporous polyacrylamide based cryogels, *Soft Matter*, 1, 303, 2005.
93. Plieva, F.M. et al., Characterization of supermacroporous monolithic polyacrylamide based matrices designed for chromatography of bioparticles, *J. Chromatog.*, A 807, 129, 2004.
94. Kato, N., Sakai, Y., and Shibata, S., Wide-range control of deswelling time for thermosensitive poly(*N*-isopropylacrylamide) gel treated by freeze-drying, *Macromolecules*, 36, 961, 2003.
95. Ozmen, M.M. and Okay, O., Superfast responsive ionic hydrogels with controllable pore size, *Polymer*, 46, 8119, 2005.

Chapter 10

Smart Boronate-Containing Copolymers and Gels at Solid–Liquid Interfaces, Cell Membranes, and Tissues

Alexander E. Ivanov, Igor Yu. Galaev, and Bo Mattiasson

CONTENTS

Abbreviations

AA	acrylamide
BCC	boronate-containing copolymer
BCG	boronate-containing copolymer gel
CCA	crystalline colloid array
DDOPBA	polymerizable derivative of 4-carboxyphenylboronic acid
DMAA	*N,N*-dimethylacrylamide
NAAPBA	*N*-acryloyl-*m*-aminophenylboronic acid
Neu5Ac	*N*-acetylneuraminic acid
NIPAM	*N*-isopropylacrylamide
NIPMM	*N*-isopropylmethacrylamide
NPAPBA	*N*-propionyl-*m*-aminophenylboronic acid
NVP	*N*-vinylpyrrolidone
PABA	poly(aniline boronic acid)
PBS	phosphate-buffered saline
PCCA	crystalline colloid array incorporated into polyacrylamide gel
PEG	polyethylene glycol
PLL	polylysine
PLL-g-(PEG;PBA)	polylysine grafted with polyethylene glycol and *m*-aminophenylboronic acid
PVA	poly(vinyl alcohol)
RBC	red blood cells
rLECs	rabbit lens epithelial cells
VPBA	4-vinylphenylboronic acid
WGA	wheat germ agglutinin

10.1 Introduction

Reversible complex formation of borate and phenylboronate ions with mono- and oligosaccharides in aqueous solutions has been well known for many years. The first reports of the complexation of saccharides with boric and phenylboronic acids appeared about 50 years ago.[1–3] These pioneering works attracted considerable attention and led to numerous fundamental and applied studies over the following decades. Affinity chromatography of sugars, nucleotides, and glycoproteins on boronate-containing supports was reviewed by Bergold and Scouten.[4] Boronate–sugar interactions were widely employed in detection of saccharides, and this field was recently reviewed by James and Shinkai.[5] Miyata et al.[6] reviewed synthesis and applications of sugar-responsive gels, including those containing chemically attached phenylboronic acids.

Among the other boronate-containing materials, water-soluble synthetic polymers with multiple phenylboronate functions (boronate-containing copolymers, BCCs) are of special interest. Synthetic soluble polymers with reactive or bioaffinity functions have long been of interest in the fields of biotechnology and medicine as carriers for conjugation of enzymes,[7,8] antibodies,[9] carbohydrates,[10] reagents for affinity precipitation of proteins,[11] and displacers for biochromatography.[12] An inherent property of the reactive polymers is their multipoint (or multivalent) interaction with biological counterparts, which results in much stronger specific binding compared with low-molecular-weight reagents with the same functional groups.[10,13] Nature uses the multivalent type of interaction, since at the cell surface, cell–cell interactions and signal transduction fundamentally depend on the formation of multiple receptor–ligand complexes. Polymer synthesis provides access to multivalent biomimicking reagents that can be used to probe or alter cellular functions by mechanisms inaccessible to natural materials.[14] Given their inherent sugar specificity, one can anticipate BCCs as promising mucoadhesives or reversible modulators of cell glycocalix, e.g., agents capable of preventing postoperative adhesions in biological tissues.[15]

Indeed, the BCCs are capable of binding not only to sugars,[16] but also to living animal cells,[17,18] the cell viability not being suppressed or affected by binding the copolymers. Due to the large number of pendant groups in the chains, the BCCs can interact with polysaccharides in a multivalent way, thus enhancing the strength of biomimetic binding.[19] The soluble copolymers, as well as the boronate-containing gels (BCGs), give a response to small changes in their aqueous environment: saccharide or saccharide-containing solute concentration, temperature, or pH. Therefore, these materials exhibit responsive or smart polymeric properties applicable in various fields of biomedical science. This chapter summarizes the responsive properties of BCGs as well as the interaction of BCCs with insoluble carbohydrates, cells, and tissues.

10.2 Boronate-Containing Copolymers as Polyelectrolytes

The BCCs are polyelectrolytes, the pendant groups of phenylboronic acid (PBA) being present in both an uncharged trigonal form and a charged boronate form (Figure 10.1) in a wide range of pH (from 6.5 to 10 for the copolymers of *N*-acryloyl-*m*-aminophenylboronic acid [NAAPBA], according to Shiomori et al.[20]). Ionization constant pK_a was reported to be 8.6 for

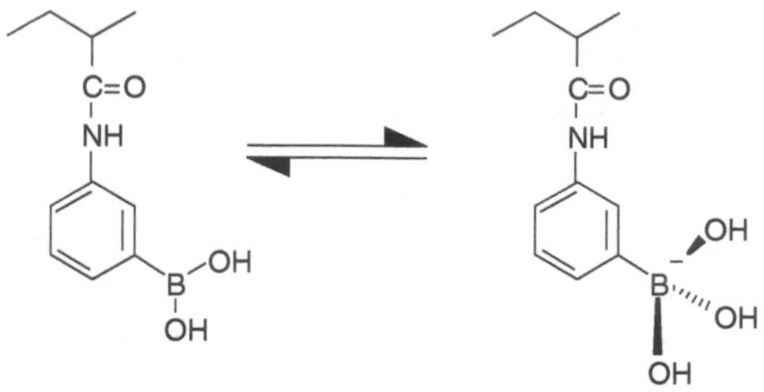

FIGURE 10.1
Equilibrium of phenylboronic acid group between an uncharged trigonal form and a charged boronate anion.

the low-molecular-weight analog of the monomer unit *N*-propionyl-*m*-aminophenylboronic acid (NPAPBA).[16] It is generally accepted that the sugar-reactive form of boric acid as well as pendant phenylboronates are anionic.[4,21] To provide the BCCs with enhanced sugar-specific reactivity at physiological pH, attempts have been made to synthesize polymerizable PBA derivatives with a lower ionization constant (pK_a).[22]

It was recently found, however, that the affinity of ring-substituted PBAs to sugars does not always increase with pH. Moreover, the pH optimum of the complex formation can be lower than the pK_a of the PBA. For example, the binding association constant of 4-bromophenylboronic acid ($pK_a = 8.3$) with glucose at pH 7.5 has a higher value compared with pH 6.5 or pH 8.5. Such results indicate that the pK_a of the boronic acid is not the only determinant of its binding to saccharides. Steric arrangement of the interacting counterparts and buffer composition are also factors affecting the complex formation.[23]

It is relevant to note that the BCC containing NAAPBA displayed a well-expressed cross-linking reactivity toward poly(vinyl alcohol) at pH 7.5, i.e., in the conditions where most of the boronate ligands were in a neutral form,[24,25] as seen in the titration curves presented in Figure 10.2. A similar reaction of galactomannan with the copolymer of acrylamide (AAm) and NAAPBA (96:4) was earlier reported to proceed at pH 7.4.[17] *N*-acetyl-neuraminic acid (Neu5Ac), an end group of many oligosaccharides, was recently reported to react with the *neutral form* of NPAPBA due to the intermolecular interaction of the boron atom with the *N*-acetyl group of Neu5Ac.[26] Therefore the sugar-specific binding of NAAPBA copolymers to the cell oligosaccharides at physiological pH seems quite possible. Indeed, a coagulation of lymphocyte suspension was induced by the NAAPBA–AAm copolymer at pH 7.4, as reported by Miyazaki et al.[17]

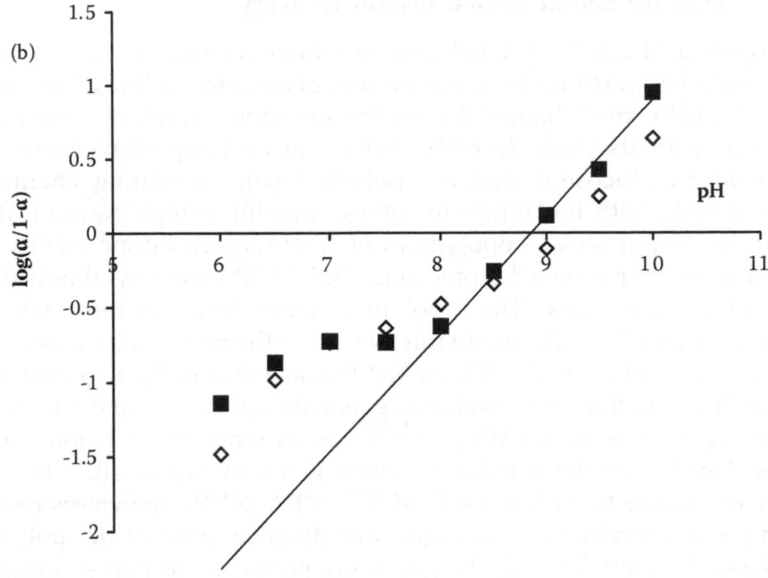

FIGURE 10.2
Titration curves of DMAA–NAAPBA copolymer in 0.15*M* sodium chloride (■) and in the absence of salt (◇). The graph charts $\log(\alpha/1-\alpha)$ as a function of pH, where α is the degree of ionization of phenylboronic acid. The linear regression is shown for the copolymer titration in the presence of sodium chloride. (From Kuzimenkova, M.V. et al., *Macromol. Biosci.*, 6, 170, 2006. With permission.)

10.3 Boronate-Containing Copolymer Gels as Components of Biosensors and Drug-Delivery Systems

Due to specific and reversible interaction with diols, polyols, sugars, and glycoproteins, phenylboronic acids (PBAs) are often employed as components of biocompatible, reversibly gelating systems. The PBA-containing polymers were shown to form gels either due to their cross-linking with a bifunctional vinyl monomer or via formation of complexes with long-chain polyols, such as polyvinyl alcohol. Both types of gels were studied as components of sensing devices for analysis of sugars, nucleotides, and nucleic acids as well as responsive, swelling matrices for self-regulated drug-delivery. In this section we consider these two types of materials in combination because of the similarity of their structures, specific reactions, and swelling/shrinking behavior resulting from changes in the environment.

10.3.1 BCG for Self-Regulated Insulin Delivery

Development of a self-regulated, biocompatible insulin delivery system is a challenging opportunity for the treatment of diabetes mellitus. The current practical treatment is limited to insulin injection, which is a significant burden on patients. Since the early 1990s, many attempts have been made to develop a glucose-responsive polymer gel, containing chemically attached PBA, with the ability to release insulin at high concentrations of glucose. In particular, copolymers of *N*-vinylpyrrolidone (NVP) with *N*-acryloyl-*m*-aminophenylboronic acid (NAAPBA) were synthesized and studied for this purpose. The copolymers were shown to form gels with poly(vinyl alcohol) (PVA) due to interaction of the pendant boronates with PVA diols in alkaline media (50m*M* NaOH), as reported by Kitano et al.[27,28] It is worth noting that PVA also forms gels with borate ions and with water-soluble copolymers of NAAPBA at pH 7.5, as reported by Ivanov et al.[25] and as depicted in the reaction schemes given in Figure 10.3. The high affinity of glucose to the boronates of NAAPBA–NVP copolymers resulted in competitive binding of the sugar and disintegration of the polymeric complexes. In a similar way, the gels were supposed to release insulin in response to increasing glucose concentration in human blood,[28] i.e., to behave as smart biomaterials. Further research has been carried out with copolymers of *N,N*-dimethylacrylamide (DMAA) with NAAPBA,[24] exhibiting good solubility in neutral and weakly acidic aqueous media, which is more acceptable for biomedical applications. In particular, tertiary copolymers of DMAA, NAAPBA, and *N,N*-dimethylaminopropyl acrylamide were shown to react with PVA four- to fivefold faster in the pH range from 5.5 to 8 compared with the binary NAAPBA–DMAA copolymers. The observed behavior was ascribed to the stabilization of the PBA–polyol complex by dimethylamino groups.

To achieve reversible and large (eight- to tenfold) variations in the gel volume needed for effective drug delivery *in vivo*, the boronate-containing gels were synthesized using *N*-isopropylacrylamide (NIPAM), which offers thermoresponsive properties to the BCCs[29] and their gels.[30] A glucose-responsive gel able to release insulin under weakly alkaline conditions (pH 9.0, 0.12*M* NaCl) at 28°C was prepared by free-radical polymerization of NIPAM with NAAPBA (10 mol%) in the presence of *N,N*′-methylene-bis-acrylamide. The gel underwent large swelling changes in response to temperature because of the thermal sensitivity of the NAAPBA–NIPAM copolymer,[30] as shown in Figure 10.4. The critical swelling temperature of the gel was dependent on the glucose concentration in contacting solution and varied from 22 to 36°C for glucose concentration changes from 0 to 5 g/l. As a consequence, the gel exhibited large volume changes caused by the changing concentration of glucose in the temperature range from 25 to 30°C. The volume changes were accompanied by the on–off regulation of insulin release well correlated with the oscillating glucose concentration in the contacting solution.[30]

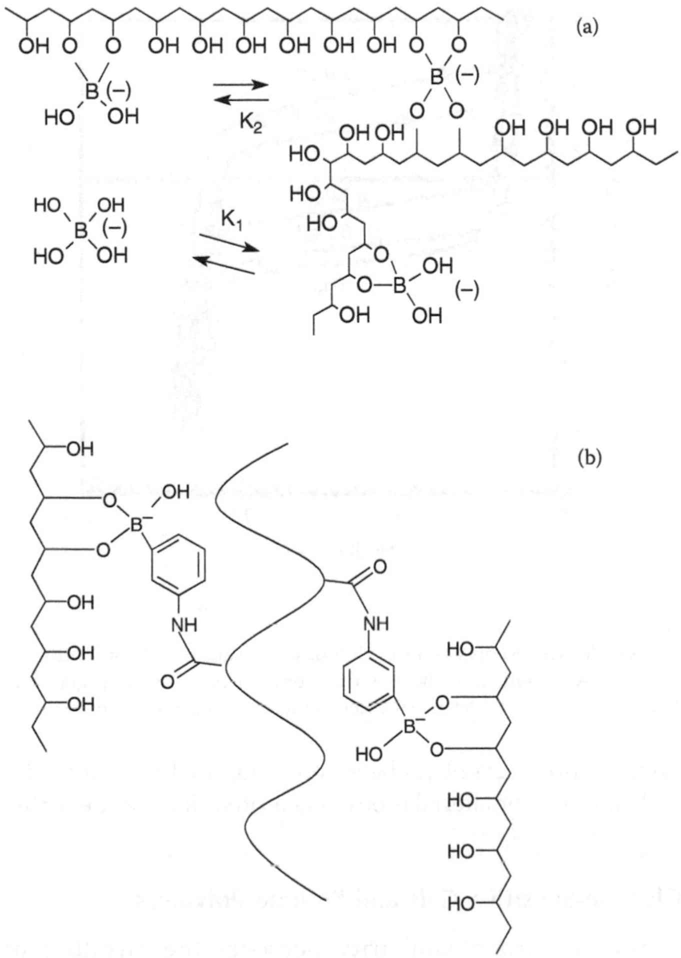

FIGURE 10.3
Scheme of PVA cross-linking with borate ions through (a) didiol complex formation and (b) with DMAA–NAAPBA copolymer through monodiol complex formation.

Further attempts to develop a BCC responsive to glucose under physiological conditions were recently made using copolymerization of *N*-isopropylmethacrylamide (NIPMM) and a polymerizable derivative of 4-carboxyphenylboronic acid (DDOPBA) having a lower pK_a (7.8) compared with that of NAAPBA (8.6). The phase transition temperature of the DDOPBA–NIPMM copolymer (10 mol% DDOPBA) turned out to be 27°C at pH 7.4 and, therefore, ≈15°C higher than that of the NAAPBA–NIPAM copolymer. This finding opened the prospect of further improvement of glucose-responsive boronate-containing gels operating close to body temperature.[22] Although preparation of a glucose-responsive gel enabling controlled release of insulin at physiological temperature, pH, ionic strength, and adequate

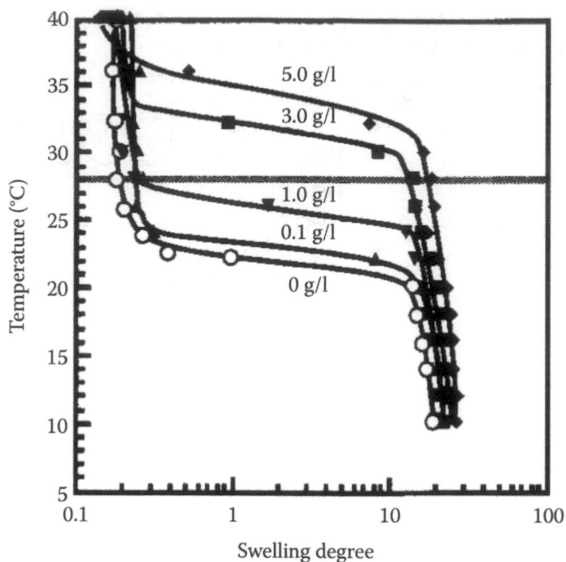

FIGURE 10.4
Swelling of NAAPBA–NIPAM gels as a function of temperature and concentration of glucose. The swelling degree was defined as the water content per mass of dried gel. (From Kataoka, K. et al., *J. Am. Chem. Soc.*, 120, 12694, 1998. Reproduced with permission.)

glucose concentration has not yet been reported, the huge biomedical importance of the desirable biomaterial is obvious motivation for the further studies.

10.3.2 Glucose-Sensitive Gels and Soluble Polymers

One can notice a certain similarity between the swelling pattern of NAAPBA–NIPAM gels[30] and the thermoprecipitation pattern of soluble NAAPBA–NIPAM copolymers,[31] observed at varied glucose concentrations (Figure 10.4 and Figure 10.5). Indeed, in the range of glucose concentration up to 3 g/l (16m*M*), the critical swelling temperature of the gel (10 mol% NAAPBA) varied in the range of ≈10°C as well as the critical flocculation temperature (CFT) of the copolymer (18 mol% NAAPBA) did. Being relatively small at 0.56m*M* glucose concentration, the shift of both critical temperatures became large in 16m*M* glucose. This similarity very probably reflects the processes of dehydration and aggregation of the copolymer chains taking place upon heating, while the hydrophilic molecules of glucose bound to the chains prevent the dehydration.

It was recently shown that the shift of the phase transition temperature (T_P) of the NAAPBA–NIPAM copolymer increases with the increasing molar portion of boronates associated with ribose moieties of adenosine.[32] The effect of glucose on the critical phenomena in the boronate-containing copolymers and gels seems to follow a similar pattern. A measurable T_P of the

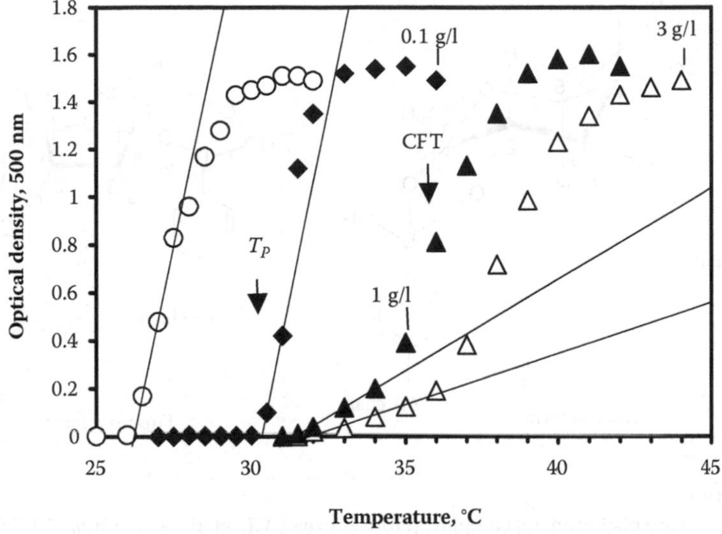

FIGURE 10.5
Temperature dependence of optical density of NAAPBA–NIPAM copolymer (0.4 mg/ml) in the absence (○) and presence of glucose at different concentrations of the sugar. Buffer: 0.1*M* glycine–NaOH, containing 0.1*M* NaCl; pH 9.2. (Redrawn from Ivanov, A.E. et al., *Biomacromolecules*, 7, 1017, 2006. With permission.)

copolymer could be registered, however, at glucose concentrations as low as 5.6μ*M*, probably due to a higher motional freedom of the free polymer chains, thus allowing for bidentate chelation of glucose by the pendant boronate groups.[20] The coordination of particular diols of glucose with the pendant boronate groups was proposed by Alexeev et al.[33] as illustrated in Figure 10.6. Formation of bidentate boronate complexes of glucose was proven by ^{1}H NMR and ^{13}C NMR spectroscopy and reported by Norrild and Eggert.[34] The high stability of bidentate glucose–PBA complexes results in efficient sugar binding at low concentrations and, therefore, in higher sensitivity of the soluble copolymer compared with the gel.

It is relevant to note that other hexoses, such as D-galactose or D-mannose, do not form bidentate complexes with boronates.[35] Their effects on thermal precipitation of the NAAPBA–NIPAM copolymer are weaker; to achieve a $T_P = 4.2$°C produced by 0.56m*M* glucose, it requires more than 25-fold higher concentrations of the above sugars.[31]

10.3.3 BCG Fixed on Solid Surfaces and Studied by Physical Methods

10.3.3.1 BCG as Sugar Sensors Based on Vinyl Polymers

Conventionally, glucose sensors utilize the enzyme glucose oxidase; however, these sensors often display poor stability as a result of long-term contact

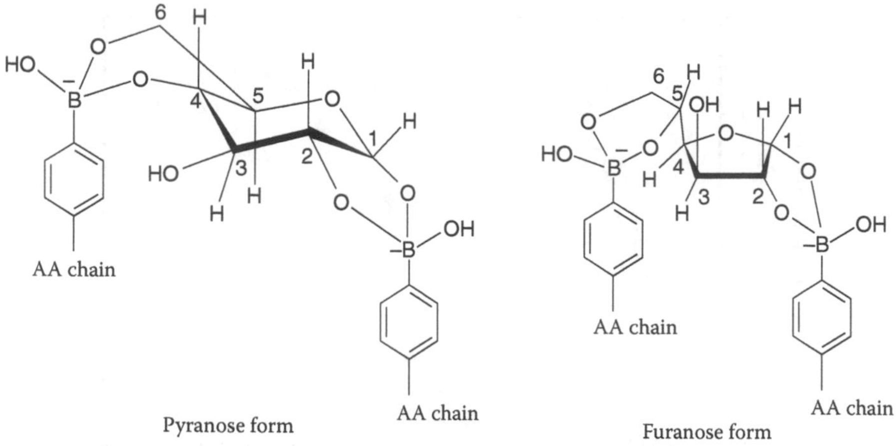

FIGURE 10.6
Proposed glucose bis(boronate) complex. (From Alexeev, V.L. et al., *Anal. Chem.*, 75, 2316, 2003. Reproduced with permission.)

with process media and problems associated with sterilization. Therefore, there is a need for stable and inexpensive sensors for use at physiological pH values in bioprocess monitoring.[36] As the most effective synthetic ligands for binding glucose in aqueous media are derived from boronic acids,[37] many attempts have been made to construct the sensors based on glucose-responsive BCG.

Swelling changes in the BCC–PVA gels that were observed upon their contact with glucose could be registered by the changes in electric current through a platinum electrode coated with the gel.[38] The swelling led to increased mobility of ion species toward the surface of the electrode. Thus, the increased current was observed upon the addition of glucose to the analyzed solution. The changes of the current were proportional to glucose concentration up to 3 g/l (16.7m*M*), the lowest estimated glucose concentration being 1.67m*M*. It is worth noting that the effect of glucose was much stronger than that of α-methyl-D-glucoside, which is known to form no complexes with phenylboronic acids[39] and very weak complexes with borate.[40] These results thus demonstrate the sugar selectivity of BCC–PVA-coated electrodes.

The boronate-containing gels (BCG) intended for glucose sensing have also been prepared by electropolymerization of AAm and NAAPBA in the presence of *N,N′*-methylene-bis-acrylamide on gold surfaces.[41] The swelling and shrinking of the gel were registered by Faradaic impedance spectroscopy, chronopotentiometry, surface plasmon resonance (SPR), and microgravimetric quartz-crystal microbalance measurements. The SPR spectra of the swollen and shrunken gels corresponded to polymer film thicknesses of 320 nm and 210 nm, respectively. The molar monomer composition of the NAAPBA–AAm gel was 9:2. All of the physical methods mentioned above have confirmed swelling of the gels in the presence of 25m*M* glucose and their shrinking upon depletion of glucose from 50m*M* HEPES buffer solution (pH 7.4). The swelling of a 210-nm-thick polymer gel took 4 to 6 min in the presence of 25m*M* glucose and was completely reversible upon depletion of glucose.

A similar type of BCG was prepared by means of polymerization carried out in the presence of nucleotides and monosaccharides, which allowed imprinting of recognition sites in the NAAPBA–AAm copolymer membranes associated with electronic transducers.[42] The gels prepared in the presence of adenosine-5′-monophosphate were able to respond on the nucleotide at its concentrations of 10^{-6} to $10^{-5}M$, whereas similar compounds such as guanosine- or cytosine-5′-monophosphates could not be detected at concentrations lower than $10^{-4}M$. The enhanced sensitivity and selectivity of the nucleotide binding to the gel was attributed to the cooperative binding interactions between the imprinted substrate and the copolymer consisting of the boronic acid ligand (the ligating site for ribose moiety) and the acrylamide units forming complementary H-bonds with other polar groups of the nucleotide.

Asher et al. have developed an ingenious physical method for quantitative measurements of gel swelling and shrinking.[43] The authors incorporated a crystalline colloid array (CCA) into a polyacrylamide hydrogel with pendant boronic acid groups (PCCA). The embedded CCA diffracted visible light and the diffraction wavelength reported on hydrogel volume. This photonic crystal-sensing material responded to glucose in low-ionic-strength aqueous solutions by swelling and red-shifting its diffraction as the glucose concentration increased. The sensing material responded to glucose and other sugars in the concentration range from 1 to 100m*M* in 2m*M* Tris-buffer, pH 8.5, although the detection limit for glucose was ≈50μ*M* in distilled water.

The response of the sensor to glucose decreased upon addition of NaCl and ceased at NaCl concentration higher than 10m*M*, presumably due to diminished osmotic pressure changes originating from the formation of charged boronate–sugar complexes at high ionic strengths and, therefore, from diminished gel-volume changes. In an effort to adapt the CCA sensor to the higher ionic strengths typical of body fluids, the same research group developed a gel containing copolymerized poly(ethylene glycol) monomethacrylate (400 Da) capable of binding sodium ions.[33] The new gel could operate in 2m*M* Tris-buffer containing 0.15*M* NaCl (pH 8.5) and gave opposite responses to glucose (shrinking) and other hexoses: mannose, galactose, and fructose (swelling), as shown in Figure 10.7. The authors concluded that glucose cross-linked the gel due to the sugar's tendency to form bidentate complexes with PBA, as seen in Figure 10.6.

Given the composition of PCCA gels, their reactivity was displayed at higher pH than similar gels described by Gabai et al.[41] It is worth noting that PCCA gels were prepared by alkaline hydrolysis of polyacrylamide gels, followed by covalent attachment of 3-amino-PBA in the presence of water-soluble carbodiimide. Numerous carboxylate functions formed at the hydrolysis

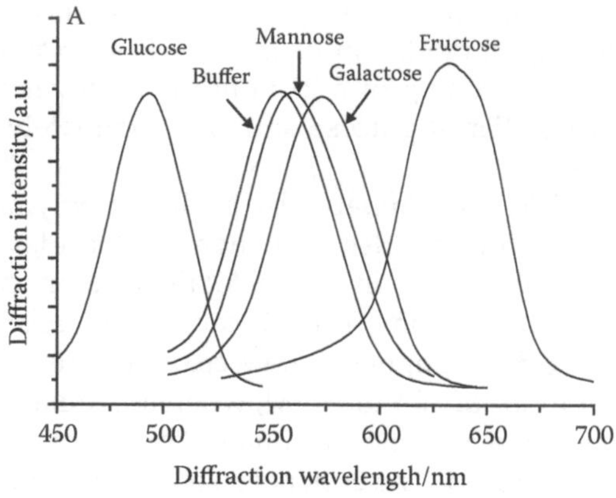

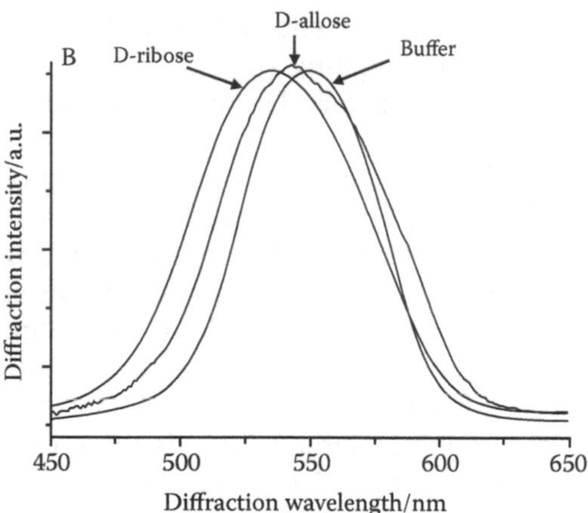

FIGURE 10.7
Response of NAAPBA–AAm–PEG sensor to (a) 4m*M* concentrations of D-glucose, D-mannose, D-galactose, and D-fructose and (b) 4m*M* concentrations of D-allose and D-ribose in 2m*M* Tris-HCl, pH 8.5, 150m*M* NaCl solution. (From Alexeev, V.L. et al., *Anal. Chem.*, 75, 2316, 2003. Reproduced with permission.)

stage might hinder ionization of the pendant PBA and, therefore, its reactivity toward sugars. The NAAPBA–AAm gels prepared by direct radical copolymerization of the monomers exhibited good reactivity toward glucose, even at pH = 7.4,[41] i.e., at a much lower value than the pK_a of the pendant phenylboronates. Similar results were recently obtained in our laboratory: it was found that physically cross-linked turbid NAAPBA–AAm gels (10 mol% NAAPBA) exhibited an optical response to glucose in 50m*M* sodium phosphate buffer in the pH range from 6.5 to 7.5. Unlike the swelling of PCCA gels or the electron-transfer resistance of the BCG fixed on gold supports,[41] the optical response of the turbid NAAPBA–AAm gel was strictly proportional to glucose concentration in the range from 1 to 50m*M*, as seen in Figure 10.8. As follows from Figure 10.8, the optical-density changes were

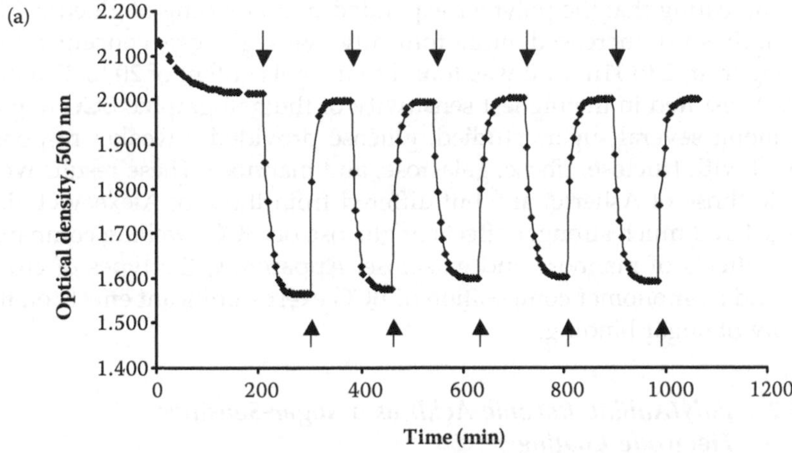

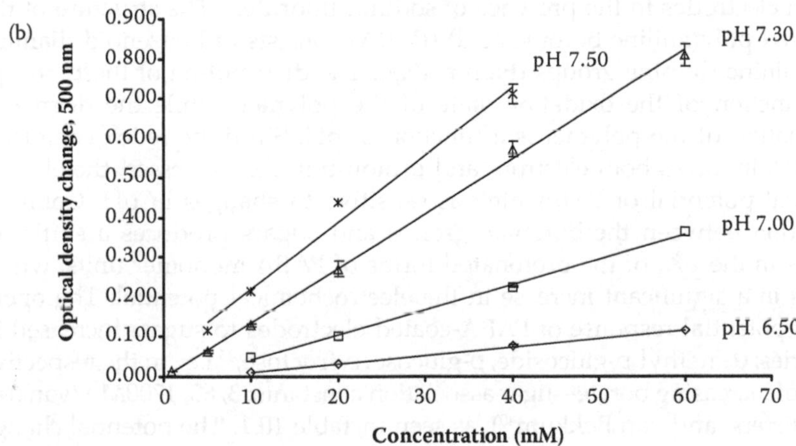

FIGURE 10.8
Response of NAAPBA–AAm gels to glucose. (a) Repeatability of the gel optical-density changes caused by 30m*M* glucose in 50m*M* sodium phosphate buffer, pH 7.3 (downward arrows), and by glucose depletion (upward arrows) by soaking the gel in the buffer. (b) pH dependence of the gel sensitivity to glucose.

completely reversible and repeatable, i.e., the starting value of the gel optical density was restored each time upon depletion of glucose.[44]

Glucose-sensitive holographic sensors based on NAAPBA–AAm gels were recently developed by Lee et al.[36] The gels containing a reflection hologram were prepared by laser irradiation of gel films impregnated by microparticles of silver bromide, followed by a photographic development of the films. The response of the hologram to glucose was studied in phosphate buffer solution at pH 7.4. The diffraction wavelength of the hologram shifted to higher values, indicating that the polymer expanded upon binding of glucose. The wavelength shift increased monotonically with glucose concentration, increasing from 2 to 11m*M*. It was found that a gel content of 20 to 25 mol% NAAPBA resulted in the highest sensitivity of the holographic BCG to glucose. Among several sugars studied, glucose provided a weaker response compared with fructose, ribose, galactose, and mannose. These results were similar to those of Asher et al.[43] but differed from those of Alexeev et al.,[33] who registered much stronger effects of glucose on BCC swelling compared with the effects of mannose and galactose. Apparently, the types of cross-linking and co-monomer composition of BCG exert significant effects on the selectivity of sugar binding.

10.3.3.2 Poly(Aniline Boronic Acid) as a Sugar-Sensitive Electrode Coating

Sugar-sensitive boronate-containing polymer films can be prepared by electropolymerization of 3-aminophenylboronic acid on the surface of glassy carbon electrodes in the presence of sodium fluoride.[45] The structure of the prepared poly(aniline boronic acid) (PABA) consists of benzenoid diamine and quinine diimine groups (Figure 10.9). The distribution of these groups is a function of the oxidation state of the polymer, while the degree of protonation of the polymer is a function of pH. Since the redox chemistry of PABA involves both electron- and proton-transfer processes, the electrochemical potential of PABA films is sensitive to changes in pH. Complex formation between the boronate groups and sugars produces a sufficient change in the pK_a of the protonated forms of PABA monomer units, which results in a significant increase in the electrochemical potential. The open-circuit potential response of PABA-coated electrodes to sugars increased in the series: α-methyl-D-glucoside, D-glucose, D-fructose,[45] i.e., in the respective order of increasing borate–sugar association constants: 3, 65, 1700M^{-1} (van den Berg, Peters, and van Bekkum[40]), as seen in Table 10.1. The potential change increased monotonically with glucose or fructose concentration in the concentration range from ≈3 to 40m*M*.

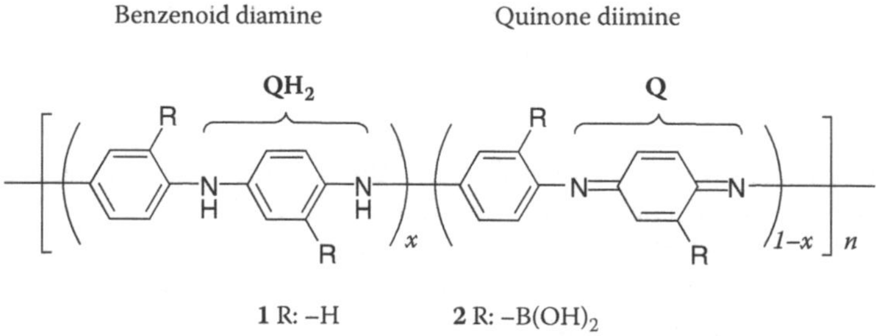

Benzenoid diamine — Quinone diimine

QH_2 — Q

1 R: –H **2** R: $-B(OH)_2$

FIGURE 10.9
Structure of poly(aniline boronic acid).

TABLE 10.1
Increase in Thermoprecipitation Temperature of NAAPBA–NIPAM Copolymer in the Presence of 0.56m*M* Sugars and Polyols (ΔT_P) and the Association Constants of Borate and Phenylboronate with Sugars and Polyols (Log K_{ass})

		Log K_{ass}	
Compounds	ΔT_P	**Borate** [a]	**Phenylboronate** [b]
D-fructose	4.5	2.82	3.6
Lactulose	3.6	2.91	…
Mannitol	4.0	3.3 [c]	3.4
D-glucose	4.2	1.80	2.0
L-arabinose	0.4	2.14	2.6
D-xylose	0.4	2.2 [d]	…
D-galactose	0.4	1.99	2.4
D-mannose	0.3	2.01	2.2
L-fucose	0.3	…	…
N-acetylneuraminic acid	2.0	…	1.3–1.5 [e]
N-acetylgalactosamine	<0.2	…	…
N-acetylglucosamine	<0.2	…	…
Sucrose	<0.2	0.86	…
Raffinose	<0.2	1.35	…
Glycerol	<0.2	1.2 [c]	1.3

[a] From Verchere and Hlaibi.[51]
[b] From Lorand and Edwards.[3]
[c] According to Lorand and Edwards.[3]
[d] According to van den Berg, Peters, and van Bekkum.[40]
[e] Calculated from data of Otsuka et al.[26]

Unlike swelling of NAAPBA–AA gels,[41] the change of PABA electropotential induced by complex formation with sugars could not be described as a first-order kinetic process. This behavior is likely due to complications associated with diffusion coupled with complexation as well as structural changes in the polymer.[45] The effective diffusion coefficient of glucose in the PABA film (0.5 μm in thickness) was estimated as 2×10^{-12} cm^2/sec at 25°C. This value is about two orders of magnitude less than that reported for methanol diffusion through polyaniline (Ball et al.[46]) and about eight orders of magnitude less than the diffusion coefficient of glucose in water at 25°C (6.7×10^{-4} cm^2/sec, as given in *CRC Handbook of Chemistry and Physics*[47]). Apparently, the electropolymerized PABA film had a very high density and low porosity.

Ma and Yang reported on saccharide sensors based on PABA-coated gold electrodes.[48] Detection of glucose, fructose, sorbitol, and mannitol was performed using impedance spectroscopy. The increase of the electron/ion-transfer resistance was proportional to the logarithm of the sugar/polyol concentration in a wide concentration range from 10^{-10} to $10^{-2}M$, allowing for detection of very low sugar concentrations. Interestingly, the effect of glucose on the electron/ion-transfer resistance of the PABA films was opposite to that observed with NAAPBA–AAm films by Gabai et al.[41] In the latter case, the resistance decreased with glucose concentration, probably due to the film swelling at high sugar concentrations (5 to 100m*M*). In contrast, swelling of PABA film brought in contact with 11μ*M* glucose was found to be very low, as detected by surface plasmon resonance (SPR),[49] so that the ion conductivity of the film could not increase due to enlargement of pores. The resistance of PABA films increased with increasing glucose concentrations, most probably due to steric effects produced by nonconductive saccharide blocking the pathways of electrons and ions.[48] On the other hand, the SPR responses to glucose and other sugars were easily detectable with the gold electrode coated with poly(4-vinylphenylboronic acid), polyVPBA, at pH > 9. The responses increased in the series glucose < mannose < galactose < fructose,[49] i.e., in the order of increasing PBA-sugar association constants (Table 10.1) at 5.6m*M* sugar concentration.

To conclude, boronate-containing gels and films prepared from vinyl monomers, such as NAAPBA or VPBA and their polymers, typically exhibit a swelling response to sugars and nucleotides taken at concentrations from 1 to 100m*M* under neutral conditions. This creates possibilities for the development of nonenzymatic sugar sensors applicable to monitoring of cell metabolism in microbial fermentation processes and cell cultures.[36]

10.4 Interaction of Boronate-Containing Copolymers with Polysaccharide Gels, Cells, and Tissues

PBA can bind to biological surfaces via diol groups that belong to the oligosaccharides present in glycoproteins and glycoconjugates. Such oligosaccharides are abundant both on the cell surface and in the extracellular matrix of the eukaryotes, and as such they represent a ubiquitous target for BCC binding.[50] Being capable of binding to the glycosylated proteins and lipids,[17,18] the BCCs seem to be promising reagents for biomedical manipulations with living cells or tissue surfaces. Given their ability to bind sugars, the BCCs can exhibit, to a certain extent, lectinlike functions.[18] They spontaneously assemble on red blood cells and sterically protect them from agglutination by lectins and by antibodies to blood groups, i.e., they modify the immunochemical properties of cells.[50] Furthermore, the BCCs readily adsorb to the surfaces of body tissues *in situ* and provide a steric barrier between such tissues, thus minimizing the tissue adhesions that typically form following surgical procedures.[15] Although the prospects for biomedical applications of BCCs are impressively wide, there are relatively few reports on the copolymer's interactions with tissues and cells. This can be partially explained by the highly proprietary character of this research area, which has recently been covered by some patent applications.[15,51,52] Another reason could be the lack of fundamental knowledge about complex formation between PBA and oligosaccharides typical of the glycoproteins and glycolipids, including mucin glycoproteins abundant at mucosal surfaces as well as the surfaces of nonadherent cells. Chemical synthesis of biocompatible, water-soluble BCCs is another problem of high importance. In Section 10.4.2 we summarize examples of biomedical applications of BCCs available from the literature and discuss them in combination with recent findings made in our laboratory.

10.4.1 Interaction of BCC with Oligosaccharides and Polysaccharide Gels

Based upon the series of sugar reactivity toward borate and phenylboronate ions (Table 10.1), it seems reasonable to predict the reactivity of BCCs toward oligosaccharides belonging to glycoproteins. One can foresee, however, a principal limitation to this approach: the stability of carbohydrate complexes with borate and phenylboronate is often determined by the presence of free glycosidic hydroxyl groups. For example, complexes of borate with maltose, which consists of two glucose moieties, are weaker than those of glucose itself, and the complexes of lactose are weaker than those of its two constituents, glucose and galactose.[53] It is apparent from ^{11}B and ^{13}C NMR studies that the furanose form of monosaccharides dominates in its borate complexes.[54] Interconversion of the pyranose and furanose forms of the sugars requires ring opening of the sugar. This will not occur if the reducing end is part of the glycosidic bond. Broadly speaking, the reactivity of glycans in glycoproteins cannot be predicted based on the reactivity of the corresponding reducing sugars.

On the other hand, there are some reports of PBA binding to aryl glycosides[55] as well as reports on the interaction of oligosaccharides with complex porphyrins bearing four covalently attached PBA groups.[56] Binding of borate to 4,6-diols of galactose moieties of galactomannan has been demonstrated by ^{11}B NMR.[57] Apparently, adsorption of BCC to polysaccharide gels exhibiting multiple sugar moieties both in the polymer chains and at nonreducing chain ends can be a tentative model for the copolymers' binding to oligo- and polysaccharides of cell membranes and mucosal surfaces.

Mannan immobilized on 4% cross-linked agarose beads was used to evaluate the relative reactivity of several BCCs prepared by simultaneous attachment of the end-group-activated poly(ethylene glycol) (PEG), mol wt = 5 kDa, and 4-formyl-PBA to polylysine (PLL), mol wt = 24 kDa, as reported by Winblade et al.[50] The prepared comblike copolymers, PLL-g-(PEG;PBA), were found to react with mannan. Mannose residues in mannans are most often connected by 1 to 2 or 1 to 6 bonds. Therefore, the 2,3- or 4,6-diols of the mannose residues should be available for interaction with PBA. Depending on the number of PEO and PBA groups in the copolymer, the chemisorbed amount increased from ≈1.5 to 5 mg/ml gel with decreasing PEO/PLL ratio for 1-mg/ml copolymer solutions in 10m*M* sodium phosphate buffer containing 0.14*M* NaCl at pH 7.4 (PBS) (Table 10.2). The high number of attached PEO chains (36 or 56 chains per PLL molecule, or 1:3

TABLE 10.2

Composition of PLL-g-(PEG;PBA) Copolymers and Their Binding to Mannan-Containing Agarose

Number of PEGs Attached per Lysine Unit	Number of PEGs Attached per PLL Backbone	Number of PBAs Attached per PLL Backbone	Copolymer Binding to Mannan-Containing Agarose (mg/ml gel)
1:21	6	41	5.3
1:9	13	65	3.7
1:6	19	99	2.4
1:3	36	100	1.4
1:2	56	123	1.4

Source: Compiled from the data of Winblade, N.D. et al., *Biomacromolecules*, 1, 523, 2000; *J. Biomed. Mater. Res.*, 59, 618, 2002.

and 1:2 PEG/lysine unit ratio, respectively) resulted in low chemisorbed amounts of the copolymers, despite the high PBA content in the same BCC. In contrast, the BCC containing only 6 PEG chains and 41 PBA groups (the PEG/lysine unit ratio of 1:21) adsorbed to the gel at markedly higher quantities under the same conditions. This was ascribed to the steric repulsion between the PEO-grafts and the gel displayed because of the high content of PEO attached to the PLL molecule.

Kuzimenkova et al.[19] have found that NAAPBA-containing copolymers of acrylamide, DMAA or NIPAM, spontaneously adsorb to 6% cross-linked agarose beads (Sepharose CL-6B) from 0.1*M* sodium phosphate of 0.1*M* sodium bicarbonate buffer, pH 7.5 to 9.2. Agarose is mainly composed of alternating D-galactose and 3,6-anhydro-L-galactose units, whereas galactose units are bound to the neighboring units via 1- and 3-hydroxyl groups. The hydroxyls in the positions 4 and 6 of the in-chain galactose units, as well as the diols of the end-groups, might be, in principle, accessible for binding to NAAPBA. The chemisorbed amount of BCC increased with pH, exhibiting unusually high values (15 mg/ml) at pH 7.5, for the copolymer of NAAPBA with acrylamide. This phenomenon was ascribed to self-association of the copolymer displayed in aqueous solutions at pH < 8. All the copolymers could be completely desorbed from the gel in the presence of fructose, a strong competitor for binding to PBA (Table 10.1 and Figure 10.10). From batch adsorption experiments carried out with NAAPBA and Sepharose CL-6B, the authors have calculated the apparent equilibrium association constant between the NAAPBA anion and the available diols of agarose gel, equal to $53 \pm 17M^{-1}$. This value is lower than that known for PBA interaction with free galactose (Table 10.1), most probably owing to unavailability of glycosidic hydroxyls in most of the sugar units in agarose. However, multivalent interaction of the pendant PBA groups with the available diols resulted in irreversible binding of the BCC to the agarose gel, in contrast to monomeric NAAPBA, which exhibited only a slight retention on the same Sepharose CL-6B column.

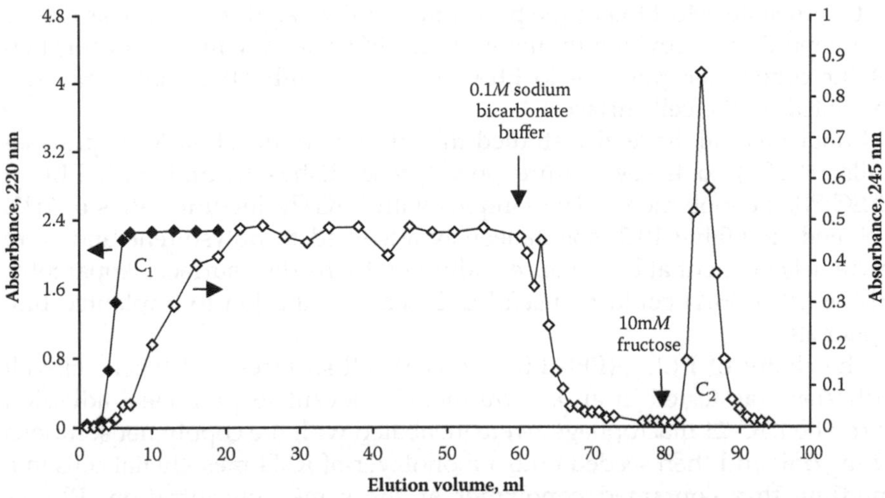

FIGURE 10.10
Frontal chromatogram of DMAA–NAAPBA copolymer (◇) and poly(*N*,*N*-dimethylacrylamide) (♦) on a Sepharose CL-6B column (1 × 2.1 cm) in 0.1*M* sodium bicarbonate buffer (pH 9.2, 22°C). The absorbance in the fractions was measured after 100-fold dilution by the buffer at 220 nm and 245 nm for poly(*N*,*N*-dimethylacrylamide) and DMAA–NAAPBA copolymer, respectively. Arrows indicate application of 10m*M* fructose in the buffer solutions. (Redrawn from Ivanov, A.E. et al., *Chem. Eur. J.*, in press, 2006. With permission.)

10.4.2 Interaction of BCC with Cells

10.4.2.1 Red Blood Cells and Epithelial Cells

Winblade et al. have shown the ability of PLL-g-(PEG;PBA) to assemble on red blood cell (RBC) surfaces and sterically protect them from agglutination by lectins (the proteins capable of multivalent recognition of sugar moieties) and by antibodies to blood groups[50] (Section 10.4.1). Wheat germ agglutinin (WGA) used in these studies is a lectin with four sites that bind *N*-acetylglucosamine residues and thus can bind to and agglutinate RBCs via the sugar residues on their surface. RBCs were incubated with the copolymer solutions (1 mg/ml) in PBS, and the maximum amount of WGA that could be added without agglutinating the cells was measured. The copolymer samples strongly interacting with mannan-containing gels (the PEO/lysine unit ratio of 1:21 and 1:9, see Table 10.2) were also able to completely prevent RBC agglutination at the highest tested concentration of WGA (0.125 mg/ml), unlike the other PLL-g-(PEG;PBA) samples. No RBC hemolysis, morphology changes, or rouleaux formation was noted in any of the agglutination experiments. The PLL-g-(PEG;PBA) copolymers alone never caused RBC agglutination in these experiments.

N-acetylglucosamine does not contain *cis*-diols and thus exhibits very weak interaction with PBA-containing polymers.[31] Therefore, strong competitive binding of the pendant PBA to the lectin receptors does not seem to be probable. Most likely, the prevention of RBC agglutination was due to steric repulsion provided by the grafted PEO chains, while chemisorption of the copolymer was due to PBA interaction with other sugar moieties exhibited on the cell surface.

Winblade et al. have also studied the attachment of rabbit lens epithelial cells (rLECs) to tissue culture polystyrene dishes treated with PLL-g-(PEG;PBA) copolymers.[58] The samples with PEG/lysine unit ratios of 1:21, 1:9, and 1:6 (Table 10.2) spontaneously adsorbed to polystyrene and were extremely efficient at blocking cell adhesion to the dish surface. Apparently, PLL-g-(PEG;PBA) could be useful in blocking adhesion to implanted biomaterials.

The ability of PLL-g-(PEG;PBA) to coat cell surfaces and block cell–cell adhesion was tested in an *in vitro* model relevant to peritoneal adhesion formation. IC-21 macrophages were incubated with the copolymer solutions (2 mg/ml) and then seeded onto a monolayer of RM4 mesothelial cells in a medium that contained copolymer at the same concentration. PLL-g-(PEG;PBA) of 1:21 and 1:9 PEG/lysine unit ratios reduced the number of adherent IC-21 cells by more than 90%. The RM4 cells remained confluent throughout the assay and showed no signs of toxic damage.

The cytotoxicity of PLL-g-(PEG;PBA) was studied with rLECs at different contact times and polymer concentrations. When applied to rLEC monolayers, PLL-g-(PEG;PBA) copolymers of 1:9 PEG/lysine unit ratio had very low toxicity, but those of 1:21 PEG/lysine unit ratio had apparent toxicity at high concentrations (>1 mg/ml) and long exposure times (4 h). In the latter case, the cells were observed to round up from the monolayer and become phase-bright in a concentration-dependent manner. When the same assay was conducted with the corresponding PLL-g-PEG polymers containing no PBA, a different kind of time- and dose-dependent toxicity was found. The cells that were killed by PLL-g-PEG became phase-dark and grainy and remained adherent to the cell-culture substrate, exhibiting a classic toxicity response clearly different from that observed for the PBA-containing copolymers.[58] This highly unusual response of the cells on the BCC treatment might be due to mitosis, which NAAPBA–AA copolymers are known to induce.[17]

10.4.2.2 Proliferating Response of Lymphocytes on BCC

Lectins are known to affect cell mitosis. They can either stimulate or inhibit proliferation of cancer cells as well as normal human epithelium cells.[59] Concanavalin A, which is a typical lectin-inducing lymphocyte proliferation, consists of four subunits, each bearing a sugar-binding site. The lectin binding to the cells results in cross-linking of glycoproteins on the plasma membrane surface of lymphocytes and induces changes in the intracellular metabolism, including inositol triphosphate metabolism and an increase in the cytoplasmatic Ca^{2+} concentration, followed by the sequential synthesis of RNA and DNA.[17] Usually, the stimulating and inhibiting effects of lectins on cell proliferation are registered by incorporation of [^{3}H]-thymidine into the cells.[59]

A dose-dependent increase in [^{3}H]-thymidine incorporation to mouse-spleen lymphocytes was achieved in the presence of NAAPBA–AAm copolymer (4 mol% AAm, mol wt = 2×10^5 g/mol) and provided evidence for the proliferative response of lymphocytes.[17] The same research group later studied the lectinlike interaction of NAAPBA–DMAA copolymers with lymphocytes in greater detail.[18] First, the sugar-specific pattern of the BCC binding to cells was confirmed by experiments carried out in the presence of sorbitol, a strong competitor for binding to PBA. The incorporation of [^{3}H]-thymidine was significantly reduced in the presence of the competitor. Second, the NAAPBA–DMAA copolymers themselves competed to bind to lymphocytes with a natural *Limax flavus* lectin having specificity to *N*-acetylneuraminic acid (Neu5Ac) moieties of carbohydrates. Interestingly, the BCC had almost no effect on the cellular binding of peanut agglutinin, a lectin with specificity to Galβ1-3GalNAc groups of carbohydrates. The authors concluded that it was the Neu5Ac-specific recognition displayed by NAAPBA–DMAA copolymers that caused their inhibitory effect on *Limax flavus* lectin binding to the cells, and not a nonspecific exclusion-volume effect of the chemisorbed polymer chains.[18]

10.4.2.3 Effect of BCC on Endothelium Cell Surface Glycocalix

Binding of BCCs to the glycocalix of endothelial cells was recently studied within the cooperative efforts of the Department of Experimental Pathology and the Department of Biotechnology at Lund University (Sweden), as reported by Szul et al.[60] and Ivanov et al.[61] Leukocyte adhesion to the endothelium is essential for leukocyte extravasation into sites of inflammation. Endothelial cell-adhesion molecules relevant for leukocyte adhesion include members of the selectin family (E-selectin and P-selectin) as well as immunoglobulin molecules and members of the mucin-type cell surface molecules, such as CD34, GlyCAM-1, or MAdCAM-1. These endothelial cell-surface-adhesion molecules can be recognized by their leukocyte counterparts, L-selectins, and integrins. As a consequence, circulating leukocytes are recruited from the bloodstream and invade the inflamed tissue.[62,63]

The cell surface exhibits a carbohydrate-rich zone called the glycocalix. Besides the mucin-type molecules mentioned above, it contains other glycoproteins and proteoglycans that extend into the extracellular space. Carbohydrate chains of the endothelial glycocalix form a matrix that resists the access of leukocytes to cell-adhesion molecules. Cross-linking of glycocalix with a multivalent polymeric reagent was supposed to influence the role of glycocalix as a regulator of leukocyte adhesion.

Adhesion assays were carried out with the mlEND1 mouse microvascular endothelial cell line seeded onto Bioptech parallel flow-chamber slides and grown to confluency. Cells were activated with interleukin IL1 for 4 h for expression of CD62E (E-selectin). The chamber was mounted on an inverse microscope with controlled humidity and heating. Sodium phosphate buffer (10m*M*), containing 0.15*M* NaCl and low amounts of $CaCl_2$ (<2m*M*) and

$MgCl_2$ (<0.3mM), pH 8.2, was pumped across the activated endothelial cell monolayer previously treated by the solutions of NAAPBA–AAm (30 mg/ml or less) in the same buffer for 30 min at room temperature. The control experiments were performed with homopolymer of acrylamide, containing no boronate groups, or with the pure buffer. Freshly isolated leukocytes (U937 monocytic cells) suspended in the phosphate buffer were injected into the flow chamber. Interaction of leukocytes could be observed in the chamber via time-lapse images that were taken with a digital camera. Images were transferred to an Openlab (Improvision) Imaging system and analyzed manually for adherent leukocytes. Quantification was performed manually from screenshots taken at defined time intervals.

Figure 10.11a and Figure 10.11b illustrate the monolayer of mlEND1 cells activated with interleukin IL1 and the same cells exposed to a flow of the buffer with suspended leukocytes, respectively. IL1-activated mlEND1 bound up to 6000 leukocytes per mm^2, making the endothelial cell monolayer almost invisible. Under these conditions, leukocytes remained spherical in shape and did not spread on the endothelial cell monolayer. Figure 10.11c illustrates the monolayer of mlEND1 cells treated with a 30-mg/ml solution of NAAPBA–AAm (13 mol% NAAPBA) for about 30 min before exposure to

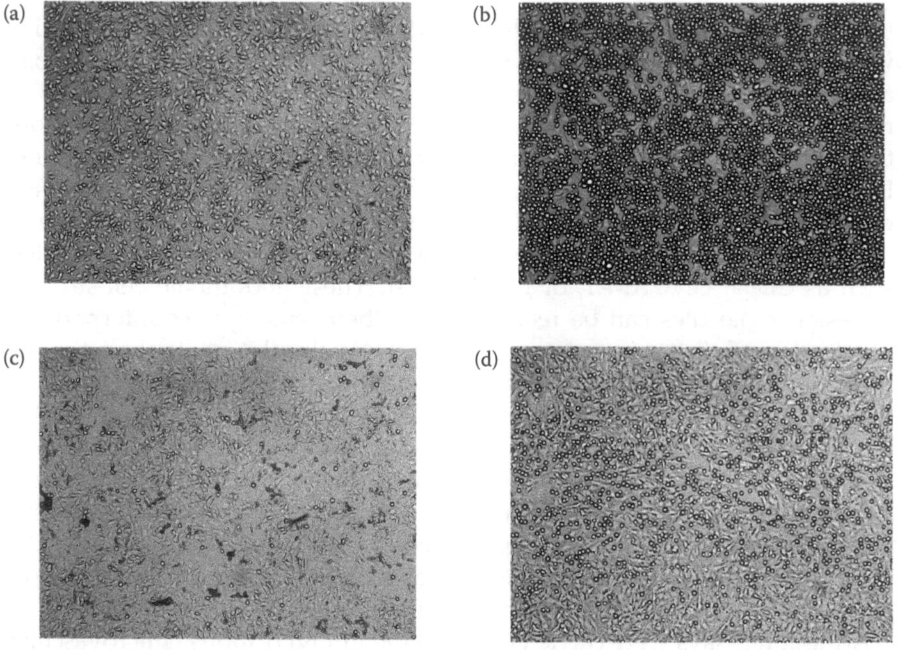

FIGURE 10.11
U937 leukocyte adhesion to mlEND1 endothelial cells: (a) monolayer of mlEND1 cells; (b) adhesion of leukocytes to the endothelial cells activated with IL-1; (c) adhesion of leukocytes to the activated endothelial cells treated with NAAPBA–AA; (d) adhesion of leukocytes to the activated endothelial cells treated with polyacrylamide.

the leukocytes. Obviously, the amount of adhered leukocytes decreased dramatically due to cross-linking of glycocalix, which restricted the accessibility of endothelial cell-adhesion molecules. On the other hand, polyacrylamide taken at a concentration of 30 mg/ml exerted a much weaker effect on leukocyte adhesion (Figure 10.11d). Interaction of the BCC with the cells was most probably due to boronate–sugar interactions between the BCC and carbohydrate-containing molecules of the cell glycocalix. Treatment of the endothelial cell monolayer with the concentrated aqueous solution of BCC had little effect on the cell viability as measured by propidium iodide cell labeling.

The reduction in quantity of adhered leukocytes depended on the copolymer concentration, as illustrated in Figure 10.12. The results obtained on the reduced cell–cell adhesion are similar to those reported by Winblade et al.[58] for the interaction between mesothelial cells and macrophages, as discussed in Section 10.4.2.1. Although the exact binding sites of the pendant boronates to the cell glycocalix are not known, it is obvious that BCC can be used for regulation of adhesion phenomena in living cells.

10.4.3 Mucoadhesive Properties of BCC and BCG

Mucosa is a layer of moist epithelial tissue lining the internal cavities in vertebrates. Almost all viral, bacterial, and parasitic agents causing infectious diseases of the intestine, respiratory, and genital tract enter or infect through mucosal membranes.[64] The mucosal membrane or the epithelial-cell-associated

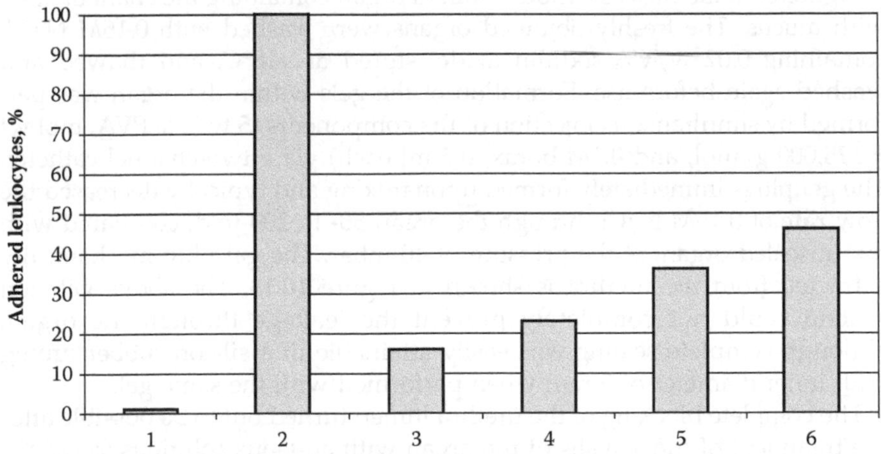

FIGURE 10.12
U937 leukocyte adhesion to monolayer of mlEND1 endothelial cells: (1) nonactivated endothelial cells; (2) endothelial cells activated with IL-1; and activated endothelial cells treated with (3) 30 mg/ml NAAPBA–AA copolymer, (4) 6 mg/ml NAAPBA–AA copolymer, (5) 3 mg/ml NAAPBA–AA copolymer, and (6) 0.6 mg/ml NAAPBA–AA copolymer in 10m*M* sodium phosphate buffer, pH 8.2.

glycocalix consists mainly of highly hydrated glycoproteins, mucins,[64] which are responsible for its rheological properties. Therefore, mucosal drug delivery and vaccination assisted by mucoadhesive polymers is a highly promising way to treat or prevent disease. Advantages are the localization at a given target site, a prolonged residence time at the site of drug absorption, and an intensified contact with the mucosa, thereby increasing the drug concentration gradient.[65]

Although some polymers exhibit considerable mucoadhesivity, they could not be regarded as a real "pharmaceutical glue." Judging by the title of one of the conceptual papers,[66] a really strong "sticky stuff" for mucosa has yet to be found. Bioadhesivity of macro dosages might be controlled by lectins, but the high cost of such bioadhesives is a limiting factor. A solution might be the BCCs exhibiting binding specificity to carbohydrates as ideal mucoadhesives. Strong interaction of BCC with mucin from porcine stomach was recently reported by Ivanov et al.[31] Apparently, such BCCs might be used to achieve formation of plugs intended to occlude channels and cavities in living mammals, the procedure being important during surgical alteration of the organs[67] or for the treatment of intracranial[68] or cerebral[69] aneurisms. Polymers such as poly(ethyl vinyl alcohol) and cellulose triacetate were used in the cited studies, although no attempts have yet been made to use BCCs for the same or similar purposes. Tight attachment of boronate-containing gels (BCG) to the inner walls of the channels and cavities might be possible, however, due to the multivalent interaction of BCCs with mucosal or endothelial surfaces.

The authors of this chapter have studied the plugging properties of the gels formed by PVA and borax ($Na_2B_4O_7{\cdot}10H_2O$) using the urethra from slaughtered male pigs as a model animal organ containing the channel lined with mucus. The freshly obtained organs were washed with 0.15*M* NaCl, containing 0.02 w/v% sodium azide, stored at –18°C, and thawed and washed again before use. Formation of the gels within the organ was performed by simultaneous injection of the components (5 w/v% PVA, mol wt = 125,000 g/mol, and 0.1*M* borax, 0.5 ml each) via a two-channel catheter. The gel plugs immediately formed upon mixing and typically decreased the flow rate of 0.15*M* NaCl through the organ 50- to 200-fold, compared with the unsealed organ, at the pressure of 40 mbar. The gel plug mechanically extracted from the urethra is shown in Figure 10.13. The above gelation system could not completely prevent the leakage through the organ, although complete sealing was easily attainable in a silicon rubber tubing with inner diameter of 5 mm when performed with the same gel.

The complete blocking of the urethra lumen turned out to be possible after the treatment of inner walls of the organ with aqueous solutions (5 w/v%, pH 8.8) of NAAPBA–DMAA or NAAPBA–AA copolymers (see Table 10.3). As shown in the table, the copolymers with a higher molar content of PBA ligands were more effective in blocking the leakage through the sealed organs. One can conceive the gel formation to follow both mechanisms shown in Figure 10.3b (near the urethra walls) and Figure 10.3a (in the gel plug), resulting in a mixed type of the gel structure. The tight attachment of

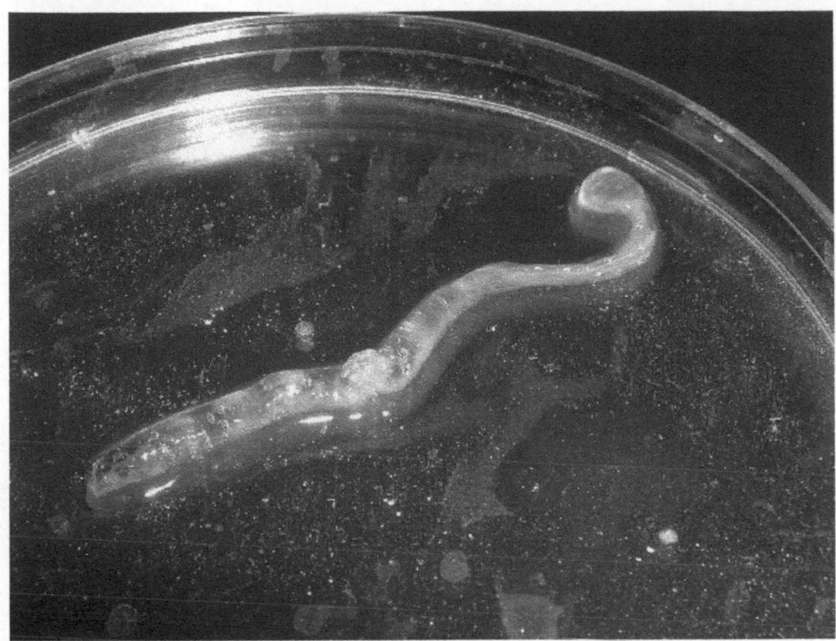

FIGURE 10.13
Cross-linked PVA–borax gel plug (1.5- to 3-mm thickness) mechanically extracted from pig urethra and placed on a 9-cm (Ø) petri dish.

the plug to the organ walls was presumably due to the multivalent interaction of the pendant boronates of BCC with mucins and other glycoproteins of mucus or epithelial glycocalix. It is relevant to note that BCC–PVA gels exhibit much higher relaxation time compared with PVA–borax gels.[25] Therefore, the shape stability of the gel formed near the walls as a result of the BCC cross-linking with PVA can be higher compared with that formed without precoating of the organ with BCC.

The plugs formed within the organs could be easily dissolved by injection of aqueous fructose (5 w/v%), followed by slight mechanical compression,

TABLE 10.3
Flow Properties of Pig Urethra Precoated with Different BCC and Plugged with a PVA-Borax Gel

BCC	NAAPBA (mol%)	Immediate Leakage (number of events)	Leakage within 10 min (number of events)	Leakage after 10 min (number of events)	No Leakage after 30 min (number of events)	Number of Trials
NAAPBA–DMAA	2.7	5	0	0	0	5
NAAPBA–DMAA	5.2	3	4	3	0	10
NAAPBA–DMAA	8.2	0	1	5	4	10
NAAPBA–AAm	12	0	1	2	9	12

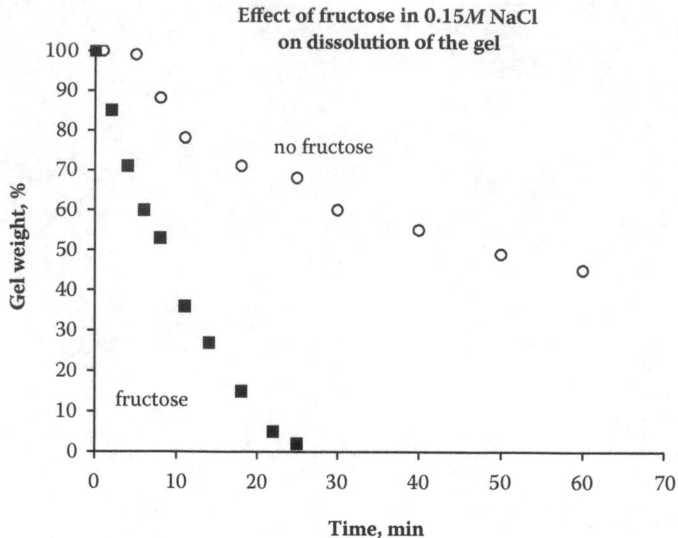

FIGURE 10.14
Dissolution of PVA–borax gel (0.7 g) in 10 ml of 0.15*M* NaCl or 1 w/v% solution of fructose in 0.15*M* NaCl. The graph depicts gel weight as a function of time.

in 3 to 5 min. PVA–borax gels could not be completely dissolved in 0.15*M* NaCl, even after 24 h at room temperature, whereas dissolution was possible in the presence of fructose, a strong competitor of PVA for binding to borate and boronate (see Figure 10.14 and Table 10.1). The association constant of PVA diols with borate was reported to be as low as $4M^{-1}$.[70] Therefore, it is possible, in principal, to form relatively stable gel plugs based on BCG adhering to mucosal epithelium. It was also possible to dissolve the plugs on demand, after the channel had been occluded for the required time.

10.5 Conclusions

Water-soluble, boronate-containing synthetic copolymers (BCCs) exhibit a strong potential for application in various fields of biotechnology and medicine. Up to now, the BCCs were mostly involved in research aimed at practical goals, such as self-regulated insulin delivery or sugar sensing. On the other hand, the knowledge of BCC binding affinity to polysaccharides and glycoproteins offers a wide range of applications in biomedical science. These include regulation of cell–cell recognition and adhesion phenomena in living organisms as well as synthesis of BCC-grafted microparticles or microdevices for controlled drug delivery. Such copolymers might also be applicable to occluding body channels and cavities in humans.

Recently, bioadhesive flat poly(methyl methacrylate) microdevices were prepared by attachment of lectins to their surfaces.[71] It looks probable that similar or better adhesion properties of microdevices can be obtained with surface-grafted BCC.[72] Therefore, comparison of the bioadhesivity displayed by lectins and BCCs attached to similar surfaces seems a reasonable endeavor. The key problem to be solved on this path is the synthesis of well-characterized biocompatible BCCs capable of binding to the glycocalix of living epithelial, endothelial, and blood cells under physiological conditions.

Acknowledgments

This research was supported by the Swedish Competence Center for Bioseparation (CBioSep), the Swedish Research Counsil (project VR621-2005-3604), the Swedish Foundation for Strategic Research (area of Chemistry and Life Sciences), and INTAS (projects 0673 and 4813).

References

1. Boeseken, J., The use of boric acid for the determination of the configuration of carbohydrates, *Adv. Carbohydr. Chem.*, 4, 189–210, 1949.
2. Kuivila, H.G., Keough, A.H., and Soboczenski, E.J., Areneboronates from diols and polyols, *J. Org. Chem.*, 19, 780–783, 1954.
3. Lorand, J.P. and Edwards, J.O., Polyol complexes and structure of the benzeneboronate ion, *J. Org. Chem.*, 24, 769–774, 1959.
4. Bergold, A. and Scouten, W.H., Borate chromatography, in *Solid Phase Biochemistry: Analytical and Synthetic Methods*, Scouten, W.H., Ed., John Wiley & Sons, London, 1983, pp. 149–187.
5. James, T.D. and Shinkai, S., Artificial receptors as chemosensors for carbohydrates, *Topics Curr. Chem.*, 218, 159–200, 2002.
6. Miyata, T., Uragami, T., and Nakamae, K., Biomolecule-sensitive hydrogels, *Adv. Drug Delivery Rev.*, 54, 79–98, 2002.
7. Ding, Z.L., Chen, G.H., and Hoffman, A.S., Synthesis and purification of thermally sensitive oligomer-enzyme conjugates of poly(*N*-isopropylacrylamide)-trypsin, *Bioconjugate Chem.*, 7, 121–125, 1996.
8. Ding, Z.L., Chen, G.H., and Hoffman, A.S., Unusual properties of thermally sensitive oligomer-enzyme conjugates of poly(*N*-isopropylacrylamide)-trypsin, *J. Biomed. Mater. Res.*, 39, 498–505, 1998.
9. Uludag, H., Norrie, B., Kousinioris, N., and Gao, T., Engineering temperature-sensitive poly(*N*-isopropylacrylamide) polymers as carriers of therapeutic properties, *Biotechnnol. Bioeng.*, 73, 510–521, 2001.
10. Mammen, M., Choi, S.-K., and Whitesides, G.M., Polyvalent interactions in biological systems: implication for design and use of multivalent ligands and inhibitors, *Angew. Chem. Int. Ed.*, 37, 2754–2794, 1998.

11. Kumar, A., Galaev, I.Yu., and Mattiasson, B., Affinity precipitation of amylase inhibitor from wheat meal by metal chelate affinity binding using Cu(II)-loaded copolymers of 1-vinylimidazole with *N*-isopropylacrylamide, *Biotechnol. Bioeng.*, 59, 695–704, 1998.
12. Ivanov, A.E., Galaev, I.Yu., Kazakov, S.V., and Mattiasson, B., Thermosensitive copolymers of *N*-vinylimidazole as displacers of proteins in immobilized metal affinity chromatography, *J. Chromatogr. A*, 907, 115–130, 2001.
13. Arvidsson, P., Ivanov, A.E., Galaev, I.Y., and Mattiasson, B., Polymer versus monomer as displacer in immobilized metal affinity chromatography, *J. Chromatogr. B: Biomed. Sci. Appl.*, 753, 279–285, 2001.
14. Cunliffe, D., Pennadam, S., and Alexander, C., Synthetic and biological polymers: merging the interface, *Eur. Polym. J.*, 40, 5–25, 2004.
15. Hubble, J.A., Elbert, D.L., and Winblade, N.D., Methods and compositions to prevent formation of adhesions in biological tissues, PCT Int. Appl. WO9910022, 1999, *C.A.* 130, 213676, 1999.
16. Kataoka, K., Miyazaki, H., Okano, T., and Sakurai, Y., Sensitive glucose-induced change of the lower critical solution temperature of poly[*N,N*-dimethylacrylamide-*co*-3-(acrylamido)phenylboronic acid] in physiological saline, *Macromolecules*, 27, 1061–1062, 1994.
17. Miyazaki, H., Kikuchi, A., Koyama, Y., Okano, T., Sakurai, Y., and Kataoka, K., Boronate-containing polymer as novel mitogen for lymphocytes, *Biochem. Biophys. Res. Commun.*, 195, 829–836, 1993.
18. Uchimura, E., Otsuka, H., Okano, T., Sakurai, Y., and Kataoka, K., Totally synthetic polymer with lectin-like function: induction of killer cells by the copolymer of 3-acrylamidophenylboronic acid with *N,N*-dimethylacrylamide, *Biotechnol. Bioeng.*, 72, 307–314, 2001.
19. Kuzimenkova, M.V., Ivanov, A.E., and Galaev, I.Yu., Boronate-containing copolymers: polyelectrolyte properties and sugar-specific interaction with agarose gel, *Macromol. Biosci.*, 6, 170–178, 2006.
20. Shiomori, K., Ivanov, A.E., Galaev, I.Yu., Kawano, Y., and Mattiasson, B., Thermoresponsive properties of sugar sensitive copolymer of *N*-isopropylacrylamide and 3-(acrylamido)phenylboronic acid, *Macromol. Chem. Phys.*, 205, 27–34, 2004.
21. Pezron, E., Richard, A., Lafuma, F., and Audebert, R., Reversible gel formation induced by ion complexation, 1: borax-galactomannan interactions, *Macromolecules*, 21, 1121–1125, 1988.
22. Matsumoto, A., Yoshida, R., and Kataoka, K., Glucose-responsive polymer gel bearing phenylboronate derivative as a glucose-sensing moiety operating at the physiological pH, *Biomacromolecules*, 5, 1038–1045, 2004.
23. Yan, J., Springsteen, G., Deeter, S., and Wang, B., The relationship among pKa, pH, and binding constants in the interactions between boronic acids and diols: it is not as simple as it appears, *Tetrahedron*, 60, 11205–11209, 2004.
24. Hisamitsu, I., Kataoka, K., Okano, T., and Sakurai, Y., Glucose-responsive gel from phenylborate polymer and poly(vinyl alcohol): prompt response at physiological pH through the interaction of borate with amino group in the gel, *Pharm. Res.*, 14, 289–293, 1997.
25. Ivanov, A.E., Larsson, H., Galaev, I.Yu., and Mattiasson, B., Synthesis of boronate-containing copolymers of *N,N*-dimethylacrylamide, their interaction with poly(vinyl alcohol) and rheological behaviour of the gels, *Polymer*, 45, 2495–2505, 2004.

26. Otsuka, H., Uchimura, E., Koshino, H., Okano, T., and Kataoka, K., Anomalous binding profile of phenylboronic acid with *N*-acetylneuraminic acid (Neu5Ac) in aqueous solution with varying pH, *J. Am. Chem. Soc.*, 125, 3493–3502, 2003.
27. Kitano, S., Kataoka, K., Koyama, Y., Okano, T., and Sakurai, Y., Glucose-responsive complex formation between poly(vinyl alcohol) and poly(*N*-vinyl-2-pyrrolidone) with pendent phenylboronic acid moieties, *Makromol. Chem. Rapid Commun.*, 12, 227–233, 1991.
28. Kitano, S., Koyama, Y., Kataoka, K., Okano, T., and Sakurai, Y., A novel drug delivery system utilising a glucose responsive polymer complex between poly(vinyl alcohol) and poly(*N*-vinyl-2-pyrrolidone) with a phenylboronic acid moiety, *J. Control. Release*, 19, 161–170, 1992.
29. Aoki, T., Nagao, Y., Sanui, K., Ogata, N., Kikuchi, A., Sakurai, Y., Kataoka, K., and Okano, T., Glucose-sensitive lower critical solution temperature changes of copolymers composed of *N*-isopropylacrylamide and phenylboronic acid moieties, *Polym. J.*, 28, 371–374, 1996.
30. Kataoka, K., Miyazaki, H., Bunya, M., Okano, T., and Sakurai, Y., Totally synthetic polymer gels responding to external glucose concentration: their preparation and application to on-off regulation of insulin release, *J. Am. Chem. Soc.*, 120, 12694–12695, 1998.
31. Ivanov, A.E., Shiomori, K., Kawano, Y., Galaev, I.Yu., and Mattiasson, B., Effects of polyols, saccharides and glycoproteins on thermoprecipitation of phenylboronate-containing copolymers, *Biomacromolecules*, 7, 1017–1024, 2006.
32. Ivanov, A.E., Galaev, I.Yu., and Mattiasson, B., Binding of adenosine to pendant phenylboronate groups of thermoresponsive polymer: a quantitative study, *Macromol. Biosci.*, 5, 795–800, 2005.
33. Alexeev, V.L., Sharma, A.C., Goponenko, A.V., Das, S., Lednev, I.K., Wilcox, C.S., Finegold, D.N., and Asher, S.A., High ionic strength glucose-sensing photonic crystal, *Anal. Chem.*, 75, 2316–2323, 2003.
34. Norrild, J.C. and Eggert, H., Evidence for mono- and bisdentate boronate complexes of glucose in the furanose form: application of 1J C-C coupling constants as a structural probe, *J. Am. Chem. Soc.*, 117, 1479–1484, 1995.
35. Takeuchi, M., Mizuno, T., Shinkai, S., Shirakami, S., and Itoh, T., Chirality sensing of saccharides using a boronic acid-appended chiral ferrocene derivative, *Tetrahedron: Asymmetry*, 11, 3311–3322, 2000.
36. Lee, M.-C., Kabilan, S., Hussain, A., Yang, X., Blyth, J., and Lowe, C.R., Glucose-sensitive holographic sensors for monitoring bacterial growth, *Anal. Chem.*, 76, 5748–5755, 2004.
37. Davis, A.P. and Wareham, R.S., Carbohydrate recognition through noncovalent interaction: a challenge for biomimetic and supramolecular chemistry, *Angew. Chem. Int. Ed.*, 38, 2978–2999, 1999.
38. Kikuchi, A., Suzuki, K., Okabayashi, O., Hoshino, H., Kataoka, K., Sakurai, Y., and Okano, T., Glucose-sensing electrode coated with polymer complex gel containing phenylboronic acid, *Anal. Chem.*, 68, 823–828, 1996.
39. Nicholls, M.P. and Paul, P.K.C., Structures of carbohydrate-boronic acid complexes determined by NMR and molecular modelling in aqueous alkaline media, *Org. Biomol. Chem.*, 2, 1434–1441, 2004.
40. van den Berg, R., Peters, J.A., and van Bekkum, H., The structure and local stability constants of borate esters of mono- and di-saccharides as studied by ^{11}B and ^{13}C NMR spectroscopy, *Carbohydr. Res.*, 253, 1–12, 1994.

41. Gabai, R., Sallacan, N., Chegel, V., Bourenko, T., Katz, E., and Willner, I., Characterization of the swelling of acrylamidophenylboronic acid–acrylamide hydrogels upon interaction with glucose by Faradaic impedance spectroscopy, chronopotentiometry, quartz-crystal microbalance (QCM), and surface plasmon resonance (SPR) experiments, *J. Phys. Chem.*, 105, 8196–8202, 2001.
42. Sallacan, N., Zayats, M., Bourenko, T., Kharitonov, A.B., and Willner, I., Imprinting of nucleotide and monosaccharide recognition sites in acrylamidephenylboronic acid–acrylamide copolymer membranes associated with electronic transducers, *Anal. Chem.*, 74, 702–712, 2002.
43. Asher, S.A., Alexeev, V.L., Goponenko, A.V., Sharma, A.C., Lednev, I.K., Wilcox, C.S., and Finegold, D.N., Photonic crystal carbohydrate sensors: low ionic strength sugar sensing, *J. Am. Chem. Soc.*, 125, 3322–3329, 2003.
44. Thammakhet, C., Ivanov, A.E., Thavarungkul, P., Kanatharana, P., Galaev, I.Yu., and Mattiasson, B., Monitoring of glucose and other carbohydrates using phenylboronate-containing copolymer gels, submitted for publication, 2007.
45. Shoji, E. and Freund, M.S., Potentiometric saccharide detection based on the pKa changes of poly(aniline boronic acid), *J. Am. Chem. Soc.*, 124, 12486–12493, 2002.
46. Ball, I.J., Huang, S.C., Wolf, R.A., Shimano, J.Y., and Kaner, R.B., pervaporation studies with polyaniline membranes and blends, *J. Membr. Sci*, 174, 161–176, 2000.
47. Lide, D.R., Frederikse, H.P.R., Eds., *CRC Handbook of Chemistry and Physics, 75th Ed.*, CRC Press, Boca Raton, 1995, p. 253.
48. Ma, Y. and Yang, X., One saccharide sensor based on the complex of the boronic acid and the monosaccharide using electrochemical impedance spectroscopy, *J. Electoanal. Chem.*, 580, 348–352, 2005.
49. Soh, N., Sonezaki, M., and Imato, T., Modification of a thin gold film with boronic acid membrane and its application to a saccharide sensor based on surface plasmon resonance, *Electroanalysis*, 15, 1281–1290, 2003.
50. Winblade, N.D., Nikolic, I.D., Hoffman, A.S., and Hubbell, J.A., Blocking adhesion to cell and tissue surfaces by the chemisorption of a poly-L-lysine-graft-(poly(ethylene glycol), phenylboronic acid) copolymer, *Biomacromolecules*, 1, 523–533, 2000.
51. Holmes-Farley, S.R., Mandeville, H.W., Dhal, P.K., Huval, C.C., Li, X., and Polomoscanik, S.C., Aryl boronate functionalized polymers for treating obesity, PCT Patent Appl. WO 03/002571, 2003, *C.A.*, 138: 90649.
52. Wang, B., Gao, X., Yang, W., Fang, H., and Yan, Y., Water soluble boronic acid fluorescent reporter compounds and methods of use thereof, PCT Int. Appl. WO 2005024416, 2005, *C.A.*, 142: 308776.
53. Verchere, J.F. and Hlaibi, M., Stability constants of borate complexes of oligosaccharides, *Polyhedron*, 6, 1415–1420, 1987.
54. Chapelle, S. and Verchere, J.-F., A ^{11}B and ^{13}C NMR determination of the structures of borate complexes of pentoses and related sugars, *Tetrahedron*, 44, 4469–4482, 1988.
55. Morin, G.T., Paugam, M.-F., Hughes, M.P., and Smith, B.D., Boronic acid mediates glycoside transport through a liquid organic membrane via reversible formation of trigonal boronate esters, *J. Org. Chem.*, 59, 2724–2728, 1994.
56. Sugasaki, A., Sugiyasu, K., Ikeda, M., Takeuchi, M., and Shinkai, S., First successful molecular design of an artificial Lewis oligosaccharide binding system utilizing positive homotropic allosterism, *J. Am. Chem. Soc.*, 123, 10239–10244, 2001.

57. Pezron, E., Richard, A., Lafuma, F., and Audebert, R., Reversible gel formation induced by ion complexation, 1: borax-galactomannan interactions, *Macromolecules*, 21, 1121–1125, 1988.
58. Winblade, N.D., Schmökel, H., Baumann, M., Hoffman, A.S., and Hubbell, J.A., Sterically blocking adhesion of cells to biological surfaces with a surface-active copolymer containing poly(ethylene glycol) and phenylboronic acid, *J. Biomed. Mater. Res.*, 59, 618–631, 2002.
59. Yu, L.G., Fernig, D.G., Smith, J.A., Milton, J.D., and Rhodes, J.M., Reversible inhibition of proliferation of epithelial cell lines by *Agaricus bisporus* (edible mushroom) lectin, *Cancer Res.*, 53, 4627–4632, 1993.
60. Szul, D., Ivanov, A., Galaev, I., and Hallmann, R., The Endothelial Cell Surface Glycocalix, a Regulator of Leukocyte Adhesion, Lund Immunology Seminar Minisymposium, Lund, Sweden, Dec. 2004.
61. Ivanov, A.E., Szul, D., Hallmann, R., Galaev, I.Yu., and Mattiasson, B., Boronate-Containing Copolymers as Carbohydrate-Specific Reagents and Regulators of Cell Adhesion, FBPS 05, 6th International Symposium on Frontiers in Biomedical Polymers, Granada, Spain, June 16–19, 2005, Book of Abstracts, P-9.
62. Hahne, M., Jäger, U., Isenmann, S., Hallmann, R., and Vestweber, D., Five TNF-inducible cell adhesion mechanisms on the surface of mouse endothelioma cells mediate the binding of leukocytes, *J. Cell Biol.*, 121, 655–664, 1993.
63. Hammel, M., Weitz-Schmidt, G., Krause, A., Moll, T., Vestweber, D., Zerwes, H., and Hallmann, R., Species-specific and conserved epitopes on mouse and human E-selectin important for leukocyte adhesion, *Exp. Cell Res.*, 269, 266–274, 2001.
64. Davis, S.S., Nasal vaccines, *Adv. Drug Delivery Rev.*, 51, 21–42, 2001.
65. Kvanvilkar, K., Donovan, M.D., and Flanagan, D.R., Drug transfer through mucus, *Adv. Drug Delivery Rev.*, 48, 173–193, 2001.
66. Lehr, C.-M., From sticky stuff to sweet receptors: achievements, limits and novel approaches to bioadhesion, *Eur. J. Drug Metab. Pharm.*, 21, 39–48, 1996.
67. Schmitt, E.E., Occlusion of channels in living mammals, PCT Int. Appl. WO 9405342 A1, 1994, *C.A.*, 120: 307569.
68. Mawad, M.E., Cekirge, S., Ciceri, E., and Saatci, I., Endovascular treatment of giant and large intracranial aneurisms by using a combination of stent placement and liquid polymer injection, *J. Neurosurg.*, 96, 474–482, 2002.
69. Piotin, M., Mandal, S., Sugiu, K., Gailloud, P., and Rufenacht, D.A., Endovascular treatment of cerebral aneurysms: an *in vitro* study with detachable platinum coils and triacetate cellulose polymer, *Am. J. Roentgenol.*, 176, 235–239, 2001.
70. Leibler, L., Pezron, E., and Pincus, P.A., Viscosity behaviour of polymer solutions in the presence of complexing ions, *Polymer*, 29, 1105–1109, 1988.
71. Tao, S.L., Lubeley, M.W., and Desai, T.A., Bioadhesive poly(methyl methacrylate) microdevices for controlled drug delivery, *J. Control. Release*, 88, 215–228, 2003.
72. Ivanov, A.E., Ahmad Panahi, H., Kuzimenkova, M.V., Nilsson, L., Bergenståhl, B., Waqif, H.S., Jahanshahi, M., Galaev, I.Yu., and Mattiasson, B., Affinity adhesion of carbohydrate particles and yeast cells to boronate-containing polymer brushes grafted onto siliceous supports, *Chem. Eur. J.*, 12, 7204–7214, 2006.

Chapter 11

Drug Delivery Using Smart Polymers: Recent Advances

Nicholas A. Peppas

CONTENTS

11.1 Introduction

Intelligent or smart polymers are biocompatible polymeric materials that can respond to the physiological or biological environment by expanding or contracting at different rates. This type of response can be used either to construct biomedical devices or to produce advanced responsive medical systems. In the pharmaceutical field, such systems have been used to develop new types of responsive or pulsatile delivery devices.

Among the various types of polymers, hydrogels are the simplest carriers to be used as intelligent systems. These are three-dimensional, hydrophilic, polymeric networks capable of imbibing large amounts of water or biological fluids.[1,2] The networks are composed of homopolymers, copolymers, or biohybrids and are insoluble in water due to the presence of chemical cross-links (tie-points, junctions) or physical cross-links such as entanglements or crystallites.[3–8] These hydrogels exhibit thermodynamic compatibility with water that allows them to swell in aqueous media.[1,2,9–11]

There are numerous applications of hydrogels in the medical and pharmaceutical sectors.[12–14] They can be used as contact lenses, membranes for biosensors, linings for artificial hearts, materials for artificial skin, and drug-delivery devices.[12–16]

Hydrogels can be neutral or ionic based on the nature of the side groups. Depending on the physical structure of the networks, they can be amorphous, semicrystalline, hydrogen-bonded structures; supermolecular structures; or hydrocolloidal aggregates.[1–8,17–21] Hydrogels can also show swelling behavior in response to conditions in the external environment. These polymers are physiologically responsive hydrogels,[22] where polymer complexes can be broken or the network can be swollen as a result of the changing external environment. They exhibit drastic changes in their swelling ratio, especially as a function of pH, ionic strength, temperature, and electromagnetic radiation.[22]

Novel materials based on polymers such as poly(hydroxyethyl methacrylate) (PHEMA), poly(*N*-isopropyl-2-acrylamide) (PNIPAAm), and poly(vinyl alcohol) (PVA) have been synthesized by various preparation techniques.[23,24] Hydrogels are also used as carriers that can interact with the mucosa lining in gastrointestinal tract, colon, vagina, nose, and other parts of the body due to their ability to prolong their residence time at the delivery location.[25] The interaction between such carriers and the glycoproteins in the mucosa is thought to occur primarily via hydrogen bonding. Therefore, materials containing a high density of carboxyl and hydroxyl groups appear to be promising candidates for this type of application.

The idea of adhesion promoters diffusing across the polymer–mucin interface has also been introduced by Peppas and Sahlin,[26] who used chains of polymerized ethylene glycol, either freely loaded in the carrier or grafted to the polymer surface, as adhesion promoters. The "stealth" properties of poly(ethylene glycol), PEG, have also been used to reduce the uptake of particulate carriers by the reticuloendothelial system.[27] PEG has also been shown to lengthen the biological half-life and to reduce the immunogenicity of high-molecular-weight substances such as adenosine deaminase (ADA) and asparaginase.[28]

11.2 Molecular Structure of Smart Polymers

The suitability of smart-polymer hydrogels as drug-delivery devices and their performance in a particular application depend to a large extent on their bulk structure. A number of excellent reviews discuss this topic in great detail.[29–32] The most important parameters used to characterize the network structure of hydrogels are the polymer volume fraction in the swollen state, $\upsilon_{2,s}$, the molecular weight of the polymer chain between two neighboring cross-linking points, $\bar{M}_c$, and the corresponding mesh size, ξ.[33]

The polymer volume fraction in the swollen state is a measure of the amount of biological fluid imbibed by the hydrogel. The molecular weight between two consecutive cross-links, which can be either chemical or physical in nature, is a measure of the degree of polymer cross-linking. It is important to note that, due to the random nature of the polymerization process itself, only average values of $\bar{M}_c$ can be calculated. The correlation distance between two adjacent cross-links, ξ, provides a measure of the space available between the macromolecular chains available for drug diffusion; again, it can be reported only as an average value. These parameters, which are related to one another, can be determined theoretically or through the use of a variety of experimental techniques. Two methods are prominent among the growing number of techniques used to elucidate the structure of hydrogels: the equilibrium swelling theory and the rubber elasticity theory.

The structure of hydrogels that do not contain ionic moieties can be analyzed by the Flory-Rehner theory.[9] This combination of thermodynamic and elasticity theories states that a cross-linked polymer gel that is immersed in a fluid and allowed to reach equilibrium with its surroundings is subject to only two opposing forces: the thermodynamic force of mixing and the retractive force of the polymer chains. At equilibrium, these two forces are equal. Equation 11.1 describes the physical situation in terms of the Gibbs free energy.

$$\Delta G_{\text{total}} = \Delta G_{\text{elastic}} + \Delta G_{\text{mixing}} \tag{11.1}$$

where $\Delta G_{\text{elastic}}$ is the contribution due to the elastic retractive forces developed inside the gel, and ΔG_{mixing} is the result of the spontaneous mixing of the fluid molecules with the polymer chains. The term ΔG_{mixing} is a measure of the compatibility of the polymer with the molecules of the surrounding fluid. This compatibility is usually expressed by the polymer–solvent interaction parameter, χ_1.[11]

Differentiation of Equation 11.1 with respect to the number of solvent molecules while keeping temperature and pressure constant results in Equation 11.2:

$$\mu_1 - \mu_{1,0} = \Delta\mu_{\text{elastic}} + \Delta\mu_{\text{mixing}} \tag{11.2}$$

where μ_1 is the chemical potential of the solvent in the polymer gel, and $\mu_{1,0}$ is the chemical potential of the pure solvent. At equilibrium, the difference between the chemical potentials of the solvent outside and inside the gel must be zero. Therefore, changes of the chemical potential due to mixing and elastic forces must balance each other. The change of chemical potential due to mixing can be expressed using heat and entropy of mixing.

The change of chemical potential due to the elastic retractive forces of the polymer chains can be determined from the theory of rubber elasticity.[11,34] Upon equaling these two contributions, an expression for determining the molecular weight between two adjacent cross-links of a neutral hydrogel prepared in the absence of a solvent can be written as

$$\frac{1}{\bar{M}_c} = \frac{2}{\bar{M}_n} - \frac{\left(\bar{\upsilon}/V_1\right)\left[\ln(1-\upsilon_{2,s}) + \upsilon_{2,s} + \chi_1 \upsilon_{2,s}^2\right]}{\left(\upsilon_{2,s}^{1/3} - \frac{\upsilon_{2,s}}{2}\right)} \tag{11.3}$$

where $\bar{M}_n$ is the molecular weight of the polymer chains prepared under identical conditions but in the absence of the cross-linking agent, $\bar{\upsilon}$ is the specific volume of the polymer, and V_1 is the molar volume of water.

Peppas and Merrill[35] modified the original Flory-Rehner theory for hydrogels prepared in the presence of water. The presence of water effectively modifies the change of chemical potential due to the elastic forces. This term must now account for the volume-fraction density of the chains during cross-linking. Equation 11.4 predicts the molecular weight between cross-links in a neutral hydrogel prepared in the presence of water.

$$\frac{1}{\bar{M}_c} = \frac{2}{\bar{M}_n} - \frac{\bar{\upsilon}/V_1 \left[\ln(1-\upsilon_{2,s}) + \upsilon_{2,s} + \chi_1 \upsilon_{2,s}^2\right]}{\upsilon_{2,r}\left[\left(\frac{\upsilon_{2,s}}{\upsilon_{2,r}}\right)^{1/3} - \left(\frac{\upsilon_{2,s}}{2\upsilon_{2,r}}\right)\right]} \tag{11.4}$$

where $\upsilon_{2,r}$ is the polymer volume fraction in the relaxed state, which is defined as the state of the polymer immediately after cross-linking but before swelling.

The presence of ionic moieties in hydrogels makes the theoretical treatment of swelling much more complex. In addition to the ΔG_{mixing} and $\Delta G_{elastic}$ in Equation 11.1, there is an additional contribution to the total change in Gibbs free energy due to the ionic nature of the polymer network, ΔG_{ionic}.

$$\Delta G_{total} = \Delta G_{elastic} + \Delta G_{mixing} + \Delta G_{ionic} \tag{11.5}$$

Upon differentiating Equation 11.5 with respect to the number of moles of solvent, keeping temperature T and pressure P constant, an expression similar to Equation 11.2 for the chemical potential can be derived.

$$\mu_1 - \mu_{1,o} = \Delta\mu_{elastic} + \Delta\mu_{mixing} + \Delta\mu_{ionic} \tag{11.6}$$

where the term $\Delta\mu_{ionic}$ is the change of chemical potential due to the ionic character of the hydrogel.

Expressions for the ionic contribution to the chemical potential have also been developed.[36–38] They exhibit strong dependencies on the ionic strength of the surrounding media and on the nature of the ions present in the solvent. Equations 11.7 and 11.8 are expressions that have been derived for swelling of anionic and cationic hydrogels, respectively, prepared in the presence of a solvent.

$$\frac{V_1}{4IM_r}\left(\frac{\upsilon_{2,s}^2}{\upsilon}\right)\left(\frac{K_a}{10^{-pH}-K_a}\right)^2 = \left[\ln(1-\upsilon_{2,s})+\upsilon_{2,s}+\chi_1\upsilon_{2,s}^2\right]$$
$$+\left(\frac{V_1}{\upsilon\bar{M}_c}\right)\left(1-\frac{2\bar{M}_c}{\bar{M}_n}\right)\upsilon_{2,r}\left[\left(\frac{\upsilon_{2,s}}{\upsilon_{2,r}}\right)^{1/3}-\left(\frac{\upsilon_{2,s}}{2\upsilon_{2,r}}\right)\right] \tag{11.7}$$

$$\frac{V_1}{4IM_r}\left(\frac{\upsilon_{2,s}^2}{\upsilon}\right)\left(\frac{K_b}{10^{pH-14}-K_a}\right)^2 = \left[\ln(1-\upsilon_{2,s})+\upsilon_{2,s}+\chi_1\upsilon_{2,s}^2\right]$$
$$+\left(\frac{V_1}{\upsilon\bar{M}_c}\right)\left(1-\frac{2\bar{M}_c}{\bar{M}_n}\right)\upsilon_{2,r}\left[\left(\frac{\upsilon_{2,s}}{\upsilon_{2,r}}\right)^{1/3}-\left(\frac{\upsilon_{2,s}}{2\upsilon_{2,r}}\right)\right] \tag{11.8}$$

where I is the ionic strength; K_a and K_b are the dissociation constants for the acid and base, respectively; and M_r is the molecular weight of the repeating unit.

11.3 Mechanical Behavior of Smart Polymers

Hydrogels subjected to a relatively small deformation will rapidly recover to their original dimension. This elastic behavior of hydrogels can be used to elucidate their structure by utilizing the rubber elasticity theory originally developed by Treloar[34] and Flory[39,40] for vulcanized rubbers and modified to polymers by Flory.[11] However, the original theory or rubber elasticity does not apply to hydrogels prepared in the presence of a solvent. Such expressions were developed by Silliman[41] and later modified by Peppas and Merrill.[42]

Here, only the form of rubber elasticity theory used to analyze the structure of hydrogels prepared in the presence of a solvent is presented, and it is left up to the reader to consult the original reference for detailed derivations.

$$\tau = \frac{\rho RT}{\bar{M}_c}\left(1-\frac{2\bar{M}_c}{\bar{M}_n}\right)\left(\alpha-\frac{1}{\alpha^2}\right)\left(\frac{\upsilon_{2,s}}{\upsilon_{2,r}}\right)^{1/3} \tag{11.9}$$

where τ is the stress applied to the polymer sample, ρ is the density of the polymer, R is the universal gas constant, T is the absolute experimental temperature, and $\bar{M}_c$ is the desired molecular weight between cross-links.

To analyze the structure of hydrogels using the rubber elasticity theory, experiments need to be performed using a tensile testing system. Interestingly, the rubber elasticity theory has been used to analyze both chemically and physically cross-linked hydrogels[43–45] as well as hydrogels exhibiting temporary cross-links due to hydrogen bonding.[46]

11.3.1 Determination of the Mesh Size

The primary mechanism of release of many drugs from hydrogels is diffusion occurring through the space available between macromolecular chains. This space is often regarded as the "pore." Depending upon the size of these pores, hydrogels can be conveniently classified as macroporous, microporous, or nonporous. A structural parameter that is often used to describe the size of the pores is the correlation length, ξ, which is defined as the linear distance between two adjacent cross-links and can be calculated using the following equation:

$$\xi = \alpha\left(\bar{r}_o^2\right)^{1/2} \tag{11.10}$$

where α is the elongation ratio of the polymer chains in any direction, and $(\bar{r}_o^2)^{1/2}$ is the root-mean-square, unperturbed, end-to-end distance of the polymer chains between two neighboring cross-links.[47] For isotropically swollen hydrogel, the elongation ratio, α, can be related to the swollen polymer volume fraction, $\upsilon_{2,s}$, using Equation 11.11.

$$\alpha = \upsilon_{2,s}^{-1/3} \tag{11.11}$$

The unperturbed end-to-end distance of the polymer chain between two adjacent cross-links can be calculated using Equation 11.12, where C_n is the Flory characteristic ratio, l is the length of the bond along the polymer backbone (for vinyl polymers 1.54 Å), and N is the number of links per chain that can be calculated by Equation 11.13.

$$\left(\bar{r}_o^2\right)^{1/2} = l(C_n N)^{1/2} \tag{11.12}$$

$$N = \frac{2\bar{M}_c}{M_r} \tag{11.13}$$

In Equation 11.13, M_r is the molecular weight of the repeating units from which the polymer chain is composed. Finally, when one combines Equations

11.10 through 11.13, the correlation distance between two adjacent cross-links in a swollen hydrogel can be obtained as follows:

$$\xi = \upsilon_{2,s}^{-1/3}\left(\frac{2C_n\bar{M}_c}{M_r}\right)^{1/2} l \qquad (11.14)$$

A detailed theoretical characterization of the network structure of the polymer carrier in terms of the correlation length, ξ, in combination with diffusion studies of model drugs and proteins, provides an invaluable insight into the very complex structure of polymer networks and aids in the design of drug-delivery carriers.[48]

To examine the effect of the molecular structure on the swelling behavior of these smart gels, we examine the behavior of cross-linked homocopolymers of poly(methacrylic acid) (PMAA, a pH-sensitive hydrogel) and poly(*N*-isopropyl acrylamide) (PNIPAAm) (a temperature-sensitive hydrogel). As shown in Figure 11.1, PNIPAAm is temperature sensitive and collapses at about 32°C, while PMAA is virtually temperature independent in the range of temperatures studied. On the other side, Figure 11.2 shows the pH dependence of the same cross-linked polymers. Clearly, PMAA is pH dependent. These studies indicate the basics of smart gel behavior.

The synergistic effects of the two phenomena can be seen in Figure 11.3, which depicts the swelling behavior of cross-linked random copolymers of

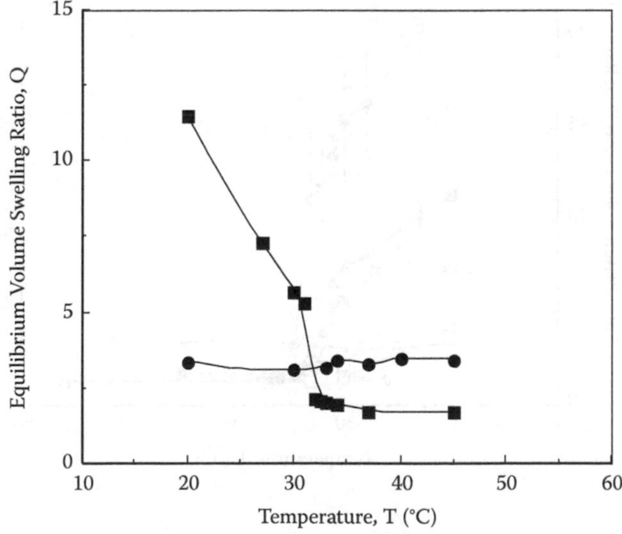

FIGURE 11.1
Equilibrium volume swelling ratio of PNIPAAm (■) and PMAA (●) homopolymer hydrogels in deionized water (pH = 6.5) as a function of temperature. Three samples were tested at each temperature; the error bars are smaller than the data symbols.

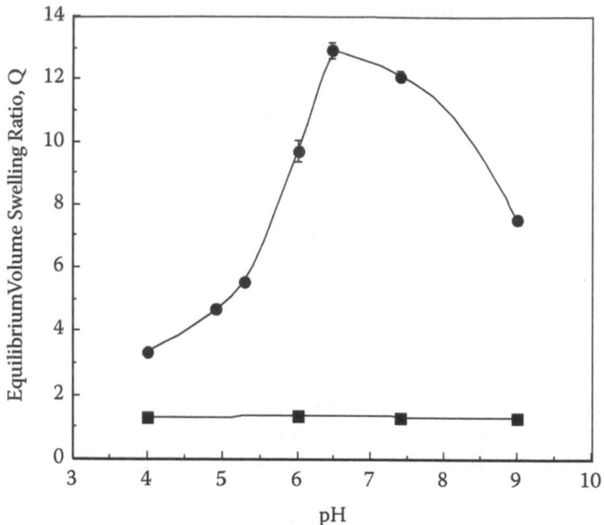

FIGURE 11.2
Equilibrium volume swelling ratio of PNIPAAm (■) and PMAA (●) homopolymer hydrogels in buffered solutions at 37°C as a function of pH. Three samples were tested at each pH value; the error bars are smaller than the data symbols in most cases.

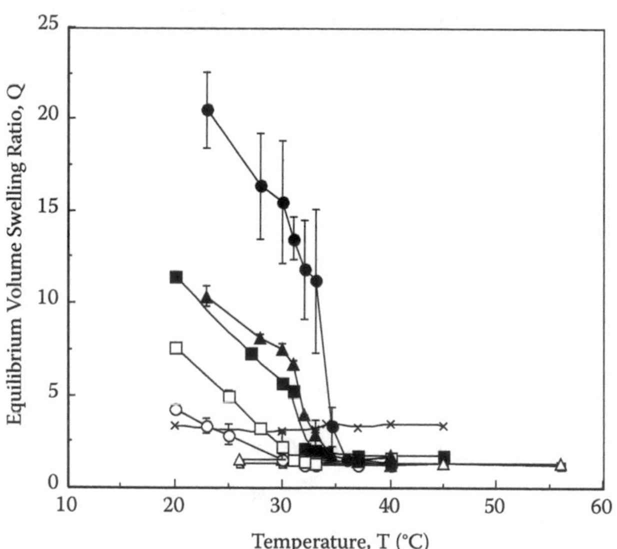

FIGURE 11.3
Equilibrium volume swelling ratio of P(NIPAAm-*co*-MAA) hydrogels in deionized water as a function of temperature. Hydrogels contained 100% NIPAAm (■), 88% NIPAAm (●), 84% NIPAAm (▲), 75% NIPAAm (□), 69% NIPAAm (○), 43% NIPAAm (Δ), 27% NIPAAm (⊠), and 0% NIPAAm (X). Data points represent the average for three hydrogel samples. In instances where error bars are not shown, the error bars are smaller than the data symbols.

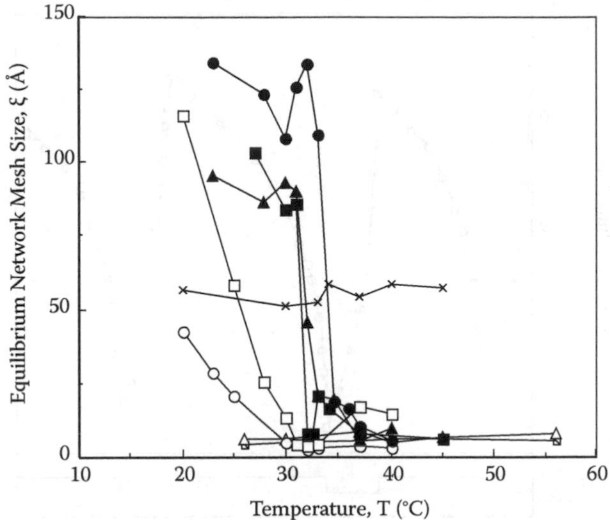

FIGURE 11.4
Equilibrium network mesh size of P(NIPAAm-*co*-MAA) hydrogels as a function of temperature. Hydrogels contained 100% NIPAAm (■), 88% NIPAAm (●), 84% NIPAAm (▲), 75% NIPAAm (□), 69% NIPAAm (○), 43% NIPAAm (Δ), 27% NIPAAm (⊠), and 0% NIPAAm (X). Data points represent the average for three hydrogel samples.

MAA and NIPAAm, henceforth designated as P(NIPAAm-*co*-MAA). Clearly, these copolymers exhibit temperature and pH sensitivity, depending on the copolymer composition. With this structural information, it is easy to apply Equations 11.12 through 11.14 to determine the equilibrium mesh size. Indeed, Figure 11.4 indicates the value of the mesh size of P(NIPAAm-*co*-MAA) networks as a function of swelling temperature.

11.3.2 Effect of Molecular Structure on Drug- and Protein-Release Behavior

One of the most important and challenging areas in the drug-delivery field is to predict the release of the active agent as a function of time using both simple and sophisticated mathematical models. The importance of such models lies in their utility during both the design stage of a pharmaceutical formulation and the experimental verification of a release mechanism.[49]

Figure 11.5 and Figure 11.6 show the possible variation of the equilibrium swelling ratio of cross-linked P(NIPAAm-*co*-MAA) networks as a function of pH and temperature, respectively. Clearly, the response of such hydrogels is significant and is very much affected by the size of the specimen, as the characteristic water diffusion time is proportional to the square of the specimen's characteristic length. Indeed, Figure 11.5 shows the dependence of the weight swelling ratio, q, on the pH change from 5 to 7.4, while Figure 11.6

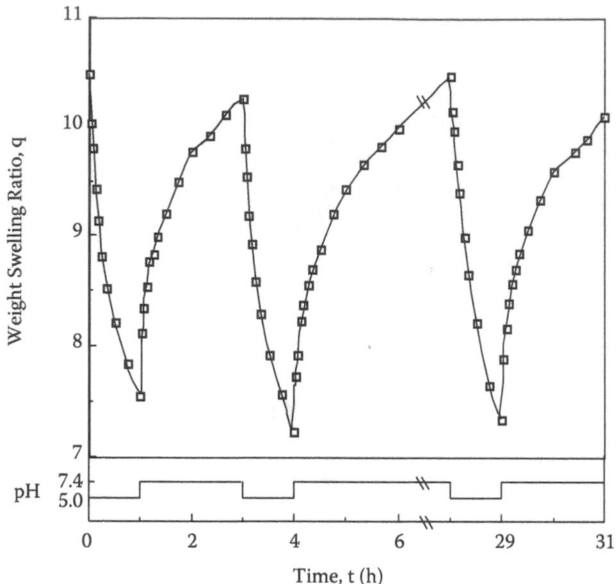

FIGURE 11.5
Weight swelling ratio of a cross-linked P(NIPAAm-*co*-MAA) hydrogel sample containing 84% NIPAAm placed successively in buffered solutions of pH 5.0 for 1 h and pH 7.4 for several cycles, all at 37°C. The sample was equilibrated at pH 7.4 prior to the experiment.

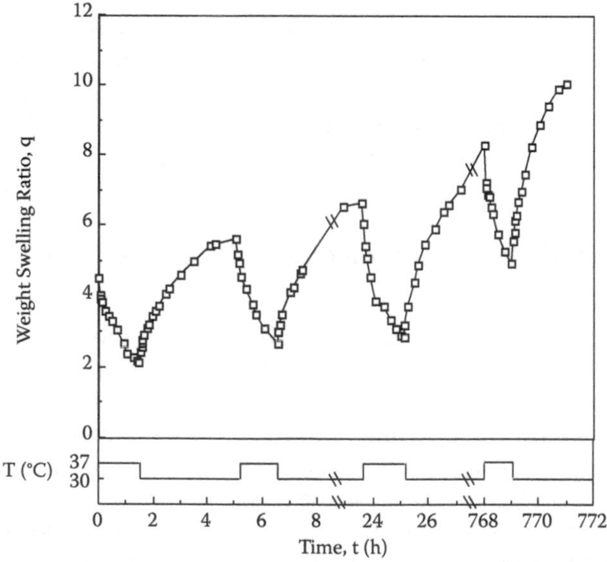

FIGURE 11.6
Weight swelling ratio of a cross-linked P(NIPAAm-*co*-MAA) hydrogel sample containing 84% NIPAAm, placed successively in deionized water, equilibrated at 37°C for approximately 1 h and 30°C for several hours. The hydrogel sample was equilibrated at 30°C prior to the experiment.

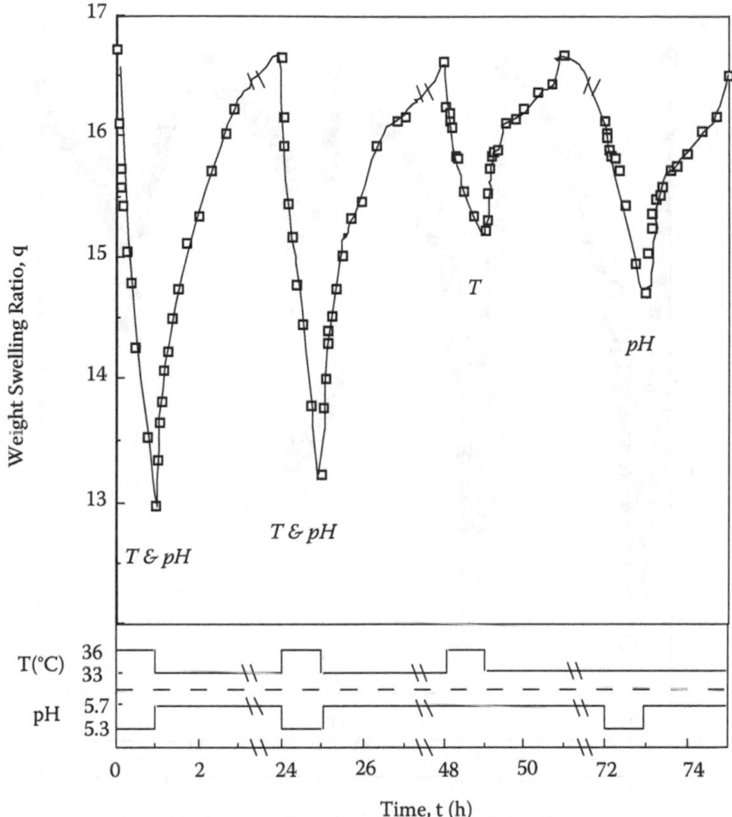

FIGURE 11.7
Weight swelling ratio of a cross-linked P(NIPAAm-*co*-MAA) hydrogel sample containing 88% NIPAAm, placed successively in buffered solutions of pH 5.3 at 36°C and pH 5.7 at 33°C, equilibrated at pH 5.3 and 36°C for approximately 1 h and pH 5.7 and 33°C for several hours. The last two peaks indicate changes in only one of the swelling parameters: temperature or pH. The hydrogel sample was equilibrated at pH 5.3 and 33°C prior to the experiment.

shows the dependence of q on the temperature from 30 to 37°C. The combined synergistic effect is clearly seen in Figure 11.7, where the swelling ratio can be varied with successive changes of temperature and pH. The associated change of the mesh size can be seen in Figure 11.8.

Such results indicate the powerful responsive behavior of such intelligent hydrogels in drug delivery. To design a particular system with a specific release mechanism, experimental data of statistical significance are compared with a solution of the theoretical model. It is therefore clear that only a combination of accurate and precise data with models accurately depicting the physical situation will provide an insight into the actual mechanism of release.

The vast majority of theoretical models is based on diffusion equations. The phenomenon of diffusion is intimately connected to the structure of the

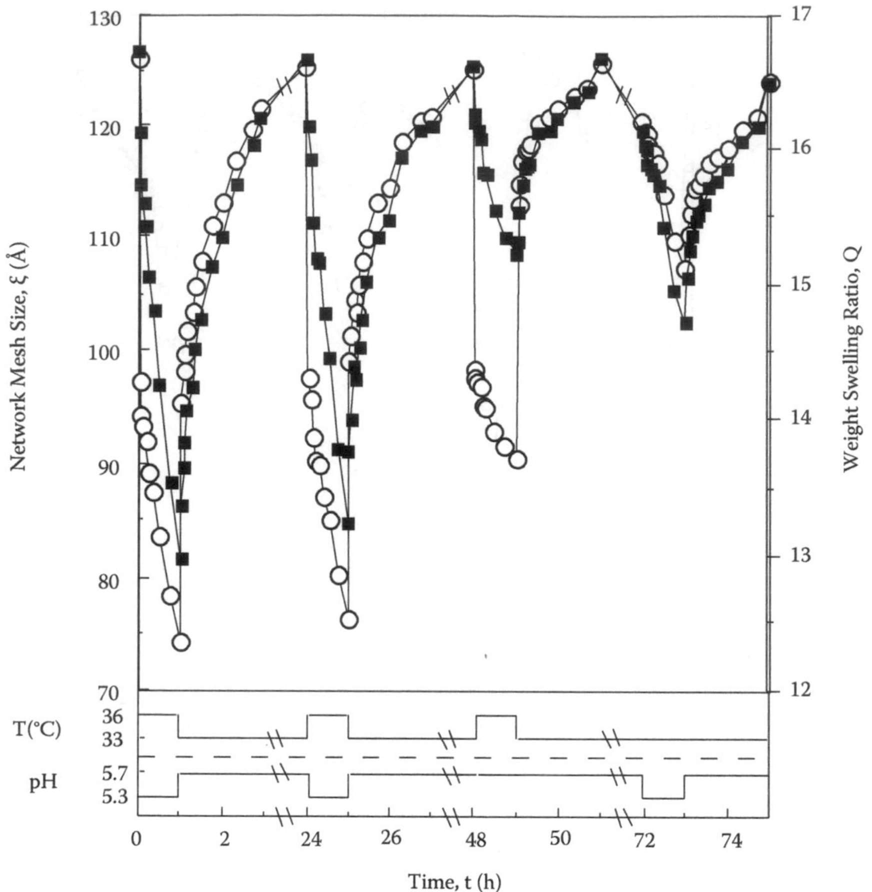

FIGURE 11.8
Network mesh size (○) and weight swelling ratio (■) of a cross-linked P(NIPAAm-*co*-MAA) hydrogel sample containing 88% NIPAAm, placed successively in buffered solutions of pH 5.3 at 36°C and pH 5.7 at 33°C, equilibrated at pH 5.3 and 36°C for approximately 1 h and pH 5.7 and 33°C for several hours. The last two peaks indicate changes in only one of the swelling parameters: temperature or pH. The hydrogel sample was equilibrated at pH 5.3 and 33°C prior to the experiment.

material through which the diffusion takes place; thus the morphology of the polymeric materials should be accounted for in a successful model. There has been a limited number of reviews addressing these aspects of controlled-release formulations.[49–51] The mechanisms of drug release offer a convenient way of categorizing controlled-release systems[52]:

Diffusion controlled
Chemically controlled
Swelling controlled

We devote a separate section to each of these mechanisms. However, given that ordinary diffusion takes place in each of these mechanisms to a certain degree, and since most of the models used are based on diffusion equations, a separate section describing the fundamentals of diffusion precedes the discussion of controlled-release systems.

The release of an active agent from a polymeric carrier consists of the movement of the drug through the bulk of the polymer. This phenomenon, known as diffusion, is to a large degree controlled by the mass-transfer limitations at the boundary between the polymer carrier and its surroundings. On a macroscopic level, the diffusion of drug molecules through the polymer carrier can be described by Fick's law of diffusion, which is mathematically stated by Equations 11.15 and 11.16 for transport in one dimension[53]

$$j_i = -D_{ip}\frac{dc_i}{dx} \tag{11.15}$$

$$\frac{\partial c_i}{\partial t} = D_{ip}\frac{\partial^2 c_i}{\partial x^2} \tag{11.16}$$

where the concentration and mass flux of species i are designated c_i, and j_i, respectively; D_{ip} is the diffusion coefficient of species i in the polymer matrix, and x and t stand for the independent variables of position and time, respectively.

Several important assumptions have been implicitly incorporated in Equations 11.15 and 11.16. First, these equations describe the release of a drug from a carrier of a thin planar geometry; equivalent equations for release from thick slabs, cylinders, and spheres have also been derived.[53] It should also be emphasized that in the above written form of Fick's law, the diffusion coefficient is assumed to be independent of concentration. This assumption, while not conceptually correct, has been largely accepted due to the computational simplicity. Finally, j_i is a flux with respect to the mass average velocity v of the system.

Initial and boundary conditions, which are necessary for solving Equations 11.15 and 11.16, allow for the appropriate description of the experimental conditions imposed upon the drug-release device. The solutions of Equations 11.15 and 11.16, subject to a number of boundary conditions that can be applied to various *in vitro* and *ex vivo* experiments, have been obtained.[53]

To improve the predictive power of the Fickian diffusion theory, a concentration-dependent diffusion coefficient is used in Equations 11.1 and 11.2. Equation 11.2 is then rewritten and solved with the appropriate boundary conditions

$$\frac{\partial c_i}{\partial t} = \frac{\partial}{\partial z}\left(D_{ip}(c_i)\frac{\partial c_i}{\partial x}\right) \tag{11.17}$$

TABLE 11.1

Diffusion Coefficients as Functions of Structural Parameters of Smart Gels

Type of Smart Polymer Carrier	Eq. No.	Form of D_{ip}
Porous	11.18	$D_{ip} = \frac{\lambda^2 v}{6}$
Porous	11.19	$D_{eff} = D_{iw} K_p K_r \frac{\varepsilon}{\tau}$
Microporous	11.20	$\frac{D_{ip}}{D_b} = (1-\lambda)^2(1+\alpha\lambda+\beta\lambda^3+\gamma\lambda^5)$
Nonporous	11.21	$D_{ip} = D_0 \exp\left\{-\frac{k}{v_f}\right\}$
Nonporous	11.22	$\frac{D_{2,1,3}}{D_{2,1}} = \varphi(q_s)\exp\left[-B\left(\frac{q_s}{V_{f,1}}\right)\left(\frac{1}{H}-1\right)\right]$
Nonporous (highly swollen)	11.23	$\frac{D_{2,1,3}}{D_{2,1}} = k_1\left(\frac{\bar{M}_c - \bar{M}_c^*}{\bar{M}_n - \bar{M}_c^*}\right)\exp\left(-\frac{k_2 r_s^2}{Q-1}\right)$

where $D_{ip}(c_i)$ is the concentration-dependent diffusion coefficient; its form of concentration dependence is affected by the structural characteristics of the polymer carrier. A selective summary of the various forms of the diffusion coefficient is provided in Table 11.1.

One of earliest approaches of estimating the diffusion coefficient through a polymer carrier is that of Eyring.[54] In this theory, diffusion of a solute through a medium is presented as a series of jumps instead of a continuous process. Therefore, in Equation 11.18 in Table 11.1, which comes from the Eyring analysis, λ is the diffusional jump of the drug in the polymer, and v is the frequency of jumping.

Fujita[55] utilized the idea of free volume in polymers to estimate the drug diffusion coefficient and arrived at an exponential dependence of the drug-diffusion coefficient on the free volume, v_f, which is given by Equation 11.21 in Table 11.1. Yasuda and Lamaze[56] refined Fujita's theory and presented molecularly based theory, which predicts the diffusion coefficients of drugs through a polymer matrix rather accurately (Equation 11.22). In their treatment, the normalized diffusion coefficient — the ratio of the diffusion coefficient of the solute in the polymer, $D_{2,1,3}$, to the diffusion coefficient of the solute in the pure solvent, $D_{2,1}$ — is related to the degree of hydration, H, and free-volume occupied by the swelling medium, $V_{f,1}$. In addition, φ is a sieving factor that provides a limiting mesh size impermeable to drugs with cross-sectional area q_s, and B is a parameter characteristic of the polymer. In Equation 11.22, the subscripts 1, 2, and 3 refer to the swelling medium, drug, and polymer, respectively.

Peppas and Reinhart[57] also developed a theoretical model based on a free volume of the polymer matrix. In their theory, they assumed the free volume of the polymer to be the same as the free volume of the solvent, and they arrived at Equation 11.23 in Table 11.1. They related the normalized diffusion coefficient to the degree of swelling, Q, the solute radius, r_s, and the molecular weight of the polymer chains. More specifically, $\bar{M}_c$ is the average molecular weight of the polymer chains between adjacent cross-links, $\bar{M}_n$ is the average molecular weight of the linear polymer chains prepared under identical conditions in the absence of the cross-linking agent, and $\bar{M}_c^*$ is the critical molecular weight between cross-links below which a drug of size r_s could not diffuse through the polymer network. In addition, k_1 and k_2 are constants related to the polymer structure. This theory is applicable to drug transport in highly swollen, nonporous hydrogels. Equations for moderately or poorly swollen[58] and semicrystalline[59] hydrogels were also developed.

Yet another approach for predicting the diffusion coefficient of a drug in a controlled-release device has been adopted from the chemical engineering field.[60] More specifically, the transport phenomena in porous rocks, ion-exchange resins, and catalysis are of very similar nature to a drug diffusing through a macro- or microporous polymer. In these types of polymers the diffusion is assumed to be taking place predominantly through the water (or bodily fluids)-filled pores. The diffusion coefficient of a drug in a polymer, D_{ip}, in Equations 11.15 and 11.16 is replaced by an effective diffusive coefficient, D_{eff}, which is defined by Equation 11.19 in Table 11.1. In Equation 11.19, ε is the porosity, or void fraction, of the polymer, which is a measure of the volume of the pores available for diffusion, and τ is the tortuosity, which describes the geometric characteristics of the pores. The term K_p is the equilibrium-partitioning coefficient, which is a parameter needed when the drug is soluble in the polymer matrix; it is the ratio of the concentration inside of the pore to the concentration outside of the pore. The term K_r describes the fractional reduction in diffusivity within the pore when the solute diameter, d_s, is comparable in size to the pore diameter d_r. Equation 11.20 in Table 11.1 is a semiempirical relation proposed by Faxén[61] for diffusion of spheres through porous media. In this equation, λ is the ratio of the drug radius, r_s, to the pore average radius, r_p; D and D_b are the diffusion coefficients of the sphere through the pore and in bulk, respectively; and α, β, and γ are constants. It is clear to see that as the size of the drug gets smaller with respect to the size of the pore, the ratio of D/D_b approaches the limit of 1.

11.3.3 Diffusion-Controlled Delivery Systems

Reservoir diffusion-controlled systems consist of a bioactive agent containing a core that is separated from the external environment by a polymer membrane. Analysis of drug release from such systems using Equation 11.15 shows that the release rate is independent of time, i.e., zero-order, for planar, cylindrical, and spherical geometry. For example, Equations 11.22

and 11.23 give the rate of drug release and the total amount of drug released for systems of spherical geometry.

$$\frac{dM_t}{dt} = \frac{4\pi D_{ip} K}{(r_e - r_i)/(r_e r_i)}(c_{i2} - c_{i1}) \tag{11.22}$$

$$M_t = \frac{4\pi D_{ip} K(c_{i2} - c_{i1})}{(r_e - r_i)/(r_e r_i)} t \tag{11.23}$$

In the above equations, D_{ip} is the concentration independent diffusion coefficient, M_t is the amount of drug released at time t, K is the drug-partition coefficient, and r_e and r_i are the external and internal radius of the sphere, respectively. Finally, c_{i1} and c_{i2} are the drug concentrations inside and outside the matrix, respectively.

A comparison of analogous equations for the planar, cylindrical, spherical, and other geometries reveals that the drug release can be manipulated by the geometry of the system. In addition, it is also clear that the amount of drug release can be controlled by the thickness of the membrane, concentration difference of the drug across the membrane, the thermodynamic characteristics of the system via the partition coefficient, and the structure of the polymer through the solute diffusion coefficient.

In matrix systems, which are also diffusion controlled, the drug can be either dissolved or dispersed throughout the network of the hydrogels. The drug release from these systems is modeled using Equation 11.16, with the concentration-dependent coefficient given by Equation 11.17 and one of the equations of Table 11.1. It is clear from solutions to Equation 11.16 that the fractional drug release obtained form these systems is proportional to $t^{1/2}$.

In smart delivery systems the release behavior is not only a function of time, but also of the external physiological conditions. Returning to the systems described in Figure 11.3 through Figure 11.8, we see now that the release of an incorporated enzyme, streptokinase, which acts as a thrombolytic enzyme, is possible only when the mesh size attains a certain value that will allow the relatively large-molecular-weight enzyme to be released. This is clear with the results of Figure 11.9 and Figure 11.10, where release of streptokinase is observed only under pH or temperature conditions that will lead to a significant mesh size expansion, as shown in Figure 11.8.

11.3.4 Chemically Controlled Delivery Systems

Chemically controlled systems consist of two major subclasses of controlled-release systems based on the mechanisms that control the drug release. In erodible systems, the drug release rate is controlled by degradation or dissolution of the polymer. In pendant chain systems, the drug is attached to

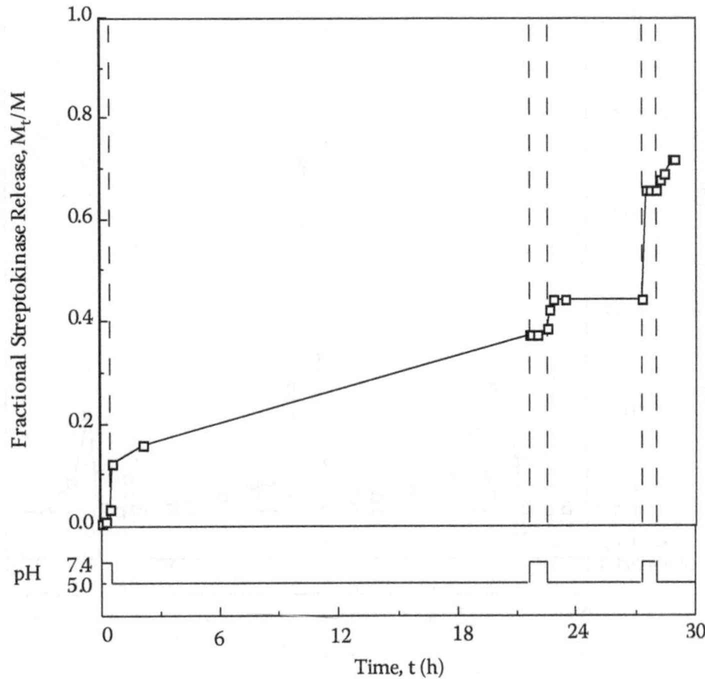

FIGURE 11.9
Fraction of streptokinase released at 37°C from P(NIPAAm-*co*-MAA) hydrogels containing 84 mol% NIPAAm upon change in pH from 5 to 7.4 for 1 h and then back to pH 5 for a longer time interval.

the polymer via hydrolytically or enzymatically labile bonds, and the drug release is controlled by the rate of degradation of the bonds.

Unlike in pendant-chain systems, ordinary diffusion takes place in erodible systems, and the actual mechanism of drug release depends on whether diffusion or erosion is the rate-controlling step. If erosion of the matrix is much slower than the diffusion of the drug through the polymer, the drug release can be analyzed using equations described in the section on diffusion-controlled systems (Section 11.3.3). In the other scenario, the drug remains incorporated in the matrix due to the low rate of diffusion; therefore, the drug release is erosion controlled. The two possible mechanisms of erosion, heterogeneous and homogeneous, can be predicted from polymer hydrophobicity and morphology. Hydrophilic polymers absorb water; therefore, erosion throughout the polymer matrix (i.e., homogeneous erosion) will take place. On the other hand, hydrophobic polymers will erode only at the surface, or heterogeneously, because water is excluded from the bulk of the matrix. In the pharmaceutical field, surface-eroding polymer matrices have received much more attention due to the fact that they exhibit near zero-order release kinetics.

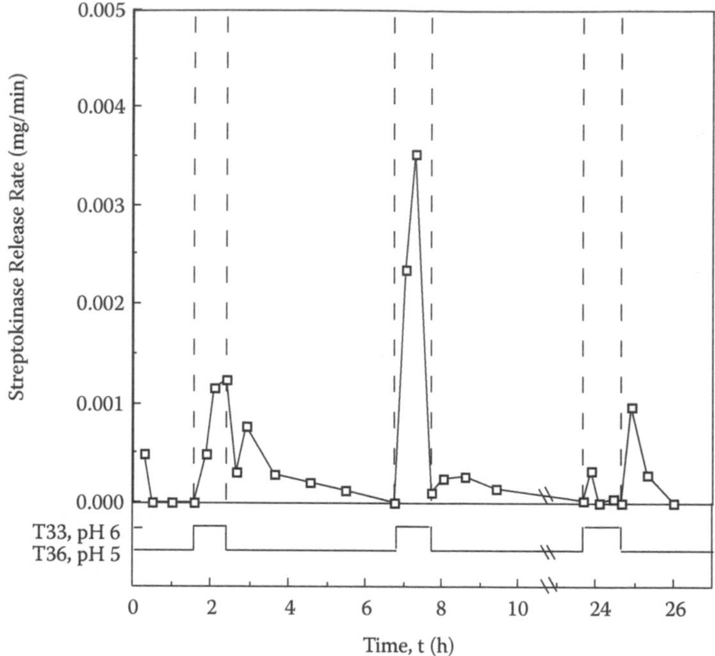

FIGURE 11.10
Release rate of streptokinase in buffered solutions from P(NIPAAm-*co*-MAA) hydrogels containing 88 mol% NIPAAm upon temperature change from 36 to 33°C and pH from 5 to 6 for 1 h and back to 36°C and pH 5 for a longer time interval.

11.3.5 Swelling-Controlled Release Systems

Swelling-controlled systems are formulations consisting of hydrophilic matrices from which the drug release is controlled by the inward flux of solvent molecules and the consequent swelling of the polymer matrix. In these systems the drugs are initially dissolved or dispersed in the glassy polymers. Upon contact with biological fluids, the polymer matrix begins to swell, and two distinct phases can be observed in the polymer: an inner glassy phase and a swollen rubbery phase. Drug molecules are able to diffuse out of the rubbery phase of the polymer. Clearly, the drug release is controlled by the velocity and position of the glassy–rubbery interface since no drug diffuses out of the glassy region of the polymer. A very important phenomenon of macromolecular relaxation takes place at the glassy–rubbery interface and significantly affects the drug release. The relative importance of macromolecular relaxation on the mechanism of drug release can be easily assessed by fitting experimental release data to Equation 11.22 and determining the exponent n.

$$\frac{M_t}{M_\infty} = kt^n \tag{11.22}$$

It is of paramount importance that Equation 11.22 be applied only to the first 60% of the total amount of drug released. Here, M_t and M_∞ are the amounts of drug released at time t and at equilibrium, respectively; k is a proportionality constant; and n is the diffusional exponent. Ritger and Peppas[62] performed analysis of the Fickian and non-Fickian diffusional behavior in terms of the value of the coefficient n.

Rigorous mathematical analysis of release behavior from this type of formulations falls into the category of moving-boundary problems. A detailed review of solutions of this type of problems is discussed by Crank.[63] Our laboratories have provided several theoretical models of drug release from swelling-controlled release devices. The first detailed model of drug transport with concentration-dependent diffusion coefficients was developed by Korsmeyer at al.[64] Lustig et al.[65] proposed a different model for drug release, based on rational thermodynamics, that also included a complete viscoelastic description of the polymer matrix and the concentration-dependent diffusion coefficient of the drug.

11.4 Smart Delivery from Complexation Polymers

There has been considerable interest in the interactions between macromolecules and the resulting polymer–polymer complexes during the past three decades. Many, if not all, of these investigations originated as studies on relatively simple model polymer systems whose ultimate goal was to shed some light on the processes occurring in the much more complex biological systems. Interpolymer complexes possess unique physical and chemical properties that are different from those of the initial components and have found applications in technology, medicine, and other fields.[66]

The unique properties of the complexes arise due to higher-order structures that are present in addition to first-order structures as a result of secondary binding forces. Interpolymer complexes can be classified based on the nature of the secondary binding forces as

Polyelectrolyte complexes
Hydrogen-bonding complexes
Stereocomplexes
Charge-transfer complexes

Polyelectrolyte complexes form due to Coulomb forces when two oppositely charged polyelectrolytes are mixed together. Mixing a Lewis acid, or proton-donating macromolecules, with a Lewis base, or proton-accepting macromolecules, results in the formation of hydrogen-bonding complexes. The formation of stereocomplexes is the result of van der Waals forces between isotactic and syndiotactic poly(methyl methacrylate), for example.

Charge-transfer complexes arise due to charge-transfer interactions between electron-accepting and electron-donating polymers.[67] A number of experimental techniques have been applied to study interpolymer complexes: potentiometry,[68] conductometry,[69] turbidimetry,[70] viscometry,[71] calorimetry,[72] sedimentation,[73] light scattering,[74] high-resolution H nuclear magnetic resonance (NMR) spetroscopy,[75] infrared,[76] and electron spectroscopy.[77]

It is the sensitivity of hydrogen bonds to their external environment that provides the pH dependence of physical properties of hydrogen-bonding interpolymer complexes that are of interest to us.

Antipina et al.[71] also investigated the complexation of poly(ethylene glycols) (PEG) with linear polymethacrylic and polyacrylic acids (PMMA and PAA) of molecular weights of 100,000 and 120,000, respectively. They used potentiometry and viscometry to examine the effect of molecular weight of the PEG chain, concentration of the polymers in solution, and pH and temperature of the medium on the complex formation. By monitoring the pH levels of solutions of the polymeric acids to which PEG chains of molecular weights of 1,000, 2,000, 3,000, 6,000, 15,000, and 40,000 were added, they found that there is a critical PEG chain length that is necessary for the complexation reaction to occur. The addition of 3000 molecular weight PEG to 0.1g/l solution of PMAA resulted in a gradual increase in the solution pH. A similar effect was observed for the PAA solutions; however, the rise in the solution pH took place upon the addition of PEG 6000, and it was less profound than for the PMAA solution. The addition of higher-molecular-weight PEGs was accompanied by steeper increase in the solution pHs, which leveled off at the polyacid/PEG molar ratio of 1. These results were in complete agreement with the results obtained by viscometry. Therefore, it was concluded that the PEG critical molecular weight needed to promote the complexation reaction with PMAA and PAA is 2000 and 6000, respectively. Additionally, the stability of the complex was suggested to be dependent on the chemical structure of the polyacid, which in turn would promote hydrophobic interactions contributing to the stability of the formed complexes.

Papisov et al.[78] performed calorimetric and potentiometric experiments to determine the thermodynamic parameters of the complex formation of PMAA and PAA with PEG. Their investigation focused on how the temperature and nature of the solvent affected the complex stability. They found that, in aqueous media, the enthalpy and entropy associated with the formation of the PMAA–PEG complex are positive, while in an aqueous mixture of methanol, both of the thermodynamic quantities become negative. The viscosities of aqueous solutions containing complexes of PMAA and PEG increase with decreasing temperature as a result of a breakdown of the complexes.

The temperature stability of the complexes seems to be dependent on the molecular weight of the PEG chain, i.e., the larger the PEG chain, the lower the temperature at which the complex dissociates. An important observation was that the complexation/decomplexation phenomena was reversible by changing the temperature of the system. The positive values of the thermodynamic parameters as well as the experimental observations clearly indicate the important role of hydrophobic interactions in the stabilization of the PMAA–PEG complexes. Since PAA is considerably more hydrophilic than PMAA, hydrophobic interactions do not play an important role in stabilizing the PAA–PEG complexes. This is demonstrated by the much lower value of ΔH of the PAA–PEG complex and the almost nonexistent effect of temperature on the stability of the PAA–PEG complex.

As mentioned above, the thermodynamic parameters change their signs when water solvent is replaced by a mixture of methanol and water. This is due to a change in the nature of stabilizing forces from hydrophobic interactions to hydrogen bonding the protons of the carboxylic groups and the ether oxygens on the PEG chains.

Miyoshi et al.[79,80] investigated interpolymer interactions, morphology, and chain dynamics of the poly(acrylic acid)–poly(ethylene oxide) (PAA–PEO) complex in the solid state using high-resolution solid-state ^{13}C NMR spectroscopy. In their study they utilized PEO and PAA with molecular weights of 20,000 and 90,000, respectively. They concluded that there are three hydrogen bonding forms of the carboxyl group in the PAA, namely

Complex form, with groups actively participating in the interpolymer hydrogen bonding with PEO chains

Dimeric form, with groups that form intramolecular hydrogen-bonding complexes among PAA molecules

Free form, with groups that are not part of the complex or dimeric forms

Philippova and Starodubtsev[81,82] have also extensively studied the complexation behavior of polyacids and PEG, especially the system of cross-linked PMAA and linear PEG. They observed that decreasing the molecular weight of PEG from 6000 to 1500 resulted in a drastic decrease in both the stability and equilibrium composition of the intermacromolecular complex. Analysis of dried polymer networks of PMAA with absorbed PEG chains by FT-IR (Fourier transform infrared) spectroscopy revealed the presence of two types of hydrogen-bonded structures: (a) dimers of methacrylic acid at absorption frequency of 1700 cm^{-1} and (b) interpolymer complexes of PMAA and PEG at 1733 cm^{-1}. In addition, they also suggested, based on the results of their studies, that the hydrogen-bonded dimer of PMAA forms preferentially to the intermacromolecular complex between the PMAA network and PEG chains.

Philippova et al.[83] utilized the pulse-field gradient (PFG) NMR method to investigate the translational mobility of linear PEG macromolecules absorbed in the loosely cross-linked PMAA hydrogels mentioned above. This study was also designed to explain why hydrogels of PMAA collapse when exposed to relatively low concentration of PEG (<5 wt%) and then reswell when immersed in solutions of high PEG concentrations (≈ 5 to 10 wt%). The results of the PFG NMR technique showed that there are two fractions of PEG macromolecules with different chain mobilities inside the collapsed gel

1. Some PEG molecules had self-diffusion characteristics similar to those of chains of the cross-linked network.
2. Some PEG molecules exhibited free-diffusion properties.

In contrast, the PEG chains inside the reswollen gels had a self-diffusion coefficient independent of time, indicating that they did not participate in interpolymer complexation with the PMAA network.

The formation of inter- and intrapolymer complexes has also been shown to affect the polymerization kinetics. For example, Ferguson et al.[84] investigated the influence of intrapolymer complexation on the kinetics of AA polymerization in the presence of copolymer matrices composed of *N*-vinylpyrrolidone (VP) and either acrylamide or styrene. They found that the polymerization rate reached a maximum in the vicinity of AA:VP = 1 for the VP–AAm matrix. This maximum in the polymerization rate was most pronounced in the presence of copolymer with the highest content of VP. Replacing the hydrophilic acrylamide with the more hydrophobic styrene in the copolymer matrix maximized the AA polymerization rate at a lower-than-equimolar ratio of AA to VP. Apparently, the hydrophilic groups of VP were interacting with the hydrophobic nucleus consisting of the styrene units in the VP–St copolymer and were thus unable to participate in the formation of the complex, unlike in the case of the VP–AAm copolymer matrix.

Bajoras and Makuska[85] investigated the effect of hydrogen-bonding complexes on the reactivities of (meth)acrylic and isotonic acids in a binary mixture of dimethyl sulfoxide and water using IR spectroscopy.[85] They demonstrated that, by altering the solvent composition, it was possible to carry out copolymerization in the azeotrope, which resulted in the production of homogeneous copolymers of definite compositions at high conversions. Furthermore, it was shown that the water–solvent fraction determines the rate of copolymerization and the reactivity ratios of the co-monomers. This in turn determines the copolymer composition.

Verhoeven et al.[86] addressed the possible effect of manufacturing conditions such as the presence of additives, namely PEG, on the polymerization of 2-hydroxyethyl methacrylate (HEMA) and MAA. The suspected complexation between the additive and the MAA monomer did not have a significant effect on the reaction ratios or on the copolymer composition and tactility, as demonstrated by ^{13}C NMR studies.

The effect of solvent on the polymerization kinetics of a system of PEG monomethacrylate and MAA was investigated by Smith and Klier.[87] In this work, a parameter of copolymer structure, namely sequence distribution in the copolymer, was estimated by determining the reactivity ratios of the monomers using ^{1}H NMR spectroscopy for two different solvents: D_2O and a mixture of ethanol and water. The paramount effect of the solvent on the polymerization process was exhibited by a profound change of r_1 and r_2 from 1.03 and 1.02 in the case of the D_2O solvent to 2.0 and 3.6 in the 50/50 wt% mixture of ethanol and water, respectively. In the context of polymer structure, this means that while the resulting polymer would have a random structure when synthesized in D_2O, it would have significantly large blocks of each co-monomer in its chains when manufactured in the ethanol–water mixture. Clearly, this would significantly affect the final properties of the polymer system. Due to their potential industrial use, these polymers were the subject of a more recent publication that characterized their molecular weight distributions and compositions.[87]

11.5 Conclusions and the Future of Smart Hydrogels

Smart polymeric networks can be prepared by (a) designing interactions between the building blocks of biocompatible networks and the desired specific ligands and then (b) stabilizing these interactions by a three-dimensional structure. Such structures must also be flexible enough to allow for diffusion of solvent and ligand into and out of the networks. Synthetic networks that can be designed to recognize and bind biologically significant molecules are of great importance and influence a number of emerging technologies. These artificial materials can be used as unique systems or incorporated into existing drug-delivery technologies that can aid in the removal or delivery of biomolecules and restore the natural profiles of compounds in the body. Such uses — as carriers for controlled and targeted drug delivery, micropatterned devices, and systems for biological recognition — have demonstrated the versatility of these biopolymeric materials, as indicated by Langer and Peppas.[88]

Of specific interest are applications requiring the patterning of vinyls, methacrylates, and acrylates during reaction, allowing for the formation of nanoscale three-dimensional structures. These micropatterned structures can be used for a host of applications, including cell adhesion, separation processes, the so-called factory-on-a-chip microscale reactors, and microfluidic devices. In recent years we have seen an explosion in the field of novel smart-polymer-based microfabricated and nanofabricated devices for drug delivery. Such devices seek to develop a platform of well-controlled functions in the micro or nano level. They include nanoparticulate systems, recognitive molecular systems, biosensing devices, and microfabricated and microelectronic devices.

In addition, biomimetic smart materials are now used to build biohybrid systems or even biomimetic materials (mimicking biological recognition) for drug delivery, drug targeting, and tissue engineering devices.[89] The synthesis and characterization of biomimetic gels and molecularly imprinted drug-release and protein-delivery systems is a significant focus of recent research. The design of a precise macromolecular chemical architecture that can recognize target molecules from an ensemble of closely related molecules has a large number of potential applications.[90]

The development of nanoparticulate smart systems for drug-delivery applications has achieved a level of sophistication never before seen in the field of drug delivery.[91] Using intelligent polymers, it is now possible to design new devices for *intelligent therapeutics*. Such systems can be employed for auto-feedback drug delivery, where the hydrogel is connected to a biosensor and responds to fast changes in external biological conditions. This concept can be used to develop novel insulin-delivery systems. Another novel use of these systems is for the release of human calcitonin. The physicochemical understanding of such hydrogels under the conditions of application is neither simple nor well developed. Considering that all of these carriers are ionic hydrogels and that several ionic and macromolecular components are involved, with their associated thermodynamically nonideal interactions, it is evident that the analysis and prediction of the swelling and drug-delivery behavior is rather complex.

References

1. Peppas, N.A. and Mikos, A.G., Hydrogel structure, in *Hydrogels in Medicine and Pharmacy,* Vol. 1, Peppas, N.A., Ed., CRC Press, Boca Raton, FL, 1986, p. 1.
2. Brannon-Peppas, L., Swelling of hydrogels, in *Absorbent Polymer Technology,* Brannon-Peppas, L. and Harland, R.S., Eds., Elsevier, Amsterdam, 1990, p. 45.
3. Peppas, N.A. and Merrill, E.W., PVA hydrogels: reinforcement of radiation-crosslinked networks by crystallization, *J. Polym. Sci., Polym. Chem. Ed.*, 14, 441, 1976.
4. Peppas, N.A. and Merrill, E.W., Differential scanning calorimetry of crystallized PVA hydrogels, *J. Appl. Polym. Sci.*, 20, 1457, 1976.
5. Peppas, N.A., Hydrogels of poly(vinyl alcohol) and its copolymers, in *Hydrogels in Medicine and Pharmacy,* Vol. 2, Peppas, N.A., Ed., CRC Press, Boca Raton, FL, 1986, p. 1.
6. Stauffer, S.R. and Peppas, N.A., Poly(vinyl alcohol) hydrogels prepared by freezing-thawing cyclic processing, *Polymer,* 33, 3932, 1992.
7. Hickey, A.S. and Peppas, N.A., Mesh size and diffusive characteristics of semicrystalline poly(vinyl alcohol) membranes prepared by freezing/thawing techniques, *J. Membr. Sci.*, 107, 229, 1995.
8. Peppas, N.A. and Mongia, N.K., Ultrapure poly(vinyl alcohol) hydrogels with mucoadhesive drug delivery characteristics, *Eur. J. Pharm. Biopharm.*, 43, 51, 1997.
9. Flory, P.J. and Rehner, J., Statistical mechanics of cross-linked polymer networks, II: swelling, *J. Chem. Phys.*, 11, 521, 1943.
10. Flory, P.J., Statistical mechanics of swelling of network structures, *J. Chem. Phys.*, 18, 108, 1950.
11. Flory, P.J., *Principles of Polymer Chemistry,* Cornell University Press, New York, 1953.
12. Ratner, B.D. and Hoffman, A.S., Synthetic hydrogels for biomedical applications, in *Hydrogels for Medical and Related Applications*, Andrade, J.D., Ed., ACS Symposium Series No. 31, ACS, Washington, DC, 1976, p. 1.
13. Peppas, N.A., *Hydrogels in Medicine*, CRS Press, Boca Raton, FL, 1986.
14. Peppas, N.A. and Langer, R., New challenges in biomaterials, *Science*, 263, 1715, 1994.

15. Park, K., *Controlled Release: Challenges and Strategies*, ACS, Washington, DC, 1997.
16. Peppas, N.A., Hydrogels and drug delivery, *Curr. Opin. Coll. Int. Sci.*, 2, 531, 1997.
17. Kabanov, V.A. and Papisov, I.M., Formation of complexes between complementary synthetic polymers and oligomers in dilute solution, *Vysokomol. Soed.*, A21, 243, 1979.
18. Bekturov, E.A. and Bimendina, L.A., Interpolymer complexes, *Adv. Polym. Sci.*, 43, 100, 1981.
19. Tsuchida, E. and Abe, K., Interactions between macromolecules in solution and intermacromolecular complexes, *Adv. Polym. Sci.*, 45, 1, 1982.
20. Klier, J. and Peppas, N.A., Structure and swelling behavior of poly(ethylene glycol)/poly(methacrylic acid) complexes, in *Absorbent Polymer Technology*, Brannon-Peppas, L. and Harland, R.S., Eds., Elsevier, Amsterdam, 1990, p. 147.
21. Bell, C.L. and Peppas, N.A., Biomedical membranes from hydrogels and interpolymer complexes, *Adv. Polym. Sci.*, 122, 125, 1995.
22. Peppas, N.A., Physiologically responsive gels, *J. Bioact. Compat. Polym.*, 6, 241, 1991.
23. Wichterle, O. and Lim, D., Hydrophilic gels for biological use, *Nature*, 185, 117, 1960.
24. Langer, R., Drug delivery and targeting, *Nature*, 392, 5, 1998.
25. Huang, Y., Leobandung, W., Foss, A., and Peppas, N.A., Molecular aspects of muco- and bioadhesion: tethered structures and site-specific surfaces, *J. Controlled Release*, 65, 63, 2000.
26. Peppas, N.A. and Sahlin, J.J., Hydrogels as mucoadhesive and bioadhesive materials: a review, *Biomaterials*, 17, 1553, 1996.
27. Stolnik, S., Illum, L., and Davis, S.S., Long circulating microparticulate drug carriers, *Adv. Drug Del. Rev.*, 16, 195, 1995.
28. Burnham, N.L., Polymers for delivering peptides and proteins, *Am. J. Hosp. Pharm.*, 51, 210, 1994.
29. Lowman, A.M. and Peppas, N.A., Hydrogels, in *Encyclopedia of Controlled Drug Delivery*, Mathiowitz, E., Ed., Wiley, New York, 1999, p. 397.
30. Ratner, B.D. and Hoffman, A.S., Synthetic hydrogels for biomedical applications, in *Hydrogels for Medical and Related Applications*, Vol. 3, Andrade, J.D., Ed., ACS, Washington, DC, 1976, p. 1.
31. Peppas, N.A. and Mikos, A.G., Preparation methods and structure of hydrogels, in *Hydrogels in Medicine and Pharmacy*, Vol. I, Peppas, N.A., Ed., CRC Press, Boca Raton, FL, 1986, p. 1.
32. am Ende, M.T. and Mikos, A.G., Diffusion-controlled delivery of proteins from hydrogels and other hydrophilic systems, in *Protein Delivery: Physical Systems*, Sanders, L.M. and Hendren, R.W., Eds., Plenum Press, Tokyo, 1997, p. 139.
33. Peppas, N.A. and Barr-Howell, B.D., Characterization of the cross-linked structure of hydrogels, in *Hydrogels in Medicine and Pharmacy*, Vol. I, Peppas, N.A., Ed., CRC Press, Boca Raton, FL, 1986, p. 27.
34. Treloar, R.G., *The Physics of Rubber Elasticity*, 2nd ed., Oxford University Press, Oxford, U.K., 1958.
35. Peppas, N.A. and Merrill, E.W., Crosslinked poly(vinyl alcohol) hydrogels as swollen elastic networks, *J. Appl. Polym. Sci.*, 21, 1763, 1977.
36. Katchalsky, A. and Michaeli, I., Polyelectrolyte gels in salt solution, *J. Polym. Sci.*, 15, 69, 1955.

37. Brannon-Peppas, L. and Peppas, N.A., Equilibrium swelling behavior of pH-sensitive hydrogels, *Chem. Eng. Sci.*, 46, 715, 1991.
38. Ricka, J. and Tanaka, T., Swelling of ionic gels: quantitative performance of the Donnan theory, *Macromolecules*, 17, 2916, 1984.
39. Flory, P.J., Rabjohn, N., and Shaffer, M.C., Dependence of elastic properties of vulcanized rubber on the degree of cross linking, *J. Polym. Sci.*, 4, 225, 1949.
40. Flory, P.J., Rabjohn, N., and Shaffer, M.C., Dependence of elastic properties of vulcanized rubber on the degree of cross linking, *J. Polym. Sci.*, 4, 435, 1949.
41. Silliman, J.E., Network Hydrogel Polymers: Application in Hemodialysis, Sc.D. thesis, Massachusetts Institute of Technology, Cambridge, MA, 1972.
42. Peppas, N.A. and Merrill, E.W., Crosslinked poly(vinyl alcohol) hydrogels as swollen elastic networks, *J. Appl. Polym. Sci.*, 21, 1763, 1977.
43. Mark, J.E., The use of model polymer networks to elucidate molecular aspects of rubberlike elasticity, *Adv. Polym. Sci.*, 44, 1, 1982.
44. Anseth, K.S., Bowman, C.N., and Brannon-Peppas, L., Mechanical properties of hydrogels and their experimental determination, *Biomaterials*, 17, 1647, 1996.
45. Patterson, K.G., Padgett, S.J., and Peppas, N.A., Microcrystallinity and three-dimensional network structure in plasticized PVC, *Colloid Polym. Sci.*, 260, 851, 1982.
46. Lowman, A.M. and Peppas, N.A., Analysis of the complexation/decomplexation phenomena in graft copolymer networks, *Macromolecules*, 30, 4959, 1989.
47. Canal, T. and Peppas, N.A., Correlation between mesh size and equilibrium degree of swelling of polymeric network, *J. Biomed. Mater. Res.*, 23, 1183, 1989.
48. Narasimhan, B. and Peppas, N.A., Molecular analysis of drug delivery systems controlled by dissolution of the polymer carrier, *J. Pharm. Sci.*, 86, 297, 1997.
49. Narasimhan, B. and Peppas, N.A., The role of modelling studies in the development of future controlled release devices, in *Controlled Drug Delivery: Challenges and Strategies*, Park, K., Ed., ACS, Washington, DC, 1997, pp. 529–557.
50. Langer, R.S. and Peppas, N.A., Chemical and physical structure of polymers as carriers for controlled release of bioactive agents: a review, *J. Macromol. Sci. Rev. Macromol. Chem. Phys.*, C23, 61, 1983.
51. Narasimhan, B., Mallapragada, S.K., and Peppas, N.A., Release kinetics: data interpretation, in *Encyclopedia of Controlled Drug Delivery*, Mathiowitz, E., Ed., Wiley, New York, 1999, p. 921.
52. Hennink, W.E., Franssen, O., van Dijk-Wolthuis, W.N.E., and Talsma, H., Dextran hydrogels for the controlled release of proteins, *J. Control Rel.*, 48, 107, 1997.
53. Crank, J., *The Mathematics of Diffusion*, 2nd ed., Oxford University Press, New York, 1975.
54. Eyring, H., Theory of rate processes, *J. Chem. Phys.*, 4, 283, 1936.
55. Fujita, H., Diffusion in polymer-diluent systems, *Fortschr. Hochpolym. Forsch*, 3, 1, 1961.
56. Yasuda, H. and Lamaze, C.E., Permselectivity of solutes in homogeneous water-swollen polymer membranes, *J. Macromol. Sci. Phys.*, B5, 111, 1971.
57. Peppas, N.A. and Reinhart, C.T., Solute diffusion in swollen membranes, I: a new theory, *J. Membr. Sci.*, 15, 275, 1983.
58. Peppas, N.A. and Moynihan, H.J., Solute diffusion in swollen membranes, IV: theories for moderately swollen networks, *J. Appl. Polym. Sci.*, 30, 2589, 1985.
59. Harland, R.S. and Peppas, N.A., Solute diffusion in swollen membranes, VII: diffusion in semicrystalline networks, *Colloid Polym. Sci.*, 267, 218, 1989.
60. Lightfoot, E.N., *Transport Phenomena and Living Systems*, Wiley, New York, 1974.

61. Faxén, H., Die bewegung einer starren kugel langs der achse eines mit zaherer flussigkeit fefullten rohres, *Arkiv. Mat. Astronom. Fys.*, 17, 27, 1923.
62. Ritger, P.L. and Peppas, N.A., A simple equation for description of solute release, I: Fickian and non-Fickian release from non-swellable devices in the form of slabs, spheres, cylinders or discs, *J. Controlled Rel.*, 5, 23, 1987.
63. Crank, J., *Free and Moving Boundary Problems*, 2nd ed., Oxford University Press, New York, 1975.
64. Korsmeyer, R.W., Lustig, S.R., and Peppas, N.A., Solute and penetrant diffusion in swellable polymers, I: mathematical modeling, *J. Polym. Sci. Polym. Phys.*, 24, 395, 1986.
65. Lustig, S.R., Caruthers, J.M., and Peppas, N.A., Theories of penetrant transport in glassy polymers, *J. Membr. Sci.*, 48, 281, 1990.
66. Bekturov, E.A. and Bimendina, L.A., Interpolymer complexes, *Adv. Polym. Sci.*, 41, 99, 1981.
67. Tsuchida, E. and Abe, K., Interactions between macromolecules in solution and intermacromolecular complexes, *Adv. Polym. Sci.*, 45, 1, 1982.
68. Bailey, F.E., Lundberg, R.D., and Callard, R.W., Some factors affecting the molecular association of poly(ethylene oxide) and poly(acrylic acid) in aqueous solutions, *J. Polym. Sci.*, A2, 845, 1964.
69. Bimendina, L.A., Roganov, V.V., and Bekturov, E.A., Hydrodynamic properties of complexes of polymethacrylic acid-polyvinylpyrrolidone in solutions, *J. Polym. Sci. Polym. Symp.*, 44, 65, 1974.
70. Sato, H. and Nakajima, A., Formation of a polyelectrolyte complex from carboxymethyl cellulose and poly(ethyleneimine), *Polym. J.*, 7, 241, 1975.
71. Antipina, A.D., Baranovskii, V., Papisov, I.M., and Kabanov, V.A., Equilibrium peculiarities in the complexing of polymeric acids with poly(ethylene glycols), *Vysokomol. Soed.*, A14, 941, 1972.
72. Biros, I., Masa, L., and Pouchly, J., Calorimetric investigation of the formation of a stereocomplex in poly(methyl methacrylate) solutions, *Eur. Polym. J.*, 10, 629, 1974.
73. Bimendina, L.A., Tleubaeva, G.S., and Bekturov, E.A., Complexing of a copolymer of maleic anhydride and methacrylic acid with poly(vinylpyrrolidone) in solutions, *Vysokomol. Soed.*, A19, 71, 1977.
74. Liquori, A.M., De Santis, S.M., and D'Alagni, M., Dilute solution properties of the stereocomplex between isotactic and syndiotactic poly(methyl methacrylate), *J. Polym. Sci.*, B4, 943, 1966.
75. Spevacek, J. and Schneider, B., HR [high resolution]–NMR study of formation and structure of the stereocomplex of poly(methyl methacrylate) in solution, *Makromol. Chem.*, 175, 2939, 1974.
76. Philippova, O.E. and Starodubtsev, S.G., Intermacromolecular complexation between poly(methacrylic acid) hydrogels and poly(ethylene glycol), *J. Mater. Sci. Pure Appl. Chem.*, 11, 1893, 1995.
77. Bakeev, N.F., Pshezhetsky, V.C., and Kargin, V.A., *Vysokomol. Soed.*, 1, 1812, 1959.
78. Papisov, I.M., Baranovskii, V., Sergieva, Y., Antipina, A.D., and Kabanov, V.A., Thermodynamics of complex formation between polymethacrylic and polyacrylic acids and polyethylene glycols: calculation of temperatures of breakdown of complexes of oligomers and matrices, *Vysokomol. Soed.*, 5, 1133, 1974.
79. Miyoshi, T., Takegoshi, K., and Hikichi, K., High-resolution solid state ^{13}C NMR study of the interpolymer interaction, morphology and chain dynamics of the poly(acrylic acid)/poly(ethylene oxide) complex, *Polymer*, 38, 2315, 1997.

80. Miyoshi, T., Takegoshi, K., and Hikichi, K., High-resolution solid-state ^{13}C nuclear magnetic resonance study of a polymer complex: poly(methacrylic acid)/poly(ethylene oxide), *Polymer*, 37, 11, 1996.
81. Philippova, O.E., Karibyants, N.S., and Starodubtzev, S.G., Conformational changes of hydrogels of poly(methacrylic acid) induced by interaction with poly(ethylene glycol), *Macromolecules*, 27, 2398, 1994.
82. Philippova, O.E. and Starodubtzev, S.G., Intermacromolecular complexation between poly(methacrylic acid) hydrogels and poly(ethylene glycol), *J. Macromol. Sci.: Pure Appl. Chem.*, 11, 1893, 1995.
83. Skirda, V.D., Aslanyan, I.Y., Philippova, O.E., Karibyants, N.S., and Khokhlov, A.R., Investigation of translational motion of poly(ethylene glycol) macromolecules in poly(methacrylic acid) hydrogels, *Macromol. Chem. Phys.*, 200, 2152, 1999.
84. Drescher, B., Scranton, A.B., and Klier, J., Synthesis and characterization of polymeric emulsifiers containing reversible hydrophobes: poly(methacrylic acid-g-ethylene glycol), *Polymer*, 42, 49, 2001.
85. Bajoras, G. and Makuska, R., Peculiarities of radical homo- and copolymerization of acrylic, methacrylic and itaconic acids in complexing solutions, *Polym. J.*, 18, 955, 1986.
86. Verhoeven, J., Peschier, L.J.C., van Det, M.A., Bouwstra, J.A., and Junginger, H.E., The physico-chemical characterization of poly(2-hydroxyethyl methacrylate-*co*-methacrylic acid), 1: effect of PEG 400 on the tacticity and reactivity ratios, *Polymer*, 30, 1942, 1989.
87. Smith, B.L. and Klier, J., Determination of monomer reactivity ratios for copolymerizations of methacrylic acid with poly(ethylene glycol) methacrylate, *J. Appl. Polym. Sci.*, 68, 1019, 1998.
88. Langer, R. and Peppas, N.A., Advances in biomaterials, drug delivery, and bionanotechnology, *AIChE J.*, 49, 2990, 2003.
89. Dillow, A.K. and Lowman, A., Eds., *Biomimetic Materials and Design: Biointerfacial Strategies, Tissue Engineering and Targeted Drug Delivery*, Dekker, New York, 2002.
90. Golumbfskie, A.J., Pande, V.S., and Chakraborty A.K., Simulation of biomimetic recognition between polymers and surfaces, *Proc. Natl. Acad. Sci.*, 96, 11707, 1999.
91. Peppas, N.A. and Byrne, M.E., New biomaterials for intelligent biosensing, recognitive drug delivery and therapeutics, *Bull. Gattefossé*, 96, 23, 2003.

Chapter 12

Polymeric Carriers for Regional Drug Therapy

Anupama Mittal, Deepak Chitkara, Neeraj Kumar, Rajendra Pawar, Avi Domb, and Ben Corn

CONTENTS

Abbreviations

BCNU	1,3-bis-(2-chloroethyl)-l-nitrosourea (BCNU, carmustine)
5-FU	5-fluorouracil
HA	hyaluronic acid
HEMA	2-hydroxyethylmethacrylate
MePEG	methoxypolyethylene glycol
PCL	poly(ε-caprolactone)
P(CPP-SA)	poly[bis(*p*-carboxy phenoxy)propane-sebacic acid]
PDLA	poly(D-lactide)
PDLLA	poly(D,L-lactide)
PEG	polyethylene glycol
PELA	poly-D,L-lactide-*co*-poly(ethylene glycol)
PGA	poly(glycolic acid)
PHA	poly(hydroxyalkanoate)s
PLA	poly(lactic acid)
PLGA	poly(D,L-lactic-*co*-glycolic acid)
PLLA	poly(L-lactide)
POE	poly(orthoester)
PSA	poly(sebacic acid)
VEGF	vascular endothelial growth factor

12.1 Introduction

Drug research has evolved and matured through several phases beginning with the botanical phase of early human civilization, through the synthetic chemistry age in the mid-20th century, and finally the biotechnology era at the dawn of the 21st century. The advent of combinatorial chemistry and advances in high-throughput screening, functional genomics, and proteomics has expanded the therapeutic armamentarium to incredible levels. This definitely calls for an aggressive search for strategies to deliver these potent molecules. Conventional routes of administration — oral ingestion as well as IV (intravenous) and IM (intramuscular) injections — distribute the drug molecule to all body parts, including targeted and nontargeted sites. This creates a burden on the whole-body system, while the requirement is only at a particular site in the body. In addition, conventionally delivered drugs become diluted in the blood and body fluids, resulting in an inadequate drug concentration at the diseased site while causing toxicity to healthy tissue. An attractive alternative for the delivery of these pharmacologically active compounds is regional/localized drug delivery.[1]

Localized drug delivery is defined as a method of delivering a drug (for its local action) from a dosage form to a particular site in the biological system where its entire pharmacological effect is desired.[2] To accomplish this targeted delivery, biodegradable and biocompatible polymeric drug-delivery systems (DDSs) are being explored extensively. After the commercial success of products such as Lupron Depot®, Zoladex®, Norplant®, and Gliadel®, interest in this field increased. One can easily outline the advantages of regional drug therapy over systemic drug delivery, the major one being high locoregional concentration of therapeutic agents with prolonged retention time. In localized delivery, a lower dose is required to fill the volume of distribution and, hence, chances of various adverse effects are reduced or completely eliminated, since the high systemic dose that otherwise would be required to achieve the therapeutic concentration at the diseased site is obviated.[2] Some of the therapeutic agents with a relatively short half-life, specifically proteins and peptides, and other biologically unstable biomolecules such as nucleic acids and oligonucleotides, can also be delivered locally with minimal loss in therapeutic activity.[3] Table 12.1 lists the various advantages of regional drug delivery.

Localized drug delivery has been most successfully exploited in the field of cancer, particularly brain tumors, neurological disorders, vascular complications, bone infections, and retinal diseases. Numerous DDSs have been explored for localized delivery, including implants and injectable systems, such as microspheres and *in situ* gelling systems. Certainly all types of diseased states favor regional therapy; however, local treatment is not feasible in all circumstances, as in the case of metastatic tumors, where the cells possess the ability to colonize other tissues, either by direct growth into adjacent tissue through invasion or by implantation into distant sites via body fluids.

Several strategies have been explored to deliver drugs to a specific site or body compartment, but delivery via polymeric carriers is one of the simplest and most successful approaches. Polymers for localized application can play structural and functional roles. Polymer science has been undergoing tremendous advancement since the last decade in terms of chemical modifications, synthesis of copolymers, and development of new delivery systems.

TABLE 12.1

Advantages of Regional Drug Therapy

Therapeutically effective dose required at the site is reduced, thus diminishing or completely eliminating adverse effects
Drugs with a relatively short half-life can also be delivered locally with minimal loss in therapeutic activity, e.g., anticancer agents
Drugs with low bioavailability can be directed directly to the required site
Patient-to-patient variability in drug pharmacokinetics is reduced, which is of significance in the case of drugs with a narrow therapeutic index
Obviates the need of premedication for drugs that show adverse effects when given systemically
Intratumoral delivery is not limited by poor blood supply caused by radiation therapy or surgery
Is advantageous for drugs with dose-dependent activity

These modifications have resulted in a range of biodegradable polymers of natural and synthetic origin with specific properties conducive to the desired applications. Many natural biodegradable polymers, such as chitin and chitosan, alginate, gelatin, hyaluronan, etc., and synthetic polymers, such as polyesters, polycaprolactone, polyorthoesters, polyanhydrides, etc., are being tailored to make them conducive for regional therapy. Synthetic polymers are preferred over natural ones, since they are presumed to be free of immunogenicity. Their physicochemical properties are also more predictable, reproducible, and easier to modify, and they can be fabricated into wafers, flexible films, or linked beads and rods to fit various diseased sites.

The versatility of polymeric carriers has generated huge interest in the scientific community and has resulted in an enormous amount of work done in the area of localized drug delivery using polymers. However, the advantages of using polymers must be weighed against the following concerns[4]:

Toxicity of the polymer and its degradation products in the body, i.e., biocompatibility

Overall cost of polymeric drug delivery systems

Problems associated with release, i.e., dose dumping or release failure

Discomfort caused by the system itself or by the means of inserting the delivery system

This chapter provides a review of both introductory as well as state-of-the-art information on the polymeric carriers used in regional drug therapy for the diseased states that are most exploited for localized treatment. Major emphasis has been placed on the biodegradable polymers of both natural and synthetic origin, focusing on their sources, designs, physicochemical properties, and a few applications. A brief mention is also made regarding the drug-delivery systems employed to accomplish the regional therapy, including microspheres, implants, and *in situ* gelling systems. The studies compiled are not encyclopedic; rather, specific examples are taken to highlight certain points.

12.2 Diseases Demanding Regional Therapy

Regional drug therapy has been extensively explored for several diseases, especially for the treatment of ailments localized in a specific organ or tissue, as enumerated in Table 12.2. Research in this field has mainly focused on cancer chemotherapy, owing to the extreme toxicity and lack of specificity of most anticancer agents. However, development of such systems has expanded in the past few years to encompass delivery systems for brain disorders, bacterial infections, and cardiovascular diseases[5]; thrombosis,

TABLE 12.2

Diseases Demanding Regional Drug Therapy and Associated Therapeutic Agents

Diseases Demanding Regional Therapy	Therapeutic Agents
Localized solid tumors, e.g., brain, head and neck, bone, breast, ovary, and colon	anticancer drugs
Chronic and acute infections of bone, soft tissue, gingiva, and the eye; and for surgical prophylaxis	antibiotics
Blood vessel and heart disorders	antithrombogenic and antiproliferative agents
Neurological disorders	neurotransmitters and neuromodulators
Chronic and acute pain	local anesthetics and anti-inflammatory agents

restenosis, and osteomyelitis[6]; local infections[7]; and glaucoma and retinal disorders[8] that are difficult to treat by systemic therapy.[1]

12.2.1 Cancer

A major hurdle in curing cancer is the ineffectiveness of systemically administered chemotherapy. Systemic therapy may be curative for some tumors, but it is rarely effective in treating most of the solid malignant tumors, especially those of the brain or liver, the major hurdle being the delivery of therapeutic agents to these solid tumors because of the impediments to drug transport that are posed by their unique ultrastructural properties.[9] The heterogeneous distribution of blood vessels combined with aberrant branching and tortuosity result in uneven and slowed blood flow within tumors. Moreover, the absence of a functional lymphatic system results in an elevated interstitial pressure that retards the convective transport of high-molecular-weight drugs. In addition to the transport barriers, a second problem in cancer therapy is that the chemo- and radiotherapeutic agents are nonselective in their action; they are cytotoxic to both the healthy cells as well as tumor cells, which leads to undesirable side effects during anticancer therapy. If the dose is decreased to alleviate undesirable side effects, this also reduces the efficacy of the therapy itself; conversely, increasing the dose, while more effective for controlling tumor growth, leads to increased toxicity of healthy tissues.[9] Localized drug delivery helps to overcome this stalemate.

Localized delivery of low-therapeutic-index drugs is found to be useful for many common solid tumors, including breast, brain, and prostate, which do not respond well to conventional systemic chemotherapy.[10] Adjuvant chemotherapy routinely follows local treatment of breast cancer, colon cancer, and rectal cancer. Localized delivery of therapeutic agents has been accomplished by implantation of drugs directly into the brain; this also helps overcome the limitation of inadequate penetration of the blood–brain barrier and poor patient compliance. In addition, the implants can be placed in specific regions of the brain, thereby avoiding undesirable distribution of

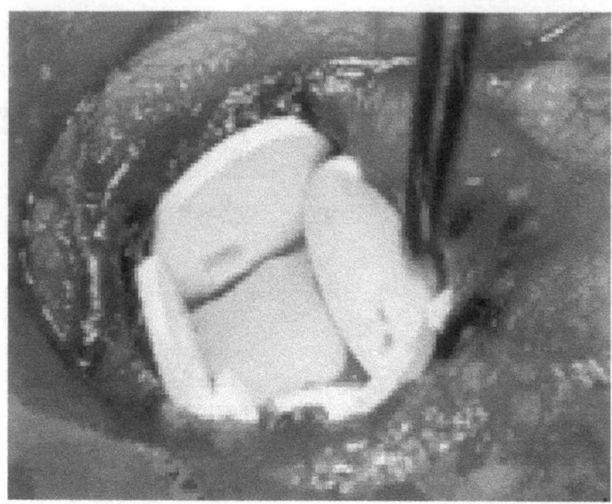

FIGURE 12.1
Gliadel wafers being inserted into the tumor cavity. (From Lesniak, M.S. and Brem, H., *Nature Rev. Drug Discov.*, 3, 499, 2004. With permission.)

the drug throughout the brain, a common problem when drugs are administered into the cerebral spinal fluid.[5] The initial studies of drug delivery to the brain for the treatment of malignant glioma by using a polyanhydride implant were performed with the chemotherapeutic alkylating agent 1,3-bis-(2-chloroethyl)-l-nitrosourea (BCNU, carmustine)[11] (Figure 12.1). Small polymer wafers loaded with 3.8 wt% BCNU were tested in rats, rabbits, and monkeys for their safety and effectiveness before use in humans.[12] Gliadel, the drug-loaded polymer wafers of carmustine in poly(bis[*p*-carboxy phenoxy] propane-sebacic acid) P(CPP-SA, 20:80) is now an FDA-approved polymeric implant for the treatment of brain tumors.[13] A polyanhydride-film-releasing cisplatin was also found to be effective in inhibiting tumor growth in nude mice. This provides a possible solution to human squamous cell carcinoma of the head and neck.[14] (See Vignette 12.1.[15])

12.1 Gliadel (polifeprosan 20 with carmustine implant)

Gliadel is a white, dime-sized wafer made up of a biocompatible polymer that contains the cancer chemotherapeutic drug, carmustine (BCNU). Gliadel wafer is designed to deliver carmustine directly into the surgical cavity created when a brain tumor is resected. Once implanted, Gliadel slowly dissolves, releasing high concentrations of BCNU into the tumor site, targeting microscopic tumor cells that sometimes remain after surgery. The specificity of Gliadel minimizes drug exposure to other areas of the body.

Rabbits implanted with wafers containing 3.85% carmustine showed no detectable levels of carmustine in the plasma or cerebrospinal fluid.

More than 70% of the copolymer degrades by three weeks. The metabolic disposition and excretion of the monomers differ. Carboxyphenoxypropane is eliminated by the kidney, and sebacic acid, an endogenous fatty acid, is metabolized by the liver and expired as CO_2 in animals.

12.2.2 Neurological Disorders

Polymeric microspheres have been employed to deliver neuroactive agents to a specific location within the brain in a controlled and prolonged manner without risk of infection from indwelling cannulae. Because of their size, these microparticles can easily be implanted by stereotaxy in discrete, precise, and functional areas of the brain without damaging the surrounding tissue. Presently, this method is most frequently applied in the fields of neuro-oncology and neurodegenerative diseases.[16] Intracerebral delivery of the cholinergic agent bethanecol as a potential treatment for Alzheimer's disease was studied by Howard et al.[17] Polymeric microspheres impregnated with bethanecol were injected into the brain of rats in which memory deficit (assessed by maze performance) had been induced by bilateral fimbria-fomix lesions. These animals demonstrated a marked improvement in performance within 10 days that lasted for the entire 40 days of testing.

A potential treatment for Parkinson's disease using injectable microspheres releasing dopamine has also been studied. Parkinson's disease is characterized by a decrease in both the brain levels of dopamine and in the number of dopaminergic neurons in the brain. Rats implanted on the ipsilateral side with dopamine microspheres exhibited contralateral rotations with an amplitude comparable with that elicited by a previously administered test dose of apomorphine (a standard experimental compound used to restore dopaminergic function), but with longer duration. Control animals injected with placebo microspheres did not show any consistent rotational behavior. It appears that direct delivery of a neurotransmitter using a biodegradable polymer to the brain can restore function for prolonged periods.[18] A novel, localized method for potential delivery of therapeutic agents to the injured spinal cord has also been investigated. The strategy consists of dispersing therapeutic agents in a polymeric solution that gels after injection into the subarachnoid space. This minimally invasive DDS provided an alternative and safe method to deliver therapeutic agents intrathecally.[19]

12.2.3 Vascular Disorders

Major complications of most of the vascular surgical procedures involve intravascular thrombosis and delayed stenosis secondary to intimal proliferation.

Restenosis of the coronary artery following coronary angioplasty induces localized injury to the vessel wall, which leads to the release of vasoactive, thrombogenic, and mitogenic factors that induce processes that cause renarrowing at the injured site. Mechanical devices designed to provide a larger lumen at the completion of the procedure and pharmacologic therapies designed to inhibit the neointimal proliferation have both been used to reduce restenosis. Debulking procedures, such as directional coronary atherectomy, rotational atherectomy, and laser angioplasty, unfortunately have not been associated with a clear-cut reduction in the incidence of restenosis. Localized drug delivery by the agency of coronary artery stents,[20] however, has reduced the restenosis rate following coronary angioplasty by approximately 50%. A drug-eluting stent has three elements: the metallic stent, a drug-carrier vehicle or coating, and a pharmacologic agent that interferes with local neointimal proliferation.[21,22] The pharmacologic agents used to interfere with local neointimal formation can be classified as antiproliferative, anti-inflammatory or immune modulating, antimigratory, antithrombotic, and prohealing agents.[21,22] There is currently a great deal of active research into evaluating different and novel pharmacologic agents that also may limit neointimal proliferation. Clinical outcomes studies comparing drug-eluting stents with bare-metal stents have shown dramatic reductions in restenosis with drug elution. The first human experience with a sirolimus drug-eluting stent (Cypher, Cordis Corp.), the Ravel trial, and finally the SIRIUS multicenter U.S. phase III trial have all demonstrated a substantial reduction in angiographic restenosis (>50% stenosis).[21,23,24] In-stent restenosis rates of zero to 3% and in-segment or vessel restenosis rates of up to only 9% were observed with the drug-eluting stents, compared with 33% in the bare-metal stent arm. These series of clinical trials led the FDA to approve the Cypher stent for clinical use (April 2003).

Drug-eluting polymeric matrices have been employed to inhibit injury-induced thrombosis and neointimal thickening. The effectiveness of heparin-releasing devices was evaluated on rats that underwent carotid artery balloon catheter injury. Thin flexible sheets comprising biodegradable polyanhydrides and polyesters loaded with heparin were evaluated *in vitro* and *in vivo*. Heparin was released at a constant concentration for one to three weeks from these films. The control group showed a significant reduction of the artery internal diameter, while the treated rats showed minimal or no proliferation of smooth muscle cells. Other agents, such as dexamethasone (a powerful anti-inflammatory agent), hirulog (a specific inhibitor of thrombin), and antisense oligonucleotides, incorporated in a polymeric system, were found to be effective in preventing smooth muscle cell proliferation following endothelial injury.[25]

12.2.4 Bone Infections

Despite the fact that many new, highly effective antibiotics have been developed in recent years, osteomyelitis (chronic bone infection) is still an important therapeutic and diagnostic problem. Osteomyelitis is a bone infection that is characterized by the presence of dead avascular bone. The current treatment consists of surgical debridement of the infected area, followed by high-dose antibiotic therapy. Four-to-six week systemic antibiotic therapy is required along with bone and soft tissue debridement in the therapy of chronic osteomyelitis. Systemic antibiotic therapy alone carries a high risk of failure, owing to the inadequate antibiotic concentrations reaching the infected tissues.[26] Infected bone located in areas of tissue that have an inadequate blood supply, particularly in patients with diabetes and with vascular insufficiency, may be particularly refractory to treatment. A logical solution to this problem is the local delivery of antibiotics for a few weeks using a biodegradable polymer.

Various polymers have been used as implant carriers for the local delivery of antibiotics.[26,27] Poly(methylmethacrylate), a nondegradable polymer carrier, was the first device to be used clinically in Europe (Septopal TM, E. Merck, Germany) for the delivery of gentamicin in bone. Since then, several biodegradable polymers, including poly(propylenefumarate), polylactide, polyanhydride, and the natural materials of collagen, gelatin, and inorganic phosphates, have been used to deliver antibiotics to infected bone. Linked beads of biodegradable polyanhydride containing gentamicin were formulated for implantation at the time of surgical treatment of osteomyelitis and soft-tissue infection. These beads degrade over a period of weeks and, in the process, release gentamicin at high local concentrations. At high concentrations, gentamicin is known to kill most organisms that are commonly associated with osteomyelitis. Various random copolyesters of 3-hydroxybutyrate with 3-hydroxyvalerate and 4-hydroxybutyrate have been used in the construction of biodegradable implantable rods for the local delivery of antibiotics (Sulperazone™ and Duocid™) in chronic osteomyelitis therapy.[28] Biodegradable microspheres of PLGA have also been manufactured containing tobramycin to treat osteomyelitis. The most effective treatment was found to be biodegradable microspheres of tobramycin plus parenteral treatment with cefazolin in a rabbit osteomyelitis model.[29] Attempts have also been made to treat the bacterial biofilm infections associated with the bone using PLGA microspheres of ofloxacin.[30] Teicoplanin microspheres and vancomycin beads of biodegradable polymer have also been explored for treatment of bone infections.[31,32]

12.2.5 Retinal Disorders

Diseases of the retina are difficult to treat with systemically administered drugs because of the blood–retinal barrier and potential systemic toxicity. The antiviral effect of ganciclovir microspheres was studied in rabbit eyes inoculated with human cytomegalovirus; vitritis, retinitis, and optic neuritis were observed to be cured without any adverse tissue reaction and minimal focal disruption of the retinal architecture.[33]

The use of steroids to treat a number of retinal diseases is gaining widespread acceptance. Intravitreal injection of corticosteroids like budesonide to treat diabetic retinopathy has proved effective, since these prevent neovascularization and microvascular alterations (by inhibiting the vascular endothelial growth factor [VEGF] expression), which are common manifestations of this ocular disease.[34] Controlled intravitreous release of retinoic acid has been studied by injecting drug-loaded microspheres against proliferative vitreoretinopathy. Sustained release of retinoic acid from PLGA microspheres was obtained *in vitro* for 40 days. A single injection of retinoic acid-loaded microspheres in suspension in BSS was effective in reducing the incidence of tractional retinal detachment after two months in a rabbit model of proliferative vitreoretinopathy.[35]

12.3 Polymeric Carriers for Regional Drug Therapy

Polymeric carriers can be broadly classified into biodegradable and nonbiodegradable, depending upon their ability to degrade over the time scale of observation. By definition, biodegradable polymers are the ones that undergo biotransformation, either enzymatically or nonenzymatically, to yield biocompatible or nontoxic by-products that are excreted via normal physiological pathways. An ideal biodegradable polymer should possess adequate mechanical properties, should be sterilizable and easily processable, and should undergo *in vivo* metabolism into such by-products that do not produce inflammation or toxic response. On the basis of source/origin, biodegradable polymers can be grouped into natural and synthetic. In the class of natural polymers, we have polysaccharides like chitin and chitosan, alginic acid, etc., while polypeptides include gelatin. On the basis of degradable linkages, synthetic polymers have been divided into various classes, i.e., polyesters, poly(ortho esters), and polyanhydrides (Figure 12.2).

12.3.1 Nonbiodegradable Polymers

Nondegradable polymers are those that are stable in biological systems. Representative polymers of this class that are used as drug carriers are polysilicones, ethylene-vinyl acetate copolymers, various acrylate-based hydrogels, and segmented polyurethane, all of which have also been used as components of implantable devices. Polymethylmethacrylate beads containing gentamicin have been approved for use in Europe.[36,37] Although this product is found to be efficacious, it suffers from the major drawback of being nondegradable and requiring subsequent removal of the beads after the antibiotic release. A thermoreversible gelling formulation of paclitaxel in poloxamer 407 has been formulated for regional delivery. Control and paclitaxel–poloxamer 407 formulations were administered intratumorally at a dose of 20 mg/kg in B16F1 melanoma-bearing mice. The initial tumor growth rate was delayed by 67% and the tumor volume doubling time was increased by 72% relative to saline control. In addition, more than 91% of the tumor-bearing animals that received paclitaxel in poloxamer 407 gel survived on day 15 post-administration as

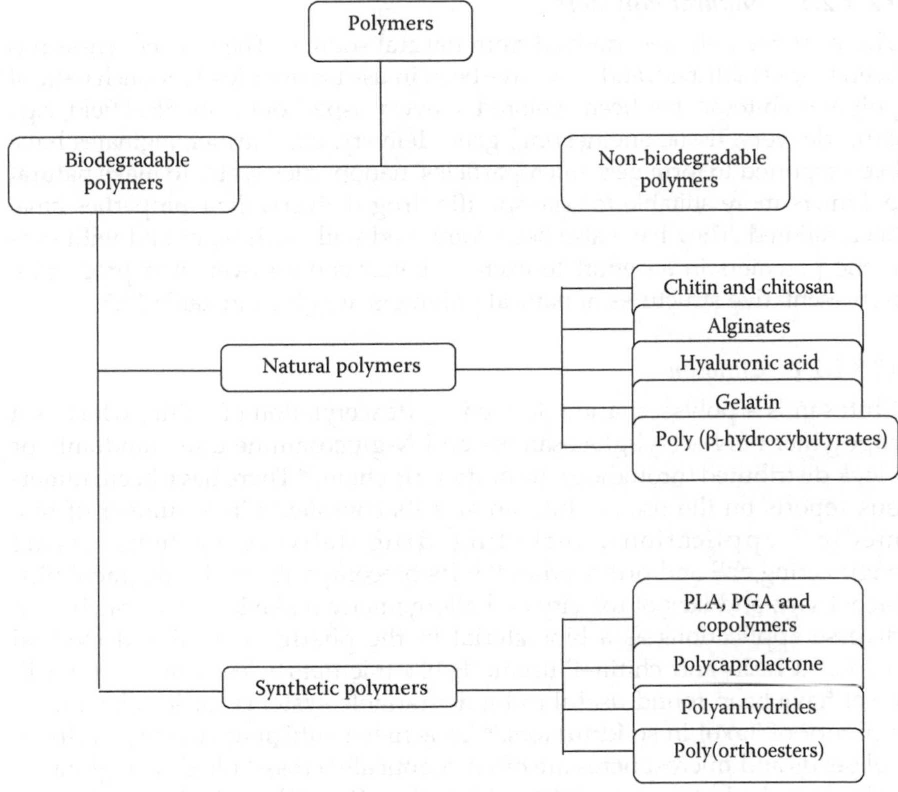

FIGURE 12.2
Classification of polymers used in regional drug therapy.

compared with 58% in the control group. The results of this study show significant benefit of paclitaxel for treatment of solid tumors when administered locally in an *in situ* gelling poloxamer 407 formulation.[38] However, the major limitation with nonbiodegradable polymeric carriers is that it is mandatory to surgically remove these from the body once the drug has been released; hence the use of these polymers has diminished considerably, especially since the advent of biodegradable polymeric carriers.

12.3.2 Biodegradable Polymers

The ever-expanding field of biodegradable polymers has gone through various stages of development. Knowledge about the properties of these polymers is required to understand the rationale that supports their applicability in the field of drug delivery, particularly as polymeric carriers for regional drug therapy, and also to discover new avenues for their use. Thus, this section gives a brief overview of the natural and synthetic polymers that have been explored as carriers in localized drug delivery.

12.3.2.1 Natural Polymers

These are the polymers derived from natural sources. Their effectiveness has been long established, and they have been in use for decades. One such natural polymer, chitosan, has been explored in every aspect of the medical field, e.g., drug delivery, tissue engineering, gene delivery, etc. Further, alginates have been reported to form gels, microparticles, nanoparticles, etc. To make natural polymers more suitable for site-specific drug delivery, their properties have been tailored. They have also been combined with each other and with synthetic polymers in an effort to exercise better control over their properties. Representative structures of natural polymers are given in Table 12.3.

12.3.2.1.1 Chitosan

Chitosan is a polysaccharide derived by deacetylation of chitin, which is a copolymer of *N*-acetyl-glucosamine and *N*-glucosamine units randomly or block distributed throughout the polymeric chain.[39] There have been numerous reports on the use of chitosan and its complexes in a number of biomedical applications, including drug-delivery systems, tissue engineering,[40–42] and orthopedics.[43,44] Its biocompatibility, biodegradability, bioactivity, and lack of toxicity and allergenicity make it very attractive for diverse applications as a biomaterial in the pharmaceutical and medical fields.[45] Chitin and chitin–Pluronic F-108 microparticles containing paclitaxel have been found useful as biodegradable systems for localized drug delivery of Taxol in solid tumors.[46] To achieve sufficient stability, chitosan gel beads and microspheres are often chemically cross-linked with glutaraldehyde and ethylene glycol diglycidyl ether. Recently, polyelectrolyte complexes have been proposed for the design of drug-delivery systems. Cationic chitosan can form complexes with nontoxic multivalent anionic counterions such as polyphosphate[47] and sodium alginate by ionic cross-linking.[48] Further, matrices of chitosan with polyglycolide have been reported and were found to have high strength and porosity.[44] Also, an injectable thermogel has been prepared by grafting poly(ethyleneglycol) onto the chitosan backbone and explored for drug release *in vitro* using bovine serum albumin as a model protein.[49]

12.3.2.1.2 Alginates

Alginates form a family of linear unbranched polysaccharides containing varying amounts of 1,4-linked β-D-mannuronic acid and α-L-glucuronic acid residues.[50] Commercial alginates are extracted primarily from three species of brown algae (kelp). These include *Laminaria hyperborea, Ascophyllum nodosum*, and *Macrocystis pyrifera*. Other sources include *Laminaria japonica, Eclonia maxima, Lesonia negrescens*, and *Sargassum* species.[50] Multivalent cations such as calcium form gels with alginates by reacting with carboxylic acid groups and bringing gluronate chains together stoichiometrically in an egg-box-like conformation, leaving mannuronate intact.[51] Yenice et al. prepared alginate beads by physically cross-linking calcium ions to sodium

TABLE 12.3
Representative Structures of Natural Polymers

Polymer	Structure
Chitosan	
Alginate	
Hyaluronic acid	
Gelatin	
Poly (hydroxyalkanoate)s	PHB PHV

alginate polymer and then evaluated them as biodegradable implants containing teicoplanin for the prevention/treatment of bone infections.[52] Further, alginate beads containing 5-fluorouracil (5-FU) have been prepared by the gelation of alginate with calcium cations, and the effect of polymer concentration and the drug loading on the release profile of 5-FU have been investigated.[53] A novel hydrogel formulation of alginate has been designed and synthesized to locally deliver the antineoplastic agents methotrexate, doxorubicin, and mitoxantrone. This novel delivery system allows for the release of single or combinations of antineoplastic agents and may find utility in localized antineoplastic agent delivery.[54]

12.3.2.1.3 *Hyaluronic Acid (HA)*

HA is a polysaccharide composed of alternating residues of the D-glucuronic acid and *N*-acetyl-D-glucosamine. It is a uniform, linear, unbranched molecule consisting of multiple identical disaccharide units[55] and is distributed throughout the extracellular matrix, connective tissues, and organs of all higher animals. Since HA is native to the body, it is nonimmunogenic and could be an ideal biomaterial for tissue engineering and drug and gene delivery. The first medical use of HA as a biomaterial was reported in 1961 as a substitute for vitreous body in retinal-reattachment surgery.[56] HA can be chemically modified to engineer robust materials with preselected mechanical properties and resorption rates as dictated by the intended clinical use. A variety of hydrophobic modifications and chemical cross-linking strategies have been explored to produce insoluble or gel-like HA materials. Liu et al. described the stability and cytotoxicity of disulfide-cross-linked thiolated HA films *in vitro*, and the persistence and biocompatibility of these films *in vivo*.[57] In view of localized drug delivery, microspheres and thin films of hyaluronan benzyl esters have been found to be the most suitable physical forms for drug and peptide delivery, since the hydrophobic, mucoadhesive properties facilitated intranasal, buccal, ocular, and vaginal delivery. High-viscosity absorbable gels made of HA and the nonsteroidal anti-inflammatory drug naproxen sodium were found to be an effective composition for preventing postsurgical tendon adhesions.[58] In fact, regulatory approval in the U.S., Canada, and Europe was granted recently for 3 wt% diclofenac in 2.5 wt% HA gel, Solaraze™, for the topical treatment of actinic keratoses, which is the third most common skin complaint in the U.S.[59] Various glucocorticoid and hyaluronic acid formulations are currently available on the market for intraarticular drug injection to treat osteoarthritis.[60] (See Vignette 12.2.[61,62])

12.2 Solaraze®

Solaraze gel contains the active ingredient, diclofenac sodium (3%), in a clear, transparent, colorless to slightly yellow gel base. Solaraze is a topical treatment for a common dermatological condition, actinic keratosis. This is an increasingly common precancerous skin condition caused by overexposure to sunlight. If untreated, it can progress to squamous cell carcinoma. This proprietary HA gel technology maximizes the concentration of the active drug (diclofenac) in the

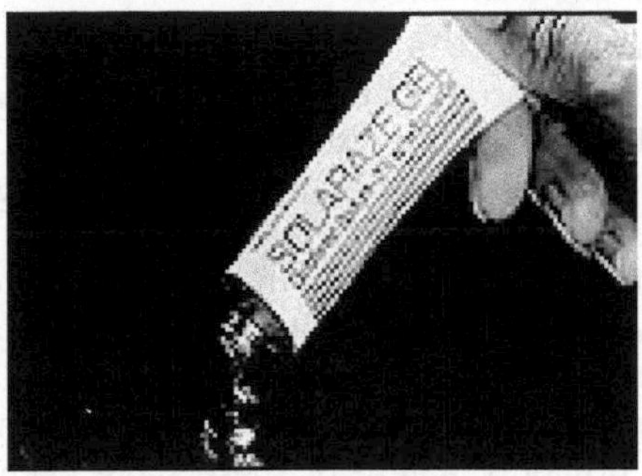

FIGURE 12.3
Solaraze gel.

upper layers of the skin. This not only reduces systemic effects, but also greatly improves the efficacy of the drug (Figure 12.3).

Patient evaluation at 30 days posttreatment showed a 90% mean decrease in target lesions vs. baseline, with 58% of patients having 100% lesion clearance and 85% of patients having 75% of lesion clearance, with both results based on target lesion number score vs. baseline. Solaraze gel was well tolerated with few side effects and did not produce serious adverse events in clinical trials.

12.3.2.1.4 *Gelatin*

Gelatin is a commonly used natural polymer derived from collagen[63] and has been extensively used for industrial, pharmaceutical, and medical applications because of its biodegradability and biocompatibility in physiological environments. In addition to controlled biodegradability, it has a proven record of clinical safety, and it can be fabricated at conditions that preserve the bioactivity of the therapeutic agent to be delivered *in vivo*. Another unique advantage is the electrochemical properties of gelatin, which can be readily modified during the processing stage in which collagen is pretreated, prior to the extraction process. A diverse range of applications has been studied for gelatin-carrier-mediated pharmaceutical drug delivery, including sustained antibiotic delivery for bone infection and repair and in cancer chemotherapy. Konishi et al. investigated the *in vivo* controlled release of cisplatin from a biodegradable hydrogel.[64] In another study it was concluded that the *in vivo* antitumor effect by dual release of cisplatin and adriamycin from a biodegradable hydrogel (prepared through the chemical cross-linking of gelatin by glutaraldehyde) synergistically enhanced their *in vivo* antitumor effect through the trans-tissue delivery.[64]

12.3.2.1.5 Poly(hydroxyalkanoate)s

Poly(hydroxyalkanoate)s (PHAs) are a group of thermoplastic polymers that are produced by a wide variety of bacteria as intracellular reserve materials. Depending upon the size of the pendant group (see Table 12.3) and the composition of the polymer, they can be varied from rigid brittle plastics to flexible plastics with good impact properties to strong tough elastomers.[65] The type of the bacterium and the growth conditions determine the chemical composition of PHAs and their molecular weight.[66,67] Analysis of the isolated PHAs reveals interesting properties such as biodegradability and biocompatibility. To date, more than 100 different monomers have been reported as PHA constituents, but only a few of these have been produced in large quantities. Research focused only recently on the application of PHAs in implants, as scaffolds in tissue engineering, or as drug carriers. A localized formulation of gentamicin has been formulated in poly(hydroxyalkanoate) polymeric carrier. It was found to be an effective preventive antibacterial modality in implant-related *Staphylococcus* infections.[68] Various random copolyesters of 3-hydroxybutyrate with 3-hydroxyvalerate and 3-hydroxybutyrate with 4-hydroxybutyrate have been used in the construction of biodegradable, implantable rods for the local delivery of antibiotics (Sulperazone™ and Duocid™) in chronic osteomyelitis therapy.[28]

12.3.2.2 Synthetic Biodegradable Polymers

Better control and flexibility over the physicomechanical properties of synthetically prepared biodegradable polymers has made them superior resources for polymer scientists and material designers. During the past five years, a battery of advancements has taken place in the area of synthetic biodegradable polymers. This category comprises a broad family of polyesters such as poly(glycolic acid) (PGA), poly(lactic acid) (PLA), and their copolymers, polycaprolactone (PCL), polyanhydrides, and polyorthoesters (POE). Representative structures of synthetic polymers are given in Table 12.4.

12.3.2.2.1 Polyesters

Polyesters belong to an exhaustively studied class of biodegradable polymers. They constitute historically the first family of synthetic condensation polymers and were investigated as part of Carothers's pioneering studies on polymerization in the 1930s.[69,70] Polyesters are currently the most widely investigated and commonly used synthetic biodegradable polymers. Since esterification is a reversible process, all the polyesters theoretically undergo degradation.

Poly(glycolic acid) (PGA) is the simplest linear, aliphatic polyester. It is the most widely used material in clinical applications, especially for sutures. PGA sutures have been available under the trade name Dexon since 1970 However, a practical limitation of Dexon is that it tends to lose its mechanical strength within 2 to 4 weeks of implantation. It has also been used in the design of internal bone-fixation devices (bone pins). These pins have been

TABLE 12.4
Representative Structures of Synthetic Polymers

Polymer	Structure
PLA	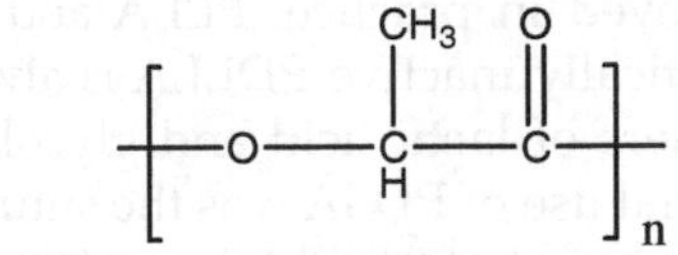 $-[O-CH(CH_3)-C(=O)]_n-$
PGA	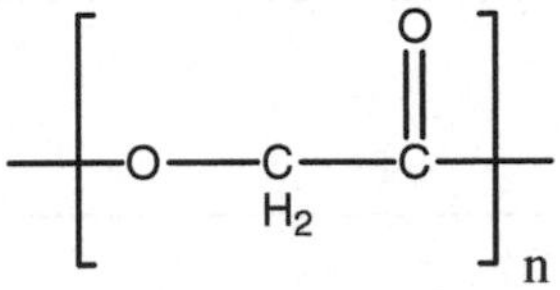 $-[O-CH_2-C(=O)]_n-$
PCL	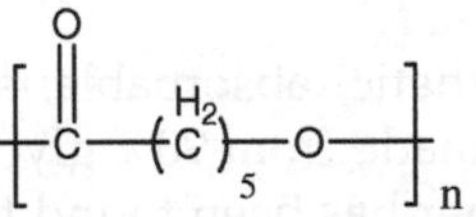 $-[C(=O)-(CH_2)_5-O]_n-$
P(CPP-SA)	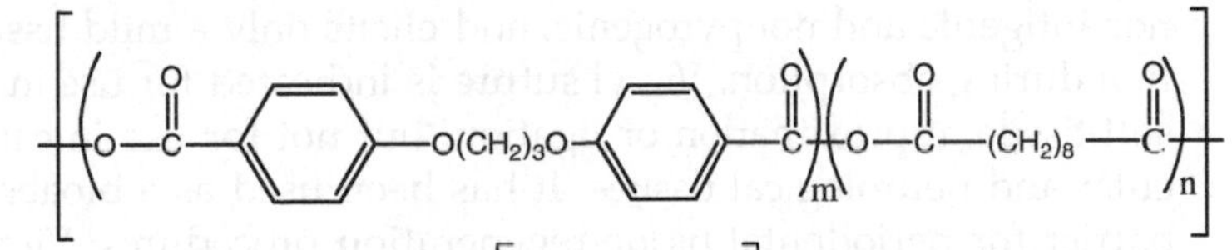 $-[(O-C(=O)-C_6H_4-O(CH_2)_3O-C_6H_4-C(=O))_m(O-C(=O)-(CH_2)_8-C(=O))]_n-$
POE I	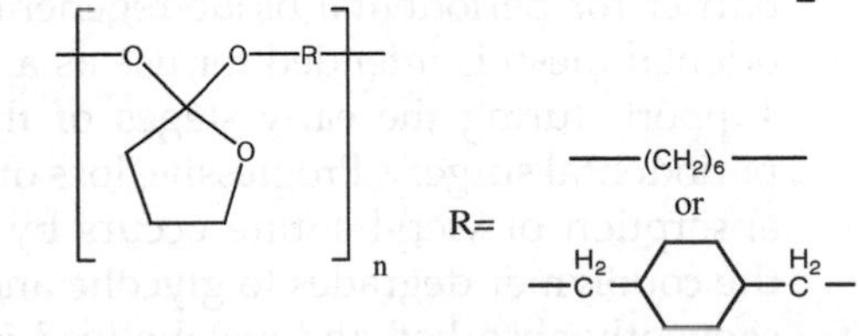 $[\ldots O-R]_n$; R= $-(CH_2)_6-$ or $-CH_2-C_6H_{10}-CH_2-$
POE II	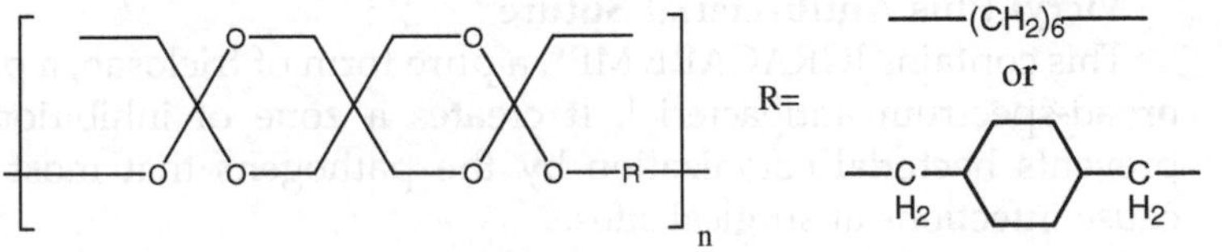 $[\ldots O-R]_n$; R= $-(CH_2)_6-$ or $-CH_2-C_6H_{10}-CH_2-$
POE III	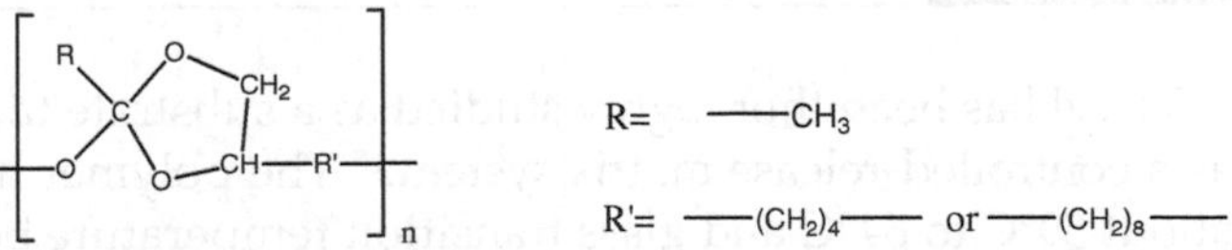 $[\ldots CH-R']_n$; R= $-CH_3$; R'= $-(CH_2)_4-$ or $-(CH_2)_8-$
POE IV	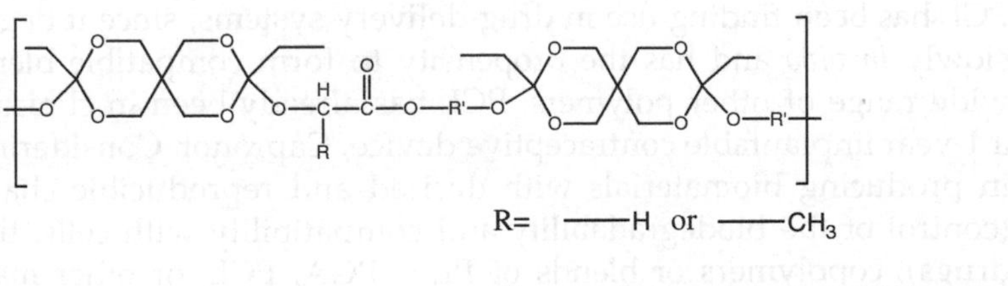 R= $-H$ or $-CH_3$; R'= $-(CH_2)_{10}-$ or $-(CH_2)_{12}-$

commercially available under the trade name Biofix. In contrast to glycolic acid, lactic acid is a chiral molecule; it exists in two stereoisomeric forms that give rise to four morphologically distinct polymers: D-form (PDLA), L-form (PLLA), and the racemic form (PDLLA). A fourth *meso*-PLA can be obtained from DL-lactide but is rarely employed in practice. PLLA and PDLA are semi-crystalline solids, while the optically inactive PDLLA is always amorphous. The entire range of copolymers of lactic acid and glycolic acid has been investigated. The first commercial use of PLGA was the suture material Vicryl (Polyglactin 910), which is composed of 8% PLLA and 92% PGA. (See Vignette 12.3.[71])

12.3 Vicryl

Vicryl (polyglactin 910) is a synthetic, absorbable, sterile surgical suture composed of a copolymer made from 90% glycolide and 10% L-lactide. Polyglactin 910 copolymer has been found to be relatively nonantigenic and nonpyrogenic, and elicits only a mild tissue reaction during absorption. Vicryl suture is indicated for use in general soft-tissue approximation or ligation, but not for use in cardiovascular and neurological tissues. It has been used as a bioabsorbable barrier for periodontal tissue regeneration procedures. Vicryl periodontal mesh is intended for use as a barrier to provide temporary support during the early stages of the healing process following periodontal surgery. Progressive loss of tensile strength and eventual absorption of vicryl suture occurs by means of hydrolysis, where the copolymer degrades to glycolic and lactic acids, which are subsequently absorbed and metabolized in the body.

Vicryl Plus Antibacterial Suture

This contains IGRACARE MP*, a pure form of triclosan, a proven broad-spectrum antibacterial. It creates a zone of inhibition that prevents bacterial colonization by the pathogens that most often cause infections at surgical sites.

Vicryl has been thoroughly studied as a substrate for biodegradation and as a controlled-release matrix system.[65] The polymer has a melting point of about 59°C to 64°C and glass transition temperature between −60 to −70°C. PCL has been finding use in drug-delivery systems, since it degrades very slowly *in vivo* and has the propensity to form compatible blends with a wide range of other polymers. PCL has already been in clinical trials for a 1-year implantable contraceptive device, Capronor. Considering the ease in producing biomaterials with desired and reproducible characteristics (control of the biodegradability and compatibility with cells, tissues, and drugs), copolymers or blends of PLA, PGA, PCL, or other materials are often used in drug-delivery systems and tissue engineering. Copolymers of polyesters with amides, imides, urethanes, anhydrides, and ethers have been prepared with the aim of improving the properties and having better control over degradation rate. PCL and its copolymers have been evaluated as delivery systems as microparticles,[72,73] polymeric paste,[74] nanoparticles,[75] micelles,[76] etc.

12.3.2.2.2 *Polyanhydrides*

Langer was the pioneer in exploiting the polyanhydrides for sustained release of drugs in controlled-release applications by recognizing their hydrolytically unstable nature and the nontoxic metabolites released by them.[77] The high hydrolytic reactivity of anhydride linkages confers versatility and controls the degradation rates of this class of bioerodible polymers. Gliadel®, an FDA-approved device of polyanhydride (P[CPP-SA 20:80]) for delivering carmustine (BCNU) for the adjuvant therapy of brain tumors, is the success story of polyanhydrides.[1,14] Polyanhydrides have also been evaluated for local delivery of antibiotics for the treatment of osteomyelitis.[6,78] Septacin® is a polyanhydride implant of a copolymer of erucic acid dimer and sebacic acid in a 1:1 weight ratio that is being developed for osteomyelitis. It contains gentamicin sulfate dispersed into a polyanhydride polymer matrix[79] (Figure 12.4).

12.3.2.2.3 *Poly(ortho esters)*

Poly(ortho esters) have been under development since 1970 after the pioneering work done by Heller and coworkers.[80] They were first designated as Chronomer and later as Alzamer[81] and have evolved through four families designated as POE I, POE II, POE III, and POE IV. One important attribute

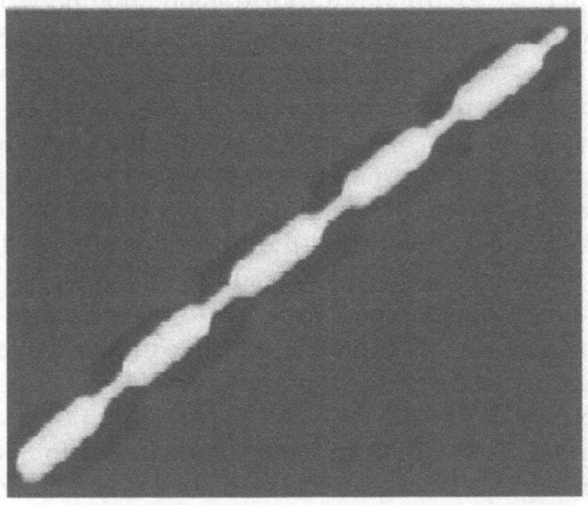

FIGURE 12.4
Molded Septacin™ beads with linkers. (From Li, L.C., Deng, J., and Stephens, D., *Adv. Drug Deliv. Rev.*, 54, 963, 2002. With permission.)

of this polymeric system is that, depending on the nature of the diols used, either (a) solid polymers are obtained that can be fabricated into desired shapes such as wafers, strands, and microspheres, or (b) viscous semisolid materials are obtained that are directly injectable, provided that their viscosity is low enough.[80] Important applications of these materials are treatment of postsurgical pain, osteoarthritis, and ophthalmic diseases as well as the delivery of proteins and DNA. Ophthalmic applications of POE can be exemplified by the use of POE III for delivery of antiproliferative agents like 5-FU and mitomycin C.[82] POE III exhibits good subconjunctival biocompatibility and efficacy to prevent the failure of trabeculectomy by providing sustained release of 5-FU.[83] Schwach-Adbellaoui et al. have investigated a bioerodible delivery system, based on POE IV and tetracycline base (TB), designed to be placed or injected into the periodontal cavity and capable of maintaining therapeutic concentrations in the gingival crevicular fluid (GCF) for up to 10 to 14 days.[83,84] *In vivo* results indicated that the POE-based formulation was well tolerated with no pain during application and no sign of irritation or discomfort after treatment. The local anesthetic agent, mepivacaine, has been incorporated into a viscous, injectable POE IV, and its potential to provide long-acting anesthesia has been explored in nonclinical models. POE was found to be a viable drug-delivery candidate for delivery of mepivacaine in the management of postsurgical pain management. The TEG-POE mepivacaine formulation is currently in phase I human clinical trials.[80] (See Vignette 12.4[85] and Figure 12.5.)

12.4 Biochronomer™ Systems

Biochronomer (AP Pharma) is a fourth-generation poly(ortho ester) prepared by the condensation of diols and a diketene acetal. Since ortho ester linkages are stable at neutral pH and become increasingly hydrolyzed as the pH at the polymer–water interface decreases, erosion rates of the polymer can be varied within very wide limits by incorporating into the polymer a short segment of poly(lactic acid) that acts as a latent acid catalyst. Upon hydrolysis of these segments, lactic acid is produced that functions as a catalyst for the hydrolysis of the ortho ester linkages in the polymer backbone.

Biochronomer technology is designed to release drugs at various implantation sites, such as under the skin, into muscle tissue, or within the peritoneal cavity. Key benefits of this technology include the ability to easily fabricate the poly(ortho ester) polymers into an array of drug-delivery forms — ranging from wafers and strands to microspheres and free-flowing injectable gels — that can be easily administered and readily accepted within the body. Biochronomer systems have potential application in a wide range of therapeutic areas, including pain management, anti-nausea, osteoarthritis, anti-inflammatory, ophthalmic diseases, bone growth, restenosis, and tissue engineering.

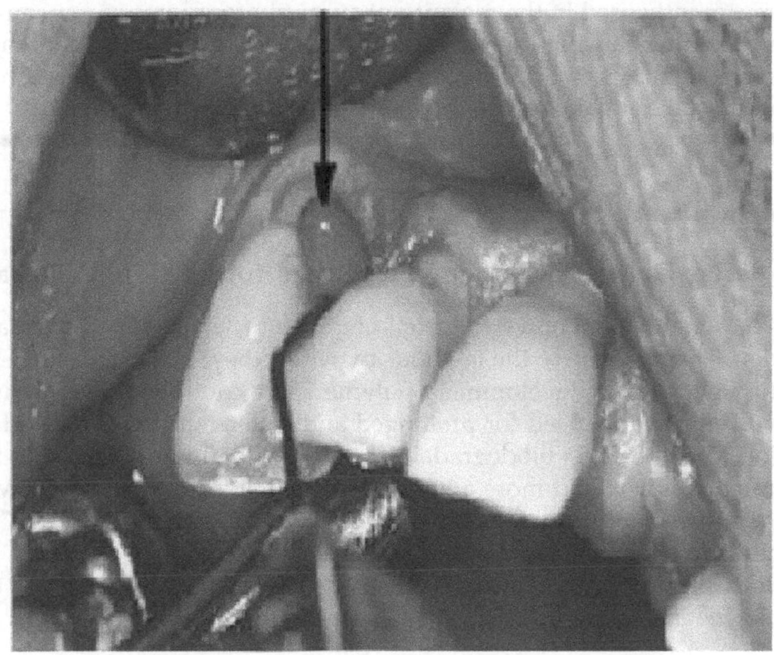

FIGURE 12.5
Typical appearance of a periodontal site after the administration of the delivery system POE/TB. (From Einmahl, S. et al., *Adv. Drug Deliv. Rev.*, 53, 45, 2001. With permission.)

The lead Biochronomer product application, APF530, is being developed for the prevention of both acute and delayed chemotherapy-induced nausea and vomiting with a single subcutaneous injection. A Phase 2 study has been completed in the U.S. in 45 patients undergoing chemotherapy for cancer with three ascending doses of this formulation (containing 5 mg, 10 mg, or 15 mg of granisetron). Importantly, measurable plasma levels over a five-day period have been observed, the time when anti-nausea therapy is most critical. In addition, the product was safe and well tolerated. It is evident that the release of granisetron from the formulation following a single subcutaneous dose provides for sustained plasma levels over the time course of both the acute and delayed phases of the condition.

The second Biochronomer product application, APF112, targets the market for postsurgical pain, which is currently estimated to be in excess of $2 billion. The treatment strategy is to provide 24 to 36 hours of postsurgical pain relief by delivering the drug, mepivacaine, directly to the surgical site. Mepivacaine is a very well-known drug for pain relief, and it has an extensive safety protocol but a short half-life for effective pain relief in its current formulation. APF112 is designed to minimize the use of opioids (morphinelike drugs), which are currently used in the majority of surgical procedures as a means of managing pain but with unpleasant side effects, including nausea, disorientation, sedation, constipation, vomiting, urinary retention, and in some situations, life-threatening respiratory depression.

12.4 Drug-Delivery Systems Employed for Regional Therapy

While newer and more powerful drugs continue to be developed, increasing attention is being given to the methods by which these active substances are administered. A new development, polymerically controlled drug delivery, has evolved from the need for prolonged and better control of drug administration. The interest in biodegradable polymers continues to grow, as their applicability in new and more advanced areas is steadily expanding. Many efforts have been put forward to widen the gap between the "efficacy" and "toxicity" by playing with new drug-delivery systems.

Various properties that make biodegradable polymers unique for their applications include

Nontoxic and nonimmunogenic nature
Complete elimination from the body
Low level of protein and cellular adsorption
Does not hamper the therapeutic effect of the drug
Solubility in a range of solvents

Biodegradable polymers can be fabricated into different delivery systems, each representing its unique advantages apart from meeting a common goal of providing controlled and local release of drugs. This section encompasses various biodegradable drug-delivery systems that have been explored so far for regional drug therapy. These include implants, microspheres, *in situ*-forming drug-delivery systems, and hydrogels.

12.4.1 Implants

Some of the systemic drug therapies are associated with side effects. To overcome these, interest in the localized delivery of therapeutic agents has increased tremendously. An attractive approach for the delivery of pharmacologically active compounds is the controlled and sustained release of active agents from a resorbable polymeric delivery system that is implanted next to the diseased tissue. The controlled localized delivery of therapeutic agents by biodegradable polymers is a method that allows exposure of a diseased area to therapeutic doses of an active agent for prolonged periods of time while avoiding high systemic doses associated with debilitating toxicities.[86]

TABLE 12.5
Implantable Drug-Delivery Products

Product	Drug	Company
Atridox	Doxycycline	CollaGenex
Eligard	Leuprolide	Sanofi
Gliadel	BCNU	MGI Pharmaceuticals Inc
Luprogel	Leuprolide	Medigene AG
Norplant	Norgestrel	Wyeth Ayerst
Profact Depot	Buserelin	Aventis
Vantas	Histerlin	Valera
Viadur	Leuprolide	ALZA
Zoladex	Goserelin	AstraZeneca

Table 12.5 lists the drugs available as implants in the market. Further, there is the possibility of removing the implant if any adverse events necessitate discontinuation of the therapy. By definition, implants are cylindrical devices of 1-mm diameter and 10 to 20 mm in length that are deposited into the subcutaneous tissue using a hollow needle.[87] Polymeric implants can be classified into two main classes: monolithic and reservoir devices. In the monolithic systems, the drug is dissolved or dispersed in an inert matrix; in the reservoir systems, the therapeutic agent forms a core surrounded by an inert polymer diffusional barrier. Combinations of both devices can also be prepared. Drugs formulated in these systems are released either by diffusion through the polymer barrier, by its erosion, or by a combination of both mechanisms.[88] Several implants have become commercially available, e.g., luteinizing hormone-releasing hormone (LHRH) agonists, such as goserelin (Zoladex™) or buserelin (DepotProfact™) in PLGA.[87] Biodegradable BCNU-loaded polyanhydride wafers consisting of 20:80 p(CPP-SA) were extensively tested for this application and are already commercially available under the brand name Gliadel®. Compared with the systemic standard therapy, the local implantation of Gliadel® was reported to improve a patient's quality of life substantially. (See Vignette 12.5.[89])

12.5 Zoladex® LA (goserelin acetate)

Zoladex LA is a depot formulation of goserelin acetate dispersed in a cylindrical rod of a biodegradable and biocompatible blend of high- and low-molecular-weight range D-L-lactide-glycolide copolymers. Each depot contains goserelin acetate equivalent to goserelin 10.8 mg. Administration of Zoladex LA every 12 weeks ensures that exposure to goserelin is maintained with no clinically significant accumulation. It has been indicated for the palliative treatment of patients with hormone-dependent advanced carcinoma of the prostate (Stage D2). Zoladex was first launched in 1987 and is currently the second-largest-selling LHRH analog in the world. It is currently available in more than 100 markets. Two controlled clinical trials were conducted with 157 patients, comparing treatment with goserelin 10.8-mg versus goserelin 3.6-mg depots. During the comparative phase, patients were randomized to receive either a single 10.8-mg depot or three consecutive 3.6-mg depots (one every 4 weeks) over this initial 12-week period. The only adverse event reported in greater than 5% of these patients during this phase was hot flushes, with the goserelin 10.8-mg group having an incidence of 47% and the goserelin 3.6-mg group having 48%. From weeks 12 to 48, all patients were treated with one Zoladex LA depot every 12 weeks. During this noncomparative phase, the following adverse events were reported in greater than 5% of patients: hot flushes (vasodilation) (63.7%), general pain (14%), gynecomastia (8.3%), pelvic pain (5.7%), bone pain (5.7%), and asthenia (5.1%).

Intraperitoneal application of BCNU solutions (three times at 20 mg BCNU/kg) had no effect on tumor growth. In contrast, the application of polymer matrices made of a physical mixture of P(CPP-SA) 20:80 with PLGA loaded with 0.25 mg BCNU slowed down the growth of tumors significantly. Implants of polyesters, especially PLGA, have been investigated for delivery of agents like 5-floxuridine paclitaxel[90,91]; antibiotics like cefazolin and gentamicin[92]; methadone[93]; heparin[94]; etc. Vogelhuber et al. investigated the efficacy of high-dosed implants. Paclitaxel-loaded implants were also studied with and without combination therapy with BCNU. The tumors responded to a local therapy with BCNU implants, even when the tumor had been fully established and had reached a size that was several times larger than its original one. Complete remission with any BCNU implant was not observed, whereas implants loaded with paclitaxel and BCNU led to a complete remission in some animals.[95] Dorta et al. prepared three types of film implants: simple monolithic discs, multilayer discs with a central monolithic layer, and multilayer discs with a central drug reservoir. The results obtained show the suitability of the proposed devices to (a) control release and avoid the burst effect with highly water-soluble drugs and (b) to modulate *in vitro* peptide release.[88] PLGA beads for long-term antibiotic delivery have also been reported.[92] Paclitaxel-loaded films, manufactured from PLGA blended with either methoxypolyethylene glycol (MePEG) or a diblock copolymer composed of poly(D,L-lactic acid)-block-methoxypolyethylene glycol, have been reported in the literature. The addition of MePEG or diblock to PLGA caused a concentration-dependent increase in the elasticity of films (due to plasticizing effects) and a decrease in glass transition temperature. Paclitaxel release from PLGA/MePEG films was very slow, with less than 5% of the encapsulated drug being released over 2 weeks. The addition of 30% diblock to paclitaxel-loaded PLGA films caused a substantial increase (five- to eightfold) in the release rate of paclitaxel.[74]

PLGA implants of ciprofloxacin have also shown some positive results against osteomyelitis.[96] In an effort to obtain a sustained fentanyl delivery with effective and precise control, a fentanyl-loaded wafer was fabricated using a PLGA oligomer by the direct-compression method. Results indicate that this constant localized-release system has the potential to provide anesthesia for a longer period than injection or topical administration.[97]

12.4.2 Microspheres

Localized drug-delivery technology using biodegradable polymers as carriers for drugs is one of the most rapidly advancing areas of pharmaceutical sciences. Microsphere-based controlled-release systems are studied to deliver a variety of therapeutics, ranging from antibiotics like gentamycin for bone infections to anticancer agents like 5-FU. They offer obvious advantages, including

- Accurate control of the release rate
- Protection and stabilization of the therapeutic agent
- Increased patient compliance by reducing the frequency of administration
- Ease of administration
- Elimination of surgical removal
- Ability to perform multipoint administration during surgery[98]
- Control of drug-release rates by manipulation of particle size, polymer degradation or erosion rates, polymer erosion mechanism (bulk vs. surface erosion), additives, etc.[99,100]

Among the biodegradable polymers, PLGA has been widely used to encapsulate, release, and deliver drugs that are both hydrophobic and hydrophilic.[101] This technology provides unique advantages over conventional delivery approaches, and many products are currently under investigation. Table 12.6 enumerates various products for regional therapy based on this technology.

Efficient and reproducible design of microspheres will require a thorough understanding of such parameters as encapsulation efficiency, surface morphology, size, etc., all of which ultimately affect the release characteristics.[102] To obtain specific drug-release patterns, a multitude of formulation and processing parameters can be varied, e.g., the average polymer molecular weight; polymer concentration; monomer ratio; drug loading; additives like PEG, fluorosilicone oil, fructose, gelatin; etc. All these factors play in conjunction with each other, and it will be difficult to assign the observed release pattern to only one factor.

Carbidopa/levodopa-loaded PLLA, PDLLA, and PLGA microspheres were designed for the intracerebral treatment of Parkinson's disease. It has been concluded that localized delivery of carbidopa/levodopa to the brain can be achieved with biodegradable microspheres, with a significant decrease of

TABLE 12.6

Microparticulate Drug-Delivery Products Using Lactide and Glycolide Polymers

Product	Drug	Company
Arestin	Minocycline	Orapharma
Decapeptyl LP	Triptorelin	Ipsen
Decapeptyl	Triptorelin	Ferring
Lupron Depot	Leuprolide	TAP
Nutropin Depot	hGh	Genentech
Risperdal Consta	Risperidone	Johnson and Johnson
Sandostatin LAR	Octreonide	Novartis
Somatuline LA	Lanreonide	Ipsen
Supercur MP	Buserelin	Aventis
TreLstar	Triptorelin	Pfizer

ipsilateral turns per minute; this therapy could be a promising approach for treatment of Parkinson's disease.[103] After loading 5-fluorouracil (5-fluoro-2,4-pyrimidinedione) into PLGA microspheres, it was determined that the formulation has the potential of increasing the long-lasting controlled-release efficacy of water-soluble anticancer drugs.[98] Microspheres of PDLLA and PLGA containing the hypnotic agent Zolpidem have also been prepared, and these have proved useful as a prolonged-release system in the treatment of insomnia.[104] Biodegradable PLGA microspheres loaded with ibuprofen destined for intraarticular administration have also shown positive results.[102]

Microparticles of PLGA have been prepared containing chlorhexidine-free base and chlorhexidine digluconate and their inclusion complex with methylated-β-cyclodextrin and hydroxypropyl-β-cyclodextrin. The system may prove useful for the localized delivery of chlorhexidine salts and other antimicrobial agents in the treatment of periodontal disease, where prolonged controlled delivery is desired.[105] Localized delivery of rifampicin-loaded microspheres of PLGA to the alveolar macrophages, the host cells for *Mycobacterium tuberculosis*, has shown a promising improvement in retarding the early development of pulmonary tuberculosis.[106]

Microsphere size has been known to affect the release kinetics. In many cases, drug release from smaller microspheres is faster due to increased surface area/volume ratio. However, the effects of microsphere size can be more complex, since microsphere size can affect the relative importance of polymer degradation in controlling release rates. The surface morphology of microspheres also influences the release kinetics, and the presence of additives has a profound influence on the formulation characteristics. A study of the effects of microparticle properties revealed that PDLLA and PLGA microparticles containing drug alone exhibited similar average diameter, whereas the ones that also contained cyclodextrins were characterized by a larger size.[104]

Surface-eroding polymers, such as polyanhydrides, can simplify the drug-release kinetics, since water penetration into the microsphere interior is minimized, and the drug release rate becomes predominantly dependent on the polymer erosion rate. A variety of factors affect drug-release rates from polyanhydride microspheres. In an effort to understand the factors affecting drug release from surface-erodible polymer devices, Berkland et al. prepared poly(sebacic acid) (PSA) microspheres loaded with three model drugs (rhodamine B, p-nitroaniline, and piroxicam), each having a variable degree of water solubility. Rhodamine, the most hydrophilic compound investigated, was localized strongly toward the microsphere surface, while the much more hydrophobic compound, piroxicam, distributed rather evenly. Rhodamine was released very quickly independent of microsphere size. The release of p-nitroaniline was more prolonged, but still showed little dependence on microsphere size. The piroxicam-loaded microspheres exhibited the most interesting release profiles, showing that release duration can be increased by decreasing microsphere size, resulting in a more uniform drug distribution. Hence, when water-soluble drugs are encapsulated with hydrophobic polymers, it may be difficult to tailor release profiles by controlling microsphere size.[99] Polyanhydrides like PSA have been combined with polyesters like PLA to tailor the polymer properties and to have a better control over release kinetics. Modi et al. have reported the synthesis and characterization of biodegradable polymers of PLA-PSA, which offers greater flexibility to alter the degradation behavior.

The microsphere drug-delivery system, based on PEG with PLA/PLGA, has been investigated with the aim of modifying formulation parameters. Poly-D,L-lactide-poly(ethylene glycol) (PELA) is a newer block copolymer of PEG with PLA. It has a hydrophilic microsegment PEG between the hydrophobic PLA segments. From the point of view of its molecular structure, PELA is expected to have many advantages over PLGA, such as controllable biodegradability, sustained release potential, friendliness to protein activity, longer blood circulation half-life, higher capability for noninvasive drug delivery, reduced amount of emulsifier to be used, and less drug adsorption to the microsphere surface.

12.4.3 *In Situ*-Forming Drug-Delivery Systems

The development of new injectable drug-delivery systems has received considerable attention over the past few years. This interest has been sparked by the advantages of these delivery systems, including ease of application, localized delivery for a site-specific action, prolonged delivery periods, decreased drug dosage with concurrent reduction in possible undesirable side effects common to most forms of systemic delivery, and improved patient compliance and comfort. These implant systems are made of biodegradable components that can be injected via a syringe into the body and, once injected, solidify to form a semisolid depot. The application is less invasive and painful than implants, where local anesthesia and surgical intervention is required. On the basis of the mechanism of solidification, these injectable implant systems are divided into four categories[107]:

Thermoplastic pastes
In situ cross-linked systems
In situ precipitation
Thermally induced gelling systems

12.4.3.1 Thermoplastic Pastes

Thermoplastic pastes are polymer systems that are injected as a hot melt (in sol phase) into the body, where they are then transformed into a semisolid depot upon cooling down to body temperature. They are characterized by their low melting point, ranging from 25 to 65°C, and low intrinsic viscosity, which is usually between 0.05 to 0.8 dl/g as measured at 25°C. Systems having intrinsic viscosity below 0.05 dl/g do not show a delayed release profile, whereas viscosity above 0.8 dl/g poses problems in administration, since the system become too viscous to pass through a needle. Monomers such as D,L-lactide, glycolide, ε-caprolactone, trimethylene carbonate, dioxanone, and ortho esters have been used to form thermoplastic pastes. They have an established track record of biocompatibility and thus are attractive starting points for development of new materials.

Block copolymers of PEG with other biodegradable polymers have been investigated. Thermoplastic biodegradable polymeric paste formulations have been prepared using triblock system composed of poly(D,L-lactide)-block–poly(ethylene glycol)-block–poly(D,L-lactide) (PLA-PEG-PLA) and studied for the delivery of Taxol™ to reduce the systemic side effects.[108] However, the noteworthy disadvantage of these systems was their high melting point (above 60°C) coupled with nonoptimal release of the incorporated drug. A melting point above 60°C causes pain at the injection site and increases the chance of necrosis and scar tissue formation.[107] In another study, MePEG has been added to PCL in an amount up to 30% by weight. PCL and its blends with MePEG have been reported to form surgical pastes. The addition of MePEG reduces the melting point of PCL, which facilitates administration at lower temperature.[109] This thermoplastic paste, with slight modification with respect to polymer composition, has been tested in human advanced prostate cancer tumors grown subcutaneously in castrated athymic mice, and promising results were obtained.[91]

A family of strong, highly flexible thermoplastic elastomers consisting of poly(ether-ester-urethane) multiblock copolymers has been synthesized by Cohn and Salomon.[110] The poly(ethylene oxide) amorphous chains performed as a molecular spring, and the crystalline PLA blocks formed strong noncovalent cross-linking domains. These highly flexible thermoplastic elastomers attained ultimate tensile strength values as high as 30 MPa and elongation-at-break levels well above 1000%.[110] Interest in these elastomers is principally focused on ocular drug delivery, treatment of periodontal diseases, and veterinary applications. Schwach-Abdellaoui et al. reported a POE-based delivery system where tetracycline base was homogeneously mixed with $POE_{70}LA_{30}$ to form a viscous formulation, whereas films were obtained by solubilizing TB and $POE_{95}LA_5$ in an organic solvent followed by drying and compression molding. These were designed to be placed or injected into the periodontal pocket and are capable of maintaining therapeutic concentrations for up to 14 days.[84]

Copolymers of ricinolic acid and sebacic acid have been shown to have thermoplastic properties. Shikanov et al. studied the effect of additives on polymer viscosity, paclitaxel release, and polymer degradation. The additives were ricinoleic acid, phospholipid, PEG 400, and PEG 2000. Addition of 20% ricinoleic acid to poly(sebacic acid-*co*-ricinolic acid) liquefied the formulation and allowed injection of the formulation containing paclitaxel via a 22-gauge needle at room temperature with no effect on paclitaxel release rate. Addition of PEG 400, PEG 2000, and phospholipid to the formulation did not affect the paclitaxel release from the formulation. *In vivo* formulations with additives (20% ricinoleic acid and PEG or phospholipid) and 5% paclitaxel content degraded faster than the formulation with only 20% ricinoleic acid and the same paclitaxel content. These results indicate that additives can be used to fine-tune the physical characteristics of the thermoplastic pastes.[111]

12.4.3.2 In Situ *Cross-Linked Systems*

Cross-linking between polymeric chains can be achieved *in situ* to form solid polymer systems or gels. This can be accomplished by different means, including free-radical reactions, usually initiated by heat or absorption of photons, or ionic interactions between small cations and polymer anions.[107] Thus, depending upon the source of cross-linking, these systems can be categorized into thermosets, photo-cross-linked gels, or ion-mediated gelation. Thermosetting polymers can initially flow and can be molded, but after heating they set into their final shape. Macromolecular networks have been formed in such systems involving covalent cross-links. Not much research has been reported in this area due to the limitations and adverse effects associated with this system.[107] Dunn et al. reported the use of biodegradable copolymers of D,L-lactide or L-lactide with ε-caprolactone using a multifunctional polyol initiator and a catalyst to form a thermosetting system for prosthetic implants and a slow-release drug-delivery system.[112] Syringeability was the obvious advantage of this system. However, this system is susceptible to burst release during the first hour, causing a high concentration of drug at the site of administration, which could result in side effects. Further, the heat released during curing may cause necrosis of the surrounding tissues. Free-radical-producing agents can also promote tumor growth.

Photo-cross-linked gels can be used advantageously in place of chemically initiated thermoset systems. In this approach, photocuring is done *in situ* by using fiber-optic cables immediately after introduction of the polymer to the desired site via injection.[107] With a photo-initiator added in advance, the gels can be photopolymerized *in situ* by exposure to UV irradiation, resulting in chemically cross-linked biodegradable hydrogels with markedly improved mechanical strength.

Succinic acid and PEG-based polyester polyol macromers and their acrylates have been synthesized by Nivasu et al., who polymerized these acrylates by exposing them to UV radiation in the presence of a photo-initiator. *In vitro* measurements of burst strength suggested that they could be useful as tissue sealants. Further *in vitro* release studies with sulfamethoxazole showed that they have the potential to be used for localized controlled delivery of antibiotics over a short period of time.[113]

Ion-mediated gelation can be achieved using natural polymers. For example, alginates can form gel in the presence of divalent ions such as calcium ions. The use of a sodium alginate-calcium chloride hydrogel to deliver combined therapeutic modalities, i.e., radiation (^{188}Re) and chemotherapy (cisplatin), in tumor-bearing animal models has been reported recently.[114] Delay in tumor growth was observed in the ^{188}Re-hydrogel group compared with the ^{188}Re and control groups, resulting in prolonged survival more than two times that of the untreated groups. Calcium performs a dual function in this system by cross-linking with sodium alginate and at the same time forming a complex with ^{188}Re, thus leading to gel formation and slow release of cisplatin coupled with trapping of ^{188}Re at the site of injection. The problems associated with the use of calcium alginates in drug-delivery devices are their potential immunogenicity and long *in vivo* degradation time.

Oster et al. have reported the use of drug-loaded nanoparticles that are capable of forming a hydrogel in the presence of ions. This ion-mediated gelation occurs because of the positive surface charge of the nanoparticles, which were formulated using a novel amine-modified branched polyester. These polymers consisted of a poly(vinyl alcohol) backbone modified with amines (diethylaminopropylamine, dimethylaminopropylamine, diethylaminoethylamine) using *N,N′*-carbonyldiimidazole linker chemistry. The hydrophilic backbones were grafted with PLGA side chains of either 10 or 20 repeating units. Degradation times from several hours to days to months could be achieved by proper alteration of the degree of amine substitution, the PLGA side-chain lengths, and the graft density of the resultant polymer.[115]

12.4.3.3 In Situ *Polymer Precipitation*

An injectable drug-delivery depot could also be obtained by causing polymer precipitation from solution. This can be achieved by solvent removal, a change in temperature, or a change in pH.[107] In the case of the solvent-removal method, an injectable implant system consists of a water-insoluble biodegradable polymer dissolved in a water-miscible, physiologically compatible solvent. Precipitation of the polymer occurs upon injection, where the solvent diffuses into the surrounding aqueous environment and water diffuses into the polymer matrix, thereby forming a solid polymeric implant *in situ*. The problem associated with this system is the burst in drug release during the first few hours due to the lag time between the injection and the formation of the solid implant. Studies have demonstrated the protein release rate from these injectable systems. Graham et al. utilized a PLGA/1-methyl-2-pyrrolidinine depot system and demonstrated that protein release rate is directly related to the phase-inversion kinetics and morphological characteristics of the formed membrane.[116] An additive that increases the phase separation rate also increases the burst effect, though the morphology remains constant. More-desirable and uniform protein release rates could also be achieved by controlling the aqueous miscibility of the depot. Phase separation occurs slowly in depots with low solvent-nonsolvent affinity, thus resulting in structures without an appreciable interconnected polymer-lean phase. Mass-transfer kinetics within the continuous, polymer-rich phase controls the protein release from these systems.[115]

The effect of Pluronic® concentration and PEO block length on the phase-inversion dynamics and subsequent *in vitro* protein-release kinetics of injectable PDLA/Pluronic depots has been reported. Increasing the Pluronic concentration and the PEO block length resulted in a decrease in the protein burst, although the release profiles still retained a typical burst-type shape. It was postulated that the hydrophilic PEO segments extend into the interconnected release pathway (polymer-lean phase), thereby reducing the number of open pores through which protein can diffuse rapidly and thus reducing the burst effect. A transition from a burst-type profile to an extended-release profile occurred by increasing the Pluronic concentration beyond a critical concentration. This result was attributed to Pluronic's leaching effect, whereby the polymer-lean phase becomes filled with the Pluronic material, consequently forming a diffusion barrier within the entire interconnected release pathway that extends the protein release and minimizes the burst.[117]

12.4.3.4 Thermally Induced Gelling Systems

Temperature is the most widely used stimulus in environmentally responsive polymer systems. The change of temperature is not only relatively easy to control, but is also easily applicable both *in vitro* and *in vivo*. These thermosensitive polymers show thermoreversible sol gel changes and are characterized by lower critical solution temperatures (LCSTs).[118,119] They are liquid at room temperature but form a gel at body temperature.

The prototype of a thermosensitive polymer is poly(*N*-isopropyl acryl amide), poly-NIPAAM, which exhibits a rather sharp LCST of approximately 32°C.[118] Unfortunately, poly-NIPAAM is not suitable for biomedical applications due to its well-known cytotoxicity. Moreover, poly-NIPAAM is nonbiodegradable.[107] Triblock poly(ethylene oxide)–poly(propylene oxide)–poly(ethylene oxide) copolymers, PEO-PPO-PEO, known as poloxamers, have shown gelation at body temperature.[118] Potential drawbacks of poloxamer gels include their weak mechanical strength, rapid erosion (i.e., dissolution from the surface), and the nonbiodegradability of PEO-PPO-PEO, which prevents the use of high-molecular-weight polymers that cannot be eliminated by renal excretion.[120]

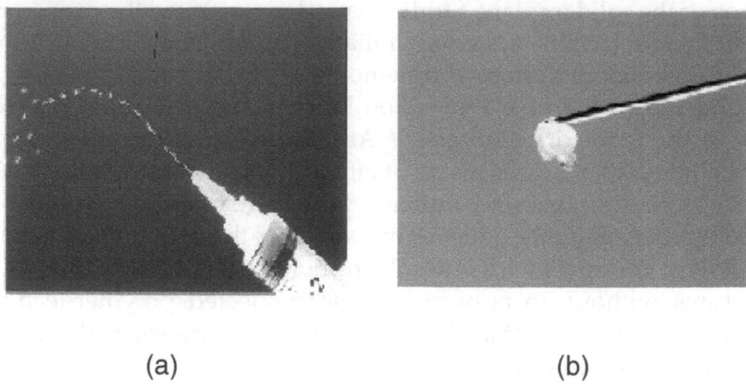

FIGURE 12.6
Sol–gel transition of thermally reversible gelling system Regel™: (a) syringibility of Regel below gelation temperature; (b) water-insoluble gel depot above gelation temperature. (From Fowers, K.D. et al., *Drug Deliv. Tech.*, 3(5), 1, 2003. With permission.)

MacroMed Inc. has developed thermosensitive biodegradable polymers based on low-molecular-weight ABA and BAB triblock copolymers, in which A denotes the hydrophobic polyester block and B denotes the hydrophilic poly(ethylene glycol) block (ReGel®). These polymers are water soluble and show reverse thermal-gelation properties.[112,121–123] The unique characteristics of ReGel center on the following two key properties: (1) ReGel is a water-soluble, biodegradable polymer at temperatures below the gel transition temperature; (2) it forms a water-insoluble gel once injected. This is consistent with a hydrophobically bonded gel state where all interactions are physical, with no covalent cross-linking. The gel forms a controlled-release drug depot with delivery times ranging from 1 to 6 weeks. (See Figure 12.6 and Vignette 12.6.[124])

12.6 OncoGel®

OncoGel is a novel intralesional injectable formulation of paclitaxel. Once injected, it maintains a high intralesional concentration of paclitaxel for six weeks. It concentrates, prolongs, and contains the cytostatic effects of paclitaxel in the tumor while avoiding the side effects associated with systemic paclitaxel products such as Taxol®. It has demonstrated an excellent safety profile in clinical trials in the U.S., with no drug-related serious adverse events. Stable disease and tumor regression were observed in a single-dose Phase I trial. Phase II trials are now underway to demonstrate the efficacy of OncoGel in the treatment of esophageal and breast cancer.

OncoGel utilizes MacroMed's ReGel® drug-delivery system. ReGel is a thermosensitive, biodegradable triblock copolymer composed of the biocompatible components PLGA and PEG. Immediately upon injection, in response to body temperature, a bioerodible gel depot is formed. This gel depot does not discharge from the needle track and remains in the tumor for a period of six weeks. Paclitaxel, a highly insoluble agent, is solubilized and stabilized within the hydrophobic domains of the ReGel system.

Each OncoGel prefilled syringe contains paclitaxel at concentrations of 1.5, 3.1, and 6.3 mg/ml of OncoGel. The injected dose is equal to 32% of the tumor volume. The product is currently packaged in 2.5-ml prefilled, single-use syringes containing 1.1 ml of OncoGel.

MacroMed expects to develop OncoGel for local tumor management in primary, neoadjuvant, and adjuvant palliative care treatment settings in paclitaxel-sensitive cancers. Initial indications selected for Phase II clinical trials include esophageal cancer, neoadjuvant breast cancer, pancreatic cancer, and brain cancer.

OncoGel has an approved inventional new drug status (IND) in the U.S. A Phase I dose-escalating clinical trial has been completed in the U.S. OncoGel is in a multinational Phase II clinical trial for esophageal cancer, and a Phase II clinical trial in the E.U. for neoadjuvancy breast cancer as DTS-301 through a license agreement with Diatos S.A. Additional Phase II trials are planned for the treatment of pancreatic and brain cancer in 2006. Market approval of OncoGel is expected in 2008.

Zentner et al. reported the release of several drugs, including paclitaxel and proteins (pGH, G-CSF, insulin, rHbsAg), from the ReGel formulation. The gel provided excellent control of the release of paclitaxel for approximately 50 days.[125] Direct intratumoral injection of ReGel/paclitaxel (OncoGel®) resulted in a slow clearance of paclitaxel from the injection site, with minimal distribution into any organ. The OncoGel treatment groups exhibited a dose response that was equal or superior to that of the systemic treatments. Paclitaxel and cyclosporine A (very hydrophobic agent) showed significantly higher solubility (enhancement of 400- to >2000-fold was achieved) when dissolved in ReGel. In addition, the stability of these drugs in aqueous cosolvent solutions was substantially improved in the presence of ReGel.

Choi and Kim used an ABA triblock copolymer of PLGA-PEG-PLGA as a drug-delivery carrier for the continuous release of human insulin. They concluded that this formulation can be used to achieve basal insulin levels over a week by a single insulin injection.[126] The aqueous-based dispersions of this polymer are free flowing at room temperature but form a gel at 37°C. The results of this study suggest that this system could be a promising platform for delivery of pDNA. Chen et al. investigated phase-sensitive and thermosensitive smart-polymer-based delivery systems for dispensing testosterone. A combination of PLA and a solvent mixture of benzyl benzoate and benzyl alcohol was used in the phase-sensitive polymer-delivery system.[127]

Bae et al. recently reported the use of a poly(caprolactone-*b*-ethylene glycol-*b*-caprolactone) (PCL-PEG-PCL) triblock copolymer aqueous solution (>15 wt%) that undergoes a temperature-dependent sol-gel-sol transition. PCL-PEG-PCL has a wider gel window of over 15 to 32 wt% and a larger gel modulus. Both PEG-PCL-PEG and PCL-PEG-PCL polymers represent important progress in the development of biodegradable thermogelling systems in that they can be lyophilized in a powder form, are easy to handle, are easy to redissolve to a clear solution, and show little syneresis through the gel phase.[128]

12.4.4 Hydrogels

The pioneering work of Wichterle and Lim in 1960 on cross-linked 2-hydroxyethylmethacrylate (HEMA) hydrogels[129] attracted the interest of biomaterial scientists. Hydrogels are hydrophilic, three-dimensional networks[119] capable of imbibing large amounts of water or biological fluids, ranging from 10 to 20% up to thousands of times their dry weight.[130] They are insoluble due to the presence of chemical (tie points, junctions) and physical cross-links such as entanglements and crystallites.[119] Many different macromolecular structures are possible for physical and chemical hydrogels. These include cross-linked or entangled networks of linear homopolymers, linear copolymers, and block or graft copolymers; polyion–multivalent ion, polyion–polyion, and H-bonded complexes; hydrophilic networks stabilized by hydrophobic domains; and interpenetrated polymer networks (IPNs) or physical blends. The compositions can be divided into natural polymer hydrogels, synthetic polymer hydrogels, and combinations of the two classes.[130]

Hennink et al. have described a biodegradable dextran hydrogel that is based on physical interactions and is particularly suitable for the controlled delivery of pharmaceutically active proteins. The unique feature of their system is that the preparation of the hydrogels takes place in an all-aqueous solution, thereby avoiding the use of organic solvents. Furthermore, chemical cross-linking agents are not needed to create the hydrogels, since cross-linking is established physically by stereocomplex formation between enantiomeric oligomeric lactic acid chains. The hydrogel is simply obtained after mixing aqueous solutions of dextran(L)-lactate (L-lactic acid oligomer grafted to dextran) and dextran(D)-lactate. Here, stereocomplex formation is used to create a hydrogel structure.[131] Hubbell et al. have developed biodegradable hydrogels from water-soluble bismacromonomers by reacting poly(ethylene glycols) with lactides or glycolides and subsequent conversion of the hydroxyl alkanoate end-groups into meth(acrylates). In another study by Kelner and Schacht, hydrogels were synthesized from degradable and nondegradable PEO bismacromonomers. The degradability was provided by the presence of a hydrolyzable segment between the main PEO chain and the methacrylate or methacrylamide groups at both PEO chain termini. The hydrolyzable segment consisted of a monomeric α-hydroxy acid or a depsipeptide. The small

FITC-dextran (4 kDa) was released rapidly from the hydrogel. The larger FITC-dextran (40 kDa) and BSA (bovine serum albumin) were retained inside the matrix, and their release rate was controlled by the degradation.[132]

These materials can be synthesized to respond to a number of physiological stimuli present in the body, such as pH, ionic strength, and temperature. Environmentally sensitive hydrogels have the ability to respond to changes in their external environment. They can exhibit dramatic changes in their swelling behavior, network structure, permeability, or mechanical strength in response to changes in the pH or ionic strength of the surrounding fluid or to temperature. Other hydrogels have the ability to respond to applied electrical or magnetic fields, or to changes in the concentration of glucose.[119]

12.5 Conclusion and Future Prospects

The plethora of therapeutically active molecules pouring into the drug-discovery pipeline and aggressive research for new drug-delivery systems have provided strong motivation to create or to search for new polymeric materials. The limitations of currently available drug therapies, particularly for diseases localized in a specific organ or tissue, e.g., solid tumors and localized infections, have provided scientists with the impetus to consider alternative means of drug administration in the form of a locally administered agent from a polymeric device. The concept of biodegradable polymers as carriers for regional therapy has gained the attention of researchers and clinicians, as exemplified by the various examples presented in this chapter.

After continuous research for more than two decades, polymer-based localized drug-delivery devices have finally reached the status of approval for marketing. For example, Gliadel®, a polyanhydride wafer to deliver BCNU, has been approved for treatment of brain tumors. Macromed has developed an *in situ* gelling system using poly(ether-esters) for use with various anticancer agents, and this system has entered the second phase of clinical trials. Given these successful applications, extensive research in both academia and industry is ongoing in an effort to develop suitable polymeric systems for regional therapy. Most of the efforts in this field are focused on cancer chemotherapy and infectious diseases, as high locoregional concentrations of therapeutic drugs are desired, and anticancer and antiinfective drugs lack regional specificity. However, these polymeric devices also have great potential for use in treating other diseases as well, for example, as carriers that can serve as long-acting drug-delivery systems, thereby ensuring that the drug is released uniformly in a predetermined fashion. The current scenario indicates that this concept is expected to mature in the near future, and many devices will become available in the market.

References

1. Jain, J.P. et al., Role of polyanhydrides as localized drug carriers, *J. Control. Release*, 103, 541, 2005.
2. Dhanikula, A.B. and Panchagnula, R., Localized paclitaxel delivery, *Int. J. Pharm.*, 183, 85, 1999.
3. Lincoff, A.M., Topol, E.J., and Ellis, S.G., Local drug delivery for the prevention of restenosis: fact, fancy and future, *Circulation*, 90, 2070, 1994.
4. Langer, R., Drug delivery and targetting, *Nature*, 392, 5, 1998.
5. Domb, A.J., Polymeric carriers for regional drug therapy, *Mol. Med. Today*, 1, 134, 1995.
6. Stephens, D. et al., Investigation of the *in vitro* release of gentamycin from a polyanhydride matrix, *J. Control. Release*, 63, 305, 2000.
7. Park, E.S., Maniar, M., and Shah, J.C., Biodegradable polyanhydride devices of cefazolin sodium, bupivacaine, and Taxol for local drug delivery: preparation and kinetics and mechanism of *in vitro* release, *J. Control. Release*, 52, 179, 1998.
8. Jampel, H.D. et al., Glaucoma filtration surgery in nonhuman primates using Taxol and etoposide in polyanhydride carriers, *Invest. Opthalmol. Vis. Sci.*, 34, 3076, 1993.
9. Chilkoti, A. et al., Targeted drug delivery by thermally responsive polymers, *Adv. Drug Deliv. Rev.*, 54, 613, 2002.
10. Chabner, B.A. et al., *Goodman and Gilman's The Pharmacological Basis of Therapeutics*, 10th ed., McGraw Hill, New York, 2001, p. 1389.
11. Lesniak, M.S. and Brem, H., Targeted therapy for brain tumors, *Nature Rev. Drug Discov.*, 3, 499, 2004.
12. Byrd, K.E. and Hamilton-Byrd, E.L., Polymer chemotherapy of neurotransmitters and neuromodulators, *Polymeric Site-Specific Pharmacotherapy*, Wiley and Sons, Chichester, 1994, 141.
13. Domb, A.J. et al., Preparation and characterisation of carmustine loaded polyanhydride wafers for treating brain tumors, *Pharm. Res.*, 16, 762, 1999.
14. Shikani, A., Eisele, D.W., and Domb, A.J., Polymer chemotherapy for squamous cell carcinoma of the head and neck, *Arch. Otolaryngol. Headneck Surg.*, 120, 1242, 1994.
15. Gliadel Wafer, http://www.mgipharma.com/wt/page/gliadel, accessed August 2006.
16. Menei, P. et al., Drug targeting into the central nervous system by stereotactic implantation of biodegradable microspheres, *Neurosurgery*, 34, 1058, 1994.
17. Howard, M.A. et al., Intracerebral drug delivery in rats with lesion-induced memory deficits, *J. Neurosurg.*, 71, 105, 1989.
18. Mason, D.W., Biodegradable poly(D,L,-lactide-*co*-glycolide) microspheres for the controlled release of catecholamines to the CNS, *Proc. Int. Symp. Control Rel. Bioact. Mater.*, 15, 270, 1988.
19. Hamann, M.C.J. et al., Novel intrathecal delivery system for treatment of spinal cord injury, *Exp. Neurol.*, 182, 300, 2003.
20. Sigwart, U. et al., Intravascular stents to prevent occlusion and restenosis after transluminal angioplasty, *N. Engl. J. Med.*, 316, 701, 1987.
21. Sousa, J., Costa, M., and Abizaid, A., Lack of neointimal proliferation after implantation of sirolimus-coated stents in human coronary arteries: a quantitative

coronary angiography and three-dimensional intravascular ultrasound study, *Circulation*, 103, 192, 2001.
22. Sousa, J.E., Serruys, P.W., and Costa, M.A., New frontiers in cardiology: drug-eluting stents, *Circulation*, 107, 2274, 2003.
23. Morice, M.-C., Serruys, P.W., and Sousa, J.E., A randomized comparison of a sirolimus-eluting stent with a standard stent for coronary revascularization, *N. Engl. J. Med.*, 346, 1173, 2002.
24. Moses, J.W., Leon, M.B., and Popma, J.J., Sirolimus-eluting stents versus standard stents in patients with stenosis in a native coronary artery, *N. Engl. J. Med.*, 349, 1315, 2003.
25. Orloff, L.A. et al., Prevention of venous thrombosis in microvascular surgery by transmural release of heparin from a polyanhydride polymer, *Surgery*, 117, 554, 1995.
26. Rens, T.J.G. and Kayser, F.H., Local antibiotic treatment in osteomyelitis and soft-tissue infections, *Int. Congr. Ser.*, 556, 51, 1980.
27. Domb, A.J. and Amselem, S., Antibiotic delivery systems for the treatment of chronic bone infections, *Polymeric Site-Specific Pharmacother.*, Wiley and Sons, Chichester, 1994, 243.
28. Türesin, F., Gürsel, I., and Hasirci, V., Biodegradable polyhydroxyalkanoate implants for osteomyelitis therapy: *in vitro* antibiotic release, *J. Biomater. Sci. Polym. Ed.*, 12, 195, 2001.
29. Yeh, H.-Y. and Huang, Y.-Y., Injectable biodegradable polymeric implants for the prevention of postoperative infection: implications for antibacterial resistance, *Am. J. Drug Deliv.*, 1, 149, 2003.
30. Habib, M. et al., Preparation and characterization of ofloxacin microspheres for the eradication of bone associated bacterial biofilm, *J. Microencapsul.*, 16, 27, 1999.
31. Venice, I. et al., *In vitro/in vivo* evaluation of the efficiency of teicoplanin-loaded biodegradable microparticles formulated for implantation to infected bone defects, *J. Microencapsul.*, 20, 705, 2003.
32. Ueng, S.W.N. et al., *In vivo* study of hot compressing molded 50:50 poly (D,L-lactide-*co*-glycolide) antibiotic beads in rabbits, *J. Orthop. Res.*, 20, 654, 2002.
33. Veloso, A.A. et al., Ganciclovir-loaded polymer microspheres in rabbit eyes inoculated with human cytomegalovirus, *Invest. Opthalmol. Vis. Sci.*, 38, 665, 1997.
34. Ciulla, T.A. et al., Corticosteroids in posterior segment disease: an update on new delivery systems and new indications, *Curr. Opin. Ophthalmol.*, 15, 211, 2004.
35. Giordano, G.G., Refojo, M.F., and Arroyo, M.H., Sustained delivery of retinoic acid from microspheres of biodegradable polymer in PVR, *Invest. Ophthalmol. Vis. Sci.*, 34, 2743, 1993.
36. Seligson, D. and Henry, S.L., Newest knowledge of treatment for bone infection: antibiotic impregnated beads, *Clin. Orthop. Rel. Res.*, 295, 2, 1993.
37. Klemm, K., The use of antibiotic-containing bead chains in the treatment of chronic bone infections, *Clin. Microbiol. Infect.*, 7, 28, 2001.
38. Amiji, M.M. et al., Intratumoral administration of paclitaxel in an *in situ* gelling Poloxamer 407 formulation, *Pharm. Dev. Technol.*, 7, 195, 2002.
39. Khor, E. and Lim, L.Y., Implantable applications of chitin and chitosan, *Biomaterials*, 24, 2339, 2003.
40. Hsieh, C.-Y. et al., Preparation of [gamma]-PGA/chitosan composite tissue engineering matrices, *Biomaterials*, 26, 5617, 2005.

41. Dhiman, H.K., Ray, A.R., and Panda, A.K., Characterization and evaluation of chitosan matrix for *in vitro* growth of MCF-7 breast cancer cell lines, *Biomaterials*, 25, 5147, 2004.
42. Dhiman, H.K., Ray, A.R., and Panda, A.K., Three-dimensional chitosan scaffold-based MCF-7 cell culture for the determination of the cytotoxicity of tamoxifen, *Biomaterials*, 26, 979, 2005.
43. Hu, Q. et al., Preparation and characterization of biodegradable chitosan/hydroxyapatite nanocomposite rods via *in situ* hybridization: a potential material as internal fixation of bone fracture, *Biomaterials*, 25, 779, 2004.
44. Wang, Y.-C. et al., Fabrication of a novel porous PGA-chitosan hybrid matrix for tissue engineering, *Biomaterials*, 24, 1047, 2003.
45. Senel, S. and McClure, S.J., Potential applications of chitosan in veterinary medicine, *Adv. Drug Deliv. Rev.*, 56, 1467, 2004.
46. Nsereko, S. and Amiji, M., Localized delivery of paclitaxel in solid tumors from biodegradable chitin microparticle formulations, *Biomaterials*, 23, 2723, 2002.
47. Gan, Q. et al., Modulation of surface charge, particle size and morphological properties of chitosan-TPP nanoparticles intended for gene delivery, *Colloids. Surf., B*, 44, 65, 2005.
48. Anal, A.K. and Stevens, W.F., Chitosan-alginate multilayer beads for controlled release of ampicillin, *Int. J. Pharm.*, 290, 45, 2005.
49. Bhattarai, N. et al., PEG-grafted chitosan as an injectable thermosensitive hydrogel for sustained protein release, *J. Control. Release*, 103, 609, 2005.
50. Gombotz, W.R. and Wee, S., Protein release from alginate matrices, *Adv. Drug Deliv. Rev.*, 31, 267, 1998.
51. Tu, J. et al., Alginate microparticles prepared by spray-coagulation method: preparation, drug loading and release characterization, *Int. J. Pharm.*, 303, 171, 2005.
52. Yenice, I. et al., Biodegradable implantable teicoplanin beads for the treatment of bone infections, *Int. J. Pharm.*, 242, 271, 2002.
53. Arica, B. et al., 5-Fluorouracil encapsulated alginate beads for the treatment of breast cancer, *Int. J. Pharm.*, 242, 267, 2002.
54. Bouhadir, K.H., Alsberg, E., and Mooney, D.J., Hydrogels for combination delivery of antineoplastic agents, *Biomaterials*, 22, 2625, 2001.
55. Andre, P., Hyaluronic acid and its use as a "rejuvenation" agent in cosmetic dermatology, *Semin. Cutan. Med. Surg.*, 23, 218, 2004.
56. Reinmuller, J., Hyaluronic acid, *Aesthetic Surg. J.*, 23, 309, 2003.
57. Liu, Y., Zheng Shu, X., and Prestwich, G.D., Biocompatibility and stability of disulfide-crosslinked hyaluronan films, *Biomaterials*, 26, 4737, 2005.
58. Miller, J.A. et al., Efficacy of hyaluronic acid/nonsteroidal anti-inflammatory drug systems in preventing postsurgical tendon adhesions, *J. Biomed. Mater. Res. Part A*, 38, 25, 1998.
59. Brown, M. and Jones, S., Hyaluronic acid: a unique topical vehicle for the localized delivery of drugs to the skin, *J. Eur. Acad. Dermatol. Venereol.*, 19, 305, 2005.
60. Gerwin, N., Hops, C., and Lucke, A., Intraarticular drug delivery in osteoarthritis, *Adv. Drug Deliv. Rev.*, 58, 226, 2006.
61. Solaraze, http://www.skyepharma.com/approved_solaraze.html, accessed August 2006.
62. Solaraze Gel, http://www.bradpharm.com/products/Doak/prescription/Solaraze.htm, accessed August 2006.

63. Lim, S.T. et al., *In vivo* evaluation of novel hyaluronan/chitosan microparticulate delivery systems for the nasal delivery of gentamicin in rabbits, *Int. J. Pharm.*, 231, 73, 2002.
64. Konishi, M. et al., *In vivo* anti-tumor effect of dual release of cisplatin and adriamycin from biodegradable gelatin hydrogel, *J. Control. Release*, 103, 7, 2005.
65. Chandra, R. and Rustgi, R., Biodegradable polymers, *Prog. Polym. Sci.*, 23, 1273, 1998.
66. Byrom, D., Polymer synthesis by microorganisms: technology and economies, *Trends Biotechnol.*, 5, 246, 1987.
67. Lee, S.Y., Bacterial polyhydroxyalkanoates, *Biotechnol. Bioeng.*, 49, 1, 1996.
68. Rossi, S., Azghani, A.O., and Omri, A., Antimicrobial efficacy of a new antibiotic-loaded poly(hydroxybutyric-*co*-hydroxyvaleric acid) controlled release system, *J. Antimicrob. Chemother.*, 54, 1013, 2004.
69. Carothers, W.H., An introduction to the general theory of condensation polymers, *J. Am. Chem. Soc.*, 51, 2548, 1929.
70. Carothers, W.H. and Arvin, G.A., Polyesters, *J. Am. Chem. Soc.*, 51, 2560, 1929.
71. Vicryl, http://www.ethicon.com/content/backgrounders/www.ethicon.com/www.ethicon.com/vicryl_epi.pdf, accessed August 2006.
72. Chen, D.R., Bei, J.Z., and Wang, S.G., Polycaprolactone microparticles and their biodegradation, *Polym. Degrad. Stab.*, 67, 455, 2000.
73. Benoit, M.-A., Baras, B., and Gillard, J., Preparation and characterization of protein-loaded poly(*o*-caprolactone) microparticles for oral vaccine delivery, *Int. J. Pharm.*, 184, 73, 1999.
74. Jackson, J.K. et al., The characterization of novel polymeric paste formulations for intratumoral delivery, *Int. J. Pharm.*, 270, 185, 2004.
75. Varela, M.C. et al., Cyclosporine-loaded polycaprolactone nanoparticles: immunosuppression and nephrotoxicity in rats, *Eur. J. Pharm. Sci.*, 12, 471, 2001.
76. Allen, C. et al., Polycaprolactone-*b*-poly(ethylene oxide) copolymer micelles as a delivery vehicle for dihydrotestosterone, *J. Control. Release*, 263, 275, 2000.
77. Rosen, H.B. et al., Bioerodible polyanhydrides for controlled drug delivery, *Biomaterials*, 4, 131, 1983.
78. Tian, Y. et al., The effect of storage temperatures on the *in vitro* properties of a polyanhydride implant containing gentamycin, *Drug Dev. Ind. Pharm.*, 28, 897, 2002.
79. Li, L.C., Deng, J., and Stephens, D., Polyanhydride implant for antibiotic delivery: from the bench to the clinic, *Adv. Drug Deliv. Rev.*, 54, 963, 2002.
80. Barr, J. et al., Postsurgical pain management with poly(ortho esters), *Adv. Drug Deliv. Rev.*, 54, 1041, 2002.
81. Heller, J. et al., Poly(ortho esters): synthesis, characterization, properties and uses, *Adv. Drug Deliv. Rev.*, 54, 1015, 2002.
82. Singh, K. et al., Trabeculectomy with intraoperative 5-fluorouracil vs. mitomycin C, *Am. J. Opthalmol.*, 123, 48, 1997.
83. Einmahl, S. et al., Therapeutic applications of viscous and injectable poly(ortho esters), *Adv. Drug Deliv. Rev.*, 53, 45, 2001.
84. Schwach-Abdellaoui, K. et al., Optimization of a novel bioerodible device based on auto-catalyzed poly(ortho esters) for controlled delivery of tetracycline to periodontal pocket, *Biomaterials*, 22, 1659, 2001.
85. Biochronomer systems, http://www.appharma.com/technology/biochronomer.html, accessed August 2006.

86. Wang, J., Wang, B.M., and Schwendeman, S.P., Characterization of the initial burst release of a model peptide from poly(D,L-lactide-*co*-glycolide) microspheres, *J. Control. Release*, 82, 282, 2002.
87. Kissel, T., Li, Y., and Unger, F., ABA-triblock copolymers from biodegradable polyester A-blocks and hydrophilic poly(ethylene oxide) B-blocks as a candidate for *in situ* forming hydrogel delivery systems for proteins, *Adv. Drug Deliv. Rev.*, 54, 99, 2002.
88. Dorta, M.J. et al., Potential applications of PLGA film-implants in modulating *in vitro* drugs release, *Int. J. Pharm.*, 248, 149, 2002.
89. Zoladex (goserelin acetate), http://www.astrazeneca.com/productbrowse/5_87.aspx, accessed August 2006.
90. Jackson, J.K. et al., Characterization of perivascular poly(lactic-*co*-glycolic acid) films containing paclitaxel, *Int. J. Pharm.*, 283, 97, 2004.
91. Jackson, J.K. et al., The suppression of human prostate tumor growth in mice by the intratumoral injection of a slow-release polymeric paste formulation of paclitaxel, *Cancer Res.*, 60, 4146, 2000.
92. Wang, G. et al., The release of cefazolin and gentamicin from biodegradable PLA/PGA beads, *Int. J. Pharm.*, 273, 203, 2004.
93. Negrin, C.M. et al., Methadone implants for methadone maintenance treatment: *in vitro* and *in vivo* animal studies, *J. Control. Release*, 95, 413, 2004.
94. Tan, L.P. et al., Effect of plasticization on heparin release from biodegradable matrices, *Int. J. Pharm.*, 283, 89, 2004.
95. Vogelhuber, W. et al., Efficacy of BCNU and paclitaxel loaded subcutaneous implants in the interstitial chemotherapy of U-87 MG human glioblastoma xenografts, *Int. J. Pharm.*, 238, 111, 2002.
96. Ramchandani, M. and Robinson, D., *In vitro* and *in vivo* release of ciprofloxacin from PLGA 50:50 implants, *J. Control. Release*, 54, 167, 1998.
97. Seo, S.-A. et al., A local delivery system for fentanyl based on biodegradable poly(L-lactide-*co*-glycolide) oligomer, *Int. J. Pharm.*, 239, 93, 2002.
98. Fournier, E. et al., Development of novel 5-FU-loaded poly(methylidene malonate 2.1.2)-based microspheres for the treatment of brain cancers, *Eur. J. Pharm. Biopharm.*, 57, 189, 2004.
99. Berkland, C. et al., Microsphere size, precipitation kinetics and drug distribution control drug release from biodegradable polyanhydride microspheres, *J. Control. Release*, 94, 129, 2004.
100. Siepmann, J. et al., Effect of the size of biodegradable microparticles on drug release: experiment and theory, *J. Control. Release*, 96, 123, 2004.
101. Sandor, M. et al., Effect of protein molecular weight on release from micron-sized PLGA microspheres, *J. Control. Release*, 76, 297, 2001.
102. Fernández-Carballido, A. et al., Biodegradable ibuprofen-loaded PLGA microspheres for intraarticular administration: effect of Labrafil addition on release *in vitro*, *Int. J. Pharm.*, 279, 33, 2004.
103. Arica, B. et al., Carbidopa/levodopa-loaded biodegradable microspheres: *in vivo* evaluation on experimental Parkinsonism in rats, *J. Control. Release*, 102, 689, 2005.
104. Trapani, G. et al., Encapsulation and release of the hypnotic agent zolpidem from biodegradable polymer microparticles containing hydroxypropyl-β-cyclodextrin, *Int. J. Pharm.*, 268, 47, 2003.
105. Yue, I.C. et al., A novel polymeric chlorhexidine delivery device for the treatment of periodontal disease, *Biomaterials*, 25, 3743, 2004.

106. Suarez, S. et al., Respirable PLGA microspheres containing rifampicin for the treatment of tuberculosis: screening in an infectious disease model, *Pharm. Res.*, 18, 1315 2001.
107. Hatefi, A. and Amsden, B., Biodegradable injectable *in situ* forming drug delivery systems, *J. Control. Release*, 80, 9, 2002.
108. Zhang, X. et al., Development of biodegradable polymeric paste formulations for Taxol: an *in vitro* and *in vivo* study, *Int. J. Pharm.*, 137, 199, 1996.
109. Winternitz, C.I. et al., Development of a polymeric surgical paste formulation for Taxol, *Pharm. Res.*, 13, 368, 1996.
110. Cohn, D. and Hotovely-Salomon, A., Biodegradable multiblock PEO/PLA thermoplastic elastomers: molecular design and properties, *Polymer*, 46, 2068, 2005.
111. Shikanov, A. et al., Poly(sebacic acid-*co*-ricinoleic acid) biodegradable carrier for paclitaxel: *in vitro* release and *in vivo* toxicity, *J. Biomed. Mater. Res. A*, 69, 47, 2004.
112. Rathi, R.C. and Zentner, G.M., U.S. patent 6004573, 1999.
113. Nivasu, V.M., Reddy, T.T., and Tammishetti, S., *In situ* polymerizable polyethyleneglycol containing polyesterpolyol acrylates for tissue sealant applications, *Biomaterials*, 25, 3283, 2004.
114. Azhdarinia, A. et al., Regional radiochemotherapy using *in situ* hydrogel, *Pharm. Res.*, 22, 776, 2005.
115. Oster, C.G. et al., Design of amine-modified graft polyesters for effective gene delivery using DNA-loaded nanoparticles, *Pharm. Res.*, 21, 927, 2004.
116. Graham, P.D., Brodbeck, K.J., and McHugh, A.J., Phase inversion dynamics of PLGA solutions related to drug delivery, *J. Control. Release*, 58, 233, 1999.
117. DesNoyer, J.R. and McHugh, A.J., The effect of Pluronic on the protein release kinetics of an injectable drug delivery system, *J. Control. Release*, 86, 15, 2003.
118. Packhaeuser, C.B. et al., *In situ* forming parenteral drug delivery systems: an overview, *Eur. J. Pharm. Biopharm.*, 58, 445, 2004.
119. Peppas, N.A. et al., Hydrogels in pharmaceutical formulations, *Eur. J. Pharm. Biopharm.*, 50, 27, 2000.
120. Ruel-Gariépy, E. and Leroux, J.C., *In situ*-forming hydrogels: review of temperature-sensitive systems, *Eur. J. Pharm. Biopharm.*, 58, 409, 2004.
121. Rathi, R.C., Zentner, G.M., and Jeong, B., U.S. patent 6117949, 2000.
122. Rathi, R.C., Zentner, G.M., and Jeong, B., U.S. patent 6201072, 2001.
123. Fowers, K.D. et al., Thermally reversible gelling materials for safe and versatile depot delivery, *Drug Deliv. Tech.*, 3(5), 1, 2003.
124. OncoGel, www.macromed.com/OncoGel%2028Jan2004v6.pdf, accessed August 2006.
125. Zentner, G.M. et al., Biodegradable block copolymers for delivery of proteins and water-insoluble drugs, *J. Control. Release*, 72, 203, 2001.
126. Choi, S. and Kim, S.W., Controlled release of insulin from injectable biodegradable triblock copolymer depot in ZDF rats, *Pharm. Res.*, 20, 2008, 2003.
127. Chen, S. et al., Triblock copolymers: synthesis, characterization, and delivery of a model protein, *Int. J. Pharm.*, 288, 207, 2005.
128. Bae, S.J. et al., Thermogelling poly(caprolactone-*b*-ethylene glycol-*b*-caprolactone) aqueous solutions, *Macromolecules*, 38, 5260, 2005.
129. Wichterle, O. and Lim, D., Hydrophilic gels in biologic use, *Nature*, 185, 117, 1960.
130. Hoffman, A.S., Hydrogels for biomedical applications, *Adv. Drug Deliv. Rev.*, 43, 3, 2002.

131. Hennink, W.E. et al., Biodegradable dextran hydrogels crosslinked by stereocomplex formation for the controlled release of pharmaceutical proteins, *Int. J. Pharm.*, 277, 99, 2004.
132. Kelner, A. and Schacht, E.H., Tailor-made polymers for local drug delivery: release of macromolecular model drugs from biodegradable hydrogels based on poly(ethylene oxide), *J. Control. Release*, 101, 13, 2005.

Chapter 13

Smart Polymers in Affinity Precipitation of Proteins

Ashok Kumar, Igor Yu. Galaev, and Bo Mattiasson

CONTENTS

Abbreviations

ATPS	aqueous two phase polymer system
ELP	elastinlike proteins
LCST	lower critical solution temperature
IMAC	immobilized metal affinity chromotography
ITC	inverse transition cycling
NIPAM	*N*-isopropylacrylamide
PAA	poly(acrylic acid)

PEC	polyelectrolyte complex
PEG	poly(ethyleneglycol)
PMAA	poly(methacrylic acid)
Poly(VI/NIPAM)	poly(vinylimidazole-*co*-*N*-isopropyl-acrylamide)
VCL	*N*-vinylcaprolactam
VI	*N*-vinylimidazole

13.1 Introduction

Development of efficient and fast purification protocols in bioseparation has always been a challenging task. The advent of recombinant DNA technology has rendered it technically feasible to produce virtually any protein of interest. However, efficient separation of the desired protein from crude mixtures is recognized as the crucial factor in deciding the commercial viability of producing that protein for biotechnological applications. All bioseparation processes include three stages:

1. The preferential partitioning of target substance and impurities between two phases (liquid–liquid or liquid–solid)
2. The mechanical separation of the phases (e.g., separation of the stationary and mobile phase in a chromatographic column)
3. Recovery of the target substance from the enriched phase

The demand in bioseparation has been to reduce the purification steps so as to increase the yields and decrease costs. The strategies in general have been (a) to use methods that eliminate clarification techniques, such as centrifugation or filtration, to remove particulate matter such as cellular debris and instead directly proceed with a separation step or (b) to use highly selective techniques such as affinity separations as early as possible in the purification process.

Precipitation of a target protein from a crude mixture has been considered as an attractive option to enrich protein and separate it from the homogeneous solution. Traditionally, precipitation of the target protein is achieved by the addition of (a) large amounts of salts, like ammonium sulfate, (b) organic solvents miscible with water, like acetone or ethanol, or (c) polymers such as polyethylene glycol (PEG).[1] Precipitation of the target protein occurs because of changing bulk parameters of the medium, the driving force being integral physicochemical and surface properties of the protein macromolecule. The macromolecular nature of protein molecules — combined with a general principle of their folding, such as amino acids that are hydrophilic at the surface and hydrophobic inside the core — makes proteins rather similar in their surface properties. Thus, high selectivity is not expected to be achieved by traditional precipitation techniques, as the selectivity of precipitation is limited to differences in the integral surface properties of protein molecules. However, this type of precipitation still remains an important operation for the laboratory- and industrial-scale recovery and purification of proteins. The technique is often used in the early stages of downstream processing for both product-stream fractionation and concentration, combining capture of the target protein with some purification. However, low selectivity remains the main limitation for the technique. Thus the introduction of high selectivity to the precipitation techniques is of great importance.

The present trend in downstream processing of proteins is the introduction of highly selective affinity steps at the early stage of the purification process using robust, biologically and chemically stable ligands.[2] Affinity precipitation is a technique that seeks to combine these features and is now considered to be a powerful technique for protein purification.[3–5] As a general rule, there are five basic steps in affinity precipitation:

1. Carrying out affinity interactions in free solution
2. Precipitation of the affinity-reagent–target-protein complex from the solution
3. Recovery of the precipitate
4. Dissociation and recovery of the target molecule from the complex
5. Recovery of the affinity reagent

This chapter discusses affinity precipitation in an attempt to introduce the highly selective affinity technique applied at the very beginning of protein purification.

Affinity precipitation methods have two main approaches that have been described in the literature[4]: precipitation with homo- and heterobifunctional ligands. The basic idea of affinity precipitation involves the use of a macroligand, i.e., a ligand with two or more affinity sites. The macroligands could be synthesized either by covalent linking of the two low-molecular-weight ligands (directly or through a short spacer) or by covalent binding of several ligands to a water-soluble polymer, thus providing two main approaches (homobifunctional and heterobifunctional modes) to affinity precipitation.

13.2 Affinity Precipitation: Homo- and Heterobifunctional Modes

In the homobifunctional mode of affinity precipitation, if the protein has a few ligand-binding sites, complexation with a macroligand results in the formation of poorly soluble large aggregates (Figure 13.1) that can be separated by centrifugation or filtration from the supernatant containing soluble impurities. Immunoprecipitation of antigens with divalent antibodies is considered to be a good illustration of this principle. However, optimal concentrations of the

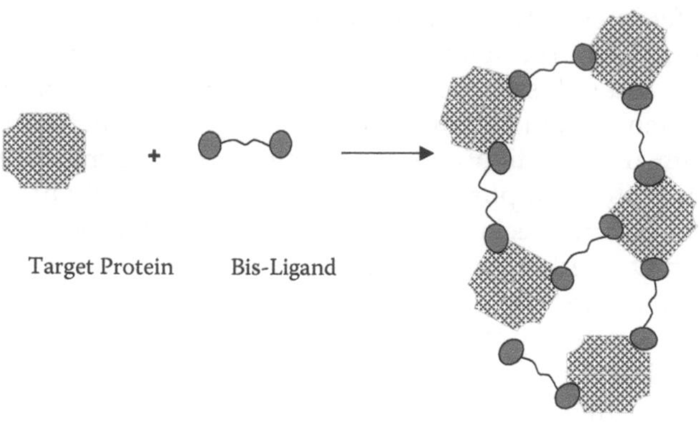

FIGURE 13.1
Homobifunctional mode of affinity precipitation.

two molecules are needed to form the aggregated complexes, and if the specific ratio is not maintained, the complex starts to dissolve.

The homobifunctional mode of affinity precipitation was introduced by Larsson and Mosbach, who performed quantitative precipitation of purified lactate dehydrogenase (tetrameric enzyme) and glutamate dehydrogenase (hexamer) from solution using bis-NAD (N^2,$N^{2'}$-adipohydrazidobis [N^6 carbonylmethyl]-NAD).[6,7] The precipitate was readily soluble in the presence of NADH (nicotinamide adenine dinucleotide reduced form), which forms stronger bonds with lactate dehydrogenase and replaces the enzyme from the complex with bis-NAD. The effectiveness of the precipitation strongly depends on the length of the spacer linking the two molecules of ligand and on the ratio of enzyme to bis derivative; an excess of the latter decreases the effectiveness of precipitation down to a completely soluble complex. With slight modifications, the homobifunctional mode of affinity precipitation has been used for the purification of other enzymes like yeast alcohol dehydrogenase. The homobifunctional ligands have to be specially synthesized for individual applications by joining two identical ligands through a spacer. Thus this mode of affinity precipitation can be used only for proteins with two or more ligand-binding centers, which significantly limits its application. An exception, however, is metal-affinity precipitation, which can exploit different metal-binding amino acid residues on the same polypeptide chain.[8] The addition of a bis-ligand at an optimum concentration creates a cross-linked network with the target protein, provided that the latter has two or more metal binding sites. The cross-linked protein–bis-ligand network precipitates from the solution eventually. The first such application was reported by Van Dam et al.,[9] who conducted model experiments in which human hemoglobin and sperm whale hemoglobin were quantitatively precipitated with bis-copper chelates. The study describes only precipitation of the pure proteins and does not present the results on the protein elution and recovery. There have been other examples where metal-affinity precipitation in homobifunctional mode has worked successfully. However, as the homobifunctional mode of affinity precipitation does not utilize smart polymers, it thus falls beyond the scope of this chapter. The reader is referred to several reviews for additional information.[3–4,10–12]

The heterobifunctional format of affinity precipitation is a more general approach wherein affinity ligands are covalently coupled to soluble–insoluble polymers to form a *macroligand* (Figure 13.2). The precipitation is induced by a component of the macroligand that is not directly involved in the affinity interactions and is represented by a soluble–insoluble polymer. The affinity ligands, for interaction with the target protein, are covalently coupled to the polymer. Thus, one part of the macroligand has the affinity for the target protein/enzyme, while the other controls the solubility of the complex. The ligand–polymer conjugate selectively binds the target protein from the crude extract and forms a complex. The phase separation of the protein–polymer complex is triggered by small changes in the environment (pH, temperature, or ionic strength), resulting in transition of polymer backbone into an insoluble state as precipitate formation. The desired protein is then dissociated from the polymer either directly from the insoluble macroligand–protein complex, or the precipitate is first dissolved and the protein dissociated from the macroligand. The ligand–polymer conjugate is reprecipitated without the protein, which remains in the supernatant in a purified form, and the ligand–polymer conjugate can be recovered and reused for another cycle[4] (Figure 13.3). The smart polymers are critical for the development of affinity precipitation of protein in a heterobifunctional mode.

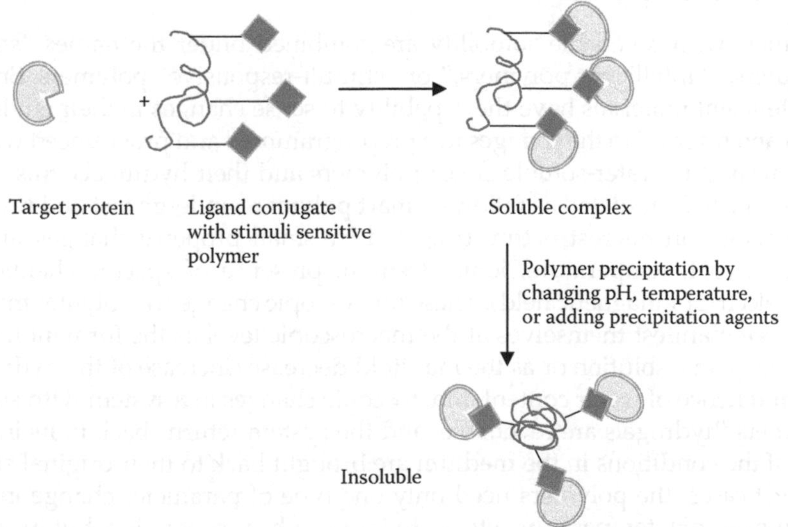

FIGURE 13.2
Heterobifunctional mode of affinity precipitation.

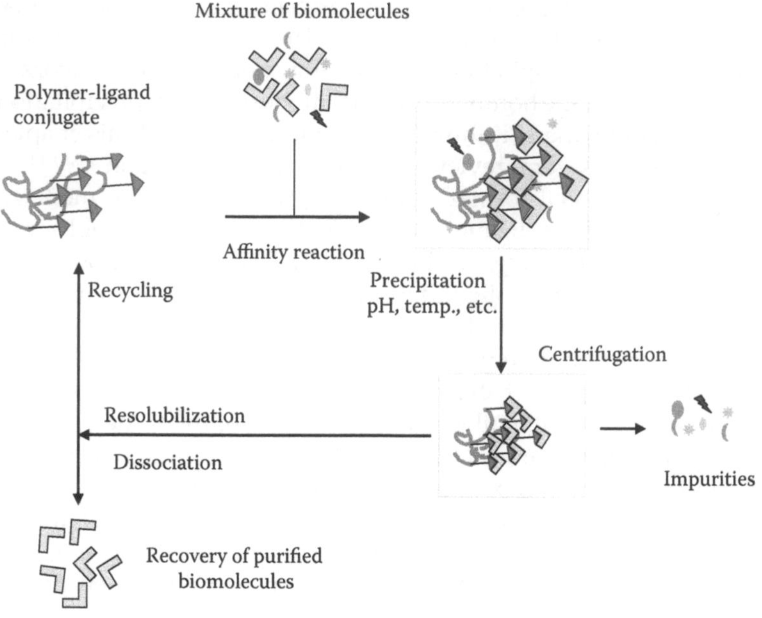

FIGURE 13.3
Schematic of the affinity precipitation of proteins.

13.3 Smart Polymers

Polymers with reversible solubility are combined under the names "smart polymers," "intelligent polymers," or "stimuli-responsive" polymers. Smart or intelligent materials have the capability to sense changes in their environment and respond to the changes in a preprogrammed and pronounced way.[13] With respect to water-soluble smart polymers and their hydrogels, this definition can be formulated as follows. Smart polymers undergo fast and reversible changes in microstructure triggered by small property changes in the medium (pH, temperature, ionic strength, presence of specific chemicals, light, electric or magnetic field). These microscopic changes of polymer microstructure manifest themselves at the macroscopic level as the formation of a precipitate in a solution or as the manifold decrease/increase of the hydrogel size and hence of water content. Macroscopic changes in a system with smart polymers/hydrogels are reversible, and the system returns back to its initial state if the conditions in the medium are brought back to their original state. In most cases, the polymers need only one type of parameter change in the medium — pH, temperature, etc. — to induce phase separation, but in some cases it may be advantageous to use more than one mode to induce precipitation.[14,15] An ideal polymer for affinity precipitation must:

- Contain reactive groups for ligand coupling
- Not interact strongly with the ligand or impurities, thereby making the ligand available for interaction with the target protein and preventing nonspecific coprecipitation of impurities
- Give complete phase separation of the polymer upon a change of property in the medium; in this respect, synthetic polymers are superior to natural polymers, which have a broader molecular-weight distribution and thereby a less-well-characterized transition
- Form polymer precipitates that are compact, thereby allowing easy separation and avoiding the trapping of impurities within a gel structure
- Be easily solubilized after the precipitate is formed; the precipitation–solubilization cycle must be repeatable many times with good recovery
- Be available and cheap

In general, all smart polymer/hydrogel systems can be divided into three groups[16,17]:

1. pH-responsive smart polymers
2. Thermoresponsive smart polymers
3. Reversibly cross-linked polymer networks

13.3.1 pH-Responsive Smart Polymers

This group consists of the polymers for which poor solvent conditions are created by decreasing the net charge of the polymer/hydrogel. In a typical pH-sensitive polymer, protonation/deprotonation events occur and impart the charge over the molecule (generally on carboxyl or amino groups). The net charge can be decreased by changing pH to neutralize the charges on the macromolecule, hence reducing the hydrophilicity (increasing the hydrophobicity) of the macromolecule and the repulsion between polymer segments. For instance, copolymers of methylmethacrylate (hydrophobic part) and methacrylic acid (hydrophilic at high pH when carboxy groups are deprotonated, but more hydrophobic when carboxy groups are protonated) precipitate from aqueous solutions upon acidification to pH around 5, while copolymers of methyl methacrylate (hydrophobic part) with dimethylaminethyl methacrylate (hydrophilic at low pH when amino groups are protonated, but more hydrophobic when amino groups are deprotonated) are soluble at low pH but precipitate at slightly alkaline conditions. The pH-induced precipitation of smart polymers is very sharp and typically requires a change in pH of not more than 0.5 units. Increasing the hydrophobicity of the pH-sensitive polymer through copolymerization with more-hydrophobic monomers results in transitions at higher pH. Figure 13.4 presents a

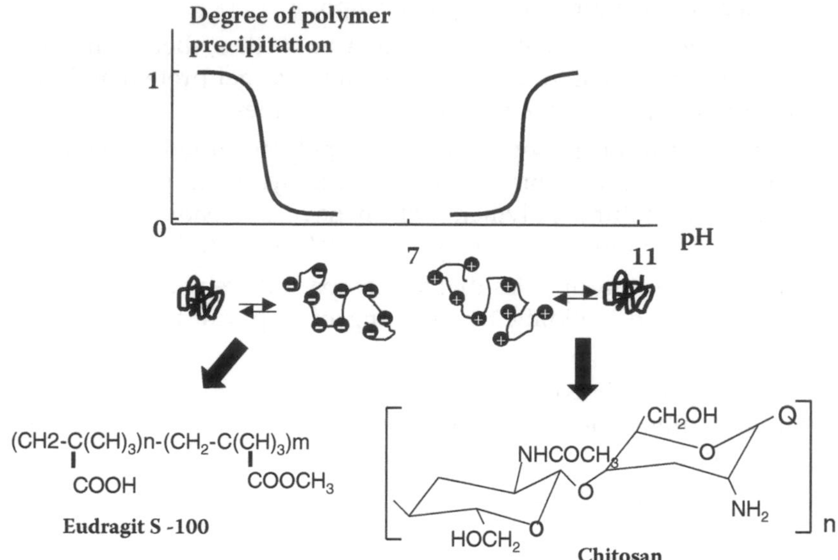

FIGURE 13.4
pH-induced soluble–insoluble behavior of synthetic (Eudragit S-100) and natural (chitosan) polymers.

pH-dependent precipitation curve of a random copolymer of methyl methacrylate and methacrylic acid (commercialized as Eudragit S-100 by Röhm Pharma GmBH, Weiterstadt, Germany) as well as chitosan, a deacetylated chitin. Hydrophobically modified cellulose derivatives with pendant carboxy groups, e.g., hydroxypropyl methyl cellulose acetate succinate, are also soluble at basic conditions but precipitate in slightly acidic media.

The charges on the macromolecule can also be neutralized by addition of an efficient counterion, e.g., a low-molecular-weight counterion or a polymer molecule with the opposite charges. The latter systems are combined under the name polyelectrolyte complexes (PEC). The complex formed by the oppositely charged polyelectrolytes, poly(methacrylic acid) (polyanion) and poly(*N*-ethyl-4-vinylpyridinium bromide) (polycation), undergoes reversible precipitation from aqueous solution at any desired pH value in the range of 4.5 to 6.5, depending on the ionic strength and polycation/polyanion ratio in the complex.[18] Polyelectrolyte complexes formed by poly(ethyleneimine) and polyacrylic acid undergo soluble–insoluble transition in an even broader pH range, from pH 3 to pH 11.[19]

There are ranges of different proteins/enzymes that have been purified successfully by affinity precipitation using pH-responsive polymers (Table 13.1). In general, a specific ligand is chemically coupled to the polymer backbone, which later binds to the target protein in solution, and the protein–polymer complex is precipitated by a change of pH, as it renders the polymer backbone insoluble. But in some cases the polymer itself has the affinity for the target protein, and

TABLE 13.1
Published Protocols of Protein Purification Using Affinity Precipitation

Target Protein	Polymer	Ligand [a]	Method Used to Precipitate Polymer	Protein Recovery (%) [b]	Purification Fold [b]	Reference
Wheat germ agglutinin	chitosan	A	ΔpH	55–70	10	Senstad and Mattiasson[42]
Lysozyme, lectins specific for *N*-acetylglucosamine	chitosan	A	ΔpH	B	B	Tyagi et al.[120]
Trypsin	chitosan	soybean trypsin inhibitor	ΔpH	93	5.5	Senstad and Mattiasson[41]
Trypsin	copolymer NIPAM and *N*-acrylhydroxysuccinimide	*p*-aminobenzamidine	ΔT	74–82	B	Nguyen and Luong[64]
Trypsin	alginate	soybean trypsin inhibitor	addition of Ca^{2+}	30–36	B	Linné-Larsson et al.[32]
Trypsin	poly(*N*-vinyl caprolactam)	soybean trypsin inhibitor	ΔT	54	28	Galaev and Mattiasson[45]
Protein A	hydroxylpropyl-methylcellulose succinate	IgG	ΔpH	50–70	4.9–5.4	Taniguchi et al.[113]
				53–90	32	Taniguchi et al.[71]
Human IgG	galactomannan	protein A	addition of tetraborate	B	B	Bradshaw and Sturgeon[37]
Alkaline protease	poly-NIPAM	*p*-aminobenzamidine	ΔT	B	B	Pécs et al.[65]
Restriction endonuclease Hind III	poly-NIPAM	phage λ DNA	ΔT	B	B	Maeda et al.[29]
Xylanase from *Trichoderma virde*	Eudragit S-100	A	ΔpH	70	4.6	Gupta and Mattiasson[4]
D-lactate dehydrogenase from *Lactobacillus bulgaricus*	Eudragit S-100	A	ΔpH	85	2.4	Guoqiang et al.[53]
Endopolygalacturonase from *Aspergillus niger*	Eudragit S-100	A	ΔpH	86	8.8	Gupta et al.[76]
Trypsin	Eudragit S-100	soybean trypsin inhibitor	ΔpH	74	1.8	Kumar and Gupta[67]
Concanavalin A	Eudragit S-100	*p*-aminophenyl-α-D-glucopyranoside	ΔpH	83–91	B	Linné-Larsson and Mattiasson[50]

(*continued*)

TABLE 13.1

Published Protocols of Protein Purification Using Affinity Precipitation

Target Protein	Polymer	Ligand [a]	Method Used to Precipitate Polymer	Protein Recovery (%) [b]	Purification Fold [b]	Reference
D-Lactate dehydrogenase from *Leuconostoc meseteroides* spp. *Cremoris*	Eudragit S-100	Cibacron blue	ΔpH	56	4.3	Shu et al.[57]
Lactate dehydrogenase and pyruvate kinase from porcine muscle	Eudragit S-100	Cibacron blue	ΔpH	63 59	7.8 4.4	Guoqiang et al.[55]
Alcohol dehydrogenase from bakers yeast	Eudragit S-100	Cibacron blue	ΔpH	40–42	8.2–9.5	Guoqiang et al.[56]
β-Glucosidase from *Trichoderma longibrachiatum*	I. chitosan II. Eudragit S-100	A	ΔpH	82–86	10–14	Agarwal and Gupta[121]
α-Amylase inhibitor from wheat	copolymer of NIPAM and vinylimidazole	immobilized Cu(II)-ions	Δ*T* and addition of salt	89	4	Kumar et al.[60]
α-Amylase inhibitors 1-1 and 1-2 from seeds of ragi (Indian finger millet)	copolymer of NIPAM and vinylimidazole	immobilized Cu(II)-ions	Δ*T* and addition of salt	84 (1-1) 89 (1-2)	13 (1-1) 4 (1-2)	Kumar et al.[60]
Monoclonal antibodies	polyelectrolyte complex formed by poly(methacrylic acid) and poly(*N*-ethyl-4-vinyl-pyridinium bromide)	antigen (inactivated glyceraldehydes-3-phosphate dehydrogenase)	ΔpH	96–98	1.4–1.8	Dainiak et al.[18]
Lactate dehydrogenase from porcine muscle	carboxymethylcellulose	Cibacron blue	addition of Ca^{2+}	87	21	Lali et al.[122]
Thermolabile α-glucosidase from *Saccharomyces cerevisiae*	poly (*N*-acryloyl-piperidine)	mMaltose	ΔT	54–68	170–200	Hoshino et al.[52]

Glucoamylase	alginate	A	addition of Ca^{2+}	81	28	Teotia[75]
His-tag β-D-galactosidase and chloramphenicol acetyltransferase	elastinlike proteins (ELP)	immobilized Ni(II)	ΔT	85	B	Kostal et al.[94]
His-tag lactate dehydrogenase (LDH)	copolymer of NIPAM and vinylimidazole	immobilized Cu(II)- and Ni(II)-ions	Δ*T* and addition of salt	82–95	3.5	Kumar et al.[62]
His-tag green fluorescent protein (GFP)	copolymer of NIPAM and vinylimidazole	immobilized Cu(II)- and Ni(II)-ions	Δ*T* and addition of salt	93.3	B	Kumar et al.[123]
Alcohol dehydrogenase	*k*-carrageenan	Cibacron blue	addition of K^+	58	13.6	Mondal et al.[124]
Pullulanase	*k*-carrageenan	A	addition of K^+	92	50	Roy and Gupta[33]
Ab-fragments	copolymer of NIPAM and vinylimidazole	immobilized Cu(II)- and Ni(II)-ions	Δ*T* and addition of salt	80–91	16–21	Kumar et al.[63]
Chitanase	Eudragit S-100	A	ΔpH	90	30	Teotia et al.[78]
Pectinase	alginate	A	addition of Ca^{2+} and microwave treated	83	20	Mondal et al.[74]
Chloramphenicol acetyl-transferase, thioredoxin, calmodulin, blue fluorescent protein	elastinlike proteins (ELP)	fusion with ELP	Δ*T*	75	B	Trabbic-Carlson et al.[98]
Carbohydrate-binding protein	elastinlike triblock protein copolymer (ELP)	carbohydrate ligands	Δ*T*	80	B	Sun[100]
Xylanase	Eudragit S-100	A	ΔpH	96	45	Roy et al.[77]
Avidin	(3-aminopropyloxy) azobenzene-poly(NIPAM)	biotin	photorespon-sive	47	B	Despond and Freitag[125]
His-tag BRCA-1 protein fragment	highly branched poly(NIPAM-*co*-GMA)	immobilized C(II)-ions	Δ*T*	B	B	Carter et al.[110]

[a] A: the polymer itself acts as a macroligand for the target protein.

[b] B: data not available.

the polymer acts as a macroligand. Chitosan was used to precipitate lysozyme or lectins wheat germ agglutinin or lysozyme or lectins, and Eudragit S-100 was similarly used as a macroligand for binding and precipitating xylanase or lactate dehydrogenase or endopolygalacturonase. pH-responsive smart polymers have been used successfully in affinity precipitation of many proteins, but the charged character of the polymer, which shows some nonspecific interactions with other proteins, has been a limiting factor in the use of these polymers.

13.3.2 Thermoresponsive Smart Polymers

This is the most important group of smart polymers and has shown tremendous potential in developing affinity-precipitation methods for protein purification. This group of smart polymers consists of uncharged polymers soluble in water due to the hydrogen bonding with water molecules. Changes in hydrophobic–hydrophilic balance are induced by increasing temperature or ionic strength. The efficiency of hydrogen bonding is reduced upon raising temperature. There is a critical temperature for some polymers at which the efficiency of hydrogen bonding becomes insufficient for continued solubility of the macromolecule, and the phase separation of polymer takes place. When the temperature of an aqueous solution of a smart polymer is raised to a point higher than its critical temperature (lower critical solution temperature [LCST] or "cloud point"), the solution separates into two phases — a polymer-enriched phase and an aqueous phase containing practically no polymer — that can be easily separated (Figure 13.5). This phase separation

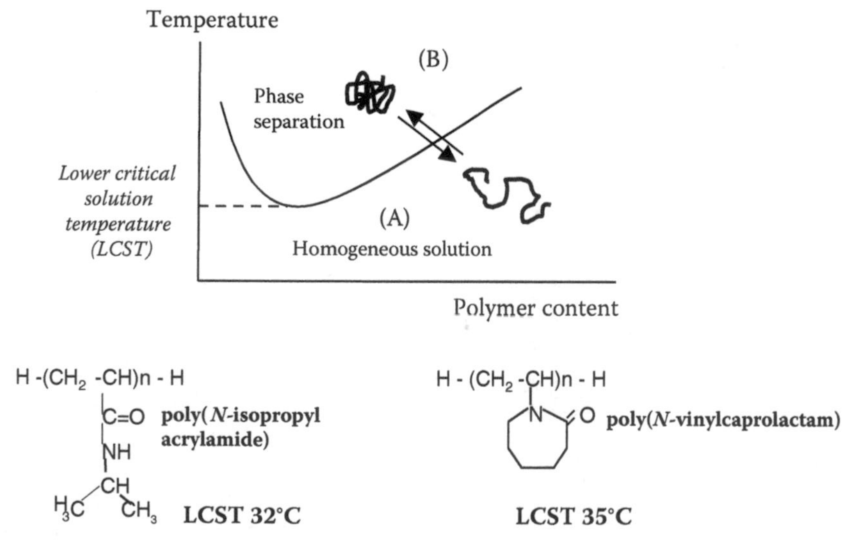

FIGURE 13.5
Temperature response for thermosensitive polymers: (a) soluble phase (below LCST); (b) insoluble phase (above LCST).

is completely reversible, and the smart polymer dissolves in water upon cooling. In aqueous solution, several *N*-substituted poly(acrylamides) undergo a thermally induced, reversible phase transition. The LCSTs of such homopolymers as poly(*N*-isopropylacrylamide), poly(*N,N*-diethylacrylamide), poly(*N*-cyclopropylacrylamide), and poly(*N*-ethylacrylamide) in distilled H_2O have been reported to be 32, 33, 58, and 74°C, respectively (Figure 13.6).[20] With present-day advanced synthetic polymer chemistry, it has been possible to design polymers with different transition temperatures ranging from 4 to 5°C for poly(*N*-vinyl piperidine) to 100°C for poly(ethylene glycol).[21] Among these homopolymers, poly(*N*-isopropylacrylamide) (PNIPAM) has received special attention because of its sharp phase-transition temperature at 32°C,[22] and various strategies for synthesizing the hydrogel and its derivatives have been described. Efforts have been directed toward altering the swelling/shrinking behavior and preparing copolymers that also respond to other stimuli. Critical insight into the deswelling mechanism was obtained through the work of Hoffman et al. on PNIPAM.[23]

In contrast to pH-sensitive smart polymers, which contain carboxy or amino groups that can be used for the covalent coupling of ligands, the thermosensitive polymers do not have inherent reactive groups. Thus copolymers containing reactive groups should be synthesized. *N*-hydroxyacrylsuccinimide[24] or glycidyl methacrylate[25] are commonly used as active co-monomers in copolymerization with NIPAM, allowing further coupling of amino-group-containing ligands to the synthesized copolymers. Other co-monomers (along with NIPAM) that have been assessed include acrylic acid, methacrylic acid, 2-methyl-2-acrylamidopropane sulfonic acid, trimethyl-acrylamidopropane sulfonic acid, trimethyl-acrylamidopropyl ammonium 3-methyl-1-vinylimidazolium iodide, sodium acrylate, sodium methacrylate, and 1-(3-sulfopropyl)-2-vinyl-pyridinium betaine.[26–28] An alternative strategy is to modify the ligand with an acryloyl group and then copolymerize the modified ligand with NIPAM.[29,30] The use of methacrylic acid, along with NIPAM, not only changes the LCST, but also makes the hydrogels responsive to both temperature and pH.[28]

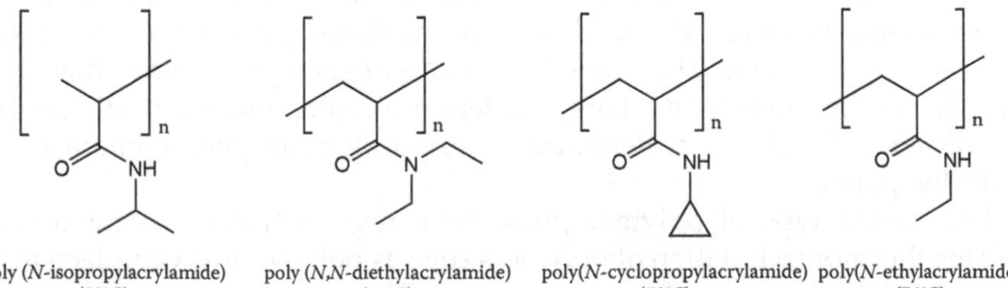

poly (*N*-isopropylacrylamide) (32°C) poly (*N,N*-diethylacrylamide) (33°C) poly(*N*-cyclopropylacrylamide) (58°C) poly(*N*-ethylacrylamide) (74°C)

FIGURE 13.6
Structures of thermoresponsive homopolymers.

13.3.3 Reversibly Cross-Linked Polymer Networks

The third group of smart polymers combines systems with reversible non-covalent cross-linking of separate polymer molecules into insoluble polymer networks. The most common systems of this group are alginate, chitosan, and *k*-carrageenan, which behave as reversibly soluble–insoluble polymers by responding to Ca^{2+}, pH, and K^+, respectively.[31–33] Others include boric acid polyols[32–36] or boric acid polysaccharides.[37] These types of polymers have found limited application as carriers in affinity precipitation, but they are more promising for the development of "smart" drug-delivery systems capable of releasing drugs in response to a signal, e.g., release of insulin when glucose concentration is increasing.[38] A polymer sensitive to pH and temperature has recently been prepared by genetic engineering.[39] Such materials are block copolymers containing repeating sequences from silk (GAGAGS) and elastin (GVGVP), where, in some cases, valine has been replaced by glutamic acid. By varying the extent of this change, the sensitivity to pH, temperature, and ionic strength can be controlled fairly precisely. Recombinant methods have also been used to design multidomain assemblies in which leucine zipper domains flank a central flexible polyelectrolyte.[40] Changes in pH/temperature stimuli trigger a sol–gel transition.

13.4 Properties of Polymer Precipitates during Affinity Precipitation

When designing a polymer for affinity precipitation, one should also consider the properties of the polymer precipitate or of the polymer-enriched phase formed. When phase separation of the polymer takes place, three extreme types of precipitate can be formed (Figure 13.7). The first type of polymer-enriched phase is usually formed after pH-induced precipitation of chitosan[41,42] or Ca-induced precipitation of alginate.[32] This type of gel has high water content and, when formed, a lot of solution is entrapped. The loose structure of such a gel makes it difficult to separate the gel from the supernatant, and the entrapped solution containing impurities decreases the purification factor. The attractive feature of such polymers is their relatively low hydrophobicity. Large protein molecules and even cells can be processed without risk of denaturing caused by hydrophobic interactions with the polymer.

The second type of polymer phase with high polymer content occurs during thermoprecipitation of polymers such as poly(*N*-vinyl caprolactam)[43] and poly(ethylene-glycol–propylene-glycol) random and block copolymers.[44] The polymer phase is easily separated from the supernatant with practically no entrapment of impurities from solution. High polymer content renders the polymer phase very hydrophobic and makes it an unfavorable

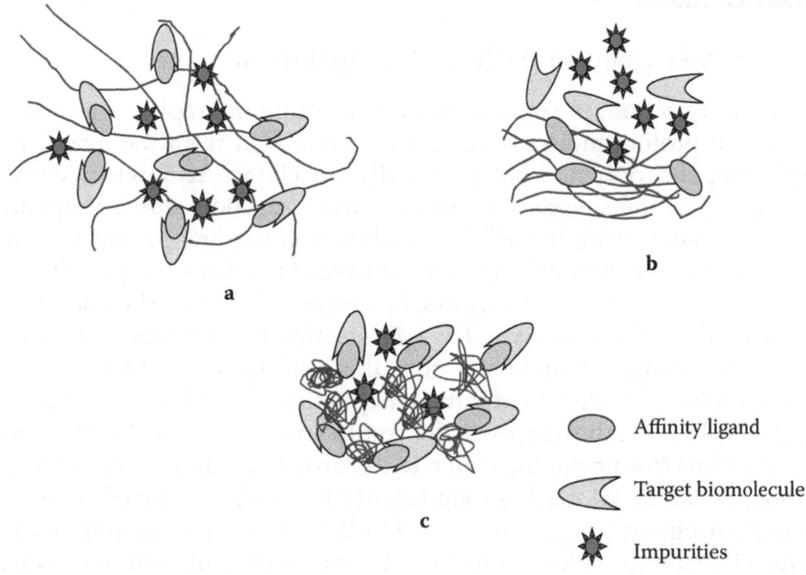

FIGURE 13.7
Types of polymer-enriched phase formed after the phase separation of the polymer during affinity precipitation: (a) loose gel with high water content; (b) compact hydrophobic phase with low water content; and (c) suspension of compact polymer.

environment for proteins. In fact, proteins are pushed off from the complex with the polymer when the polymer phase is formed. High-affinity ligands with strong binding toward the target protein are required to achieve specific coprecipitation of the protein with the polymer.[43,45] On the other hand, this property, disadvantageous for affinity precipitation, could be quite useful in another mode of protein purification, namely partitioning in aqueous two-phase systems (ATPSs).[46] Some thermoreactive polymers are able to form ATPSs with dextrans. When a specific ligand is coupled to the polymer, the target protein partitions preferentially into the aqueous phase formed by this polymer. After mechanical separation from the dextran phase, the target protein can be recovered by thermoprecipitation of the polymer. The latter forms a compact polymer-enriched phase, while the purified target protein remains in the supernatant. The high hydrophobicity of the polymer phase results in small coprecipitation of the target protein with the polymer, especially when the conditions (pH, ionic strength) are also changed to reduce specific protein–polymer interactions.[47,48]

The third case appears to be the most preferable for affinity precipitation. Compact particles of aggregated polymer are easily separated from the supernatant, accompanied by only minimal entrapment of the supernatant and impurities. On the other hand, the ligands with bound target protein are exposed to the solution and hence are in a comfortable environment.

13.5 Ligands Used in Affinity Precipitation

Affinity precipitation, in contrast to affinity chromatography, is a one-plate process. The protein molecule, once it is dissociated from the ligand upon precipitating the polymer, has practically no chance to interact with the ligand again, as the polymer is removed from solution after precipitation. Hence, to be successful, the affinity precipitation requires stronger protein–ligand interactions than affinity chromatography, where the protein molecule, once dissociated from the ligand, has several chances to be bound again as it moves along the column matrix with its high concentration of available ligands. The rough estimation is that the binding constant for a single ligand–protein interaction in affinity precipitation should be at least $10^{-5}M$.[43] Ligand coupling to a polymer usually results in a 100- to 1000-fold decrease in affinity. Thus free ligand must bind to the protein of interest with constants of about 10^{-7} to $10^{-8}M$. Such strong affinity is mostly restricted to systems with protein ligands, e.g., antibodies. On the other hand, flexibility of the polymer chain allows interaction of few ligands with multi-subunit proteins, and this multipoint attachment promotes binding strength, thereby also making relatively weak interactions useful. This fact allows exploitation of relatively weak-binding pairs, such as sugar–lectin, for the efficient affinity precipitation of tetrameric lectin, Concanavalin A, using a pH-sensitive conjugate of Eudragit S-100 with *p*-aminophenyl-*N*-acetyl-D-galactosamine[49] or *p*-aminophenyl-α-D-glucopyranoside.[50]

The affinity precipitation is designed to be applied at the first stages of a purification protocol dealing with crude unprocessed extracts. Thus the ligands used should be robust to withstand both harmful components present within the crude extracts and also precipitating/eluting agents added from outside. Sugar ligands, for example, have been used successfully for the bioseparation of lectins.[49–52] Similarly, triazine dyes, which are robust affinity ligands for many nucleotide-dependent enzymes, were successfully used in conjugates with Eudragit S to purify dehydrogenases from various sources by affinity precipitation.[53–57] Chelating groups charged with metal ions have shown tremendous potential in affinity precipitation of proteins having His-tag at the surface or of natural proteins with inherent amino acid residues that show binding for metal ions.[12,58–63] *p*-Aminobenzamidine ligands were used for affinity precipitation of proteolytic enzymes,[64,65] and boronates were used for binding to glycoproteins or to many sugar molecules.[66]

Protein ligands offer good selectivity and the high binding strength needed for affinity precipitation, provided that they are stable enough under conditions of polymer precipitation and elution. Such proteins include soybean trypsin inhibitor,[45,67,68] immunoglobulin G,[3,69,70] antibodies,[71] and protein antigen for binding monoclonal antibodies.[17] Other complex structures have even been used as ligands in affinity precipitation. Poly(NIPAM) conjugate with $(dT)_8$ was used for a model separation of a complementary oligonucleotide $(dA)_8$ from a mixture of $(dA)_8$ and $(dA)_3(dT)(dA)_4$. Restriction endonuclease Hind III was isolated using poly(NIPAM) conjugate with λ-phage DNA,[29] and C-reactive protein was isolated using poly(NIPAM) conjugate with *p*-amino-phenylphosphorylcholine.[25]

The selectivity of affinity precipitation for protein purification depends on the nature of the polymer as well as the ligands used. Precipitation of non-charged thermosensitive polymers shows very low nonspecific coprecipitation of proteins compared with pH-sensitive polymers, where ionic interactions can contribute to the unwanted protein–polymer interactions. It has been shown that coupling of specific ligands to the ionic polymers significantly decreases the nonspecific interaction of the proteins to the polymer–ligand conjugate compared with the native polymer.[72,73] When polyelectrolyte complexes precipitate, the charges of one polyanion are compensated by the charges of the oppositely charged polyanion. Strong cooperative interactions of oppositely charged polyions resulted in efficient displacement of nonspecifically bound proteins; the amount of nonspecifically bound protein does not exceed a few percentage points of that present in solution.[18] At the same time, the affinity of the proteins to nonmodified polymer could be harnessed for purification of the proteins. Endopolygalacturonase, glucoamylase, and pectinase were purified by coprecipitation with alginate.[74–76] Xylanase and chitanase have been purified by coprecipitation with Eudragit S-100[77,78] and, similarly, pullulanase has been coprecipitated using *k*-carrageenan.[34]

13.6 Application of Affinity Precipitation

Affinity precipitation of proteins using smart polymers emerged in the early 1980s. Since then it has evolved as a technique capable of simple, fast, and efficient purification of a variety of proteins. However, the affinity-precipitation technique has remained underexploited by researchers, as their description is limited to journals devoted to biochemical engineering/biotechnology. By now, these techniques have shown sufficient promise to be adopted and routinely used by biochemists and molecular biologists. A number of examples are now available in which the affinity-precipitation technique has been used for protein purification. Table 13.1 summarizes published protocols of protein isolation using the affinity-precipitation technique. The authors consider it relevant to discuss here some of their recent works and the work of others in this area as model examples.

13.6.1 Metal Chelate Affinity Precipitation

The concept of using metal chelating in affinity techniques, such as immobilized metal-affinity chromatography (IMAC), was a breakthrough introduction.[79] The IMAC technique has a wide application in protein purification, particularly when dealing with recombinant proteins.[8,80] This offers a number of important advantages over other "biospecific" affinity techniques for protein purification, particularly with respect to ligand stability, protein loading, and recovery.[8] The technique is generally based on the selective interaction between metal ions like Cu(II) or Ni(II), which are immobilized on the solid support and on the electron donor groups on the proteins. The amino acids histidine, cysteine, tryptophan, and arginine have strong electron donor groups in their side chains, and the presence of such exposed residues is an important factor for IMA-binding properties.[81] In the recombinant proteins, polyhistidine tag (His-tag) fused to either the N- or C-terminal end of the protein has become the selective and efficient separation tool for application in IMAC separation. At present, it is one of the most popular and successful methods used in molecular biology for the purification of recombinant proteins. The widespread application of the metal-affinity concept has also been extended by adopting the technique in a nonchromatographic format such as metal chelating affinity precipitation.[12,58,62,63] This separation strategy makes metal-affinity methods simpler and more cost effective when the intended applications are for large-scale processes.

By combining the versatile properties of metal affinity with affinity precipitation, the technique presents enormous potential as a simple and selective separation strategy. The first such application was reported by Van Dam et al.,[9] when human hemoglobin and sperm whale hemoglobin were quantitatively precipitated in model experiments with bis-copper chelates. In another study, Lilius et al.[82] described the purification of genetically engineered galactose dehydrogenase with polyhistidine tail by metal-affinity precipitation. The histidines functioned as the affinity tail, and the enzyme could be precipitated when the bis-zinc complex with ethylene glycol-bis-(β-aminoethyl ether) *N*,*N*,*N'*,*N'*-tetraacetic acid, EGTA $(Zn)_2$, was added to the protein solution. However, in general, the application of affinity precipitation with homobifunctional ligands has been quite limited.[83] The requirement of a multibinding functionality of the target protein and the low precipitation rate restricts the use of this type of affinity-precipitation process.[9,84,85] The concentration dependence and the risk of terminal aggregate formation further complicates its use.[84]

In metal chelate affinity precipitation, metal ligands are covalently coupled to the reversible soluble–insoluble polymers (mainly thermoresponsive polymers) by radical copolymerization. The copolymers carrying metal-chelating ligands are charged with metal ions, and the target protein binds the metal-loaded copolymer in solution via the interaction between the histidine on the protein and the metal ion. The complex of the target protein with copolymer is precipitated from the solution by increasing the temperature in the presence of NaCl, whereas impurities remain in the supernatant and are discarded after the separation of precipitate. The precipitated complex is solubilized by reversing the precipitation conditions, and the target protein is dissociated from the precipitated polymer by using imidazole or EDTA as eluting agent. The protein is recovered from the copolymer by precipitating the latter at elevated temperature in the presence of NaCl.

Traditionally, polydentate carboxy-containing ligands like iminodiacetic acid (IDA) or nitrilotriacetic acid (NTA) have been quite successful in IMAC for metal-chelate-mediated purification of proteins.[86] The ligands coordinate well with the metal ion and still leave coordinating sites on the metal ion available for binding the target protein. Such ligands, however, show some limitations in metal chelate affinity precipitation when copolymerized with NIPAM.[60] The introduction of highly charged co-monomers (at neutral conditions) like IDA or NTA into the polymer results in a drastic decrease in the efficiency of precipitation with temperature compared with the behavior of NIPAM homopolymer.

The breakthrough in this direction came when a new ligand, imidazole, was successfully incorporated into NIPAM and the copolymer achieved efficient precipitation.[58] Copolymers of *N*-vinylimidazole (VI) with *N*-isopropylacrylamide (NIPAM) can be synthesized by radical copolymerization in aqueous solution, where VI concentrations up to 25 mol% can be incorporated in the copolymer in aqueous solution. The incorporation of VI in the copolymers is determined using ^{1}H NMR (Figure 13.8). Clear separation of the peaks from imidazole protons and from *N*-isopropyl amide protons is observed. Copolymers of poly(VI-NIPAM) with 15 and 24 mol% of VI are precipitated by heating up to 60°C. Incorporation of relatively hydrophilic imidazole moieties in poly(VI-NIPAM) copolymer hindered the hydrophobic interactions of the poly(NIPAM) segments. This results in higher precipitation temperatures in poly(VI-NIPAM) copolymers compared with homopolymer of poly(NIPAM) and depends upon the VI content in the copolymer.[12,60,61] Poly(VI-NIPAM) copolymers with up to 33 mol% VI content

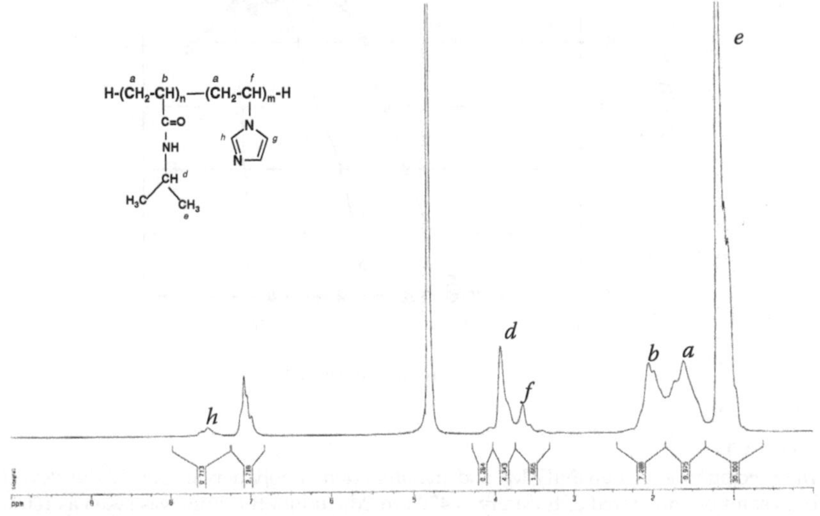

FIGURE 13.8
500-MHz ^{1}H NMR spectrum of poly(VI-NIPAM) (D_2O, 25°C; 24 mol% vinylimidazole).

in the reaction mixture could be precipitated from the solution by heating, and beyond that concentration, the precipitation of the copolymer becomes difficult.[60] When loaded with metal ions like Cu(II) or Ni(II), the hydrophilicity of the copolymer is increased even more (metal ions induce more positive charges),[12] and it prevents the copolymers from precipitating at all, even when increasing the temperature up to 70°C.

For complete precipitation of metal-loaded copolymers, one should incorporate conditions that promote hydrophobic interactions and reduce mutual repulsions caused by similar charges. The increase in ionic strength, and hence the decrease in charge repulsion, by adding NaCl facilitates precipitation of metal-bound copolymers. At relatively moderate salt concentrations of 0.4*M* NaCl, the metal-bound copolymers can be precipitated quantitatively below 30°C (Figure 13.9). The Cu(II)- or Ni(II)-poly(VI-NIPAM) precipitated completely around 20°C and 28°C, respectively, which is lower than the phase transition temperature (32°C) of poly(NIPAM). Above these temperatures, the decrease in the turbidity measured is actually due to the flocculation of the polymer precipitates (which allows some transmittance) rather than polymer dissolution. The efficient precipitation of these metal-loaded copolymers in the presence of NaCl at mild temperatures is very convenient for metal-affinity precipitation and allows its application for purification of a wide range of proteins. High salt concentrations do not interfere with protein–metal ion–chelate interaction,[87] while the possibility of nonspecific binding of foreign proteins to the polymer is reduced at high salt concentrations.

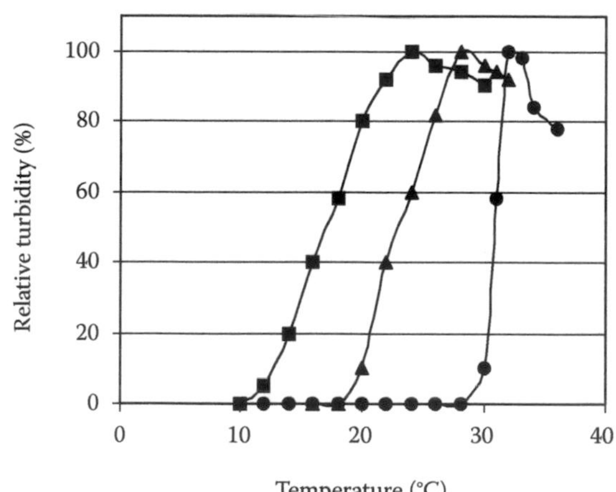

FIGURE 13.9
Thermoprecipitation of poly(NIPAM) and metal-loaded copolymers of poly(VI-NIPAM) from aqueous solution monitored as turbidity at 470 nm. Maximum turbidity was taken as 100% and relative turbidities were calculated from that. Polymer concentration = 1.0 mg/ml. The figure shows poly(NIPAM) precipitation (●), Ni(II)-poly(VI-NIPAM) precipitation (▲), and Cu(II)-poly(VI-NIPAM) precipitation (■) at different temperatures in presence of 0.4*M* NaCl. The VI concentration was 15 mol% and 24 mol% for the Cu(II) and Ni(II) copolymers, respectively.

Imidazole is a monodentate ligand in Cu-complexes. Up to four imidazoles bind to one Cu(II) ion, and the log *K* (where *K* is association constant, M^{-1}) for each imidazole ligand decreases from $\log K_1 = 3.76$ for binding the first imidazole ligand to $\log K_4 = 2.66$ for binding the fourth imidazole ligand.[88] The binding of a single imidazole ligand to a Cu(II)-ion in solution is much weaker compared with the binding of tridentate IDA ($\log K = 11$).[89] On the other hand, when a Cu(II)-ion forms a complex with four imidazole ligands, the combined binding constant $\log K = \log K_1 + \log K_2 + \log K_3 + \log K_4 =$ 12.6 to 12.7. The strength of this complex is close to that of a Cu(II)-ion complex with poly(*N*-vinylimidazole) (poly-VI), $\log K = 10.64$ to 14.21,[88,90] and comparable with the binding of tridentate IDA ligand, $\log K_4 = 5.5$ to 6.0. When coupled to solid matrices, imidazole ligands are spatially separated, and the proper orientation of the ligands to form a complex with the same Cu(II)-ion is unlikely, and the imidazole ligands are not used for IMA-chromatography.[58] In solution, a flexible polymer like poly(VI-NIPAM) can adopt a solution-phase conformation where two to three imidazole ligands are close enough to form a complex with the same Cu(II)-ion, providing significant strength of interaction (Figure 13.10). It is clear that not all available coordination sites of the metal ion are occupied by imidazole ligands

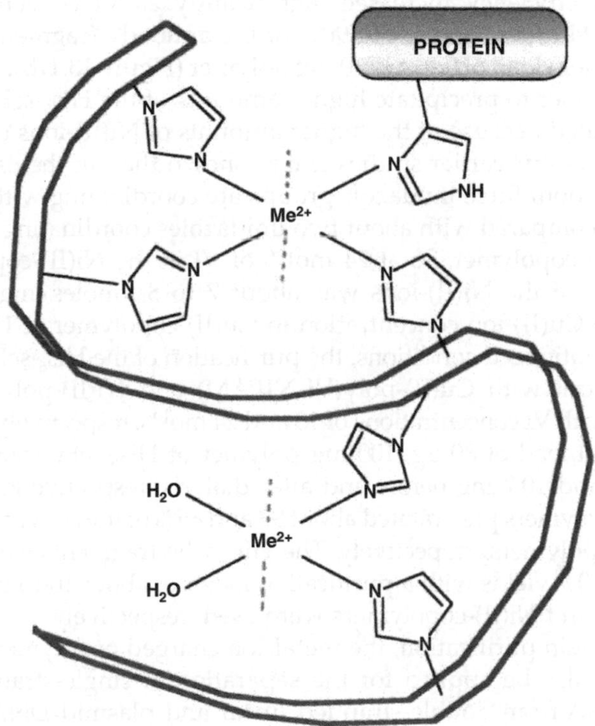

FIGURE 13.10
Smart-polymer–metal conjugate for protein purifications.

of the polymer. The unoccupied coordination sites of the metal ion could be used for complex formation with the protein molecule via histidine residues on its surface.

The properties of these metal copolymers make them attractive as soluble–insoluble affinity macroligands for metal chelate affinity precipitation of proteins. Many proteins containing natural metal-ion binding residues and recombinant proteins containing His-tag residues have been purified using metal chelate affinity precipitation (Table 13.1).

In a recent study, purification of extracellularly expressed six histidine-tagged single-chain Fv-antibody fragments (His_6-scFv fragments) from recombinant *Escherichia coli* cell culture broth was performed. Quantitative precipitation of the His_6-scFv fragments occurred at different loads of the cell supernatant (0.43 to 1.07 mg total protein) with 10 mg Cu(II)-poly(VI-NIPAM) (15 mol% VI) (Figure 13.11a). Up to a 1.07-mg protein load, the precipitation of the His_6-scFv fragments was more than 95% (activity), whereas the total protein binding was around 16 to 18%. Some decrease in precipitation efficiency at higher protein loads was observed. The bound His_6-scFv fragments were recovered almost completely (>95%) by elution with 50m*M* EDTA buffer, pH 8.0. On the other hand, the precipitation efficiency with Ni(II)-poly(VI-NIPAM) at 15 mol% VI was lower. However, the precipitation efficiency increased significantly as VI concentration was increased to 24 mol%, and about 90% of the antibody fragments were captured at a protein load of 0.9 mg/10 mg polymer (Figure 13.11b). The capacity of this copolymer to precipitate higher amounts of the His_6-scFv fragments can be explained because of the higher amounts of Ni(II)-ions on the copolymer. In one of our earlier studies it was shown that, in the case of Ni(II)-copolymers, about three imidazole groups are coordinating with each metal ion.[12] This is compared with about two imidazoles coordinating with Cu(II)-ions in Cu(II)-copolymer. So at 24 mol% of VI in the Ni(II)-copolymer, the concentration of the Ni(II)-ions was about 7 to 8 μmoles/mg copolymer, equivalent to Cu(II)-ion concentration in Cu(II)-copolymer at 15 mol% VI.

Under the optimized conditions, the purification of the His_6-scFv fragments was carried out with Cu(II)-poly(VI-NIPAM) and Ni(II)-poly(VI-NIPAM) copolymers with VI concentrations of 15 and 24 mol%, respectively (Table 13.2). With an initial load of 80 μg/100 mg polymer of His_6-scFv fragments (total protein 36.4 and 10.7 mg before and after dialysis, respectively), 5 ml of the 2% metal copolymers precipitated about 93 and 85% of the activity with Cu(II)- and Ni(II)-copolymers, respectively. The His_6-scFv fragments were recovered with 91 and 80% yields with a purification factor of about 16-fold and 21-fold when Cu(II)- and Ni(II)-copolymers were used, respectively.

Besides protein purification, the metal-ion charged copolymer of poly(VI-NIPAM) can also be applied for the separation of single-stranded nucleic acids like RNA from double-stranded linear and plasmid DNA by affinity precipitation.[91] The separation method is based on the interaction of metal ions with the aromatic nitrogens in exposed purines in single-stranded nucleic acids.[92]

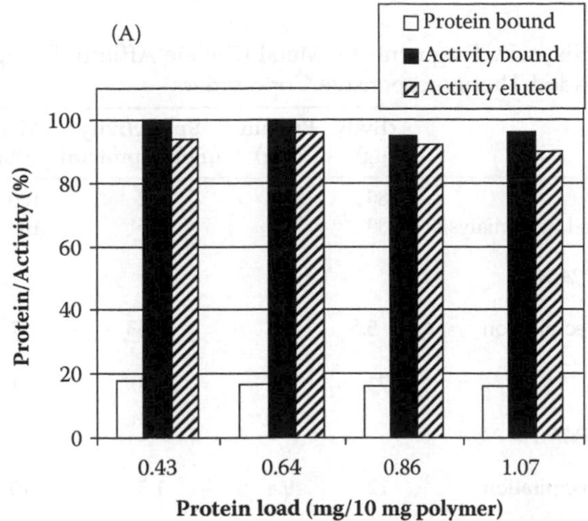

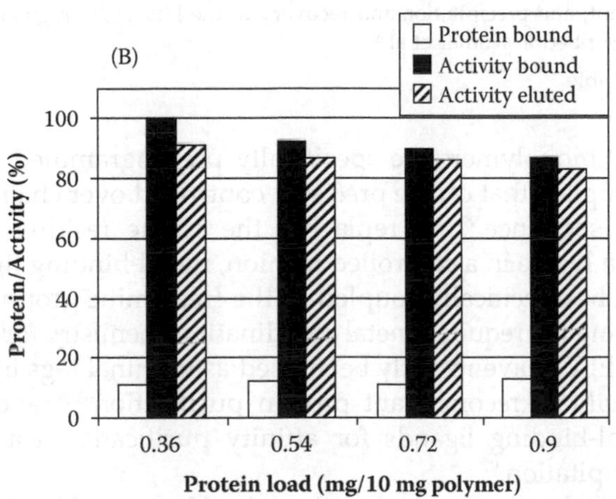

FIGURE 13.11
Affinity precipitation and recovery of His_6-scFv fragments with (a) Cu(II)-poly(VI-NIPAM) and (b) Ni(II)-poly(VI-NIPAM) copolymers at different protein loads. The copolymers contained 15 mol% and 24 mol% VI for the Cu(II) and Ni(II) copolymers, respectively. The 0.5-ml volume of the 2% copolymer solutions were augmented with 0.2 to 0.5 ml of the dialyzed cell supernatant.

13.6.2 Elastinlike Polymers for Affinity Precipitation

Elastinlike proteins (ELPs) are biopolymers consisting of the repeating pentapeptide, VPGVG. They behave similarly to poly(NIPAM) polymers and have been shown to undergo reversible phase transitions within a wide range of conditions.[93,94] Unlike the statistical nature of step- and chain-polymerization

TABLE 13.2

Purification of His$_6$-scFv Fragments by Metal Chelate Affinity Precipitation Using Metal-Loaded Thermoresponsive Copolymers

Steps	Activity (μg)	Protein (mg)	Sp. Activity (μg/mg protein)	Yield (%)	Fold Purification
Cell supernatant	80	36.4	2.2	100	1
Supernatant loaded after dialysis	80	10.7	7.5	100	3.4
Cu(II)-poly(VI-NIPAM)					
Unbound after precipitation and wash	5.5	8.8	0.63	7	NA [a]
Eluate	73	2.1	34.8	91	16
Ni(II)-poly(VI-NIPAM)					
Unbound after precipitation and wash	12	9.4	1.3	15	NA [a]
Eluate	64	1.4	45.7	80	21

Note: 5 ml of the 2% metal-copolymer solution was added to 5 ml of the dialyzed cell supernatant, and precipitation and recovery of the His$_6$-scFv fragments were carried out as described in Kumar et al.[63]

[a] NA: Not applicable.

reactions, ELP biopolymers are specifically preprogrammed within a synthetic gene template that can be precisely controlled over chain length, composition, and sequence.[93] By replacing the valine residue at the fourth position with a lysine in a controlled fashion, metal-binding ligands such as imidazole can be specifically coupled to the free amine group on the lysine residues, creating the required metal coordination chemistry for metal-affinity precipitation. ELPs have recently been used as terminal tags in recombinant systems to facilitate recombinant protein purification[95,96] and been conjugated to metal-binding ligands for affinity purification via temperature-triggered precipitation.[97]

The metal-affinity purification method for His-tagged proteins based on temperature-triggered precipitation of chemically modified ELP biopolymers was demonstrated.[97] ELPs with repeating sequences of ([VPGVG]$_2$ [VPGKG] [VPGVG]$_2$)$_{21}$ were synthesized, and the free amino groups on the lysine residues were modified by reacting with imidazole-2-carboxyaldehyde to incorporate the metal-binding ligands into the ELP biopolymers. Biopolymers charged with Ni(II) were able to interact with a His-tag on the target proteins based on metal coordination chemistry. Purifications of two His-tagged enzymes, β-D-galactosidase and chloramphenicol acetyltransferase, were used to demonstrate the application of metal-affinity precipitation using this new type of affinity reagent. The bound enzymes were easily released by addition of either EDTA or imidazole, and over 85% recovery was observed in both cases. The recovered ELPs were reused with no observable decrease in the purification performance. The feature of ELPs to reversibly precipitate above their transition temperature was exploited as a general method for the purification of His-tagged proteins. The principle of the single-step metal-affinity method is based on coordinated ligand bridging between the modified ELPs and the target proteins. This has been the first report exploiting the features of ELP for protein purification based on metal-affinity purification. The capability of modulating purification conditions by simple temperature triggers and the low cost of preparation will probably make the ELP-based metal-affinity precipitation a useful method in the future, not only for protein purification, but also for diverse applications in bioseparation such as DNA purification and environmental remediation.[94]

ELPs were also used to purify proteins from *Escherichia coli* culture when proteins are expressed as a fusion with an ELP. Nonchromatographic purification of ELP fusion proteins, termed inverse transition cycling (ITC), exploits the reversible soluble–insoluble phase transition behavior by inducing precipitation that is imparted by the ELP tag.[98] Purification of recombinant proteins is described by fusing a target protein with an intein and an elastinlike polypeptide that only requires NaCl, dithiothreitol, and a syringe filter to isolate the target protein from *E. coli* lysate. This tripartite fusion system enables rapid isolation of the target protein. The elastinlike polypeptide tag imparts reversible phase transition behavior to the tripartite fusion so that the fusion protein can be selectively aggregated in cell lysate by the addition of NaCl. The aggregates are isolated by microfiltration and resolubilized by reversal of the phase transition in low-ionic-strength buffer. After resolubilizing the fusion protein, the intein is activated to cleave the target protein from the elastin–intein tag, and the target protein is then isolated from the elastin–intein fusion by an additional phase transition cycle.[99]

Another interesting example has been a one-pot affinity precipitation purification of carbohydrate-binding protein reported by Sun et al.[100] By designing thermally responsive glyco-polypeptide polymers, which were synthesized by selective coupling of pendant carbohydrate groups to a recombinant elastinlike triblock protein copolymer (ELP), glyco-affinity precipitation purification of carbohydrate-binding protein was demonstrated. The thermally driven inverse transition temperature of the ELP-based triblock polymer is maintained upon incorporation of carbohydrate ligand, which was confirmed by differential scanning calorimetry and ^{1}H NMR spectroscopy experiments. As a test system, lactose-derivatized ELP was used to selectively purify a galactose-specific binding lectin through simple temperature-triggered precipitation at a high level of efficiency. This strategy takes the advantages of both solution-phase capturing and releasing and solid-phase separation of target protein, which facilitate a simple and highly efficient protein purification and further identification (Figure 13.12).

13.6.3 Affinity Precipitation Using Polyelectrolytes

Many separation procedures by affinity precipitation discussed above are based on the use of smart polymers that can respond to temperature or pH.

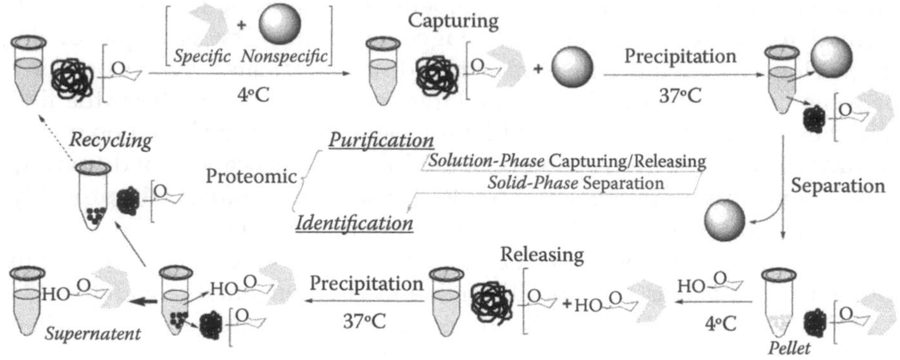

FIGURE 13.12
Schematic illustration of one-pot glycol-affinity precipitation purification. (Reproduced from Sun, X.-L. et al., *J. Proteome Res.*, 4, 2355, 2005. With permission.)

However, in many situations these systems work within a limited range of conditions under which polymer precipitation occurs, which in turn considerably limits the number of proteins that can be purified by affinity precipitation using such polymers. Polyelectrolyte complexes have recently emerged as useful polymeric materials that overcome this limitation and are considered to be useful for affinity precipitation. Both natural and synthetic oppositely charged polyelectrolytes interact with each other to form complexes. Such polycomplexes (PECs) attain properties that are different from those of the initial components. PECs are very stable with respect to dissociation and are also able to participate in interpolyelectrolyte reactions.[101] The solubility of nonstoichiometric complexes formed with an excess of one of the PEC components is regulated by the composition of the complex as well as by pH and ionic strength.[102] Reversible precipitation–dissolution of PECs could be achieved at different pH values, depending on the ionic strength and composition of the complex. For example, PECs formed by poly(ethyleneimine) and poly(acrylic acid) undergo phase transitions in a broader pH range, from pH 3 to 11.[19]

The electrostatic nonspecific binding of polyelectrolytes to proteins can result in the formation of soluble complexes,[103] complex coacervation,[104] or the formation of amorphous precipitates.[105] Conditions favoring the formation of compact insoluble protein–polyelectrolyte complex aggregates are selected for protein separation and purification. A number of polycations and polyanions, including poly(ethyleneimine), poly(diallyl dimethyl ammonium chloride), carboxymethylcellulose, dextran sulfate, poly(acrylic acid), and poly(methacrylic acid), were used for separation by nonspecific precipitation of a number of proteins at different pH values and in the presence of different salt concentrations.[106–108]

However, the main problem encountered in nonspecific polyelectrolyte precipitation of proteins is the selectivity of polymer–protein interaction.

The introduction of affinity interactions in the system by coupling an affinity ligand to one of the components makes the polyelectrolyte precipitation very specific for the target protein. This was shown when lactate dehydrogenase from beef heart extract was specifically purified using Cibacron blue 3GA coupled to positively charged polymer poly(ethyleneimine).[19] The polyelectrolyte complex was formed by adding negatively charged polymer, poly(acrylic acid). The precipitated complex with the target protein was removed from the solution containing impurities. LDH (lactate dehydrogenase) was obtained with a yield of 85% and a purification factor of 11-fold. In another example, when antigen (inactivated glyceraldehyde-3-phosphate dehydrogenase) from rabbit was covalently coupled to a polycation, the resulting complex was used for purification of antibodies from 6G7 clone specific for the protein. The crude extract was incubated with the polymer complex, and the precipitation of the latter was carried out at 0.01*M* NaCl and pH 4.5, 5.3, 6.0, and 6.5 using complexes with polycation/polyanion ratio of 0.45, 0.35, 0.2, and 0.15, respectively. Purified antibodies were eluted at pH 4.0, where polyelectrolyte complexes of all compositions used were insoluble. Quantitative recoveries were achieved under optimal conditions of polycation/polyanion ratios of 0.35 and 0.45 at respective precipitation pH levels of 6.0 and 6.5. Precipitated polyelectrolyte complexes could be dissolved at pH 7.3 and used repeatedly.[18]

The successful affinity precipitation of antibodies in the above example indicates that the ligand is exposed to the solution. This fact was used to develop a new method of producing monovalent Fab fragments of antibodies containing only one binding site. Traditionally, Fab fragments are produced by proteolytic digestion of antibodies in solution followed by separation of Fab fragments. In the case of monoclonal antibodies against inactivated subunits of glyceraldehyde-3-phosphate dehydrogenase, digestion with papain resulted in significant damage of binding sites of the Fab fragment. Proteolysis of monoclonal antibodies in the presence of the antigen–polycation conjugate followed by (a) precipitation induced by addition of polyanion, poly(methacrylic) acid, and pH shift from 7.3 to 6.5 and (b) elution at pH 3.0 resulted in 90% immunologically competent Fab fragments. Moreover, the papain concentration required for proteolysis was ten times lower than that for free antibodies in solution. The antibodies bound to the antigen–polyelectrolyte complex were digested by papain to a lesser extent, suggesting that binding to the antigen–polycation conjugate not only protected binding sites of monoclonal antibodies from proteolytic damage, but also facilitated the proteolysis, probably by exposing antibody molecules in a way convenient for proteolytic attack by the protease.[109]

13.6.4 Other Applications

Other types of metal-chelating polymers for affinity precipitation of proteins were recently obtained by synthesizing highly branched copolymers of NIPAM and 1,2-propanediol-3-methacrylate (glycerol monomethacrylate

[GMA]), poly(NIPAM-*co*-GMA), using the technique of reversible addition-fragmentation chain transfer (RAFT) polymerization employing a chain-transfer agent that allows the incorporation of imidazole functionality in the polymer chain ends. The LCST of the copolymers can be controlled by the amount of hydrophobic and GMA comonomers incorporated during copolymerization procedures. The copolymers demonstrated LCST below 18°C and were successfully used to purify a His-tagged BRCA-1 protein fragment (a protein implicated in breast cancer) by affinity precipitation.[110,111]

We have studied the rheological properties of the polymer phase formed after pH-induced precipitation of Eudragit S-100. Neutralization of the negatively charged carboxy groups results in an increase in the hydrophobicity of the polymer molecules. The latter aggregate in response to the hydrophobic interactions to form a three-dimensional network with high viscosity and strong shear-thinning behavior. Upon a further decrease in pH and an increase in the hydrophobicity of the polymer molecules, the latter rearrange from the three-dimensional network into a suspension of particles of aggregated polymer chains. This rearrangement is accompanied by a significant drop in viscosity and much less pronounced shear-thinning behavior.[48] Quite a few successful affinity-precipitation procedures were reported when using this polymer[52–56,68,69] or a similar polymer, Eudragit L-100.[112] An alternative pH-sensitive polymer used successfully for affinity precipitation is hydroxypropylmethylcellulose acetate succinate.[71,113]

An interesting example of the use of poly(*N*-acryloylpiperidine) terminally modified with maltose for affinity precipitation of thermolabile α-glucosidase has also been demonstrated.[52] Use of the polymer with extremely low LCST (soluble below 4°C and completely insoluble above 8°C) made it possible to use the technique for purification of thermolabile α-glucosidase from cell-free extract of *Saccharomyces cerevisiae*, achieving 206-fold purification with 68% recovery.

Affinity precipitation is readily combined with other protein isolation techniques, e.g., partitioning in aqueous two-phase polymer systems.[54,68–70,78] Partitioning of protein complexed with ligand–polymer conjugate is usually directed to the upper hydrophobic phase of aqueous two-phase polymer systems formed by PEG and dextran/hydroxypropyl starch, whereas most of the proteins present in crude extracts or cell homogenates partition into the lower hydrophilic phase. Then the precipitation of the protein–polymer complex is promoted by changing pH. Trypsin was purified using a conjugate of soybean trypsin inhibitor with hydroxypropylcellulose succinate acetate,[68] lactate dehydrogenase, and protein A using a conjugate of Eudragit S-100 with the triazine dye Cibacron blue[54] and immunoglobulin G.[69,70] The approach was also successfully demonstrated by the purification of microbial xylanases, pullulanases, wheat germ α-amylase, and sweet potato α-amylase[114,115] and by the purification of lectins from wheat germ, potato, and tomato. Other attractive extensions of this approach have been to separate animal cells by crafting the smart macroaffinity ligands by coupling an antibody (against a cell-surface protein) to a smart polymer.[116] Combination of partitioning with affinity precipitation improves both the yield and the purification factor and allows easier isolation of protein from particulate feed streams.

A new concept has recently been introduced where macroaffinity-ligand-facilitated three-phase partitioning (MLFTPP) converts three-phase partitioning (TPP) into a more selective and predictable technique for bioseparation of proteins using smart affinity ligands.[117,118] In this method, a water-soluble polymer is floated as an interfacial precipitate by adding ammonium sulfate and tertiary butanol. The polymer (appropriately chosen), in the presence of an enzyme for which it shows affinity, selectively binds to the enzyme and floats as a polymer–enzyme complex. By using this approach, purification of wheat germ agglutinin and wheat germ lipase (94% activity recovery and 27-fold purification in the case of wheat germ agglutinin and 99% activity recovery and 40-fold purification in the case of the lipase) was achieved by using chitosan as a macroaffinity ligand.[119]

13.7 Conclusion

Precipitation is a well-established technique for separating biomolecules from reaction fluids. On a large scale, the process shows a great promise; however, the selectivity of the process remains a great challenge. Affinity precipitation of proteins using smart polymers emerged in the early 1980s, and since then it has matured to a technique capable of simple, fast, and efficient purification of a variety of proteins. To be successful, affinity precipitation requires a robust ligand with affinity for a target protein; the ligand should be coupled to a polymer backbone that renders the conjugate a property of reversible solubility in response to small changes of pH, ionic strength, or temperature. The design of a polymer for affinity precipitation of proteins should address a few points, such as the completeness and sharpness of the transition from soluble to insoluble state and the properties of the polymer phase formed (nonspecific interaction with proteins, entrapment of the supernatant, structure and rheological behavior). With a proper choice of the polymer backbone (poly-NIPAM, poly[*N*-vinyl caprolactam], poly[ethylene oxide–propylene oxide]) as well as the ligand (positively or negatively charged) and environmental conditions (pH, ionic strength), one can achieve efficient polymer precipitation and the desired structure of the polymer phase. Polymers that are pH sensitive, like Eudragit and chitosan, are commercially available, whereas thermoresponsive polymers based on *N*-isopropylacrylamide are straightforward to synthesize in aqueous solutions. Thus protein purification using affinity precipitation constitutes a potentially very attractive mode of operation when designing protein-purification protocols.

References

1. Scopes, R.K., *Protein Purification: Principles and Practice*, Springer-Verlag, New York, 1994.
2. Kaul, R. and Mattiasson, B., Secondary purification, *Bioseparation*, 3, 1, 1992.
3. Mattiasson, B. and Kaul, R., Affinity precipitation, in *Molecular Interactions in Bioseparations*, Ngo, T., Ed., Plenum Press, New York, 1993, pp. 469–477.
4. Gupta, M.N. and Mattiasson, B., Affinity precipitation, in *Highly Selective Separations in Biotechnology*, Street, G., Ed., Blackie Academic and Professional, London, 1994, pp. 7–33.
5. Galaev, I.Yu. and Mattiasson, B., New methods for affinity purification of proteins: affinity precipitation, a review, *Biochemistry (Moscow)*, 62, 571, 1997.
6. Larsson, P.-O., Flygare, S., and Mosbach, K., Affinity precipitation of dehydrogenases, *Methods Enzymol.*, 104, 364, 1984.
7. Larsson, P.-O. and Mosbach, K., Affinity precipitation of enzymes, *FEBS Lett.*, 98, 333, 1979.
8. Arnold, F.H., Metal-affinity separations: a new dimension in protein processing, *Bio/Technology*, 9, 151, 1991.
9. Van Dam, M.E, Wuenchell, G.E., and Arnold, F.H., Metal affinity precipitation of proteins, *Biotechnol. Appl. Biochem.*, 11, 492, 1989.
10. Hilbreg, F. and Freitag, R., Protein purification by affinity precipitation, *J. Chromatogr. B*, 790, 79, 2003.
11. Mattiasson, B., Kumar, A., and Galaev, I.Yu., Affinity precipitation of proteins: design criteria for an efficient polymer, *J. Mol. Recog.*, 11, 211, 1998.
12. Kumar, A., Galaev, I.Yu., and Mattiasson, B., Metal chelate affinity precipitation: a new approach to protein purification, *Bioseparation*, 7, 185, 1999.
13. Gisser, K.R.C. et al., Nickel-titanium memory metal: a "smart" material exhibiting a solid-state phase change and superelasticity, *J. Chem. Educ.*, 71, 334, 1994.
14. Guoqiang, D. et al., Alternative modes of precipitation of Eudragit S-100: a potential ligand carrier for affinity precipitation of proteins, *Bioseparation*, 5, 339, 1995.
15. Galaev, I.Yu. and Mattiasson, B., Smart polymers and what they could do in biotechnology and medicine, *Trends Biotechnol.*, 17, 335, 1999.
16. Galaev, I.Yu., Gupta, M.N., and Mattiasson, B., Use smart polymers for bioseparations, *Chemtech*, 26, 19, 1996.
17. Galaev, I.Yu. and Mattiasson, B., Affinity precipitation of proteins using smart polymers, in *Smart Polymers for Bioseparation and Bioprocessing*, Galaev, I.Yu. and Mattiasson, B., Eds., Taylor & Francis, London, 2002, pp. 55–77.
18. Dainiak, M.B. et al., Affinity precipitation of monoclonal antibodies by nonstoichiometric polyelectrolyte complexes, *Bioseparation*, 7, 231, 1999.
19. Dissing, U. and Mattiasson, B., Polyelectrolyte complexes as vehicles for affinity precipitation of proteins, *J. Biotechnol.*, 52, 1, 1996.
20. Idziak, I. et al., Thermosensitivity of aqueous solutions of poly(*N,N*-diethylacrylamide), *Macromolecules*, 32, 1260, 1999.
21. Galaev, I.Yu. and Mattiasson, B., Thermoreactive water-soluble polymers, nonionic surfactants, and hydrogels as reagents in biotechnology, *Enzyme Microb. Technol.*, 15, 354, 1993.
22. Schild, H.G., Poly(*N*-isopropylacrylamide): experiment, theory and applications, *Prog. Polym. Sci.*, 17, 163, 1992.

23. Park, T.G. and Hoffman, A.S., Estimation of temperature-dependent pore size in poly(*N*-isopropylacrylamide) hydrogel beads, *Biotechnol. Prog.*, 10, 82, 1994.
24. Liu, F. et al., Development of a polymer-enzyme immunoassay method and its application, *Biotechnol. Appl. Biochem.*, 21, 257, 1995.
25. Mori, S., Nakata, Y., and Endo, H., Purification of rabbit C-reactive protein by affinity precipitation with thermosensitive polymer, *Prot. Expr. Purif.*, 5, 151, 1994.
26. Xue, W. and Hamley, I.W., Thermoreversible swelling behaviour of hydrogels based on *N*-isopropylacrylamide with a hydrophobic comonomer, *Polymer*, 43, 3069, 2002.
27. Solpan, D., Duran, S., and Guven, O., Synthesis and properties of radiation-induced acrylamide-acrylic acid hydrogels, *J. Appl. Polym. Sci.*, 86, 357, 2002.
28. Lee, W.-F. and Shieh, C.H., pH-thermoreversible hydrogels, II: synthesis and swelling behaviours of *N*-isopropylacrylamide-*co*-acrylic acid-*co*-sodium acrylate hydrogels, *J. Appl. Polym. Sci.*, 73, 1955, 1999.
29. Maeda, M. et al., Modification of DNA with poly(*N*-isopropylacrylamide) for thermally induced affinity precipitation, *Reactive Funct. Polym.*, 21, 27, 1993.
30. Umeno, D., Mori, T., and Maeda, M., Single stranded DNA-poly(*N*-isopropylacrylamide) conjugate for affinity precipitation separation of oligonucleotides, *Chem. Commun.*, 14, 1433, 1998.
31. Charles, M., Coughlin, R.W., and Hasselberger, F.X., Soluble-insoluble enzyme catalysis, *Biotechnol. Bioeng.*, 16, 1553, 1974.
32. Linné-Larsson, E. et al., Evaluation of alginate as a carrier in affinity precipitation, *Biotechnol. Appl. Biochem.*, 16, 48, 1992.
33. Roy, I. and Gupta, M.N., *k*-Carrageenan as a new smart macroaffinity ligand for the purification of pullulanase, *J. Chromatogr. A*, 998, 103, 2003.
34. Wu, K.-Y.A. and Wisecarver, K.D., Cell immobilization using PVA crosslinked with boric acid, *Biotechnol. Bioeng.*, 39, 447, 1992.
35. Kokufuta, E. and Matsukawa, S., Enzymatically induced reversible gel-sol transition of a synthetic polymer system, *Macromolecules*, 28, 3474, 1995.
36. Kitano, S. et al., Preparation of glucose-responsive polymer complex system having phenylboronic acid moiety and its application to insulin-releasing device, in *Proceedings of the First International Conference on Intelligent Materials*, Takagi, T., Takahashi, K., Aizawa, M., and Miyata, S., Eds., Technomic, Lancaster, PA, 1993, pp. 383–388.
37. Bradshaw, A.P. and Sturgeon, R.J., The synthesis of soluble polymer-ligand complexes for affinity precipitation studies, *Biotechnol. Techniq.*, 4, 254, 1990.
38. Lee, S.J. and Park, K., Glucose-sensitive phase-reversible hydrogels, in *Hydrogels and Biodegradable Polymers for Bioapplications*, Ottenbrite, R.M., Huang, S.J., and Park, K., Eds., American Chemical Society, Washington, DC, 1996, pp. 2–10.
39. Nagarsekar, A. et al., Genetic engineering of stimuli-sensitive silk elastin-like protein block copolymers, *Biomacromolecules*, 4, 602, 2003.
40. Petka, W.A. et al., Reversible hydrogels from self-assembling artificial proteins, *Science*, 281, 389, 1998.
41. Senstad, C. and Mattiasson, B., Affinity precipitation using chitosan as a ligand carrier, *Biotechnol. Bioeng.*, 33, 216, 1989.
42. Senstad, C. and Mattiasson, B., Purification of wheat germ agglutinin using affinity flocculation with chitosan and a subsequent centrifugation or flotation step, *Biotechnol. Bioeng.*, 34, 387,1989.

43. Galaev, I.Yu. and Mattiasson, B., Affinity thermoprecipitation: contribution of the efficiency of ligand-protein interaction and access of the ligand, *Biotechnol. Bioeng.*, 41, 1101, 1993.
44. Lu, M. et al., Ucon-benzoyl dextran aqueous two-phase systems: protein purification with phase component recycling, *J. Chromatogr. B.*, 680, 65, 1996.
45. Galaev, I.Yu. and Mattiasson, B., Affinity thermoprecipitation of trypsin using soybean trypsin inhibitor conjugated with a thermo-reactive polymer, poly(*N*-vinyl caprolactam), *Biotechnol. Techniq.*, 6, 353, 1992.
46. Alred, P.A. et al., Application of temperature-induced phase partitioning at ambient temperature for enzyme purification, *J. Chromatogr. A.*, 659, 289, 1994.
47. Harris, P.A., Karlström, G., and Tjerneld, F., Enzyme purification using temperature induced phase formation, *Bioseparation*, 2, 237, 1991.
48. Franco, T.T. et al., Aqueous two-phase system formed by thermoreactive vinyl imidazole/vinyl caprolactam copolymer and dextran for partitioning of a protein with a polyhistidine tail, *Biotechnol. Techniq.*, 11, 231, 1997.
49. Linné-Larsson, E. et al., Affinity precipitation of Concanavalin A with *p*-amino-α-D-glycopyranoside modified Eudragit S-100, I: initial complex formation and build-up of the precipitate, *Bioseparation*, 6, 273, 1996.
50. Linné-Larsson, E. and Mattiasson, B., Isolation of Concanavalin A by affinity precipitation, *Biotechnol. Techniq.*, 8, 51, 1994.
51. Linné-Larsson, E. and Mattiasson, B., Evaluation of affinity precipitation and a traditional affinity chromatography for purification of soybean lectin, from extracts of soya flour, *J. Biotechnol.*, 49, 189, 1996.
52. Hoshino, K. et al., Preparation of a new thermo-responsive adsorbent with maltose as a ligand and its application to affinity precipitation, *Biotechnol. Bioeng.*, 60, 568, 1998.
53. Guoqiang, D., Kaul, R., and Mattiasson, B., Purification of *Lactobacillus bulgaricus* D-lactate dehydrogenase by precipitation with an anionic polymer, *Bioseparation*, 3, 333, 1993.
54. Guoqiang, D., Kaul, R., and Mattiasson. B., Integration of aqueous two-phase extraction and affinity precipitation for the purification of lactate dehydrogenase, *J. Chromatogr. A*, 668, 145, 1994.
55. Guoqiang, D. et al., Affinity precipitation of lactate dehydrogenase and pyruvate kinase from porcine muscle using Eudragit bound Cibacron blue, *J. Biotechnol.*, 37, 23, 1994.
56. Guoqiang, D. et al., Affinity precipitation of yeast alcohol dehydrogenase through metal ion promoted binding with Eudragit bound Cibacron blue 3GA, *Biotechnol. Prog.*, 11, 187, 1995.
57. Shu, H.-C. et al., Purification of the D-lactate dehydrogenase from *Leuconostoc mesenteroides* ssp. Cremoris using a sequential precipitation procedure. *J. Biotechnol.*, 34, 1, 1994.
58. Galaev, I.Yu. et al., Imidazole a new ligand for metal affinity precipitation: precipitation of Kunitz soybean trypsin inhibitor using Cu(II)-loaded copolymers of 1-vinylimidazole with *N*-vinylcaprolactam or *N*-isopropylacrylamide, *Appl. Biochem. Biotechnol.*, 68, 121, 1997.
59. Galaev, I.Yu., Kumar, A., and Mattiasson, B., Metal-copolymer complexes of *N*-isopropylacrylamide for affinity precipitation of proteins, *J. Mol. Sci. Pure Appl. Chem.*, A36, 1093, 1999.

60. Kumar, A., Galaev, I.Yu., and Mattiasson, B., Affinity precipitation of α-amylase inhibitor from wheat meal by metal chelate affinity binding using Cu(II)-loaded copolymers of 1-vinylimidazole with *N*-isopropylacrylamide, *Biotechnol. Bioeng.*, 59, 695, 1998.
61. Kumar, A., Galaev, I.Yu., and Mattiasson, B., Isolation and separation of α-amylase inhibitors I-1 and I-2 from seeds of ragi (Indian finger millet, *Eleusine coracana*) by metal chelate affinity precipitation, *Bioseparation*, 7, 129, 1998.
62. Kumar, A. et al., Metal chelate affinity precipitation: purification of $(His)_6$-tagged lactate dehydrogenase using poly(vinylimidazole-*co*-*N*-isopropylacrylamide) copolymers, *Enzyme Microb. Technol.*, 33, 113, 2003.
63. Kumar, A. et al., Purification of histidine-tagged single chain Fv-antibody fragments by metal chelate affinity precipitation using thermo-responsive copolymers, *Biotechnol. Bioeng.*, 84, 495, 2003.
64. Nguyen, A.L. and Luong, J.H.T., Synthesis and application of water-soluble reactive polymers for purification and immobilization of biomolecules, *Biotechnol. Bioeng.*, 34, 1186, 1989.
65. Pécs, M., Eggert, M., and Schügerl, K., Affinity precipitation of extracellular microbial enzymes, *J. Biotechnol.*, 21, 137, 1991.
66. Shiomori, K. et al., Thermoresponsive properties of sugar sensitive copolymer of *N*-isopropylacrylamide and 3-(acrylamido)phenylboronic acid, *Macromol. Chem. Phys.*, 205, 27, 2004.
67. Kumar, A. and Gupta, M.N., Affinity precipitation of trypsin using soybean trypsin inhibitor linked Eudragit S-100, *J. Biotechnol.*, 37, 185, 1994.
68. Chen, J.-P. and Jang, F.-L., Purification of trypsin by affinity precipitation combining with aqueous two-phase extraction, *Biotechnol. Techn.*, 9, 461, 1995.
69. Kamihira, M., Kaul, R., and Mattiasson, B., Purification of recombinant protein A by aqueous two-phase extraction integrated with affinity precipitation, *Biotechnol. Bioeng.*, 40, 1381, 1992.
70. Mattiasson, B. and Kaul, R., One-pot protein purification by process integration, *Bio/Technology*, 12, 1087, 1994.
71. Taniguchi, M. et al., Purification of staphylococcal protein A by affinity precipitation using a reversibly soluble-insoluble polymer with human IgG as a ligand, *J. Ferment. Bioeng.*, 68, 32, 1989.
72. Kumar, A. et al., Effect of polymer concentration on recovery of the target proteins in precipitation methods, *Biotechnol. Techn.*, 8, 651, 1994.
73. Kumar, A. and Gupta, M.N., An assessment of nonspecific adsorption to Eudragit S-100 during affinity precipitation, *Mol. Biotechnol.*, 6, 1, 1996.
74. Mondal, K., Sharma, A., and Gupta, M.N., Three phase partitioning of starch and its structural consequences, *Carbohydrate Polymers*, 56, 355, 2004.
75. Teotia, S. et al., One-step purification of glucoamylase by affinity precipitation with alginate, *J. Mol. Recog.*, 14, 295, 2001.
76. Gupta, M.N., Guoqiang, D., and Mattiasson, B., Purification of endopolygalacturonase by affinity precipitation using alginate, *Biotechnol. Appl. Biochem.*, 18, 321, 1993.
77. Roy, I. et al., Simultaneous refolding/purification of xylanase with microwave treated smart polymer, *Biochim. Biophys. Acta, Proteins and Proteomics*, 1742, 179, 2005.

78. Teotia, S., Lata, R., and Gupta, M.N., Chitosan as a macroaffinity ligand: purification of chitinases by affinity precipitation and aqueous two-phase extractions, *J. Chromatogr.* A, 1052, 85, 2004.
79. Porath, J. et al., Metal chelate affinity chromatography: a new approach to protein fractionation, *Nature*, 258, 598, 1975.
80. Sulkowski, E., Purification of proteins by IMAC, *Trends Biotechnol.*, 3, 1, 1985.
81. Hemdan, E.S. and Porath, J., Interaction of amino acids with immobilized nickel iminodiacetate, *J. Chromatogr.*, 323, 255, 1985.
82. Lilius, G. et al., Metal affinity precipitation of proteins carrying genetically attached polyhistidine affinity tails, *Eur. J. Biochem.*, 198, 499, 1991
83. Gupta, M.N. et al., Affinity precipitation of proteins, *J. Mol. Recog.*, 9, 356, 1996.
84. Flygare, S. et al., Affinity precipitation of dehydrogenases, *Anal. Biochem.*, 133, 409, 1983.
85. So, L.L. and Goldstein, I.J., Protein–carbohydrate interaction, IV: application of the quantitative precipitin method to polysaccharide–Concanavalin A interaction, *J. Biol. Chem.*, 242, 1617, 1967.
86. Porath, J., Immobilized metal affinity chromatography, *Protein Expr. Purif.*, 3, 263, 1992.
87. Porath, J. and Olin, B., Immobilized metal ion affinity adsorption and immobilized metal ion affinity chromatography of biomaterials: serum protein affinities for gel-immobilized iron and nickel ions, *Biochemistry*, 22, 1621, 1983.
88. Liu, K.J. and Gregor, H.P., Metal-polyelectrolyte, X: poly-*N*-vinylimidazole complexes with zinc(II) and with copper(II) and nitrilotriacetic acid, *J. Phys. Chem.*, 69, 1252, 1965.
89. Todd, R.J., Johnson, R.D., and Arnold, F.H., Multiple-site binding interactions in metal-affinity chromatography, I: equilibrium binding of engineered histidine-containing cytochromes c, *J. Chromatogr.*, 662, 13, 1994.
90. Gold, D.H. and Gregor, H.P., Metal–polyelectrolyte complexes, VIII: the poly-*N*-vinylimidazole–copper(II) complex, *J. Phys. Chem.*, 64, 1464, 1960.
91. Balan, S. et al., Metal chelate affinity precipitation of RNA and purification of plasmid DNA, *Biotechnol. Lett.*, 25, 1111, 2003.
92. Murphy, J.C. et al., Nucleic acid separations utilizing immobilized metal affinity chromatography, *Biotechnol. Prog.*, 19, 982, 2003
93. Urry, D.W. et al., Protein-based materials with a profound range of properties and applications: the elastin T_t hydrophobic paradigm, in *Protein Based Materials*, McGrath, K. and Kaplan, D., Eds., Birkhauser, Boston, 1997, pp. 133–177.
94. Kostal, J., Mulchandani, A., and Chen, W., Tunable biopolymers for heavy metal removal, *Macromolecules*, 34, 2257, 2001.
95. Meyer, D.E. and Chilkoti, A., Purification of recombinant proteins by fusion with thermally responsive polypeptides, *Nat. Biotechnol.*, 17, 1112, 1999.
96. Shimazu, M., Mulchandani, A., and Chen, W., Thermally triggered purification and immobilization of elastin-OPH fusion, *Biotechnol. Bioeng.*, 81, 74, 2003.
97. Stiborova, H. et al., One-step metal-affinity purification of histidine-tagged proteins by temperature-triggered precipitation, *Biotechnol. Bioeng.*, 82, 605, 2003.
98. Trabbic-Carlson, K. et al., Expression and purification of recombinant proteins from *Escherichia coli*: comparison of an elastin-like polypeptide fusion with an oligohistidine fusion, *Protein Sci.*, 13, 3274, 2004.
99. Ge, X. et al., Self-cleavable stimulus responsive tags for protein purification without chromatography, *J. Am. Chem. Soc.*, 127, 11228, 2005.

100. Sun, X.-L. et al., One-pot glyco-affinity precipitation purification for enhanced proteomics: the flexible alignment of solution-phase capture/release and solid-phase separation, *J. Proteome Res.*, 4, 2355, 2005.
101. Izumrudov, V.A., Galaev, I.Yu., and Mattiasson, B., Polycomplexes-potential for bioseparation, *Bioseparation*, 7, 207, 1999.
102. Kabanov, V.A., Basic properties of soluble interpolyelectrolyte complexes applied to bioengineering and cell transformations, in *Macromolecular Complexes in Chemistry and Biology*, Dubin, P.L., Ed., Springer-Verlag, Berlin-Heidelberg, 1994, pp. 151–174.
103. Izumrudov, V.A., Competitive reactions in solutions of protein-polyelectrolyte complexes, *Ber. Bunsenges. Phys. Chem.*, 100, 1017, 1996.
104. Dubin, P.L. et al., Coacervation of polyelectrolyte-protein complexes, in *Ordered Media in Chemical Separations*, Hinze, W. and Armstrong, D., Eds., American Chemical Society, Washington, DC, 1987, pp. 162–169.
105. Nguyen, T.Q., Interactions of human haemoglobin with high-molecular-weight dextran sulfate and diethylaminoethyl dextran, *Makromol. Chem.*, 187, 2567, 1986.
106. Wang, Y., Jy, G., and Dubin, P.L., Protein separation via polyelectrolyte coacervation: selectivity and efficiency, *Biotechnol. Progress*, 12, 356, 1996.
107. Zaman, F., Kusnadi, A.R., and Glatz, C.E., Strategies for recombinant protein recovery from canola by precipitation, *Biotechnol. Prog.*, 15, 488, 1999.
108. Kiknadze, E.V. and Antonov, Yu.A., Use of polyelectrolytes for isolation of trypsin inhibitor from industrial waste of alfalfa leaf protein fractionation, *Appl. Biochem. Microbiol.*, 34, 508, 1998.
109. Dainiak, M.B. et al., Production of Fab fragments of monoclonal antibodies using polyelectrolyte complexes, *Anal. Biochem.*, 277, 58, 2000.
110. Carter, S. et al., Highly branched stimuli responsive poly([*N*-isopropylacrylamide]-*co*-[1,2-propandiol-3-methacrylate])s with protein binding functionality, *Macromol. Biosci.*, 5, 373, 2005.
111. Carter, S., Hunt, B., and Rimmer, S., Highly branched poly(*N*-isopropylacrylamide)s with imidazole end groups prepared by radical polymerization in the presence of a styryl monomer containing a dithioester group, *Macromolecules*, 38, 4595, 2005.
112. Chern, C.S., Lee, C.K., and Chen, C.Y., Biotin-modified submicron latex particles for affinity precipitation of avidin, *Colloids Surf. B Biointerfaces*, 7, 55, 1996.
113. Taniguchi, M., Tanahashi, S., and Fujii, M., Purification of staphylococcal protein A by affinity precipitation: dissociation of protein A from the adsorbent with chemical reagents, *J. Ferment. Bioeng.*, 69, 362, 1990.
114. Teotia, S. and Gupta, M.N., Free polymeric bioligands in aqueous two phase affinity extractions of microbial xylanases and pullulanase, *Protein Expr. Purif.*, 22, 484, 2001.
115. Teotia, S. and Gupta, M.N., Reversibly soluble macroaffinity ligand in aqueous two phase separation of enzymes, *J. Chromatogr. A*, 923, 275, 2001.
116. Kumar, A., Kamihara, M., and Mattiasson, B., Two-phase affinity partitioning of animal cells: implications of multipoint interactions, in *Methods for Affinity-Based Separation of Proteins/Enzymes*, Gupta, M.N., Ed., Birkhauser Verlag, Basel, Switzerland, 2002, pp. 163–180.
117. Sharma, A., Roy, I., and Gupta, M.N., Affinity precipitation and macroaffinity ligand facilitate three-phase partitioning for refolding and simultaneous purification of urea-denatured pectinase, *Biotechnol. Prog.*, 20, 1255, 2004.

118. Mondal, K. et al., Macroaffinity ligand-facilitated three-phase partitioning (MLFTPP) of α-amylases using a modified alginate, *Biotechnol. Prog.*, 19, 493, 2003.
119. Teotia, S., Mondal, K., and Gupta, M.N., Integration of affinity precipitation with partitioning methods for bioseparation of chitin binding lectins, *Food Bioproducts Process.*, 84, 37, 2006.
120. Tyagi, R. et al., Chitosan as an affinity macroligand for precipitation of *N*-acetyl glucosamine binding proteins/enzymes, *Isol. Purif.*, 2, 217, 1996.
121. Agarwal, R. and Gupta, M.N., Sequential precipitation with reversibly soluble insoluble polymers as a bioseparation strategy: purification of α-glucosidase from *Trichoderma longibrachiatum*, *Protein Expr. Purif.*, 7, 294, 1996.
122. Lali, A. et al., Carboxymethyl cellulose as a new heterofunctional ligand carrier for affinity precipitation of proteins, *Bioseparation*, 7, 195, 1999.
123. Kumar, A. et al., Binding of Cu(II)-poly(*N*-isopropylacrylamide/vinylimidazole) copolymer to histidine-tagged protein: a surface plasmon resonance study, *Langmuir*, 16, 865, 2003.
124. Mondal, K., Roy, I., and Gupta, M.N., *k*-Carrageenan as a carrier in affinity precipitation of yeast alcohol dehydrogenase, *Protein Express. Purific.*, 32, 151, 2003.
125. Despond, A. and Freitag, R., Light-responsive bioconjugates as novel tools for specific capture of biologicals by photoaffinity precipitation, *Biotechnol. Bioeng.*, 91, 583, 2005.

Chapter 14

Hydrogels in Microfluidics

Jaisree Moorthy

CONTENTS

14.1 Introduction

Over the last 15 years, many researchers have worked to downsize the laboratory workspace into what is popularly referred to as "lab on a chip" devices.[1] These devices promise efficient methods for diagnosis,[2] drug delivery (and discovery),[3] genomics,[4] proteomics,[5] and cellular studies.[6,7] Efficient transport and manipulation of small volumes (nanoliters to microliters) of fluids is a necessity in these devices; hence they are also called microfluidic devices. Nature has evolved some of the best designs for microscale transport: the circulatory system composed of the heart, artery, and veins is one such example. In these systems, the key features are the "softness" of the materials and the intrinsic control of the various components. Hydrogel materials are both soft and responsive, which makes them ideal candidates in designing biomimetic components (e.g., valves, sensors) for use in microfluidic devices. This chapter describes various applications of hydrogels in microfluidics.

14.2 Microfluidics

The key aspects of microfluidics are (a) understanding the physics of small volumes of fluid in confined regions and (b) the efficient manipulation of fluid through micron-sized conduits (microchannels). The forces that are dominant at the microscale can be counterintuitive.[8] For example, viscous forces dominate at the microscale, as opposed to inertial forces at the macroscale. Therefore, the fluid (and particles therein) in a microwell is analogous to a swimming pool filled with honey.[9] The fluid flow in microchannels is described as laminar flow, as it is free of fluctuations and is predictable. Particles in adjacent laminar streams redistribute via diffusion. This phenomenon has been used in developing separation and sensing (T-sensor) techniques.[10] Another advantage is in creating concentration gradients, since the particles diffuse slowly over time. Such a high degree of control on spatiotemporal parameters has been useful in performing experiments that would not be possible with traditional methods. For example, characterizing biological events such as signaling, chemotaxis, and cell growth has been easier with microfluidics systems.[11] It is important to note that the scale of such biological phenomena matches that of the microfluidic systems used to study them. Surface tension is a force that dominates at such small scales. This force is the result of cohesive interactions between liquid molecules at a liquid–gas or liquid–liquid interface. The capillary forces resulting from the liquid–air interface are strong enough to move fluid through micron-sized conduits.[12] Interfaces between immiscible and miscible fluids have been harnessed for a variety of functions, such as generating droplets[13] in microchannels and creating a virtual wall.[14] Atencia and Beebe have provided a useful review of controlled interfaces in microchannels.[15]

In going from a large vessel to a microfluidic device, the surface-to-volume ratio increases dramatically, and many more particles in the volume are exposed to the channel wall. A high ratio of surface area to volume is beneficial in detection because more particles can make contact with the detector. Quick heat dissipation is another consequence of a high ratio of surface area to volume. This is advantageous in electrokinetically driven systems, where joule heating is common. However, fast dissipation of heat could be a problem when high temperatures are desired, e.g., polymerase chain reaction (PCR), inside the device, requiring the researcher to use ingenious designs.

For a more complete view of fluid phenomena at the microscale, the reader is referred to publications by Beebe et al.,[16] Stone et al.,[17] and Squires and Quake.[18]

14.2.1 Microfluidic Devices

Microfluidic devices are created using microfabrication techniques that provide a basis for mass production. These systems have the capability to offer point-of-care diagnostics and to perform a large number of assays simultaneously in a cost-efficient manner for high-throughput studies.[19] The essential components of a microfluidic device include valves and pumps to transport fluid efficiently through microchannels, filters for sample preparation, and sensors for detection. The past decade has seen much progress in the production of microfluidic devices and components for diverse applications. The materials used initially were silicon and glass, as traditional microfabrication methods can be used to shape these materials into devices.[20] However, the long fabrication process and the difficulty in obtaining good seals (for valves) have limited the use of these materials. Soft fabrication with poly(dimethyl siloxane) (PDMS) helped in overcoming some of these challenges.[21,22] However, external methods such as pneumatic, electric, and magnetic signals were required to control (e.g., open and close valve) the microfluidic components. With the advent of hydrogels, smart and autonomous components could be created,[23] because these materials can sense the local environment and undergo phase transition, such as an abrupt swelling. The following section reviews additional properties of these materials.

14.3 Hydrogels

Hydrogels are polymers with high water content. The polymer chains are cross-linked into a network to form a porous sieve-like structure that can be used to separate particles based on size (e.g., gel electrophoresis). The hydrophilic nature of the polymer inside the hydrogel preserves the integrity of cells and biomolecules. For this reason, hydrogels are used extensively in tissue engineering and drug delivery.[24] A subclass of hydrogels includes polymers that can undergo phase transition[25,26] in the presence of stimuli such as pH, temperature, electric or magnetic fields, and specific chemicals (or biomolecules). The most common phase transition behavior used in microfluidic devices is the ability of the hydrogel to change its volume abruptly in the presence of these stimuli. The volume expansion, which can be as high as 1000-fold, can produce a mechanical force, resulting in transduction from a chemical stimulus to a mechanical action (as in muscles), and this force has been exploited in designing valves and pumps.[27]

The chemical composition of the polymer matrix determines the type of stimuli to which the hydrogel can respond. The extent of swelling is limited by the elasticity of the polymer chains and the cross-link density. Swelling/shrinking behavior stems from events occurring at the molecular level, which results in the transport of water and ions. The rate of swelling depends on the distribution of ionizable chemical groups and is limited by diffusion. Therefore, the time scales for the volume change will depend on the distance traveled (by ions and water molecules), which in turn depends on the initial size of the hydrogel, the diffusion constant of the species, and pore size. These parameters are taken into consideration when designing microfluidic components and optimizing for performance (Figure 14.1).

The performance of a component is usually evaluated by the response time and the mechanical stability of the material. Hydrogels are soft materials and are capable of repeatedly undergoing a swelling/shrinking process. However, hydrogel components are usually fabricated along with other rigid materials that are not able to withstand repeated cycles (greater than ten). This mismatch results in delamination of channels and other complications. An advantage of micron-sized hydrogel structures is that the diffusion is rapid and the response time is usually within a few minutes. It is interesting to note that plants and fungi use a swelling/shrinking mechanism for moving small parts (micrometer to millimeter range). They display a range of response times, from 0.1 to ≈100,000 sec.[28] Mimicking this mechanism will be useful in making devices for quick analysis (e.g., PCR) and also for slow processes, such as in drug delivery and in cell culture.

14.3.1 Fabrication

Hydrogels are commonly prepared by polymerization of monomers or by cross-linking polymers. These processes (polymerization and cross-linking)

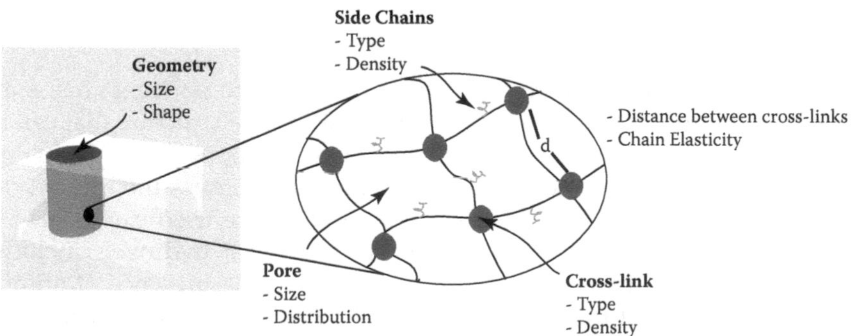

FIGURE 14.1
Parameter space governing the property of hydrogel. The geometry of the final component, pore size of the matrix, chain elasticity, cross-link density, and distribution of ionizable chemical groups, determines the response time. The chemical groups on the side chains and at the cross-linking positions determine the type of the stimuli.

can be initiated by free-radical reactions, condensation, or by using an enzyme. In free-radical polymerization, the reaction is initiated using specific molecules (initiator) that form radicals in the presence of light, temperature, or a redox mechanism.

In photopolymerization, the hydrogel structures are fabricated from a prepolymer mixture consisting of a monomer, cross-linker, and a photoinitiator. The mixture is introduced into a cavity of a microfluidic channel and irradiated with UV light through a transparency mask (Figure 14.2A). The exposed regions are polymerized into three-dimensional (3-D) structures. After removing the unpolymerized mixture, the channel and the hydrogel structures are washed thoroughly. Since the hydrogel components are fabricated *in situ*, there is flexibility in choosing an appropriate location inside the labyrinth of microchannels.

In thermally initiated polymerization, a heat source is required. There is less control in defining size and location for this method. Mostly, thermal polymerization can be performed by photopolymerization by replacing the initiator in the prepolymer mixture. Table 14.1 summarizes the various components of the prepolymer mixture. One can engineer specific functionality by choosing an appropriate composition. For example, hydrogel composed of purely 2-hydroxyethylmethacrylate (HEMA) is nonresponsive. With the addition of acrylic acid (AA), the hydrogel is responsive to pH, i.e., it swells under basic conditions. Another way to affect the properties of the hydrogel structure is to change the rate of polymerization. In photopolymerization, this is achieved by increasing the intensity of UV radiation. At increased rate, a large number of small polymer chains are formed. The response time of such a structure is improved (i.e., shorter time required to swell), while the mechanical strength of the structure is compromised.

Photopolymerization has worked very well in making devices for diagnostics[2] and protein studies.[29] However, the radicals formed from the initiators are toxic to the cells. Cross-linking between polymer chains avoids this problem. Temporary hydrogel networks can be formed by combining calcium ions and alginate.[30] The electrostatic interaction between the doubly positive ion and the negative charges on the polymer chain helps in maintaining the 3-D architecture. Using a T-junction channel, hydrogel structures can be formed via interfacial polymerization by transporting these reagents in adjacent streams (Figure 14.2B,C). Recently, Paguirigan and Beebe have demonstrated that gelatin can be cross-linked with transglutaminase enzyme to form microfluidic channels.[31] These channels were fabricated by pouring a mixture of gelatin and the enzyme into a silicone mold. After incubation, the gelatin was cross-linked into a functional device. Since gelatin is a common protein found in the extracellular matrix, these channels are biocompatible and are a suitable environment for adherent cell cultures.

Gradient hydrogels have been fabricated by combining microfluidic phenomena and photopolymerization (Figure 14.2D). These hydrogels have internal spatial gradients of properties such as porosity due to different cross-linking densities or co-monomer concentrations. The hydrogels are created

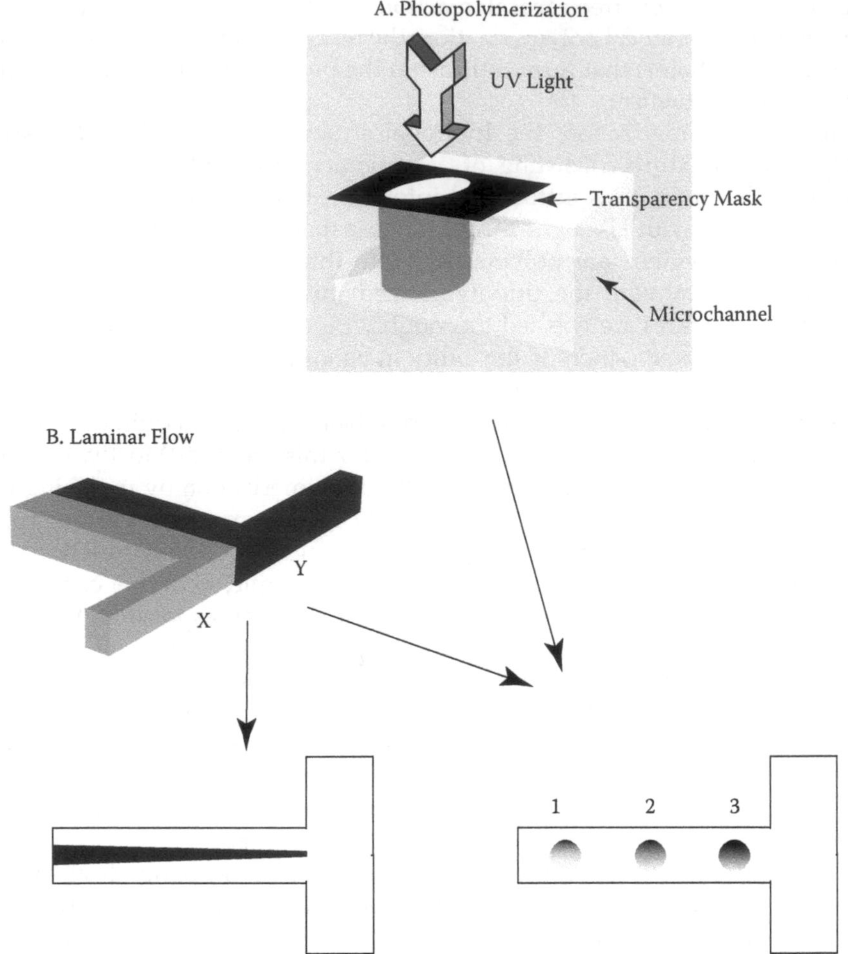

FIGURE 14.2
Fabrication of microfluidic components. (A) Photopolymerization: in this method, the microchannel is filled with a prepolymer mixture (Table 14.1) and irradiated through a transparency mask. The regions exposed are polymerized (shown in gray). (B) Laminar flow: streams flowing adjacent form an interface. The particles in the streams mix slowly by diffusion. (C) Interfacial polymerization: this occurs when the adjacent flow streams contain prepolymer mixture. (D) Combination: gradient hydrogels can be fabricated by combining microfluidics and photopolymerization. These hydrogels have a spatial distribution in their property (pore size or responsiveness).

by first making a gradient in one of the prepolymer constituents by using a T-junction (Figure 14.2B) or a gradient generator[32] and then polymerizing the resulting structure. For example, a gradient in monomer concentration (HEMA and AA) was generated by flowing streams containing different prepolymer compositions to form hydrogel with inhomogeneous properties.[33]

TABLE 14.1

List of Prepolymer Constituents and Their Function during Polymerization or in the Final Hydrogel Structure

Constituent	Function
Monomer	Forms the main framework and gives primary properties such as surface energy
Cross-linker	Added for mechanical strength, durability, and responsiveness
Co-monomer	Can be added for additional properties, e.g., responsiveness
Initiator	Initiates the polymerization process
Co-initiator	Mediates transfer of energy from initiator to monomer
Porogen	Provides additional properties such as pores for the filter in emulsion photopolymerization

In the presence of the stimulant (pH), asymmetry in swelling and in the final structure was observed. Burdick et al. have used a gradient generator to fabricate polyethylene glycol hydrogels with different ligand concentrations[34] for cell studies. Zaari et al. have fabricated gradient polyacrylamide hydrogels with different cross-linking densities[35] for use in studying cell–matrix interaction. Thus, by combining microfluidics and photopolymerization, hydrogels with a spatial gradient in structure and responsiveness can be produced, which is not possible using traditional fabrication methods.

14.4 Applications

The applications of hydrogels in microfluidic devices are reviewed in this section. The ability of hydrogels to respond to changes in the local environment has been used to make sensors. Shape changes of the hydrogel due to swelling have been used to design biomimetic valves. Physical change can generate mechanical forces, and these have been harnessed to create pumps, clutches, and lenses and have been used in active assembly. Hydrogel structures composed of pores and polymer chains have been used for separation and in sample preparation.

14.4.1 Sensor

The performance of a sensor is described by its specificity and sensitivity to detect molecules in a sample. Sensors based on various factors such as mass, fluorescence, and binding forces have been developed. While each of these display reliable specificity and a high degree of sensitivity, they are hard to integrate in microfluidic devices due to fabrication complexities. Moreover, they require another component for reporting, and integration of this component could be cumbersome. The display component could be electronically complex, bulky, and require a power supply. In contrast, hydrogel-based sensors undergo a perceptible physical change that provides visual evidence of the detection. Thus, a single unit essentially performs both the sensor and display functions. A hydrogel sensor can be created by designing a cross-linker that is responsive to a specific chemical or a biomolecule.

Dissolvable hydrogel structures containing a disulfide bond as a cross-linker (*N,N'*-cystamine-bisacrylamide) have been fabricated[33] and studied experimentally and computationally[36] (Figure 14.3A). The hydrogel is dissolved in the presence of a reducing agent such as DTT (dithioerythritol) and glutathione. The factors that affect the rate of dissolution are: the initial diameter of the hydrogel, concentration of the reducing agent, cross-link density, and temperature. Using a similar approach, hydrogels cleavable by enzymes have been designed by incorporating a peptide substrate as cross-linker. A chymotrypsin-sensitive hydrogel was designed by using a tetramer

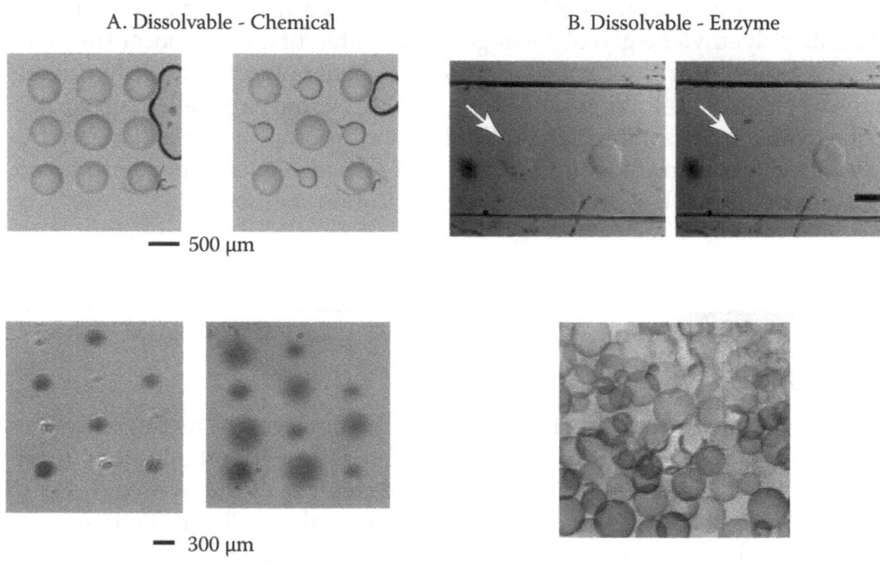

FIGURE 14.3
Detection components. (A) Dissolvable hydrogel obtained by using *N,N'*-cystamine-bisacrylamide as the cross-linker. In the presence of a reducing agent, the cross-links are compromised, resulting in degradation of the hydrogel. (From Moorthy, J. and Beebe, D.J., *Anal. Chem.*, 75, 292A, 2003. Copyright, American Chemical Society. Adapted with permission.) (B) By incorporating Cys-Tyr-Lys-Cys (CYKC) as a cross-linker, the hydrogel can be dissolved (shown by white arrows) in the presence of chymotrypsin enzyme. (From Plunkett, K.N., Berkowski, K.L., and Moore, J.S., *Biomacromolecules*, 6, 632, 2005. Copyright, American Chemical Society. Adapted with permission.) (C) pH-sensitive dyes were immobilized into poly(HEMA) hydrogel to form the combinatorial display. (From Moorthy, J. and Beebe, D.J., *Lab Chip*, 2, 76, 2002. Copyright, RSC. Adapted with permission.) (D) Using biotin–avidin interaction, antibody was adsorbed on agarose beads to function as a "readout" in immunoassay. (From Moorthy, J. et al., *Electrophoresis*, 25, 1705, 2004. Copyright, Wiley Interscience. Adapted with permission.)

(Cys-Tyr-Lys-Cys) as a cross-linker[37] (Figure 14.3B), and a botulinum neurotoxin-sensitive hydrogel was obtained by incorporating a 19-mer (NH_2-Cys-Ser-Asn-Lys-Thr-Arg-Ile-Asp-Glu-Ala-Asn-Gln-Arg-Ala-Thr-Lys-{Nle}Lys-Cys-COOH) as cross-linker.[38] In both these approaches, cysteine is included in the sequence so that the peptide can be coupled to a photopolymerizable methacrylamide cross-linker by a disulfide exchange protocol. The degree of degradation depends on the cross-link density of the hydrogel, i.e., less-cross-linked structures degrade faster. The ability to create peptide–hydrogel hybrids allows for programming the sensitivity of the sensor by varying the length (or sequence) of the peptide and the response time by adjusting the cross-link density of the hydrogel. By using antigen–antibody binding as a cross-linker, Miyata et al. have developed a hydrogel that expands in the presence of the antigen.[39] This change occurs because the free antigen molecules compete for the antibody, which weakens the cross-linking.

Another way to make a hydrogel sensor is to immobilize chemicals (or biomolecules). Here the high-volume support of the hydrogel is employed to make "readouts." A sensor and display for local pH has been fabricated by immobilizing pH-sensitive dyes inside HEMA hydrogels by mixing dye powder in the prepolymer mixture.[40] The hydrogel posts changed color based on the local pH. Using sequential photopolymerization of two different types of hydrogels, a combinatorial display was fabricated (Figure 14.3C). Proteins have also been immobilized inside the polyacrylamide hydrogel in a similar approach to detect antigen molecules[41] or for characterizing proteins (proteomics).[42] The matrix of the hydrogel provides a "natural" environment for proteins. Another approach is to immobilize the protein "on" the hydrogel. In this method, protein is allowed to diffuse or adsorb onto polymerized hydrogel. More specific immobilization of proteins on hydrogel was achieved by incorporating biotin–avidin interaction.[2] Here, agarose hydrogel beads containing streptavidin were incubated with antibody containing biotin functional groups. Since each avidin can bind to four biotins, a dense coverage of antibody molecules was achieved. The beads were loaded into the device and were used to detect specific antigens (Figure 14.3D).

14.4.2 Valve

Valves are required to control fluid flow, to open or shut certain microchannels, and to control flow direction (check valve). This function requires a moving part, which is usually achieved by incorporating electric, piezoelectric, or magnetic capabilities. However, the forces from these methods do not scale well and involve high-energy expenditure. Moreover, these forces require rigid materials that are difficult to seal. Taking a lesson from nature, one-way check valves have been fabricated that mimic the sponge-like tissue material involved in the water-transport mechanism in trees[43] and the shape of the valve found in the veins. The bi-strip valve was fabricated by combining a responsive hydrogel with a nonresponsive hydrogel.[44] In the presence of a stimulus, in this case a pH change, the responsive hydrogel swells and bends the structure. The shape favors flow in only one direction (Figure 14.4A). Kim et al. have used an arrowhead-shaped hydrogel to achieve the same function.[45] When the hydrogel swells, the structure is elongated and the head of the valve makes contact with the valve neck, allowing flow in only one direction, as shown in Figure 14.4B. Being soft, the seal obtained using hydrogel is improved over rigid materials. Since

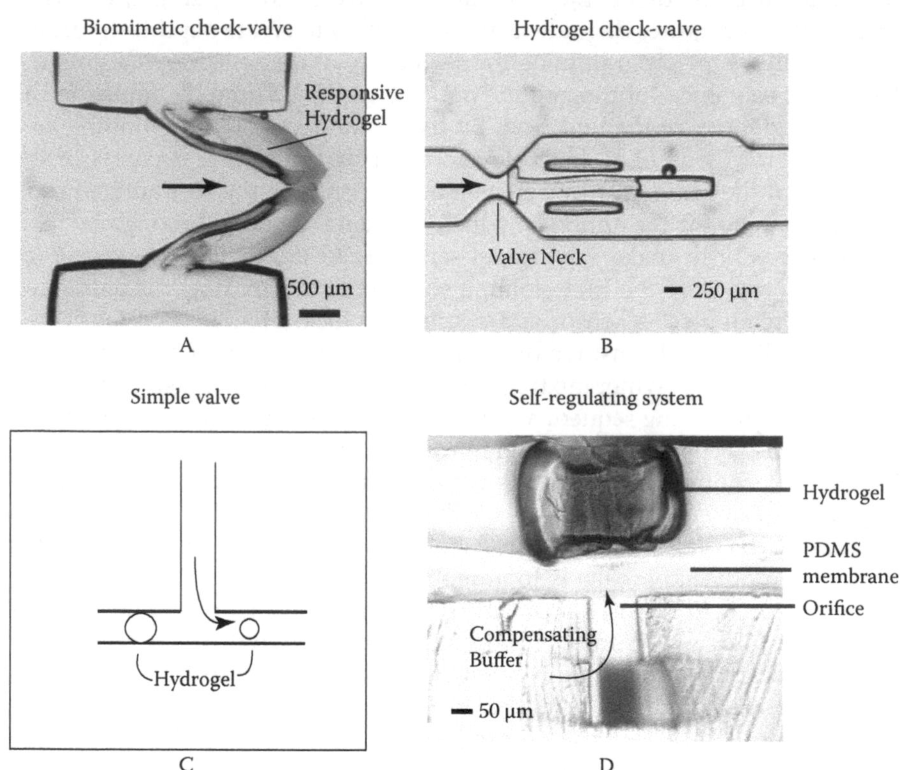

FIGURE 14.4
Valves. (A) Biomimetic check valve: integration of responsive and nonresponsive hydrogels forms the bi-strip. In the presence of stimulus, the structure swells, forming the curved shape that allows flow in only one direction (shown by arrow). (From Yu, Q. et al., *Appl. Phys. Lett.*, 78, 2589, 2001. Copyright, American Institute of Physics. Adapted with permission.) (B) Hydrogel check valve: formed by shaping the hydrogel as an arrow. The structure swells to close the valve neck and allows flow in the direction shown. (From Moorthy, J. and Beebe, D.J., *Anal. Chem.*, 75, 292A, 2003. Copyright, American Chemical Society. Adapted with permission.) (C) Simple valve: can be fabricated as a hydrogel cylinder placed in the path of the channel. When the hydrogel swells, the channel is blocked. (From Beebe, D.J. et al., *Nature*, 404, 588, 2000. Reprinted with permission from Macmillan Publishers.) (D) Self-regulating system: fabricated by coupling hydrogel swelling to a flexible membrane. (From Moorthy, J. and Beebe, D.J., *Anal. Chem.*, 75, 292A, 2003. Copyright, American Chemical Society. Adapted with permission.) The system was used to maintain a controlled pH environment. Fluctuations in pH caused the hydrogel to swell or shrink and thus control the flow of compensating buffer through the orifice.

hydrogel can be made to respond to various stimuli, the valve can be opened or closed on demand, even by specific molecules in the flowing fluid.

A simpler valve design (Figure 14.4C) includes a cylinder at the intersection of the microchannels that swells in the presence of pH[46] or temperature[47] change. Light-activated valves have been fabricated by introducing particles (gold colloids and nanoshells) in temperature-sensitive hydrogel before polymerization.[48] The trapped particles absorb energy at specific wavelengths (532 nm and 832 nm) and release energy as heat, causing the polymer chains to respond.

The force exerted by a swelling hydrogel can move soft materials such as a PDMS membrane. This energy conversion, from chemical to mechanical, has been utilized to make self-regulated valves (Figure 14.4D). Here, the swelling of the hydrogel is coupled to the displacement of a PDMS membrane. The membrane is actuated to open and close an orifice. A self-regulated system for maintaining a constant pH in the device was designed[49] using a pH-sensitive hydrogel (HEMA based). Fluctuation in pH caused the hydrogel to swell or shrink and thus open or close to allow compensating buffer to flow into the system. This design has been used to make a drug delivery device as well.[50] The phenylboronic-based hydrogel used in the device is sensitive to glucose and pH, so that oscillations in concentration cause release of the drug. Apart from the stimuli, the other parameters of importance include size and thickness of the membrane, channel depth, and shape of the orifice. An elaborate description of the parameter space can be found in papers by Liu[51] and Sanghoon.[52]

14.4.3 Pump

Pumps or dispensers are required to initiate fluid flow through the microchannels. By coupling a responsive hydrogel with a flexible membrane, dispensers have been created for use in protein therapeutics.[53] The design is similar to the self-regulated valve described in the previous section. An array of pH-sensitive hydrogels is fabricated over a flexible membrane. When actuated, the membrane is deformed, causing fluid to be dispensed from the reservoir placed below the membrane. A flow rate of 2 µl/min over 12 h was demonstrated for use in an open-loop infusion system. In another design, a hydrogel microvalve was also incorporated in the system.[54] By fabricating different inlet and outlet flow streams, the actuations of the hydrogel array and the valve were separated from each other and also from the reservoir. Depending on the state of the valve (open or closed), fluid is dispensed or pressure is stored in the "pumping chamber" (Figure 14.5A,B). Precise control on the volume dispensed, and at the specified time, was possible, making the system a good prototype for drug delivery. By incorporating hydrogel responsive to a specific target molecule (e.g., glucose), a closed-loop infusion system can be realized.

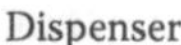

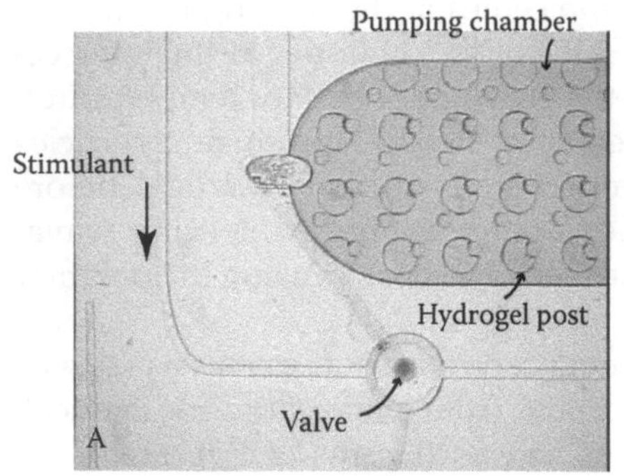

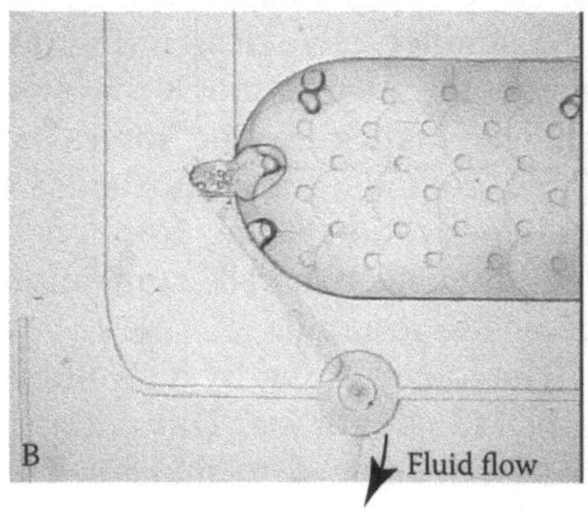

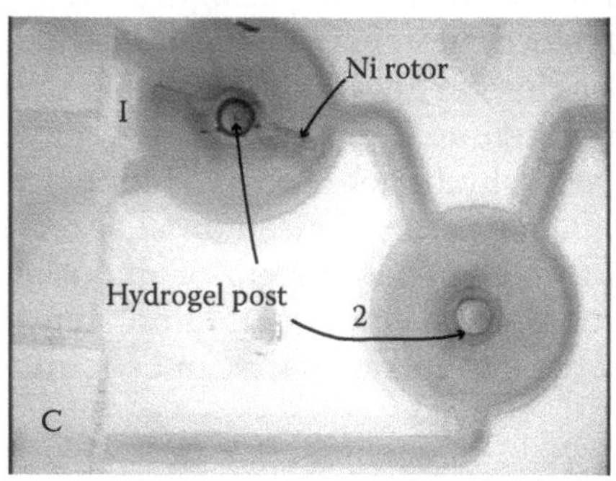

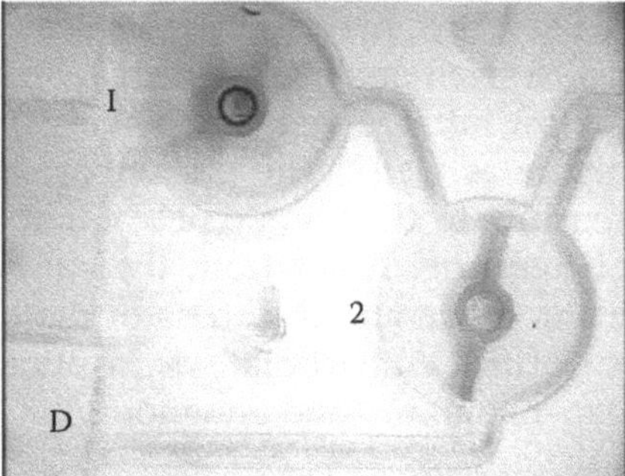

FIGURE 14.5
An array of hydrogel posts in the pumping chamber is coupled with a flexible membrane. When stimulated, the hydrogels swell and cause the membrane to squeeze fluid out through the outlet. A microvalve was included to control the release of fluid from the chamber. The stimulant for the valve is separate from that of the hydrogel posts. (A) Valve is closed and the hydrogel posts are not stimulated. (B) Valve is open and the hydrogel posts are swollen. (Figures 14.5A,B are from Eddington, D.T. and Beebe, D.J., *J. Microelectromech. Syst.*, 13, 586, 2004. Copyright, IEEE. Adapted with permission.) An array of programmable micromixers containing a Ni rotor and responsive hydrogel posts was fabricated. In the presence of appropriate stimuli, the hydrogel swells to stop the rotation of the Ni rotor. The hydrogels in 1 and 2 respond to different pH. (C) Hydrogel post in 1 is swollen. (D) Hydrogel post in 2 is swollen. (Figures 14.5C,D are from Agarwal, A.K. et al., *J. Microelectromech. Syst.*, 14, 1409, 2005. Copyright, IEEE. Adapted with permission.)

14.4.4 Clutch

One of the characteristics of laminar flow is diffusion-limited mixing between adjacent flow streams (Figure 14.2B). The downside of this phenomenon is that larger particles take a longer time to mix. The upside is that concentration gradients can be generated easily for use in detection, separation, and cellular studies. Thus, depending on the application, quick mixing or a concentration gradient may be desired. By using hydrogel cylinders as clutches, Agarwal et al. have designed systems wherein the control to "mix" or "not mix" is provided by the pH or temperature of the sample.[55] Here, a nickel rotor was used as a "stirrer"; in the presence of a rotating magnetic field, the structure rotated to mix the fluid streams. A temperature or pH-responsive hydrogel post was photopolymerized at the center of the nickel rotor. In the presence of the stimulant, the post expands, applying pressure on the rotor and causing it to stop rotating, even in the presence of the magnetic field. The authors have also fabricated an array of micromixers that can be controlled by different stimuli by choosing appropriate hydrogels (Figure 14.5C,D). They have also extended this concept in realizing programmable micropumps. These pumps use the magnetic field and nickel rotor to create a pressure difference to drive fluid flow. By incorporating responsive hydrogel posts in the microstructure, the pump could be stopped in the presence of the stimulus. Such programmable micromixers and micropumps allow for the design of generic microfluidic devices for different applications and yet provide specific control mechanisms.

14.4.5 Lens

We are able to focus on objects at different distances because the ciliary muscles in our eyes control the shape of the lens and hence the focal distance. Taking inspiration from nature, and coupling a responsive hydrogel with a pinned liquid–liquid interface, a smart lens has been fabricated.[56] The device consists of a hydrogel ring fabricated below the aperture. The ring encloses water and forms an interface with oil (Figure 14.6A,B). Microfluidic channels were constructed to transport stimulant to the hydrogel jacket. In the presence of the stimulus (temperature or pH), the hydrogel underwent a transformation in its physical size, causing a volume change of the enclosed water and thus affecting the geometry of the interface and the focal length. The hydrogel functions like the ciliary muscles in modulating the focal length of the microlens. This approach was extended to incorporate two microlenses controlled with different stimuli in a manner analogous to the compound eyes of insects. This system makes use of increased surface tension at the microscale and used local environmental conditions as an input to produce an optical output.

14.4.6 Active Assembly

Reconfiguring the layout of microfluidic devices to route fluid through different parts of the device would be valuable for many applications. Toward this goal, a "mobile" valve was designed wherein a polymer plug was moved to open and close certain parts of the channel network.[57] This was accomplished by applying a high pressure to the plug. By using responsive hydrogel, this task was achieved by stimulating the hydrogel to expand and close

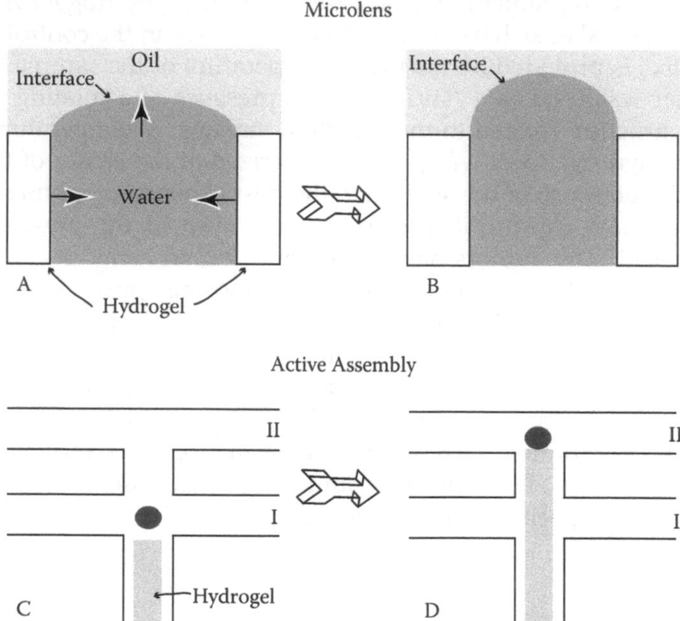

FIGURE 14.6
(A) Concept design of microlens. A hydrogel ring encloses water that forms an interface with oil. In the presence of stimulant, the hydrogel ring swells, causing a change in the geometry of the interface and thus affecting the focal length, as shown in B. (C,D) Concept design of active assembly. The circular object placed in channel I is moved to channel II when the hydrogel bar swells and elongates.

certain parts of the channel network in an "active" manner.[58] The mechanical force generated during expansion was also utilized to move objects from one place to another (Figure 14.6C,D). For example, after treatment with the sample, the detector was moved into an adjoining channel for subsequent steps, thus reducing background interference. Since the assembly process can be initiated by conditions in the local environment, one can produce devices where subsequent reaction steps are determined by the presence of specific molecules in the sample or the product.

14.4.7 Separation

The cross-linked structure of the hydrogel forms a sieving environment that has proven to be useful in separating biomolecules based on size. The pore size in the hydrogel can range from a few nanometers to tens of microns. In biological studies, gel electrophoresis uses polyacrylamide and agarose gels to separate proteins and DNA based on their size. The nano-sized pores and noninteracting environment of the gel make them ideal for this application. Burns and coworkers have down-sized gel electrophoresis onto a small chip.[59] Polyacrylamide hydrogel was photopolymerized inside the microchannel and, under the influence of an electric field, separation of DNA was demonstrated. Using a similar approach, separation of proteins in their native state was achieved in a polyacrylamide hydrogel by Herr et al.[60] This system was used to detect tetanus toxin and antibody by quantifying binding between the antigen and appropriate antibody via gel electrophoresis. These advances are promising, as they could lead to the development of large-scale, high-throughput separation of small sample volumes. This type of brute force approach will be required to characterize the information content in genomics and proteomics. An advantage of using hydrogels over fabrication of nano-posts is that the pore size can be "tuned" by starting with different prepolymer compositions. An interesting application of nanoporous hydrogel has been to enhance the sensing capability of the T-sensor.[61] This sensor is based on the principle that free and bound proteins will diffuse to different degrees; thus, in the presence of an antigen, there is a distinct change in diffusion of the biomolecules due to formation of an antigen–antibody complex. Inside the hydrogel, the diffusion of the biomolecules is greatly reduced, and the reduction is dependent on the size of the biomolecules. Thus better resolution was achieved, allowing for more accurate detection.

Macroporous hydrogels have been fabricated inside microchannels[62] by emulsion photopolymerization of a mixture containing monomer, cross-linker, photo-initiator, and porogen (e.g., water). In this method, agitating the mixture creates an emulsion composed of monomer droplets in a water medium. The emulsion mixture is then photopolymerized inside the microchannel to form a contiguous polymer network surrounded by interconnected paths or pores (Figure 14.7A). The pore size and its distribution are controlled by many parameters: composition of the emulsion mixture (e.g., concentration of cross-linker), polymerization technique, and the surface energy of the channel walls. This breadth in parameter space allows for "tuning" of the material as per the needs of the application. A HEMA-based porous monolith was fabricated inside a microchannel and used to prepare a sample for diagnosis by separating the cells from the proteins (Figure 14.7B). This separation is often required to prepare a body sample (e.g., whole blood) for diagnosis. Centrifuges are customarily used to perform this step in a laboratory setting, but centrifugal forces do not scale well with size, making them inefficient for working with small volumes of fluid. Using the porous filter, efficiency of separation was comparable with that of centrifugation, with the added advantages of working with small volumes and using no external power supply.

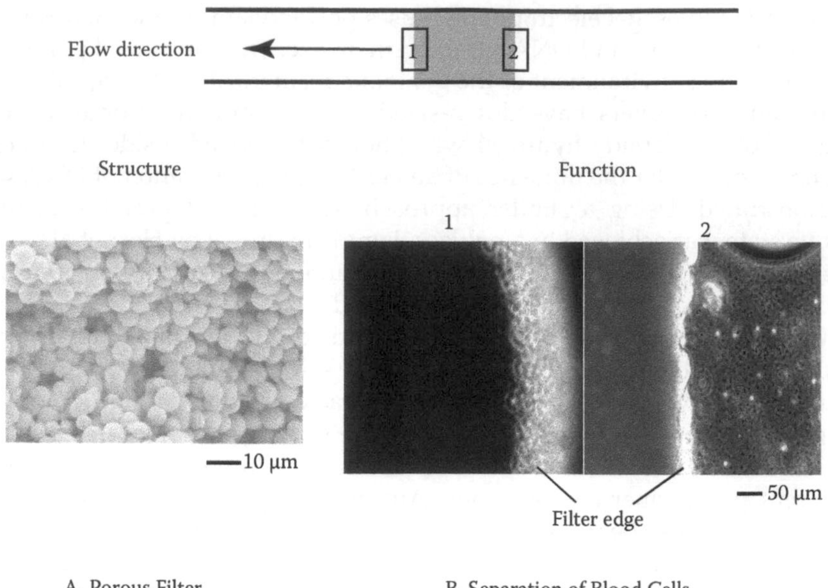

FIGURE 14.7
Porous filter for use in separation. (A) SEM image showing polymer beads formed from emulsion photopolymerization. (From Moorthy, J. et al., *Electrophoresis*, 25, 1705, 2004. Copyright, Wiley Interscience. Adapted with permission.) (B) The filter was used to remove blood cells from whole blood. The image labeled as 2 shows the filter region before separation; the white and black dots are blood cells. The image labeled as 1 shows the region after separation. Removing blood cells from whole blood is often required to prepare the biological sample for protein analysis. (From Moorthy, J. and Beebe, D.J., *Lab Chip*, 3, 62, 2003. Copyright, RSC. Adapted with permission.)

14.5 Integrated Device

In the previous section, we reviewed various applications of hydrogels in the production of microfluidic components. In a typical microfluidic device for use in diagnostic or high-throughput studies, various functions are performed sequentially or in a parallel manner. Thus, more than one component must be incorporated to form an integrated system. Earlier problems in developing such systems were due to material mismatch and complications in fabrication steps. Moreover, each system was designed for a specific assay, and since there is a great deal of variation in the assays, the process was cumbersome. Beebe et al. introduced the "microfluidic tectonics" platform[63] wherein the components are considered analogous to "Lego blocks" and fabricated *in situ* (via photopolymerization) at any location in the device. This flexibility allowed for the construction of a device specific to an assay by arranging the components in an *ad hoc* manner. Using this

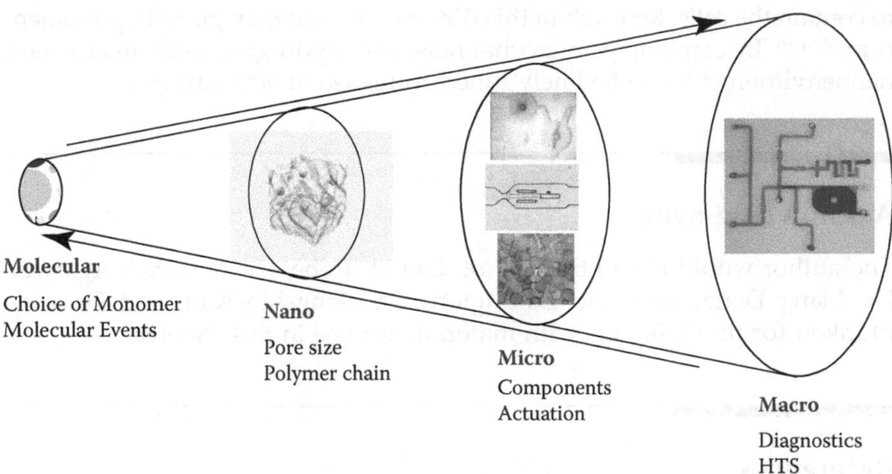

FIGURE 14.8
Multiscale hierarchical relationship of hydrogel functionality. Molecular events result in changes at the nanoscale (polymer chains), which are harnessed to create components at the microscale. Integrating the individual parts leads to a functional microfluidic device at the macroscale. (From Moorthy, J. et al., *Electrophoresis*, 25, 1705, 2004. Copyright, Wiley Interscience. Adapted with permission.)

platform, an integrated device was fabricated for detecting botulinum neurotoxin from whole blood.[2] The device consisted of hydrogel components such as a porous filter for sample preparation and one-way check valves to control flow direction. Agarose beads coated with antibody were used for detection and as a substrate for performing the immunoassay. Since the behavior of the hydrogel is controlled by choosing appropriate prepolymer mixtures, the properties of the components can be incorporated to match the need. Hydrogel functionality is essentially multiscale; molecular events result in changes at the nanoscale (polymer chains), and these are harnessed to create components at the microscale, and integrating the individual parts leads to a functional microfluidic device at the macroscale (Figure 14.8). Such hierarchies are common motifs in the complexity of biological systems.

14.6 Conclusion

In this chapter we have explored the use of hydrogels in manipulating fluid flow and in sample preparation and detection. These functions have been instrumental in making autonomous systems. As microfluidic devices find more applications in cellular studies, hydrogels are also likely to be more widely used. The polymer chain network provides a biomimetic framework to contain the cells. Research in this direction has already picked up momentum.[30,64–66] In combining microchannels and hydrogels, both micro- and nanoenvironments can be finely tuned to realize *in vivo* settings.

Acknowledgment

The author would like to thank Prof. David Beebe, Dr. Abhishek Agarwal, Dr. Liang Dong, Dr. David Eddington, Dr. Dongshin Kim, and Dr. Kyle Plunkett for providing me with materials for use in this chapter.

References

1. Mitchell, P., Microfluidics: downsizing large-scale biology, *Nat. Biotechnol.*, 19, 717–721, 2001.
2. Moorthy, J. et al., Microfluidic tectonics platform: a colorimetric, disposable botulinum toxin enzyme-linked immunosorbent assay system, *Electrophoresis*, 25, 1705–1713, 2004.
3. Dittrich, P.S. and Manz, A., Lab-on-a-chip: microfluidics in drug discovery, *Nat. Rev. Drug Discov.*, 5, 210–218, 2006.
4. Chou, H.P. et al., A microfabricated device for sizing and sorting DNA molecules, *Proc. Natl. Acad. Sci. USA*, 96, 11–13, 1999.
5. Hansen, C. and Quake, S.R., Microfluidics in structural biology: smaller, faster ... better, *Curr. Opin. Struct. Biol.*, 13, 538–544, 2003.
6. Walker, G.M., Zeringue, H.C., and Beebe, D.J., Microenvironment design considerations for cellular scale studies, *Lab Chip*, 4, 91–97, 2004.
7. Wheeler, A.R. et al., Microfluidic device for single-cell analysis, *Anal. Chem.*, 75, 3581–3586, 2003.
8. Brody, J.P. et al., Biotechnology at low Reynolds numbers, *Biophys. J.*, 71, 3430–3441, 1996.
9. Purcell, E.M., Life at low Reynolds number, *Am. J. Phys.*, 45, 3–11, 1977.
10. Weigl, B.H. and Yager, P., Microfluidic diffusion-based separation and detection, *Science*, 283, 346–347, 1999.
11. Takayama, S. et al., Patterning cells and their environments using multiple laminar fluid flows in capillary networks, *Proc. Natl. Acad. Sci. USA*, 96, 5545–5548, 1999.
12. Walker, G.M. and Beebe, D.J., A passive pumping method for microfluidic devices, *Lab Chip*, 2, 131–134, 2002.
13. Tice, J.D. et al., Formation of droplets and mixing in multiphase microfluidics at low values of the Reynolds and the capillary numbers, *Langmuir*, 19, 9127–9133, 2003.
14. Zhao, B., Moore, J.S., and Beebe, D.J., Surface-directed liquid flow inside microchannels, *Science*, 291, 1023–1026, 2001.
15. Atencia, J. and Beebe, D.J., Controlled microfluidic interfaces, *Nature*, 437, 648–655, 2005.

16. Beebe, D.J., Mensing, G.A., and Walker, G.M., Physics and applications of microfluidics in biology, *Annu. Rev. Biomed. Eng.*, 4, 261–286, 2002.
17. Stone, H.A., Stroock, A.D., and Ajdari, A., Engineering flows in small devices microfluidics towards a lab-on-a-chip, *Annu. Rev. Fluid. Mech.*, 36, 381–411, 2004.
18. Squires, T.M. and Quake, S.R., Microfluidics: fluid physics at the nanoliter scale, *Rev. Mod. Phys.*, 77, 977–1026, 2005.
19. Thorsen, T., Maerkl, S.J., and Quake, S.R., Microfluidic large-scale integration, *Science*, 298, 580–584, 2002.
20. Wise, K.D. and Najafi, K., Microfabrication techniques for integrated sensors and microsystems, *Science*, 254, 1335–1342, 1991.
21. McDonald, J.C. and Whitesides, G.M., Poly(dimethylsiloxane) as a material for fabricating microfluidic devices, *Acc. Chem. Res.*, 35, 491–499, 2002.
22. Quake, S.R. and Scherer, A., From micro- to nanofabrication with soft materials, *Science*, 290, 1536–1540, 2000.
23. Moorthy, J. and Beebe, D.J., Organic and biomimetic designs for microfluidic systems, *Anal. Chem.*, 75, 292A–301A, 2003.
24. Hoffman, A.S., Hydrogels for biomedical applications, *Adv. Drug Deliv. Rev.*, 43, 3–12, 2002.
25. Osada, Y. and Ross-Murphy, S.B., Intelligent gels, *Sci. Am.*, 268, 82–87, 1993.
26. Kuhn, W. et al., Reversible dilation and contraction by changing the state of ionization of high-polymer acid networks, *Nature*, 165, 514–516, 1950.
27. Eddington, D.T. and Beebe, D.J., Flow control with hydrogels, *Adv. Drug Deliv. Rev.*, 56, 199–210, 2004.
28. Skotheim, J.M. and Mahadevan, L., Physical limits and design principles for plant and fungal movements, *Science*, 308, 1308–1310, 2005.
29. Khoury, C. et al., Tunable microfabricated hydrogels: a study in protein interaction and diffusion, *Biomed. Microdevices*, 5, 35–45, 2003.
30. Braschler, T. et al., Gentle cell trapping and release on a microfluidic chip by *in situ* alginate hydrogel formation, *Lab Chip*, 5, 553–559, 2005.
31. Paguirigan, A. and Beebe, D.J., Gelatin based microfluidic devices for cell culture, *Lab Chip*, 6, 407–413, 2006.
32. Dertinger, S.K. et al., Generation of gradients having complex shapes using microfluidic networks, *Anal. Chem.*, 73, 1240–1246, 2001.
33. Yu, Q., Moore, J.S., and Beebe, D.J., Dissolvable and asymmetric hydrogels as components for microfluidic systems, in *Micro Total Analysis Systems*, Baba, Y., Shoji, S., and van den Berg, A., Eds., Kluwer Academic, Nara, Japan, 2002, pp. 712–714.
34. Burdick, J.A., Khademhosseini, A., and Langer, R., Fabrication of gradient hydrogels using a microfluidics/photopolymerization process, *Langmuir*, 20, 5153–5156, 2004.
35. Zaari, N. et al., Photopolymerization in microfluidic gradient generators: microscale control of substrate compliance to manipulate cell response, *Adv. Mater.*, 16, 2133–2137, 2004.
36. Chatterjee, A.N. et al., Mathematical modeling and simulation of dissolvable hydrogels, *J. Aerosp. Eng.*, 16, 55–64, 2003.
37. Plunkett, K.N., Berkowski, K.L., and Moore, J.S., Chymotrypsin responsive hydrogel: application of a disulfide exchange protocol for the preparation of methacrylamide containing peptides, *Biomacromolecules*, 6, 632–637, 2005.

38. Plunkett, K. et al., BoNT responsive hydrogels as sensors in microchannels, in *Ninth International Conference on Minituarized Systems for Chemistry and Life Sciences*, Jensen, K.F., Han, J., Harrison, D.J., and Voldman, J., Eds., Transducer Research Foundation, Boston, 2005, pp. 295–297.
39. Miyata, T., Asami, N., and Uragami, T., A reversibly antigen-responsive hydrogel, *Nature*, 399, 766–769, 1999.
40. Moorthy, J. and Beebe, D.J., A hydrogel readout for autonomous detection of ions in microchannels, *Lab Chip*, 2, 76–80, 2002.
41. Thomas, G. et al., Hydrogel-immobilized antibodies for microfluidic immunoassays: hydrogel immunoassay, *Methods Mol. Biol.*, 321, 83–95, 2006.
42. Mirzabekov, A. and Kolchinsky, A., Emerging array-based technologies in proteomics, *Curr. Opin. Chem. Biol.*, 6, 70–75, 2002.
43. Zwieniecki, M.A., Melcher, P.J., and Holbrook, N.M., Hydrogel controls of xylem hydraulic resistance in plants, *Science*, 291, 1059–1062, 2001.
44. Yu, Q. et al., Responsive biomimetic hydrogel valve for microfluidics, *Appl. Phys. Lett.*, 78, 2589–2591, 2001.
45. Kim, D. and Beebe, D.J., *In-situ* fabricated micro check-valve utilizing the spring force of a hydrogel, in *Proceedings of the Micro Total Analysis Systems*, Northrup, M.A., Jensen, K.F., and Harrison, D.J., Eds., Transducers Research Foundation, Squaw Valley, CA, 2003, pp. 527–530.
46. Beebe, D.J. et al., Functional hydrogel structures for autonomous flow control inside microfluidic channels, *Nature*, 404, 588–590, 2000.
47. Wang, J. et al., Self-actuated, thermo-responsive hydrogel valves for lab on a chip, *Biomed. Microdevices*, 7, 313–322, 2005.
48. Sershen, S.R. et al., Independent optical control of microfluidic valves formed from optomechanically responsive nanocomposite hydrogels, *Adv. Mater.*, 17, 1368–1372, 2005.
49. Eddington, D.T. et al., An organic self-regulating microfluidic system, *Lab Chip*, 1, 96–99, 2001.
50. Baldi, A. et al., A hydrogel-actuated environmentally sensitive microvalve for active flow control, *J. Microelectromech. Syst.*, 12, 613–621, 2003.
51. Liu, R.H., Yu, Q., and Beebe, D.J., Fabrication and characterization of hydrogel-based microvalves, *J. Microelectromech. Syst.*, 11, 45–53, 2002.
52. Lee, S. et al., Control mechanism of an organic self-regulating microfluidic system, *J. Microelectromech. Syst.*, 12, 848–854, 2003.
53. Eddington, D.T. and Beebe, D.J., Development of a disposable infusion system for the delivery of protein therapeutics, *Biomed. Microdevices*, 7, 223–230, 2005.
54. Eddington, D.T. and Beebe, D.J., A valved responsive hydrogel microdispensing device with integrated pressure source, *J. Microelectromech. Syst.*, 13, 586–593, 2004.
55. Agarwal, A.K. et al., Programmable autonomous micromixers and micropumps, *J. Microelectromech. Syst.*, 14, 1409–1421, 2005.
56. Dong, L. et al., Adaptive liquid microlenses activated by stimuli-responsive hydrogels, *Nature*, 442, 551–554, 2006.
57. Hasselbrink, E.F., Shepodd, T.J., and Rehm, J.E., High-pressure microfluidic control in lab-on-a-chip devices using mobile polymer monoliths, *Anal. Chem.*, 74, 4913–4918, 2002.

58. Kim, D., Mohanty, S.K., and Beebe, D.J., Active assembly methods for microfluidic systems, in *Proceedings of the Micro Total Analysis Systems*, Laurell, T., Nilsson, J., Jensen, K.F., Harrison, D.J., and Kutter, J.P., Eds., Royal Society of Chemistry, Malmö, Sweden, 2004, pp. 36–38.
59. Brahmasandra, S.N. et al., Electrophoresis in microfabricated devices using photopolymerized polyacrylamide gels and electrode-defined sample injection, *Electrophoresis*, 22, 300–311, 2001.
60. Herr, A.E. et al., On-chip native gel electrophoresis-based immunoassays for tetanus antibody and toxin, *Anal. Chem.*, 77, 585–590, 2005.
61. Hatch, A., Garcia, E., and Yager, P., Diffusion-based analysis of molecular interactions in microfluidic devices, *Proc. IEEE*, 92, 126–139, 2004.
62. Moorthy, J. and Beebe, D.J., *In situ* fabricated porous filters for microsystems, *Lab Chip*, 3, 62–66, 2003.
63. Beebe, D.J. et al., Microfluidic tectonics: a comprehensive construction platform for microfluidic systems, *Proc. Natl. Acad. Sci. USA*, 97, 13488–13493, 2000.
64. Nguyen, K.T. and West, J.L., Photopolymerizable hydrogels for tissue engineering applications, *Biomaterials*, 23, 4307–4314, 2002.
65. Tan, W. and Desai, T.A., Microfluidic patterning of cellular biopolymer matrices for biomimetic 3-D structures, *Biomed. Microdevices*, 5, 235–244, 2003.
66. Luo, Y. and Shoichet, M.S., A photolabile hydrogel for guided three-dimensional cell growth and migration, *Nat. Mater.*, 3, 249–253, 2004.

Index

A

B

C

D

E

F

G

H

L

M

N

O

P

Q

R

T

U

V

W

X

Z

Second Edition

SMART POLYMERS

Applications in Biotechnology and Biomedicine

Second Edition

SMART POLYMERS

Applications in Biotechnology and Biomedicine

Milton Keynes UK
Ingram Content Group UK Ltd.
UKHW021908071024
449327UK00022B/1637